Food Hygiene and Toxicology in Ready-to-Eat Foods

Food Hygiene and Toxicology in Ready-to-Eat Foods

Parthena Kotzekidou
Department of Food Science and Technology,
Faculty of Agriculture,
Aristotle University of Thessaloniki,
Thessaloniki, Greece

ELSEVIER

AMSTERDAM • BOSTON • HEIDELBERG • LONDON
NEW YORK • OXFORD • PARIS • SAN DIEGO
SAN FRANCISCO • SINGAPORE • SYDNEY • TOKYO
Academic Press is an imprint of Elsevier

Academic Press is an imprint of Elsevier
125 London Wall, London EC2Y 5AS, UK
525 B Street, Suite 1800, San Diego, CA 92101-4495, USA
50 Hampshire Street, 5th Floor, Cambridge, MA 02139, USA
The Boulevard, Langford Lane, Kidlington, Oxford OX5 1GB, UK

British Library Cataloguing-in-Publication Data
A catalogue record for this book is available from the British Library.

Library of Congress Cataloging-in-Publication Data
A catalog record for this book is available from the Library of Congress.

ISBN: 978-0-12-801916-0

Publisher: Nikki Levy
Acquisition Editor: Patricia Osborn
Editorial Project Manager: Karen Miller
Production Project Manager: Melissa Read
Designer: Matthew Limbert

Typeset by MPS Limited, Chennai, India

Contents

PART II MICROBIOLOGICAL HAZARDS

List of Contributors

Daniela Abela
Department of Food Studies and Environmental Health, Faculty of Health Sciences, University of Malta, Msida, Malta

Mohammad Al Mamun
Infectious Diseases Epidemiology Section, Department of Public Health, General Directorate of Health Affairs in Tabuk Region, Ministry of Health, Tabuk, Kingdom of Saudi Arabia

Adriana Pavesi Arisseto
Faculty of Food Engineering, State University of Campinas, Campinas, Sao Paulo, Brazil

María José Cardador
Department of Analytical Chemistry, Campus of Rabanales, University of Córdoba, Córdoba, Spain

Leonardo Casini
Department of Agricultural, Food and Forestry Systems, University of Florence, Florence, Italy

Alejandro Castillo
Texas A&M Center for Food Safety, Department of Animal Science, Texas A&M University, College Station, TX, United States

Rafael Djalma Chaves
Department of Food Science, Faculty of Food Engineering, University of Campinas Rua Monteiro Lobato, Campinas, Sao Paulo, Brazil

Caterina Contini
Department of Agricultural, Food and Forestry Systems, University of Florence, Florence, Italy

Palmira De Bellis
Institute of Sciences of Food Production, National Research Council, Bari, Italy

Mercedes Gallego
Department of Analytical Chemistry, Campus of Rabanales, University of Córdoba, Córdoba, Spain

Ramandeep Garg
Department of Food Studies and Environmental Health, Faculty of Health Sciences, University of Malta, Msida, Malta

Sara E. Gragg
Department of Animal Sciences and Industry, Kansas State University, Olathe, KS, United States

Elizabeth M. Grasso-Kelley
Department of Food Science and Nutrition, Illinois Institute of Technology, Chicago, IL, United States

Norbert U. Haase
Department of Safety and Quality of Cereals, Federal Research Institute of Nutrition and Food, Max Rubner-Institut, Detmold, Germany

Juan José Rodríguez Herrera
Seafood Microbiology and Technology Group, Marine Research Institute, Agencia Estatal Consejo Superior de Investigaciones Científicas, Vigo, Spain

Jacob R. Jenott
Department of Animal Sciences and Industry, Kansas State University, Olathe, KS, United States

Wanchun Jin
Department of Bacteriology, Nagoya University Graduate School of Medicine, Showa-ku, Nagoya, Aichi, Japan

Kimon Andreas G. Karatzas
Department of Food & Nutritional Sciences, School of Chemistry, Food & Pharmacy, University of Reading, Reading, United Kingdom

Justin J. Kastner
Department of Diagnostic Medicine and Pathobiology, Kansas State University, Manhattan, KS, United States

Parthena Kotzekidou
Department of Food Science and Technology, Faculty of Agriculture, Aristotle University of Thessaloniki, Thessaloniki, Greece

Paola Lavermicocca
Institute of Sciences of Food Production, National Research Council, Bari, Italy

Ivan Leguérinel
Université de Brest, Laboratoire Universitaire de Biodiversité et Ecologie Microbienne, Quimper, France

Sonia Marín
Food Technology Department, XaRTA-UTPV, Agrotecnio Center, University of Lleida, Lleida, Spain

Rafael Chacon Ruiz Martinez
Department of Food Science, Faculty of Food Engineering, University of Campinas Rua Monteiro Lobato, Campinas, Sao Paulo, Brazil

Nanci Edid Martínez-Gonzáles
Departamento de Farmacobiología, Universidad de Guadalajara, Guadalajara, Jalisco, Mexico

Bertrand Matthäus
Department of Safety and Quality of Cereals, Federal Research Institute of Nutrition and Food, Max Rubner-Institut, Detmold, Germany

Leena Maunula
Department of Food Hygiene and Environmental Health, Faculty of Veterinary Medicine, University of Helsinki, Helsinki, Finland

Jeanne-Marie Membré
Institut National de la Recherche Agronomique, UMR1014 Sécurité des Aliments et Microbiologie, Nantes, France; L'Université Nantes Angers Le Mans, Oniris, Nantes, France

Hülya Ölmez
TÜBİTAK MRC Material Institute, Gebze, Kocaeli, Turkey

Juan M. Oteiza
Center for Research and Technical Assistance to Industry, Centenario, Neuquén, Argentina

Ranju Paudyal
Department of Food & Nutritional Sciences, School of Chemistry, Food & Pharmacy, University of Reading, Reading, United Kingdom

Ljerka Prester
Institute for Medical Research and Occupational Health, Zagreb, Croatia

Antonio J. Ramos
Food Technology Department, XaRTA-UTPV, Agrotecnio Center, University of Lleida, Lleida, Spain

Zuliana Razali
Institute of Biological Sciences and Centre for Research in Biotechnology for Agriculture, Faculty of Science, University of Malaya, Kuala Lumpur, Malaysia

Ana Carolina Bortolossi Rezende
Department of Food Science, Faculty of Food Engineering, University of Campinas Rua Monteiro Lobato, Campinas, Sao Paulo, Brazil

Lucy J. Robertson
Department of Food Safety and Infection Biology, Norwegian University of Life Sciences, Oslo, Norway

Monyca Dias Rocha
Department of Food Science, Faculty of Food Engineering, University of Campinas Rua Monteiro Lobato, Campinas, Sao Paulo, Brazil

Caterina Romano
Department of Agricultural, Food and Forestry Systems, University of Florence, Florence, Italy

Anderson de Souza Sant'Ana
Department of Food Science, Faculty of Food Engineering, University of Campinas Rua Monteiro Lobato, Campinas, Sao Paulo, Brazil

Vicknesha Santhirasegaram
Institute of Biological Sciences and Centre for Research in Biotechnology for Agriculture, Faculty of Science, University of Malaya, Kuala Lumpur, Malaysia

Gabriele Scozzafava
Department of Agricultural, Food and Forestry Systems, University of Florence, Florence, Italy

Angelo Sisto
Institute of Sciences of Food Production, National Research Council, Bari, Italy

Chandran Somasundram
Institute of Biological Sciences and Centre for Research in Biotechnology for Agriculture, Faculty of Science, University of Malaya, Kuala Lumpur, Malaysia

Carlo Spanu
Department of Veterinary Medicine, University of Sassari, Sassari, Italy

Brijesh Tiwari
Department of Food Biosciences, Teagasc Food Research Centre, Dublin, Ireland

Tanvir Chowdhury Turin
Department of Family Medicine, Department of Community Health Sciences, Cumming School of Medicine, University of Calgary, Calgary, Canada

Daniel A. Unruh
Department of Animal Sciences and Industry, Kansas State University, Olathe, KS, United States

Vasilis Valdramidis
Department of Food Studies and Environmental Health, Faculty of Health Sciences, University of Malta, Msida, Malta

Francesca Valerio
Institute of Sciences of Food Production, National Research Council, Bari, Italy

Carl-Henrik von Bonsdorff
Department of Food Hygiene and Environmental Health, Faculty of Veterinary Medicine, University of Helsinki, Helsinki, Finland

Keiko Yamada
Department of Bacteriology, Nagoya University Graduate School of Medicine, Showa-ku, Nagoya, Aichi, Japan

Preface

Ready-to-eat (RTE) foods are very attractive to consumers looking for convenient meals as they provide a source of readily available and nutritious meals. Thus, the production of RTE foods is growing as a result of lifestyle changes and consumers' increased demand for convenience foods. According to the Codex Alimentarius Commission, RTE foods include any food (including beverages) consumed in its raw state or any food handled, processed, mixed, cooked, or otherwise prepared into a form in which it is normally consumed without further processing. The guidelines from the European Commission see RTE food as "food intended by the producer or manufacturer for direct human consumption without the need for cooking or other processing effective to eliminate or reduce to an acceptable level micro-organisms of concern." RTE foods prepared and/or sold by street vendors are recognized as potential vehicles of microbial foodborne pathogens, if they are mishandled during processing and preparation.

Globalization of the food trade is considered to be able to contribute to a more widespread dissemination of hazards affecting food safety. RTE foods infected by pathogenic bacteria, viruses, or mycotoxins may have acute health effects on many affected consumers. Minimal processing applied nowadays to achieve safety, while maintaining other quality attributes, increases the chance of survival of foodborne pathogens and sporeformers. Some bacteria form toxins as byproducts while growing in foods and cause severe food poisoning. Viruses in fresh produce and seafood products can also cause food-based morbidity and mortality.

Consumers are interested in foods that are safe from a toxicological point of view. The nature, sources, and formation of toxic substances in foods, as well as the mechanism of toxic effects, are of importance in order to avoid harmful effects to the consumers. Also, it is important to determine the safety limits of unintended chemicals contaminating foods during various stages of RTE food production. A number of environmental contaminants, that are persistent, can cause problems due to their accumulation through the food chain. Foods contaminated by residues of chemicals that are used as food additives, sanitizing agents, cleaning compounds, and in food packaging may cause chronic health effects to consumers, which are projected over their entire lifetime.

As the public expectation of food safety is high, the efforts to keep RTE foods safe for the benefit and well-being of consumers should be enhanced. This book reviews the challenges associated with microbial and toxicological hazards and provides strategies to minimize the risks associated with the consumption of RTE foods. It is the result of the collaboration of 51 scientists from 27 universities and research centers. My deepest appreciation and thanks go to the chapter authors, who contributed knowledge, expertise, and a significant amount of their time to bring this book to life. Without their excellent contributions, this book would never have been published. I am also grateful to staff members of Academic Press, Ms Karen R. Miller, Ms Jackie Truesdell, and Ms Carrie Bolger for their helpful advice and assistance during the preparation and completion of this book.

Parthena Kotzekidou

PART

I

Trends in food habits

CHAPTER

Food habits and the increase in ready-to-eat and easy-to-prepare products

1

Caterina Contini, Caterina Romano, Gabriele Scozzafava and Leonardo Casini

Department of Agricultural, Food and Forestry Systems, University of Florence, Florence, Italy

INTRODUCTION

Ready-to-eat and easy-to-prepare (RTE and ETP) meals include a wide range of foods whose composition, importance, and image have transformed over the decades, reflecting the technological developments and changes in uses and customs that have characterized different epochs (Scholliers, 2015). In this regard, the precursors of RTE and ETP meals can be traced back to the products that long ago were made to confront long journeys, such as dried or smoked products, which were easy to preserve and required no particular preparation. With the introduction of glass containers in the 1800s, food preservation took a step forward. It was only 50 years later, however, that food preservation on an industrial scale received a boost from the use of tin, which facilitated food transport and its diffusion over increasingly vaster areas. With the development of industrialized processes, in the 1900s we witnessed the supply of more complex products in tins, which made for a greater variability of assortments, as in the case of sardines in tomato sauce. The preparation of a complete dish was achieved in the early 1950s with TV dinners, made up of a main course, always accompanied by a dressing and a side dish, and packaged in a container that could be heated directly in the oven. The evolution of RTE and ETP meals further developed in the late 1980s with the diffusion of microwave ovens, and the use of these products was no longer limited to particular or emergency situations but instead officially entered the dietary habits of families (Scholliers, 2015). We are currently witnessing yet another transformation of these products in the sense of an expansion of their ranges, accompanied by a strong increase in demand (Scholderer and Grunert, 2005).

To get a better grasp of these phenomena, studies on consumer behavior have analyzed the characteristics of RTE and ETP meals in relation to the preferences of individuals and the evolution of these preferences. This chapter intends to

P. Kotzekidou (Ed): Food Hygiene and Toxicology in Ready-to-Eat Foods.
DOI: http://dx.doi.org/10.1016/B978-0-12-801916-0.00001-7

confront these themes by analyzing the studies reported in the literature, and answering three key questions: "Is there a common denominator that characterizes the different RTE and ETP meals?," "What are the motives behind their growing demand?," and "Is there a typical profile of individuals who choose to consume these types of products?" These three questions will be developed in the following sections and constitute the paper's framework. In the conclusions section we shall discuss the principal arguments in light of possible future scenarios for RTE and ETP meals.

IS THERE A COMMON DENOMINATOR IN RTE AND ETP MEALS?

The answer to the first question is yes: despite their multifaced nature, it is possible to qualify RTE and ETP meals on the basis of several traits common to all the products belonging to this category. The literature underlines, in particular, that their definition involves two key dimensions. The first concerns those activities for which RTE and ETP meals can determine a benefit for the consumer, while the second concerns the type of benefit associated with the product. Regarding the former aspect, the authors propose to consider all the phases tied to the consumption of food, which include purchase, preparation, and cleaning up, inasmuch as RTE and ETP meals influence, though in different ways, each of these activities (Brown and McEnally, 1993; Darian and Cohen, 1995). As for the type of benefit, the advantages of these foods are to be sought in their convenience in terms of time, physical energy, and mental effort (Brunner et al., 2010).

These considerations permit us to extrapolate the common denominator of RTE and ETP meals, and thus arrive at the following definition: "RTE and ETP meals are food products that help save time, physical energy, and mental effort in all food activities compared to what is required for meals prepared from raw materials." This characteristic represents an answer to the demands of an increasingly more time-constrained society in which fewer opportunities to prepare meals lead to a progressive substitution of the homemade with the more convenient (Ryan et al., 2002). In this context, convenience becomes a decisive attribute of the consumer's preferences and a determining factor in food choice on a par with other product attributes such as taste, price, or health traits (Candel, 2001; Grunert, 2006).

WHAT ARE THE MOTIVES BEHIND THE GROWING DEMAND FOR RTE AND ETP MEALS?

Numerous studies have observed a growing demand for convenience foods in various countries (Casini et al., 2015; Rischke et al., 2015; Smith et al., 2013). On the macroeconomic level, the reasons for the success of RTE and ETP meals are traced to the growth and technological progress that have increased the consumers' assets and facilitated the conservation and transformation of foods (Jabs and Devine, 2006). Other factors can be identified in family structure and organization. These include the woman's participation in the world of work, the extension of work hours (Buckley et al., 2007; Senauer et al., 1991; Traill, 1997), and the growth of single-member units, which have played a decisive role in increasing the demand for quickly prepared foods (Khan, 2000; Swoboda and Morschett, 2001; Verlegh and Candel, 1999).

While the aforementioned elements may constitute some of the most macroscopic factors that can be correlated to the increase in convenience food consumption, for a proper understanding of the phenomenon we must shift our attention from the objective characteristics of consumers and available resources to the subjective aspects. In this ambit, the literature points out two major fields of analysis. The first concerns beliefs, intended as the individual's perception of the outcomes of a certain behavior, the expectations of the "others" and the capability to perform a particular behavior (Ajzen, 1991, 2011). The other field concerns the value system and the ways individuals pursue personal values (de Boer et al., 2004; Brunsø et al., 1996; Schwartz, 1992).

The answer to the initial question as to the success of convenience foods therefore requires a closer examination of these concepts. The two following sections will describe the behaviors of consumers that emerge from the studies conducted by applying theories based on beliefs and values, in view of reaching an answer capable of explaining the growing demand for RTE and ETP meals.

BELIEFS, ATTITUDES, AND BEHAVIOR

In the field that analyses the connection between behavior and beliefs, the study of convenience orientation has been confronted using the cognitive model known as the theory of planned behavior (TPB), which proposes a structured approach to explain the behavior of individuals (Ajzen, 1991, 2011).

At the basis of the model are the beliefs, shaped into behavioral, normative, and control beliefs (Ajzen, 1985). The first are defined as the expected outcomes of a particular behavior. They are weighed on the basis of a personal evaluation of these outcomes, generating the attitude towards the behavior, which can be positive or negative.

Normative beliefs concern the perception of the expectations of family members, friends, and other important referents. Along with personal motivations in satisfying the expectations of others, they are responsible for the social pressure to adopt (or not to adopt) a particular behavior.

Finally, control beliefs express the perception the individual has of his/her own assets in terms of resources, skills, and opportunities to adopt a particular behavior. The individual's evaluation of the power of the aforesaid resources to facilitate or oppose the behavior determines the perceived behavioral control. According to the TPB, the connection between beliefs and behavior is represented

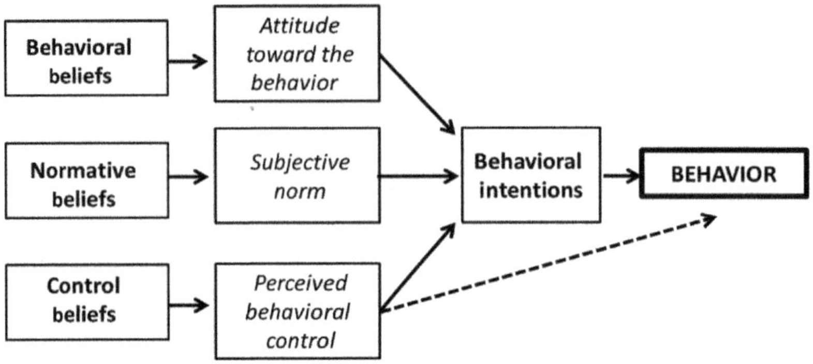

FIGURE 1.1

Schematic representation of the theory of planned behavior.

by intention, defined as the individual's inclination to adopt a particular behavior. The consequences of intention on behavior are moderated by the perceptions of behavioral control, in the sense that a favorable intention produces the behavior only if the perceived behavioral controls favor the behavior (Fig. 1.1).

The studies conducted in the ambit of RTE and ETP meals have analyzed the factors that can influence purchase behaviors; these factors can be traced to all three categories of beliefs, though the behavioral beliefs and therefore attitude proved to represent the best predictor of the behavioral intention to buy convenience foods (Olsen et al., 2010; Mahon et al., 2006). The importance of attitude on behavior has also been verified by Scholderer and Grunert (2005), who pointed out a stricter form of mediation between attitude and behavior, showing how all effects of perceived resource constraints (perceived behavioral controls) on choice were completely mediated by consumers' attitudes. The literature highlights that among the principal behavioral beliefs emerge value for money, time-saving, taste, healthiness, and weight control. In particular, value for money has been found to exert a positive influence on the attitude to purchase these products (Brunner et al., 2010; Buckley et al., 2007; de Boer et al., 2004; Ryan et al., 2002), along with time-saving (Buckley et al., 2007), while taste, healthiness, and weight control can carry a negative influence (Olsen, 2012).

While the behavioral beliefs described here have a fundamental role on purchase behavior, studies have pointed out the influence of normative and control beliefs as well. Among the former, social pressure can constitute a restraint to consumption by generating a sense of guilt for the limited care dedicated to preparing food, both in terms of time and of mental effort (Costa et al., 2007; Olsen et al., 2010; Olsen, 2012). In this same ambit, other elements that limit the consumption of RTE and ETP meals are tied to the regret for the loss of family meal-time and of the culinary traditions to pass down to one's children (Dunn et al., 2011). The influence of these factors on the consumption of convenience foods is variable.

On the territorial level, these differences appear to be attributable to a different market exposure. In particular, in countries where there is a high penetration of these products, such as Holland, nonsignificant results are observed for the effect of subjective norm on the intention to consume ready-to-eat meals (Buckley et al., 2007; Olsen et al., 2010).

Among the control beliefs, time pressure is identified as a key factor in encouraging the intention to purchase convenience foods (de Boer et al., 2004; Ryan et al., 2002; Verlegh and Candel, 1999). In terms of skills, scarce cooking ability is associated with a greater use of RTE and ETP meals (Brunner et al., 2010; Buckley et al., 2007). Moreover, in terms of opportunities, a different stance towards these products is associated with their variety, quantity, and diffusion (de Boer et al., 2004; Sloan et al., 1984). Likewise, the consumption of convenience foods is connected with the technological development and utilization of appliances suitable for RTE and ETP meals, such as microwave ovens or freezers (Buckley et al., 2007).

VALUE SYSTEM AND FOOD-RELATED LIFESTYLES

In analyzing the values system to understand the consumer's behavior, values are defined as the higher-order reasons that guide the individual in evaluating stimuli, and represent the most abstract level of cognition (Schwartz, 1992). In other words, they are the desirable end-states that orient behavior. These are universal categories that, as such, do not depend on a specific situation but can instead vary from individual to individual, depending on the relative importance he/she attributes to them.

Schwartz (1992) indicates 10 value domains, the achievement of each of which can determine an effect that is congruous or in opposition to the other domains. These relationships are summed up into two bipolar dimensions represented by self-enhancement versus self-transcendence, and openness to change versus conservation. Self-enhancement expresses power and achievement, while self-transcendence represents universalism and benevolence. On the other hand, openness to change represents values such as stimulation and self-direction, while conservation embraces tradition, conformity, and security. Finally, the hedonism value shares parts of both openness to change and self-enhancement (Fig. 1.2). Therefore, those who consider security as the most important value, for example, will be more mindful of food safety, while those who prefer stimulation will look for novelty and excitement, even to the detriment of security. Likewise, the individuals who mainly pursue benevolence will devote attention to the influence of their choices on the wellbeing of the society, while those who consider hedonism more important will attribute greater weight to appeal, relegating considerations of a social nature to second place.

In order to study the influence of personal values in the ambit of food-related behavior, a set of cognitive categories has been proposed, which goes under the name of food-related lifestyles (Brunsø et al., 2004; Scholderer et al., 2004). This construct makes it possible to relate the higher values to the way individuals employ food and eating. By utilizing this approach, studies have pointed out how

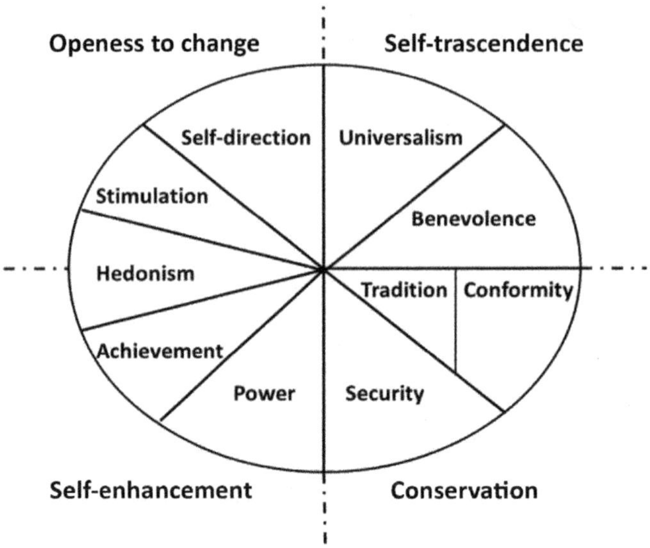

FIGURE 1.2

Schwartz value system.

the convenience orientation is determined by different value categories, depending on the socio-territorial context considered. In Germany, in fact, it is associated with the power domain, which includes values such as social status and prestige. In Spain, an association of convenience foods with the search for an exciting life is underlined, and the convenience orientation is more connected with stimulation, which reflects boldness, novelty, and challenge in life (Brunsø et al., 2004). On the other hand, in Greece, Botonaki and Mattas (2010) have divided the convenience orientation into its various constituent elements, shedding light on how dealing with cleaning up in a convenient way proves to be correlated with power, while the preference for takeaways is associated with openness to change. Finally, the orientation towards purchase and consumption is principally correlated to the achievement domain, which concerns values such as personal success, competence, and ambition in the professional field.

THE MOTIVATIONS OF SUCCESS

The increase in available income, along with several changes that have occurred in the organizational structure of families have constituted, in many countries, the basis for a development of RTE and ETP foods consumption. The actual growth in demand, however, can be explained more thoroughly only by also considering the changes that have occurred in the beliefs and values systems of individuals, which are in turn connected to changes in supply.

The analysis of the results of applying the TPB points out, first of all, the importance of the increase of time pressure as a determinant of the phenomenon. Furthermore, the reduction of cooking skills represents the other factor that has contributed to increasing the purchasing behavior of these products. The increased interest in convenience foods that has evolved has, in turn, generated major improvements in their supply in terms of quality, quantity, and distribution. In particular, the development of new products associated with a better taste and healthiness may have contributed to improving the expected behavioral outcomes and, consequently, encouraged a positive attitude towards these products, further boosting their consumption. Moreover, the greater market penetration of convenience foods may have attenuated the negative social pressure against the consumption of these products, which are increasingly less associated with a lack of care and attention towards family and guests in moments of conviviality (Buckley et al., 2007). Again from the perspective of supply, the increase in the number of places where convenience foods are available, and the variety of the foods distributed has improved the access to these foods, thereby encouraging their consumption.

Another factor that has acted positively on the perceived behavioral controls is represented by the diffusion of new technologies that in facilitating the conservation and preparation of convenience foods have enhanced their capability to reduce the time and effort associated with meals and, in turn, also contribute to improving consumers' attitudes towards them.

As far as the value system is concerned, the analysis of the literature presents a variegated picture with various aspects, which allows us to add an important element to answer the question as to the motives that have led to an increase in the demand for convenience foods: the increased interest in RTE and ETP meals has been encouraged by a significant shift in the value system. In particular, there has been a passage from the dimension of self-transcendence to that of self-enhancement and a greater inclination towards change, to the detriment of the dimension of conservation, associated with harmony and the stability of family and society (Cerulo, 2008; Sharma and Sharma 2010).

Lastly, combining the TPB and the approach of food-related lifestyles permits us to offer a more exhaustive answer to the question that opened this section: the motives behind the growing demand for RTE and ETP meals can be traced to a combined change in the behavioral, normative, and control beliefs, and of the way the consumer evaluates them. In particular, while on one hand we witness a transformation of the characteristics of convenience foods that improve their use and image, on the other hand, there is a shift in the value system that leads to a change in the opinion of the individual, as well as in that of society overall. Thus the individual is faced with an increased availability, variety, and quality of convenience foods, which make these food products increasingly more accessible and respondent to one's own tastes and lifestyles. On the other hand, the change in the system of individual and collective values affects the evaluation of the expected outcomes and the subjective norms, increasing the consumer's interest and, at the same time, lowering the barriers to consumption.

IS THERE A TYPICAL PROFILE OF INDIVIDUALS WHO CHOOSE TO CONSUME RTE AND ETP FOODS?

The studies on the consumer profile of RTE and ETP foods, though conducted in different contexts, have pointed out characteristics shared by the purchasers of this category of products. They are, in particular, prevalently male, single, young people (Brunner et al., 2010), people without children (Buckley et al., 2007; Candel, 2001) and with longer working hours (Candel, 2001). This set of sociodemographic and objectively observable traits proves to be coherent with the picture of the system of beliefs and values described in the previous paragraphs. Integrating the results of the studies on sociodemographic profiles with the values and perceptions represents the most comprehensive way to acquire an effective understanding of the most significant convenience-oriented market segments.

The characteristic of having a job commitment for many hours of the day is certainly tied to the control belief of time pressure. People with a higher role overload (Brunner et al., 2010; Buckley et al., 2007; de Boer et al., 2004; Candel, 2001) can perceive a greater time pressure, which could be translated into a more marked inclination to consume convenience foods. On the other hand though, this relationship is not essentially one of cause and effect, inasmuch as the inclination towards convenience foods can be determined by many other elements that affect time pressure, in addition to or even in opposition to the length of working hours, such as engaging in other high-involvement activities or daily travel time. Proper profiling would therefore require integrating the work time characteristic with other elements capable of better defining the presence of the individual's time pressure condition, though with the awareness that time pressure has marked features of subjectivity and can thus vary from one individual to another, even under the same objective conditions.

Age proves to be another important sociodemographic trait in defining the inclination towards convenience foods. In this regard, young age can be connected with fewer skills, low cooking involvement, and other values positively correlated to the consumption of these products. In fact, young people present a greater individualism (Danesi, 2012), which shows in the habit of breaking meal times, eating alone, snacking, and in the limited involvement of family members in meal activities, factors that characterize convenience-oriented consumers (Jabs and Devine, 2006). Furthermore, young people also show more openness to novelty and a greater inclination towards change (Ebner et al., 2006), all elements that characterize this category of consumers (de Boer et al., 2004).

Analogous considerations can also be made for the characteristics of gender, living alone, and presence of children. In particular, males generally prove to have fewer cooking skills, single persons show less inclination to preparing a proper meal when eating alone (Brunner et al., 2010), and families without children are associated with being less sensitive towards the health-related characteristics of food (Grunert and Wills, 2007), and thus prove more open to foregoing freshness in favor of convenience (Buckley et al., 2007).

The relationships described above point out how it is possible to outline a profile of the "typical" convenience food consumer by integrating the objectively verifiable elements, such as demographic characteristics, with the values and beliefs of individuals, as explanatory factors of the connection between the observable characteristics and the preferences of consumers.

DISCUSSION AND CONCLUSIONS

Our study permits us to select several explanatory elements of the role RTE and ETP foods assume in the scenario of consumption. Two principal aspects emerge in the ambit of values: the first is curiosity to try new ways of eating; the second is the individualist food behavior in which the personal interest prevails over that of the family in decisions concerning consumption. In the ambit of beliefs, we can note the importance several factors have on consumer behavior, first and foremost, the control belief of time pressure, which acts as a catalyst in the choice process of RTE and ETP foods. As far as behavioral and normative beliefs are concerned, a positive role is played by the quality traits of the supply and by market penetration, which respectively affect the attitude towards convenience products and loosen the normative barriers connected with social pressure.

It is not easy to foresee the future scenario of RTE and ETP meals, insofar as the development of these products is a function of the structural relationships between objective factors—such as time, skills, or the availability of products—and subjective factors concerning their perception (Scholderer and Grunert, 2005).

The reasons for the recent popularity of RTE and ETP meals, tied to the improvement of the quality characteristics of supply and to changes in the value system and lifestyles, in any event allow us to glimpse a future scenario of further expansion in the food habits of these products, at least for some categories of individuals.

The conclusions that the literature has come to in terms of profiling the convenience-oriented consumer cite, as fundamental characteristics, time pressure, youth, and scarce cooking skills, which thus offers several important elements to foresee the future of these foods. The segment of young people, which currently shows a marked interest in these products, in the future will constitute the group of consumers with the greatest purchasing power, and will therefore be able to also exert a growing influence on overall food demand. The supply of quickly prepared foods with high contents of quality and healthiness, though more expensive, could therefore be an interesting strategy to develop.

From the public viewpoint, in this framework, a growing diffusion of RTE and ETP foods can determine the advisability of effective food education campaigns and greater attention to the rules that regulate the production of these foods, both in terms of preparation techniques and of ingredients. In this regard, mandatory labeling constitutes a fundamental tool to permit informed consumption.

The public administration's adoption of this orientation could bring about a synergy between the needs for complete and correct information, the improvement of supply characteristics, and the consumer's greater willingness to purchase. The guarantee offered by the regulations could indeed provide the consumer with a tool to bypass several barriers that, as we have seen, negatively affect beliefs and, at the same time, reward the companies that offer the best-quality products.

REFERENCES

Ajzen, I., 1985. From intentions to actions: a theory of planned behaviour. In: Kuhl, J., Beckmann, J. (Eds.), Action Control: From Commitment to Behaviour. Springer, Heidelberg, pp. 11–39.

Ajzen, I., 1991. The theory of planned behaviour. Organ. Behav. Hum. Decision Processes 50, 179–211.

Ajzen, I., 2011. The theory of planned behaviour: reactions and reflections. Psychol. Health 26, 1113–1127.

Bernués, A., Ripoll, G., Panea, B., 2012. Consumer segmentation based on convenience orientation and attitudes towards quality attributes of lamb meat. Food Qual. Preference 26, 211–220.

Botonaki, A., Mattas, K., 2010. Revealing the values behind convenience food consumption. Appetite 55, 629–638.

Brown, L.G., McEnally, M.R., 1993. Convenience: definition, structure, and application. J. Mark. Manage. 2, 47–56.

Brunner, T.M., van der Horst, K., Siegrist, M., 2010. Convenience food products. Drivers for consumption. Appetite 55, 498–506.

Brunsø, K., Grunert, K.G., Bredahl, L., 1996. An analysis of national and cross-national consumer segments using the food-related lifestyle instrument in Denmark, France, Germany and The United Kingdom. MAPP, Working Paper No. 35. The Aarhus School of Business, Aarhus.

Brunsø, K., Scholderer, J., Grunert, K.G., 2004. Testing relationships between values and food-related lifestyle: results from two European countries. Appetite 43, 195–205.

Buckley, M., Cowan, C., McCarthy, M., 2007. The convenience food market in Great Britain: convenience food lifestyle, CFL segments. Appetite 49, 600–617.

Casini, L., Contini, C., Marone, E., Romano, C., 2013. Food habits. Changes among young Italians in the last 10 years. Appetite 68, 21–29.

Casini, L., Contini, C., Romano, C., Scozzafava, G., 2015. Trends in food consumptions: what is happening to generation X? Br. Food J. 117, 705–718.

Candel, M., 2001. Consumer's convenience orientation towards meal preparation. Conceptualization and measurement. Appetite 36, 15–28.

Cerulo, K.A., 2008. Social relations, core values, and the polyphony of the American experience. Sociol. Forum 23, 351–362.

Costa, A.I.A., Dekker, M., Beumer, R.R., Rombouts, F.M., Jongen, W.M.F., 2001. A consumer oriented classification system for home meal replacements. Food Qual. Preference 12, 229–242.

Costa, A.I.A., Schoolmeester, D., Dekker, M., Jongen, W.M.F., 2007. To cook or not to cook: a means-end study of the motivations behind meal choice. Food Qual. Pref. 18, 77–88.

Danesi, G., 2012. Commensality in French and German young adults: an ethnographic study. Hospitality Soc. 1, 153–172.

Darian, J.C., Cohen, J., 1995. Segmenting by consumer time shortage. J. Consum. Mark. 12, 32–44.

de Boer, M., McCarthy, M., Cowan, C., Ryan, I., 2004. The influence of lifestyle characteristics and beliefs about convenience food on the demand for convenience foods in the Irish market. Food Qual. Preference 15, 155–165.

Dunn, K.I., Mohr, P., Wilson, C.J., Wittert, G.A., 2011. Determinants of fast-food consumption. An application of the theory of planned behaviour. Appetite 57, 349–357.

Ebner, N.C., Freund, A.M., Baltes, P.B., 2006. Developmental changes in personal goal orientation from young to late adulthood: from striving for gains to maintenance and prevention of losses. Psychol. Aging 21, 664–678.

Grunert, K.G., 2006. Future trends and consumer lifestyles with regard to meat consumption. Meat Sci. 74, 149–160.

Grunert, K.G., Wills, J.M., 2007. A review of European research on consumer response to nutrition information on food labels. J. Public Health (Bangkok) 15, 385–399.

Jabs, J., Devine, C.M., 2006. Time scarcity and food choices: an overview. Appetite 47, 196–204.

Khan, Y., 2000. Meal solution trends. World Food Ingredients, April/May, 16–23

Lusk, J.L., Briggeman, B.C., 2009. Food values. Am. J. Agric. Econ. 91, 184–196.

Mahon, D., Cowan, C., McCarthy, M., 2006. The role of attitudes, subjective norm, perceived control and habit in the consumption of ready meals and takeaways in Great Britain. Food Qual. Preference 17, 474–481.

Marshall, J.J., Duxbury, L., Heslop, L.A., 1995. Coping with household stress in the 1990s: who uses "convenience foods" and do they help? Adv. Consum. Res. 22, 729–734.

Olsen, N.V., 2012. The convenience consumer's dilemma. Br. Food J. 114, 1613–1625.

Olsen, N.V., Sijtsema, S.J., Hall, G., 2010. Predicting consumers' intention to consume ready-to-eat meals. The role of moral attitude. Appetite 55, 534–539.

Rischke, R., Kimenju, S.C., Klasen, S., Qaim, M., 2015. Supermarkets and food consumption patterns: the case of small towns in Kenya. Food. Policy 52, 9–21.

Ryan, I., Cowan, C., McCarthy, M., Sullivan, C., 2002. Food-related lifestyle segments in Ireland with a convenience orientation. J. Int. Food Agribusiness Mark. 14, 29–47.

Scholderer, J., Grunert, K.G., 2005. Consumers, food and convenience: the long way from resource constraints to actual consumption patterns. J. Econ. Psychol. 26, 105–128.

Scholderer, J., Brunsø, K., Bredahl, L., Grunert, K.G., 2004. Cross-cultural validity of the food-related lifestyles instrument, FRL within Western Europe. Appetite 42, 197–211.

Scholliers, P., 2015. Convenience foods. What, why, and when. Appetite 94, 2–6.

Schwartz, S.H., 1992. Universals in the content and structure of values: theoretical advances and empirical tests in 20 countries. Adv. Exp. Soc. Psychol. 25, 1–65.

Senauer, B., Asp, E., Kinsley, J., 1991. Food Trends and the Changing Consumer. Eagle Press, Saint Paul, MN.

Sharma, S., Sharma, M., 2010. Globalization, threatened identities, coping and well-being. Psychol. Stud. 55, 313–322.

Sloan, E., Leone, L.C., Powers, M., McNutt, K.W., 1984. Changing consumer lifestyles. Food Technol. 38, 99–103.

Smith, L.P., Ng, S.W., Popkin, B.M., 2013. Trends in US home food preparation and consumption: analysis of national nutrition surveys and time use studies from 1965–1966 to 2007–2008. Nutr. J. 12, 45.

Swoboda, B., Morschett, D., 2001. Convenience-oriented shopping: a model from the perspective of consumer research. In: Frewer, L., Riskvik, E., Schifferstein, H. (Eds.), Food, People and Society: A European Perspective of Consumers' Food Choices. Springer, Berlin, pp. 177–196.

Traill, B., 1997. Structural changes in the European food industry: consequences for innovation. In: Traill, B., Grunert, K.G. (Eds.), Product and Process Innovation in the Food Industry. Chapman & Hall, London, pp. 38–60.

Verlegh, P.W.J., Candel, M.J.J.M., 1999. The consumption of convenience foods: reference groups and eating situations. Food Qual. Preference 10, 457–464.

CHAPTER

Safety of street foods

2

Mohammad Al Mamun[1] and Tanvir Chowdhury Turin[2]

[1]Infectious Diseases Epidemiology Section, Department of Public Health, General Directorate of Health Affairs in Tabuk Region, Ministry of Health, Tabuk, Kingdom of Saudi Arabia

[2]Department of Family Medicine, Department of Community Health Sciences, Cumming School of Medicine, University of Calgary, Calgary, Canada

STREET FOODS: DEFINITION AND BASIC CHARACTERISTICS

Nowadays, an estimated 2.5 billion people consume street foods in some form every day (Fellows and Hilmi, 2011; Kraig and Sen, 2013), and street foods are considered as the most dominant form of public dining (den Hartog et al., 2006; Kraig and Sen, 2013). Street foods are very familiar and sold as quick meals or snacks from daybreak until night in almost every corner of the world (Fellows and Hilmi, 2011; Tinker, 1997). Street foods are widely obtainable in both urban (den Hartog et al., 2006; Fellows and Hilmi, 2011; INFOSAN, 2010; WHO, 1996) and rural (Fellows and Hilmi, 2011; WHO, 1996) settings, and offer an easy substitute to home-cooked food (Winarno and Allain, 1991).

Street foods are low-priced meals compared with restaurant foods (Fellows and Hilmi, 2011; INFOSAN, 2010; Kraig and Sen, 2013; WHO, 1996; Winarno and Allain, 1991), and that's why these foods are easily accessible and affordable for large numbers of underprivileged people in many areas of poverty, particularly in many developing countries (den Hartog et al., 2006; FAO, 2007; Fellows and Hilmi, 2011; INFOSAN, 2010; Kraig and Sen, 2013; WHO, 1996). Street food is also a convenient source of attractive food for tourists and the economically privileged (FAO, 2007; INFOSAN, 2010; Kraig and Sen, 2013; Tinker, 1997; WHO, 1996). Street foods are an important source of daily calories and nutritional requirements for many people (den Hartog et al., 2006; FAO, 2007; Fellows and Hilmi, 2011; INFOSAN, 2010; Kraig and Sen, 2013; Tinker, 1997; WHO, 1996; Winarno and Allain, 1991). The nutritional value of street foods depends on their ingredients and methods of preparation (den Hartog et al., 2006; Fellows and Hilmi, 2011).

Street foods have been available for thousands of years, and reflect traditional local and regional cultures (Fellows and Hilmi, 2011; INFOSAN, 2010; Kraig and Sen, 2013; Tinker, 1997; Winarno and Allain, 1991). Cultural, ethnic

P. Kotzekidou (Ed): Food Hygiene and Toxicology in Ready-to-Eat Foods.
DOI: http://dx.doi.org/10.1016/B978-0-12-801916-0.00002-9

and religious diversities influence the type and nature of such foods (Fellows and Hilmi, 2011). These foods are prepared and consumed based on local traditional knowledge, and usually stick to the seasonality of farm production (Fellows and Hilmi, 2011). The street food sector plays an important role in providing employment opportunities for millions of men and women with inadequate education or skills and limited capital investment (den Hartog et al., 2006; INFOSAN, 2010; Roever, 2014; WHO, 1996; Winarno and Allain, 1991). Furthermore, this sector has become a significant part of the food distribution system, especially for midday meals in both developing and developed countries (INFOSAN, 2010).

In 1986, the Food and Agriculture Organization of the United Nations (FAO) defined the term "street foods" as:

> *a wide range of ready-to-eat foods and beverages sold and sometimes prepared in public places, notably streets*
>
> **Winarno and Allain (1991).**

This definition of street foods was accepted by the FAO Regional Workshop on Street Foods in Asia, held in Yogyakarta, Indonesia, during Nov. 3–7, 1986.

In 1988, an FAO Expert Consultation on Street Foods (held in Yogyakarta, Indonesia, during Dec. 5–9, 1988) defined "street foods" as follows:

> *Street foods are ready-to-eat foods and beverages prepared and/or sold by vendors and hawkers especially in streets and other similar public places*
>
> **den Hartog et al. (2006), Draper (1996), FAO (1989), FAO/WHO-CAC (1999), Fellows and Hilmi (2011), INFOSAN (2010).**

Although several attempts have been made to define "street foods," this is the most widely cited definition so far.

The World Health Organization (WHO) has defined "street-vended foods" or its equivalent "street foods" as:

> *foods and beverages prepared and/or sold by vendors in streets and other public places for immediate consumption or consumption at a later time without further processing or preparation*
>
> **WHO (1996).**

This definition includes fresh fruits and vegetables that are vended for immediate consumption.

Tinker (1997) defined street food as:

> *any minimally processed food sold on the street for immediate consumption*
>
> **Tinker (1997).**

In order to make it clear, Tinker refined her definition of street foods by including fresh fruits that are sold on the streets (eg, one or two bananas, not a large bunch of bananas), and by excluding nonnutritious chewing foods (eg, chewing gum, or chewing "paan," ie, betel leaf with betel nut, lime, and spices) (Tinker, 1997).

While Tinker concentrated on minimally processed foods for immediate consumption, the diversity of street food production and consumption has been emphasized in another definition:

> *Street foods are minimally processed to highly processed foods that are sold on streets and other public places, consumed on the spot and/or ready to take home or delivered to the work place, including catering activities that can serve celebrations such as weddings*
>
> **Fellows and Hilmi (2011).**

In an encyclopedia of food and culture, Kraig and Sen (2013) stated an ordinary definition of the term "street food" as:

> *a food prepared by a vendor and sold from an open-air stand, cart, truck, or perhaps a market stall*
>
> **Kraig and Sen (2013).**

The key feature of street foods mentioned in the definitions above is the retail location "on the street." Typically, street foods are sold on the street from push-carts, bicycles, baskets, balance poles, or from stalls or shops that do not have four permanent walls (Draper, 1996; Fellows and Hilmi, 2011; Tinker, 1997). This retailing method differentiates street food vendors from other formal food service establishments, such as cafés, "takeaways," "chopbars," and restaurants (Draper, 1996; Fellows and Hilmi, 2011). Vendors' stalls are usually situated in the open air or under a roof that is easily accessible from the street (Winarno and Allain, 1991). Because street foods are often sold in a place where people gather for shopping or recreation purposes, the places where these foods are served may include not only streets, but also outdoors and enclosed markets where individuals vend different commodities (Kraig and Sen, 2013). Amusement venues, such as carnivals, fairs, athletic events, footpaths, around factories and offices, bus and train stations, schools and colleges, and crowded residential areas, are some places for selling street foods (den Hartog et al., 2006; Kraig and Sen, 2013). The types of street food vending operations may vary in different regions of the globe. Vendors can be mobile or ambulatory vendors, for example, simply on foot or using bicycles, pushcarts, vans, or stationary vendors that sell from a fixed stall which may be a permanent or semipermanent structure (den Hartog et al., 2006; Draper, 1996; Fellows and Hilmi, 2011; INFOSAN, 2010; WHO, 1996).

Street foods exist in a limitless variety (den Hartog et al., 2006; Draper, 1996; Fellows and Hilmi, 2011; INFOSAN, 2010; Kraig and Sen, 2013; WHO, 1996; Winarno and Allain, 1991) comprising meals, drinks, and snacks (Draper, 1996; INFOSAN, 2010; Winarno and Allain, 1991), while Kraig and Sen categorized actual street foods as snacks, junk food, and basic foodstuffs in their *Encyclopedia of Food and Culture* (Kraig and Sen, 2013). Ranging from simple snacks to elaborate dishes, street-vended foods may include foods as diverse as meat, fish, fruits, vegetables, grains, cereals, frozen products, and all sorts of beverages (den Hartog et al., 2006; WHO, 1996). Street foods also show extreme diversity in terms of raw materials or ingredients, methods of preparation, processing, and consumption (Draper, 1996; Fellows and Hilmi, 2011; INFOSAN, 2010; Winarno and Allain, 1991).

Street foods can be produced in several ways: (1) many are centrally processed foods produced by formal commercial industries or manufacturers; (2) others are processed foods within the street food trade, either by the individual vendor him/herself or their families, or by another small-scale entrepreneur (Draper, 1996; Kraig and Sen, 2013). There are differences in the spots where street foods are prepared, such as small-scale food factories or traditional workshops, vendor's homes, markets, or streets (INFOSAN, 2010). The final preparation of street foods occurs when the customer places an order for the meal. The food is prepared instantly from premade ingredients and served in an appropriate manner, which can be consumed on the spot or on the move, or taken away (Kraig and Sen, 2013; Tinker, 1997; Winarno and Allain, 1991).

Tinker (1997) introduced the term "invisible street foods" as: "ready-to-eat food that was carried through the streets by the preparer rather than the purchaser, for eating at home or office, and was not for sale on the street" (Tinker, 1997). This type of street foods is prepared in or served from the basement of an office building or from a nearby home, and delivered to the workers on a regular basis. Because the foods were not vended from a visible shop or cart, they were termed "invisible" (Tinker, 1997).

SAFETY OF STREET FOODS: A MAJOR CONCERN

Despite many potential advantages, it is recognized that street foods may pose significant public health risks (WHO, 1996). Microbiological contamination is the most important health hazard associated with street foods, while the use of unpermitted chemical additives, pesticide residues, transmission of parasites, and environmental contamination have also been considered as other possible health risks (Abdussalam and Kaferstein, 1993; Arambulo et al., 1994). Many factors are responsible for these health problems, such as lack of basic infrastructure and services (eg, potable water supplies), general lack of factual knowledge about the microbiological status of street foods, insufficient resources for inspection and laboratory analysis, poor knowledge of street vendors in basic food safety measures, and inadequate public awareness of hazards posed by certain street foods (WHO, 1996). Moreover, it is too difficult to control the large numbers of street food vending operations due to their diversity, mobility, and temporary nature (WHO, 1996). Because food is susceptible to contamination at any phase of the food chain from origin to consumption (Kraig and Sen, 2013; Tinker, 1997; WHO, 1996), the safety of street foods is a point of major concern, particularly in overcrowded streets and public places (den Hartog et al., 2006; Draper, 1996; FAO, 2007).

MICROBIOLOGICAL CONTAMINATION OF STREET FOODS

Microbiological contamination of street foods is a threat to consumers' health in many parts of the world (FAO, 2009; INFOSAN, 2010). Foodborne pathogens can cause serious health hazards including food poisoning and other foodborne diseases (FAO, 2009). High levels of total coliforms were identified in certain street-vended food samples by laboratory analysis (Al Mamun et al., 2013b; INFOSAN, 2010). Bacterial contamination of street foods can also occur by a number of pathogenic bacteria such as *Salmonella* spp., *Shigella sonnei*, *Staphylococcus aureas*, *Clostridium perfringens*, *Vibrio cholerae* and *Escherichia coli* (Ghosh et al., 2007; Hanashiro et al., 2005; Kraig and Sen, 2013). The risk of microbiological contamination is primarily dependent on the type of food, the method of preparation, and the manner by which it is preserved before consumption (Draper, 1996; FAO, 2009; INFOSAN, 2010; WHO, 1996).

The major risks usually remain in low-acid foods (for instance dumplings, rice, porridges and steamed doughs, raw vegetables, fruits, and grains) (Fellows and Hilmi, 2011; WHO, 1996). Animal foods (such as fish, meat, egg, milk) also contain low-acid components (Fellows and Hilmi, 2011), and these foods are frequently contaminated with pathogens before they are prepared into foods and cooked (Fellows and Hilmi, 2011; WHO, 1996). Bacteria may release toxins into these products that are not destroyed by cooking and may cause food poisoning (Fellows and Hilmi, 2011). Moreover, the dairy, egg, fish, and meat products support quicker bacterial growth compared to cereal and bakery products with a low moisture content, products with high concentration of sugars, salt, or acidulants, and several fermented food products (Draper, 1996). Acidic foods such as fruit juices, pickles, chutneys, and fermented dairy products are less likely to be contaminated, because the growth of pathogenic bacteria is not possible in these foods (Fellows and Hilmi, 2011). Snacks that are fried or baked during preparation are considered as safe foods, although some snacks with high water content may lead to potential bacterial contamination (Winarno and Allain, 1991).

From the microbiological point of view, street foods that are just cooked or served hot (eg, noodles, meat balls) are considered to be safe foods (Draper, 1996; FAO, 2007; Fellows and Hilmi, 2011; INFOSAN, 2010; Tinker, 1997; Winarno and Allain, 1991), whereas cold meals that are kept on display at ambient temperature for a long period of time (>6 h) after they are cooked or before they are consumed, are categorized as "high-risk" foods (Draper, 1996; Fellows and Hilmi, 2011; Kraig and Sen, 2013; Tinker, 1997; Winarno and Allain, 1991). The problems are intensified in tropical countries where foods are converted into excellent culture media for bacterial growth due to high ambient temperature and humidity (Kraig and Sen, 2013). Most meals containing peanut sauce or coconut milk are considered to be in the high-risk category (Winarno and Allain, 1991). Dishes containing raw ingredients, foods that are inadequately processed, and foods/beverages served with ice are also high-risk food items

(Arambulo et al., 1994; WHO, 1996). Because freezing is not a safe process for the removal of biological hazards, contaminated ice may introduce hazards to street foods and beverages (WHO, 1996).

Water is widely used almost in all street food vending operations, and it is frequently: (1) used for drinking purposes; (2) used for washing of food ingredients; (3) incorporated into food as a component; (4) used in the processing of food; and (5) used in the washing of hands, equipment, utensils, and containers (WHO, 1996). Because water is susceptible to be contaminated with microbiological hazards (Tinker, 1997; WHO, 1996), the safety of street foods is directly associated with the quality of the water available to street food vendors (Winarno and Allain, 1991). The supply of water of acceptable quality and in sufficient quantities for drinking, washing, cleaning, and other operations is the most critical challenge for street food vendors (den Hartog et al., 2006; FAO, 2009; Fellows and Hilmi, 2011; Kraig and Sen, 2013; Winarno and Allain, 1991). Inadequate access to safe water or using contaminated water may lead to the occurrence of foodborne diseases among the consumers (FAO, 2007; Fellows and Hilmi, 2011).

Numerous factors are associated with microbiological contamination of street foods. Unhygienic practices during food preparation (Abdussalam and Kaferstein, 1993; Draper, 1996; Fellows and Hilmi, 2011; Kraig and Sen, 2013), handling (Abdussalam and Kaferstein, 1993; Draper, 1996; INFOSAN, 2010; Kraig and Sen, 2013; WHO, 1996) and transport (WHO, 1996), vendors' poor personal hygiene (Abdussalam and Kaferstein, 1993; Draper, 1996; FAO, 2009; Fellows and Hilmi, 2011; Kraig and Sen, 2013), inadequate processing or preparation of foods (Arambulo et al., 1994; FAO, 2007; WHO, 1996), and inadequate storage facilities of prepared food (Abdussalam and Kaferstein, 1993; Draper, 1996; Kraig and Sen, 2013; WHO, 1996) may lead to pathogen growth and toxin formation. If food handlers suffer from specific diseases, microbiological hazards present on their skin or in their respiratory tract, intestine, and feces may contaminate foods (WHO, 1996). Moreover, cross-contamination can occur by food handlers after handling raw materials (WHO, 1996).

Poor environmental sanitation of the street food vending location is another significant contributory factor for microbiological contamination (Abdussalam and Kaferstein, 1993; Draper, 1996; FAO, 2007, 2009; Fellows and Hilmi, 2011; INFOSAN, 2010; Kraig and Sen, 2013). Unhealthy surroundings (eg, proximity of drains and public discharge sites) (FAO, 2009; Tinker, 1997), insufficient facilities for waste disposal (Abdussalam and Kaferstein, 1993; Draper, 1996; FAO, 2009; Fellows and Hilmi, 2011; Kraig and Sen, 2013), and inappropriate accumulations of garbage, dirty plates or utensils that attract insects, flies, rodents, or birds (Fellows and Hilmi, 2011; Kraig and Sen, 2013; Tinker, 1997) are contributing factors to microbial growth and increased risk of contamination of street foods. The design, infrastructure, and maintenance of vending units, utensils, and equipment are also essential issues for microbiological safety of street foods (INFOSAN, 2010; WHO, 1996). Several reasons for microbiological contamination of street foods are shown in Fig. 2.1.

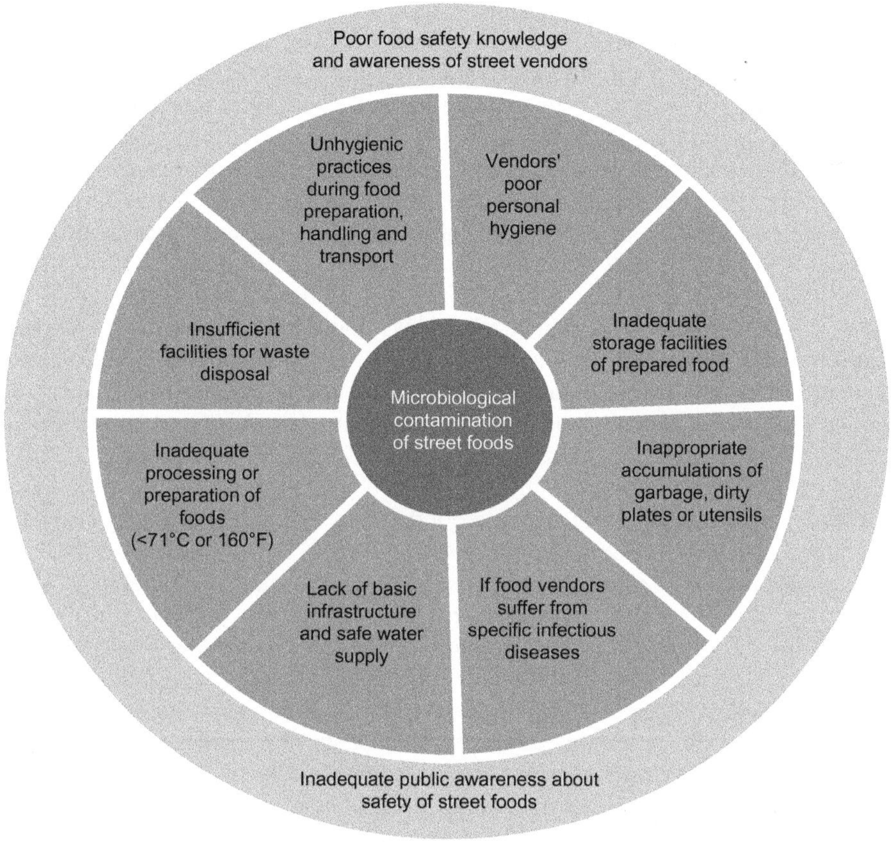

FIGURE 2.1

Several reasons for microbiological contamination of street foods.

CHEMICAL AND PHYSICAL HAZARDS IN STREET FOODS

In addition to the microbiological hazards described above, street foods can be contaminated with high levels of toxic chemicals. Various unpermitted nonfood-grade chemical additives, such as the coloring agents, rhodamine B, methanal yellow, or other textile dyes, are widely used by street food vendors in some countries (Draper, 1996; FAO, 2007, 2009; Fellows and Hilmi, 2011; INFOSAN, 2010; Winarno and Allain, 1991). These colorants are added to hide the poor quality of cheap food ingredients (Draper, 1996). In the same way, unapproved synthetic sweeteners are frequently used to adulterate beverages sold on the street (Fellows and Hilmi, 2011; Winarno and Allain, 1991). Although the proper uses of salt, spices, nitrates, and sugar are appropriate techniques for prevention of food spoilage, street food vendors purchase cheap ingredients containing unauthorized chemical preservatives or additives from illegal suppliers to keep the prices of their food items down (Draper, 1996; FAO, 2007).

A common toxicological problem with street foods that contain groundnuts (peanuts) is contamination with aflatoxins at levels above 30 μg/kg (ppb), the safety margin set by FAO/WHO guidelines (Fellows and Hilmi, 2011; Winarno and Allain, 1991). Pesticide residues higher than authorized levels have been identified in several street foods, particularly in vegetable-based products (Draper, 1996; FAO, 2009; Fellows and Hilmi, 2011; INFOSAN, 2010; Winarno and Allain, 1991). However, similar to street foods, home-cooked meals are likely to contain similar concentrations of aflatoxins and pesticide residues (Fellows and Hilmi, 2011; Winarno and Allain, 1991). Chemical contamination of street foods by heavy metals (FAO, 2009; INFOSAN, 2010) (for instance, lead ranging from 1.0 to 9.63 ppm (Fellows and Hilmi, 2011; Winarno and Allain, 1991)) and mycotoxins (FAO, 2009; INFOSAN, 2010) has also been reported. Using inappropriate equipment and utensils by food vendors may cause leaching of hazardous chemicals into food items, particularly into acidic food and beverages (WHO, 1996).

Physical hazards such as dust, dirt, and vehicular exhaust fumes may enter into street foods from the surrounding environment (den Hartog et al., 2006; INFOSAN, 2010). Physical contamination may also occur by food handlers wearing jewelry, bandages, or by careless food-handling practices (WHO, 1996). However, the public health threat from street foods is not only determined by the concentration of various additives and contaminants in a food item, but also by the cumulative daily intake of a certain additive or contaminant throughout the consumers' meal (Winarno and Allain, 1991).

ENSURING AND IMPROVING SAFETY OF STREET FOODS

Whether the street food products are prepared and served on the street or sold as prepackaged foods, they must be safe to eat, and not cause a threat to consumers' health. This safety should be maintained in every stage of the food chain from the farm to the fork, that is, from growing the food following proper good agricultural practices (GAPs), through handling and transport, to storage of ingredients, preparation and processing of the foods, and packaging or serving (Fellows and Hilmi, 2011).

Preparation and processing should ensure that growth of pathogens, production of toxic chemicals, and the introduction of physical hazards are prevented, and foods are not recontaminated (WHO, 1996). The most effective way of killing dangerous microorganisms from street food is appropriate cooking to 71°C (160°F) for long enough, while serving and consuming of hot foods immediately after cooking can also reduce the risk of foodborne diseases caused by pathogens

(Fellows and Hilmi, 2011; INFOSAN, 2010; Kraig and Sen, 2013; Tinker, 1997). Similarly, reheating food to 71°C (160°F) may eliminate many of the bacteria (Kraig and Sen, 2013). Moreover, proper use of utensils and equipment can prevent cross-contamination from raw materials (WHO, 1996). Uniform and universal supply of good-quality ingredients is also essential (Draper, 1996).

One of the most cost-effective options for reducing health risks posed by street foods is greater awareness of vendors of the basic food safety principles and measures (INFOSAN, 2010). At the point of sale, street food vendors should be aware of good personal hygiene and cleanliness, appropriate washing methods of utensils using clean and potable water, not allowing waste to accumulate nearby to prevent contamination by insects, birds, and rodents, and also protecting the food from environmental pollution (Fellows and Hilmi, 2011).

Many interventions are recommended to ensure the safety of street foods, such as the provision of adequate safe water, sanitation and refuse disposal facilities (den Hartog et al., 2006; Draper, 1996; Tinker, 1997), education and training programs for vendors (Draper, 1996; INFOSAN, 2010; Tinker, 1997; WHO, 1996), the infrastructural improvement of vending units and equipment for food preparation and storage (Draper, 1996; WHO, 1996), and raising food safety awareness among consumers (WHO, 1996; Winarno and Allain, 1991). Several approaches to ensure and improve safety of street foods are summarized in Fig. 2.2.

EDUCATION AND TRAINING OF STREET FOOD VENDORS

Tinker (1997) defined a street food vendor as "one selling ready-to-eat foods from a place having no more than three permanent walls" (Tinker, 1997). This definition straightforwardly indicates the poor socioeconomic condition of street food vendors, and concurrently it has been observed across many countries of the world that the majority of street food vendors are self-employed (Tinker, 1997), women-predominant (Draper, 1996; Tinker, 1997), middle-aged (Roever, 2014; Tinker, 1997), and illiterate or undereducated (Al Mamun et al., 2013a; INFOSAN, 2010; Roever, 2014; Tinker, 1997; WHO, 1996). Unfortunately, street food vendors have to work under unhygienic conditions with little or no infrastructure support (INFOSAN, 2010), and most of them are deficient in food safety knowledge and awareness (Al Mamun et al., 2013a; INFOSAN, 2010; WHO, 1996).

Because the use of appropriate food-handling techniques can prevent the majority of food-related illnesses and deaths (INFOSAN, 2010), education and training of street food vendors on safe handling and preparation of food and good hygienic practice may provide a cost-effective strategy to improve the safety of street-vended foods (INFOSAN, 2010; Tinker, 1997; WHO, 1996). Training and education programs should be customized according to the condition of the local street food safety hazards (WHO, 1996). Certification may be a useful tool to identify the vendors who are trained, and it may encourage consumers to consider

FIGURE 2.2

Several approaches to ensure and improve the safety of street foods.

the safety of street foods by selecting the proper vendor during their purchase (WHO, 1996). However, despite having a basic knowledge or training on food hygiene, food vendors may not bring this knowledge into practice unless several essential amenities, such as safe water supply and refuse disposal facilities, are sufficiently available (den Hartog et al., 2006).

WHO'S "FIVE KEYS TO SAFER FOOD" IN STREET FOOD OPERATIONS

The WHO's "Five Keys to Safer Food" (WHO, 2006, 2015) should be adopted for the street food sector to ensure good hygienic practices among the street food vendors, and these keys should be used as the basis for training of street food vendors worldwide (INFOSAN, 2010). Each key message containing a number of specific measures should be tailored to the conditions of the street foods operation, the food items being vended and the specific factors of the food vendors themselves (INFOSAN, 2010). The principles of the "Five Keys to Safer Food" addressing the street food sector are summarized below (INFOSAN, 2010; Kraig and Sen, 2013):

1. *Keep clean:* Keep vending unit and location clean. Wash your hands during food preparation and before handling food. Practice good personal hygiene.
2. *Separate raw and cooked:* Make sure that raw foods, particularly raw meat, poultry, and fish, are not in contact with cooked foods. Use separate equipment and utensils for handling raw and cooked foods.
3. *Cook thoroughly:* Make sure foods, especially meat, poultry, eggs, and seafood, have been thoroughly cooked to at least 70°C.
4. *Keep food at safe temperatures:* Keep cooked foods steaming hot until served. Do not leave cooked foods at ambient temperature for more than 2 h. Store foods either hot (best above 60°C) or cold (best below 5°C).
5. *Use safe water and raw materials:* If you are not sure about the safe water supply, boil it before adding it to food. Ensure that food ingredients and additives are authorized, and from reliable sources.

PROGRAMS, GUIDELINES, AND MANUALS FOR SAFETY OF STREET FOODS

International organizations, particularly the FAO and the WHO, have developed several programs, guidelines, and manuals to assist national and municipal authorities in ensuring and improving the safety of street-vended foods worldwide (FAO, 2009; FAO/WHO-CAC, 1999; INFOSAN, 2010; WHO, 1996, 2006, 2012, 2015). To provide guidance for governments, the WHO's *Codex Alimentarius Commission* prepared regional guidelines on the design of control measures for street-vended foods (FAO/WHO-CAC, 1999), and the FAO developed a training manual about the code of good hygienic practices for the preparation and sale of street-vended foods (FAO, 2009). Moreover, the WHO published a structured report on essential safety requirements for street-vended foods (WHO, 1996). Although many countries have already adopted these guidelines by implementing appropriate programs related to street foods, some sections of the guidelines are difficult to implement, particularly in regard to the provision of infrastructure for street food operations (INFOSAN, 2010).

In addition, there are specific training and educational materials on safe practices for street food preparation and handling issues, such as Good Agricultural Practices (GAPs), Good Handling Practices (GHPs), Good Manufacturing Practices (GMPs), and Hazard Analysis and Critical Control Point (HACCP) systems. These "Good Practices" approaches are initiated from the farm and end at the point of sale or consumption in order to ensure the maintenance of safety, hygiene, and quality of street food products at all levels of the food chain (Fellows and Hilmi, 2011).

IMPLEMENTATION OF HACCP APPROACH IN STREET FOOD SECTOR

The Codex Alimentarius Commission has recognized the HACCP system as the most cost-effective and flexible approach to ensure food safety at all stages of the food supply (WHO, 1996). Therefore, it is recommended for the relevant authorities to conduct HACCP studies to identify and integrate critical control measures into strategies for improving the safety of street foods (Arambulo et al., 1994; Draper, 1996; WHO, 1996). In this method, a number of checklists are used to identify and characterize the potential hazards at each critical point of the food chain where risk occurs, and to establish priorities for intervention and control (Draper, 1996; Kraig and Sen, 2013; WHO, 1996). The outcome of the HACCP analysis can be utilized for setting priorities, formulating interventions, and identifying the needs of vendors and customers for education and training (Draper, 1996; WHO, 1996). However, change in attitude by policy- and decisionmakers is a prerequisite to implement the HACCP approach in the street food sector, and at the same time training on HACCP application for the technical and inspection staff is a big challenge (WHO, 1996). Details of the HACCP analysis procedures including its seven principal activities related to street food sector are described in a WHO document titled "Essential Safety Requirements for Street-Vended Foods" (WHO, 1996).

REGULATION, REGISTRATION, AND LICENSING OF STREET FOOD VENDORS

Regulation is a common approach for controlling the street food sector worldwide (WHO, 1996). Legislation and regulatory control of street foods may differ from country to country (Draper, 1996; Kraig and Sen, 2013), and even from city to city within the same country (Kraig and Sen, 2013). Many countries have already adopted a number of ordinances or laws to regulate street food operations (INFOSAN, 2010; Kraig and Sen, 2013; WHO, 1996), while several countries have no specific regulation or control system at all (INFOSAN, 2010; Jayasuriya, 1994; Kraig and Sen, 2013). When such regulations exist, enforcement is a big challenge due to a wide diversity of the street food operations in terms of the large number of vendors and processors, the types of vending units, the type of food sold, and the vendors' mobility (Fellows and Hilmi, 2011; INFOSAN, 2010). In some countries, regulations are not enforced or enforced irregularly because of indifference or a shortage of resources and inspection staff (Fellows and Hilmi, 2011; Kraig and Sen, 2013).

Given that adulteration of foods is a major consumer concern in many countries (Draper, 1996), there is a necessity for legislation and control by government intervention (Draper, 1996; WHO, 1996). Although the principal purpose of regulations addressing the street food sector is to promote health and food safety (Kraig and Sen, 2013; Winarno and Allain, 1991), these regulations cover several issues, such as authorization or licensing to operate the production and sale of street foods, composition and presentation of food products, offenses and penalties, and institutions and officials responsible for food surveillance and control (Kraig and Sen, 2013). In some instances, regulations introduce fees to street food vendors for occupation of public space, causing traffic congestion, having access to clean water, electricity, and waste disposal systems (Fellows and Hilmi, 2011). Regulations should be realistic and achievable, so that these can be easily enforced to the street food vendors (Winarno and Allain, 1991). Regulations should be formulated and customized following an integrated consultation with vendors and consumers (WHO, 1996), particularly consumers' needs should be taken into consideration (Winarno and Allain, 1991). However, restricting street food operations or setting impossible requirements in the legislations may lead vendors to perform unhygienic procedures secretly (Winarno and Allain, 1991).

Many urban authorities have already started registration and licensing systems for street food vendors, while a considerable proportion of vendors still remain unlicensed (Draper, 1996; WHO, 1996). Easier registration or licensing processes, as well as fair inspections by relevant authorities, enable vendors to ensure safer street food (Fellows and Hilmi, 2011; WHO, 1996; Winarno and Allain, 1991). Simultaneously, the authorities may get several advantages from registering or licensing vendors; for example, identifying the vendors and the types of food sold, raising revenue, and creating an opportunity to provide training for food handlers in food safety (WHO, 1996). In order to facilitate smooth licensing and renewal, the authorities should simplify licensing requirements, keep the process transparent, omit unnecessary requirements, make the issuance or renewal of a license depending upon food safety awareness and a commitment to produce safe food (WHO, 1996).

AWARENESS RAISING AMONG STREET FOOD CONSUMERS

Street food consumers around the world represent all ages (den Hartog et al., 2006) and various socioeconomic classes (den Hartog et al., 2006; Tinker, 1997; Winarno and Allain, 1991). These consumers may include a variety of occupations with different income levels, such as business professionals, market traders, tourists, taxi drivers, laborers, construction workers, students, and housewives (Tinker, 1997). It should be kept in mind that consumers are the individuals who bear the consequences of eating unsafe street foods (WHO, 1996). Because consumers are usually attracted by convenience and low prices of any commodity, they may ignore the importance of hygiene or sanitation of street foods (Winarno and Allain, 1991). Moreover, many customers do not have proper understanding or awareness regarding the association between contaminated foods and foodborne diseases (WHO, 1996; Winarno and Allain, 1991).

In order to improve consumers' awareness about the safety of street-vended foods, relevant health authorities must come forward with a variety of educational approaches including mass media, national seminars, and community health education (FAO/WHO-CAC, 1999; WHO, 1996). In addition, incorporation of food safety messages into school textbooks will create opportunities for increasing awareness of children regarding street food safety (Al Mamun et al., 2013a; WHO, 1996). Consumers can always utilize their purchasing power to rectify vendors' unhygienic food handling practices by providing a powerful encouraging force to vendors (WHO, 1996). Consumers' organizations may also play a vital role to improve the safety of street foods in association with government authorities, vendors' associations, and food scientists (Winarno and Allain, 1991).

REFERENCES

Abdussalam, M., Kaferstein, F.K., 1993. Safety of street foods. World Health Forum 14, 191–194.

Al Mamun, M., Rahman, S.M., Turin, T.C., 2013a. Knowledge and awareness of children's food safety among school-based street food vendors in Dhaka, Bangladesh. Foodborne Pathog. Dis. 10, 323–330.

Al Mamun, M., Rahman, S.M., Turin, T.C., 2013b. Microbiological quality of selected street food items vended by school-based street food vendors in Dhaka, Bangladesh. Int. J. Food Microbiol. 166, 413–418.

Arambulo III, P., Almeida, C.R., Cuellar, J., Belotto, A.J., 1994. Street food vending in Latin America. Bull. Pan Am. Health Organ. 28, 344–354.

den Hartog, A., van Staveren, W.A., Brouwer, I.D., 2006. Food Habits and Consumption in Developing Countries: Manual for Field Studies. Wageningen Academic Publishers, Wageningen.

Draper, A., 1996. Street Foods in Developing Countries: The Potential for Micronutrient Fortification. London School of Hygiene and Tropical Medicine, London.

FAO, 1989. Food and Nutrition Paper 46, Report of an Expert Consultation on Street Foods, Yogyakarta, Indonesia, December 5–9, 1988. Food and Agriculture Organization of the United Nations (FAO), Rome.

FAO, 2007. Promises and challenges of the informal food sector in developing countries. Food and Agriculture Organization of the United Nations (FAO), Agricultural Economics and Engineering Department, University of Bologna, Italy, and Department of Sociology and Anthropology, University of Ottawa, Canada, Rome.

FAO, 2009. Good Hygienic Practices in the Preparation and Sale of Street Foods in Africa: Tools for Training. The Food and Agriculture Organization of the United Nations (FAO), Rome.

FAO/WHO-CAC, 1999. Revised Regional Guidelines for the Design of Control Measures for Street-Vended Foods in Africa (CAC/GL 22R—1999). FAO/WHO Codex Alimentarius Commission.

Fellows, P., Hilmi, M., 2011. Selling Street and Snack Foods. Rural Infrastructure and Agro-Industries Division, Food and Agriculture Organization of the United Nations (FAO), Rome.

Ghosh, M., Wahi, S., Kumar, M., Ganguli, A., 2007. Prevalence of enterotoxigenic *Staphylococcus aureus* and *Shigella* spp. in some raw street vended Indian foods. Int. J. Environ. Health Res. 17, 151–156.

Hanashiro, A., Morita, M., Matté, G.R., Matté, M.H., Torres, E.A., 2005. Microbiological quality of selected street foods from a restricted area of São Paulo city, Brazil. Food Control 16, 439–444.

INFOSAN, 2010. Basic Steps to Improve Safety of Street-Vended Food (INFOSAN Information Note No. 3/2010—Safety of Street-Vended Food). International Food Safety Authorities Network (INFOSAN), Geneva.

Jayasuriya, D., 1994. Street food vending in Asia: some policy and legal aspects. Food Control 5, 222–226.

Kraig, B., Sen, C.T., 2013. Street Food Around the World: An Encyclopedia of Food and Culture. ABC-CLIO, LLC, Santa Barbara, CA.

Roever, S., 2014. Informal Economy Monitoring Study Sector Report: Street Vendors. Women in Informal Employment Globalizing and Organizing (WIEGO), Cambridge, MA.

Tinker, I., 1997. Street Foods: Urban Food and Employment in Developing Countries. Oxford University Press, Inc., New York, NY.

WHO, 1996. Essential Safety Requirements for Street-Vended Foods, revised ed. Food Safety Unit, Division of Food and Nutrition, World Health Organization, Geneva.

WHO, 2006. Five Keys to Safer Food Manual. Department of Food Safety, Zoonoses and Foodborne Diseases, World Health Organization, Geneva.

WHO, 2012. Regional Consultation on Safe Street Foods: Bangkok, Thailand, June 20–23, 2011. World Health Organization Regional Office for South-East Asia, Food and Agriculture Organization Regional Office for Asia and the Pacific, and Institute of Nutrition, Mahidol University, Thailand, New Delhi.

WHO, 2015. Food Safety: The Five Keys to Safer Food Programme. World Health Organization, Geneva.

Winarno, F., Allain, A., 1991. Street foods in developing countries: lessons from Asia. Food, Nutr. Agric. (FAO) 1, 11–18.

PART

Microbiological hazards

II

CHAPTER

Factors influencing microbial safety of ready-to-eat foods

3

Parthena Kotzekidou

Department of Food Science and Technology, Faculty of Agriculture, Aristotle University of Thessaloniki, Thessaloniki, Greece

INTRODUCTION

Changing lifestyles, including convenience foods, for on-the-go consumption has led to a broad range of emerging microbiological issues in food safety. In addition, today's consumer has become increasingly concerned about the safety of food. "Food safety" refers to the content of various microbiological and chemical elements in food. In the context of work carried out by United Nations agencies and international organizations "food safety is the assurance that food will not cause harm to the consumer when it is prepared and/or eaten according to its intended use" (Codex Alimentarius, FAO/WHO, 2001). The FAO recognizes that the responsibility for the supply of safe, healthy and nutritious food is shared by all involved, from primary production to final preparation and consumption. In a global world trade context, a "holistic food chain approach" recognizes that the responsibility for supplying safe and nutritious food lies with all involved in food production (Burlingame and Pineiro, 2007). Consumers have become more insistent that foods available in the marketplace should be of good quality, safe, and not pose risks to them. Thus, in order to ensure this further, the responsibility of all those involved is high. Knowledge of the factors that are important in producing good-quality and safe foods is a prerequisite.

The safety of ready-to-eat (RTE) foods depends on the use of appropriate raw materials, the processing operation, and other parameters indicated in Table 3.1. Food composition, the variation in the level of pathogen injury, and the composition and level of the competitive microbiota as well as combinations of pH, water activity (a_w), and low-temperature storage after processing to limit the growth of foodborne pathogens affect food safety. In developing countries, poor food handling and sanitation practices, inadequate food safety laws, weak regulatory systems, lack of financial resources to invest in safer equipment, and lack

P. Kotzekidou (Ed): Food Hygiene and Toxicology in Ready-to-Eat Foods.
DOI: http://dx.doi.org/10.1016/B978-0-12-801916-0.00003-0

Table 3.1 Factors Affecting Safety of RTE Foods

Main Factors	Subfactors
Product category	• Fresh produce (raw fruit and vegetables ready for consumption) • Multi-ingredient preparations consisting of cooked and uncooked foods • Cooked foods
Product formulation	• Water activity (a_w) (NaCl, sucrose) • pH (lactic acid, acetic acid) • Antimicrobials and preservatives • Indigenous microbiota • Epiphytic microbiota (fruit, vegetables) • Microstructure (liquid vs structured foods) • Scattered availability of nutrients • Edible coating
Storage temperature	• Cold storage (≤4°C) • Hot-held products (>60°C)
Packaging	• Vacuum-packed • Modified-atmosphere packed • "Smart packaging"
Shelf-life	• 2–5 days • >5 days
Food preparation surfaces (food contact surfaces, cutting boards)	• Biofilm formation • Disinfection
Hands of food handlers	• Use of disposable gloves • Bare hand contact with RTE foods • Frequency of hand washing • Dishcloths
Cross-contamination	• Mishandling of RTE foods • Irrigation water (fresh produce) • Slicing of foodstuffs • Utensils (knives, serving spoons)

of education for food handlers could also contribute to increased foodborne infections and intoxications (Tessema et al., 2014).

Companies implement food safety management systems to address microbial and chemical safety of their products (CAC, 2003), and in many countries this is even a legal requirement (EC, 2004; FDA, 2011). Additionally, many voluntary certifications (such as GlobalGAP—the European retailers' private collective standard; SQF, Safe Quality Foods; BRC, British Retail Consortium's global food safety standard; IFS, International Food Standard) and benchmarked national standards (eg, KenyaGAP, ChinaGAP, FlandriaGAP) are widely adopted as a license to trade. Another quality assurance standard developed on a national level is ICQM-Integrated Chain Quality Management in Belgium (www.vegaplan.be). The ICQM Standard applies specifically to agricultural crop production and horticulture and describes the minimum requirements for producers and workers on good practices to gain access to the high-value fresh produce market. GlobalGap is the European retailers' private collective standard set and also acts as an organization for benchmarking other voluntary standards in agricultural production (including fresh produce production) around the globe, enabling certification of GAPs (www.globalgap.org). An alternative organization, SQF (Safe Quality Foods), initially developed in Australia in the early 1990s and currently owned and managed by the Food Marketing Institute in the United States, has elaborated the SQF 1000 Code for primary production as a quality assurance standard for certification of GAPs (www.sqfi.com).

ENVIRONMENTAL FACTORS AFFECTING SAFETY OF RTE FOODS

Global warming and climate changes may have a potential negative impact on water- and foodborne diseases, which are caused by microbiological agents. Climate changes mainly impact on the contamination sources and pathways of microorganisms in the food chain. Temperature increases and changes in rainfall patterns have an impact on the persistence and patterns of occurrence of bacteria, viruses, parasites, and fungi and the patterns of their corresponding foodborne diseases. During recent years there has been a significant shift from sporadic cases of *Vibrio parahaemolyticus* towards large outbreaks attributed to the consumption of oysters harvested from northern waters linked to higher mean water temperatures (Tirado et al., 2010).

Any noticeable rise in ambient temperature resulting from climatic change will increase the risk of food poisoning and food spoilage unless the cold-chain is extended and improved. On the other side, energy is required to maintain the cold-chain and the generation of this energy contributes to CO_2 production and climatic change. The limited available data suggest that the cold-chain accounts for approximately 1% of CO_2 production globally. However, this is likely to increase if global temperatures increase significantly (James and James, 2010).

Soil is a natural environment for a variety of human pathogens. Zoonotic pathogens survive longer in moist clay-based soils at lower temperatures in the presence of manure. Fields that contain animal manure are more likely to be contaminated with enteric pathogens because of their ability to survive in soils for months or years (Doyle and Erickson, 2008). The maximum survival of *Escherichia coli* O157:H7, *Listeria monocytogenes*, *Salmonella*, and *Campylobacter jejuni* in soil ranges from 45 to 100 days and survival was found to be in the order *Campylobacter* $<$ *Listeria* $<$ *Salmonella* $<$ *E. coli* O157:H7 (Nicholson et al., 2005).

Among different climate conditions, rainfall is associated with the largest increase in bacterial contamination of surface waters by foodborne pathogens, such as *L. monocytogenes*. Especially, seafood farms have potentially greater risk of *Listeria* contamination because brooks, rivers, and other surface waters enter such farms after heavy rainfall (Miettinen and Wirtanen, 2005).

The quality of irrigation water and the type of irrigation system influence the microbial safety of fresh produce (fresh fruits and vegetables). Flood and spray irrigation represent the greatest risk, because contaminated water can be directly deposited onto the edible leaves of produce. When the crop is irrigated with contaminated water close to harvest and/or contamination is transferred by cutting implements during the harvesting process, internalization of human pathogens into the inner plant tissue can occur. Solar irradiation is suggested to reduce the levels of pathogenic microorganisms in irrigation water and ensure safety. Water disinfectant treatments include chlorination, use of peroxyacetic acid, and UV treatment. Ozonation has also been described as a possible disinfection treatment for irrigation water (Suslow et al., 2003). The technology that is the best fit for the performance of the water treatment process depends upon physicochemical and microbial parameters associated with the water to be treated. The selection of technology relates also to aspects including capital and operational costs, complexity of the technology, required monitoring, and safety issues (Van Haute et al., 2015).

Extreme weather conditions could lead to a high rate of fungal spoilage and production of mycotoxins. The most widespread mycotoxins are metabolites produced by species of the genera *Aspergillus, Penicillium*, and *Fusarium*. Due to the public health and economic implications, the most commonly legislated mycotoxins are the aflatoxins, ochratoxin A, deoxynivalenol, fumonisin, zearalenone, and patulin (Tirado et al., 2010).

High temperatures may increase the rate of development of pathogens, leading to larger populations. Temperature and rainfall are positively associated with the number of salmonellosis cases (Zhang et al., 2010). A time series analysis study on human salmonellosis in several European countries showed that, in general, cases of salmonellosis increased by 5–10% for each 1°C increase in weekly temperature, for ambient temperatures above about 5°C. Similarly, a relationship between increasing temperature and salmonellosis reports has been reported in Canada and Australia (Tirado et al., 2010). However, the mechanisms underlying the seasonality in foodborne disease are a complex interplay of different factors, including human behavior and consumption patterns, pathogen prevalence in the animal reservoir, and pathogen environmental survival patterns.

Human norovirus infections, easily transferred via RTE foods prepared by infected food handlers, are seasonal; the number of laboratory-confirmed isolates and reported outbreaks of norovirus infection increases markedly during the winter months. Studies of the detection of norovirus in oysters also show higher rates of detection during the winter months. Norovirus is extremely difficult to remove from contaminated shellfish and consuming raw molluscs, such as oysters, involves potential exposure to norovirus. The public should be aware of the potential risks of norovirus associated with the consumption of raw shellfish such as oysters (HPA, 2010).

Extreme weather conditions (such as heavy rain, hail, or strong winds) influence the microbial safety of fresh produce; they affect plant physiology, tissue structure, and microbial ecology, and consequently plant tissue susceptibility for contamination and internalization with foodborne pathogens (Uyttendaele et al., 2015). On the other hand, low humidity, high temperatures, exposure to UV, and wind-mediated drying of the surface contribute to bacterial death on fresh produce. Survival of foodborne pathogens on fresh produce is significantly enhanced once the protective epidermal barrier has been broken either by physical damage, such as punctures or bruising, or degradation by plant pathogens or spoilage organisms (bacteria or fungi). These conditions can also promote the multiplication of human pathogens, especially at ambient temperatures. Injured cells and released cell fluids provide a nourishing environment for microbial growth. Especially, internalization of enteric pathogens in stomata provides not only a protective niche, but also a source of nutrients in the apoplastic fluid. In addition, internalized into the inner tissue, foodborne pathogens are protected from the postharvest washing process. Further internalization can occur during washing, especially if warm produce is submerged in cold water. Particularly, *E. coli* O157:H7 and *Salmonella* are adapted to survive in the plant environment. The above pathogens have surface epitopes that can bind to plant structures, such as stomata, and aid attachment. In addition, *Salmonella* is attracted to and metabolizes nutrients, which are contained within the apoplastic fluid of plants (Warriner and Namvar, 2010).

SOURCES OF MICROBIAL CONTAMINATION OF RTE FOODS

Changes in agricultural practices, such as increased use of improperly composted manure, may serve as a reservoir of bacterial pathogens. Intensive livestock husbandry and aquaculture may lead to the prophylactic use of antibiotics, potentially leading to residues of these compounds in animal-edible products as well as development of multiple-resistant pathogenic bacteria. *Listeria monocytogenes* is not only a human foodborne pathogen, but also causes disease in farm ruminants (goats, sheep, and cattle) with sheep being more susceptible to *L. monocytogenes* than cattle. Physical proximity between production animal and dairy processing facilities provides a higher risk for introduction of *L. monocytogenes* into the dairy production process, because it can be shed by ruminants with and without clinical symptoms. As *L. monocytogenes* is ubiquitous, it can persist in many other environments, such as retail operations, smoked seafood processing plants, and urban environments. Stringent sanitation programs, as well as measures to prevent and control persistent *Listeria* contamination in niches in a plant are thus critical to assuring the microbial safety of RTE products (Ho et al., 2007).

Food preparation surfaces may be reservoirs for bacterial contamination of RTE foods and they must be washed and disinfected on a regular basis to prevent bacterial aggregation and dissemination. In the case of contaminated preparation areas, due to the fecal–oral route of transmission and the possibility of less than optimal hand washing, consumers may be exposed to foodborne pathogens. Frequency and concentrations of bacteria found on preparation surfaces show inadequate cleaning processes that could lead to biofilm formation, which facilitates their survival on these surfaces. Biofilm formation is a bacterial mode of growth and survival which protects bacteria from stressful environmental conditions such as drying and cleaning procedures of food surfaces and the food processing environment. The ability of foodborne pathogens (ie, *Salmonella, L. monocytogenes, Yersinia enterocolitica, Bacillus cereus, Cronobacter sakazakii*, Shiga-toxin-producing *E. coli*) to colonize different inert food contact surfaces to form biofilm and the potential to act as a continuous source of bacterial contamination, which may cause postprocessing contamination, underlines the importance of different strategies for prevention of biofilm formation. The biofilm formation on preparation surfaces by the enteric pathogens, like *E. coli* O157:H7 and *Salmonella*, is facilitated by the bacterial fimbriae or flagella. Even if human pathogens cannot produce homogeneous biofilms, they may become entrapped in heterogeneous biofilms produced by nonpathogenic bacteria, making them much more resistant to stress conditions (Uyttendaele et al., 2015). The use of an effective disinfectant for contact surfaces is important.

The formation of biofilms on leaves of fresh vegetables (ie, spinach, lettuce, Chinese cabbage, celery, leek, basil, parsley, and endive leaves) has been observed (Morris and Monier, 2003). On fresh produce, biofilms occur as groups of bacterial cell aggregate in exopolysaccharide materials that serve to protect the cells from environmental stress, including desiccation and bactericidal agents. Between 30% and 80% of the total bacterial population on a leaf surface is located in biofilms and has an increased survival rate. Bacterial harborage in stomata, surface imperfections, and biofilm formation by bacteria on plant surfaces complicate efforts to consistently sanitize contaminated produce.

Fresh-cut fruit and vegetables generally consist of washed, cut, treated with sanitizing agents, and packaged products stored under refrigerated conditions. Disinfection is one of the most critical processing steps in fresh-cut vegetable production, and affects the quality, safety, and shelf-life of the end product. Washing contributes to removal of dirt, pesticide residues, and microorganisms that are responsible for quality loss. Also, washing serves to precool cut produce and remove cell exudates that may support microbial growth. The fresh-cut industry has used chlorine as one of the most effective sanitizers to assure product safety. However, there is a trend in eliminating chlorine from the disinfection process, because there are concerns about its efficacy on the produce. There are also concerns regarding the environmental and health risks that are associated with the formation of carcinogenic halogenated disinfection byproducts. Taking into account that chlorine-derived products have greater potential for the disinfection of vegetables when organic matter is eliminated, modern produce washers should be designed to assist with the disinfection process by incorporating different stages, such as showers, to remove fluids and exudates from cut surfaces before the disinfection. However, the last stage prior to packaging should be the disinfection step, which requires very low doses of disinfecting agent to achieve good results (Gil et al., 2009). Conventional sanitizing methods are ineffective for removing internalized bacteria. Thus, improved sanitizing treatments should be developed (Olaimat and Holley, 2012). The World Health Organization (WHO) categorized leafy green vegetables as the highest priority in terms of fresh produce safety from a global perspective, as the most common produce items associated with outbreaks were greens-based salads and leafy vegetables (spinach, lettuce, and lettuce mixes or salads), herbs, sprouted seeds, tomatoes, and berries (World Health Organization, 2008).

When RTE foods come into contact with raw contaminated vegetables through cutting boards and stainless steel utensils, *Salmonella* survives. Salmonellosis associated with the consumption of raw or lightly heat-treated vegetables is due to poor hygienic practices in food service premises as shown by cross-contamination of RTE foods, food preparation surfaces, and utensils (Little et al., 2011). Slicing is also a risk factor in cross-contamination of RTE products, as blades are able to disseminate high numbers of *Salmonella*. RTE products are frequently sold in delicatessens and fast-food restaurants, where stainless steel blade retail slicers are commonly used in meal preparations. Most likely underreported, cross-contamination scenarios involving equipment and utensils such as boards, knives, and slicers are likely to occur when good hygiene practices are not applied by personnel at retail levels. Slicing of food, such as RTE meat products, can enhance the risk of cross-contamination by contact with contaminated surfaces. Apart from slicing, cutting, handling, or packaging operations contribute to an increase in cross-contamination risk. In this sense, improving cleaning and disinfection procedures through hand washing, glove changing, and daily sanitation of cutting utensils and display cabinets is essential (Carrasco et al., 2012).

Food ingredients like spices, which are applied to RTE foods after the final food manufacturing pathogen reduction step, account for at least 70% of foodborne illnesses (CDC, 2010). Contamination of spices with pathogens is significant and may contribute to the foodborne illness burden. This is due to the long shelf-life of spices and the ability of pathogens to survive in spices for long periods of time. *Salmonella* can survive in dried spices and other low-moisture foods for prolonged periods. A number of studies throughout the world have found significant microbial contamination of spices at the point of retail. *Salmonella enterica* subspecies *enterica* and *Bacillus* spp. are the most common bacterial pathogens linked to spice-associated outbreaks (Van Doren et al., 2013).

According to the World Health Organization (WHO, 1992), 25% of foodborne outbreaks are closely associated with cross-contamination events involving deficient hygiene practices, contaminated equipment during slicing, contamination via food handlers, or preparation surfaces. When foodborne pathogen survival is allowed under cross-contamination and recontamination conditions, several factors, such as inadequate temperatures during food preparation, survival through undercooking or food handling, become critical.

HURDLES AFFECTING MICROBIAL GROWTH IN RTE FOODS

In RTE foods the growth and/or survival of pathogens (and their spores) are affected by the intrinsic properties (eg, a_w, pH, presence of preservatives) as well as the applied inactivation technology (eg, sterilized, pasteurized, probability on post-contamination). RTE products are considered as moderate or high risk when the final product has a water activity (a_w) >0.6 and/or a pH > 4.2, and contains no intrinsic antimicrobials. Regarding the high-risk situation, it is crucial that the product has a very high water activity (a_w > 0.98) and/or a pH 6.5–7.5, and/or is sensitive to post-contamination (Luning et al., 2011).

The most important hurdles that should be taken into account to improve microbial food safety are temperature, a_w, and pH. For example, Shiga-toxin-producing *E. coli* cannot grow at temperatures <8°C, pH < 4.5, and a_w of <0.95 (ICMSF, 1996). *Salmonella* cannot grow at temperature <5.2°C, pH < 3.8, and a_w of <0.95, while *L. monocytogenes* cannot grow at temperature <− 0.4°C, pH < 4.39, and a_w of <0.92 (ICMSF, 1996). The behavior of these pathogens, when subjected to a combination of low pH, reduced a_w, and various temperatures, is crucial for the safety of RTE foods. *Campylobacter* is the principal cause of bacterial gastroenteritis in the United Kingdom and its control in poultry meat is a major public health strategy for the prevention of campylobacteriosis. *Campylobacter* is killed through proper cooking, and avoiding contamination of RTE foods by raw poultry is of importance (Little et al., 2011).

Refrigeration is important in maintaining the safety and quality of food. It stops or reduces the rate at which changes occur in food. These changes can be microbiological (growth of microorganisms), physiological (eg, ripening, senescence, and respiration), biochemical (eg, browning reactions, lipid oxidation, and pigment degradation) and/or physical (eg, moisture loss, shriveling). Ensuring that temperatures are kept at or below 4°C throughout the cold-chain is essential for microbial safety. Survival of *Salmonella* and *E. coli* O157:H7 for extended periods in chilled conditions may be of concern. Psychrotrophic organisms, such as *L. monocytogenes*, are capable of growth at low temperatures. However, reducing the storage temperature (≤4°C) will significantly reduce the growth rate.

Improper cooling or "slow cooling," of food has been identified as a food safety problem in retail food service operations. The FDA Model Food Code requires that food should be cooled from 57.2°C to 21.1°C (135°F to 70°F) within 2 h and from 57.2°C to 5°C (135°F to 41°F) within a total of 6 h. The FDA defines cooling as a critical control point that is essential in preventing foodborne illness outbreaks in retail foodservice operations (FDA, 2009).

Listeria monocytogenes is a pathogen which has the ability to survive in stress conditions such as low temperature, high salt concentration, and low pH. Taking into account that *L. monocytogenes* is a psychrotrophic microorganism, refrigerated RTE foods must be kept at safe temperatures (ie, ≤4°C) to minimize the growth of the pathogen, especially in the case of products with a long shelf-life (González et al., 2013). Listeriosis is normally related to the consumption of refrigerated minimally processed RTE food, which is one of the major food vehicles of listeriosis outbreaks (EFSA, 2007). As listeriae are ubiquitous organisms able to multiply at refrigeration temperatures and under anaerobic conditions (as facultatively anaerobic bacteria with psychrotrophic and mesophilic features), they are of major concern, especially in contaminated RTE meat, poultry, and seafood products. However, RTE cheese and dairy products have been linked to multiple human listeriosis cases (Ho et al., 2007).

Raw milk, which is contaminated with foodborne pathogens and introduced into dairy processing plants, constitutes a risk to human health if used unpasteurized for the production of some types of cheeses or in cases of cross-contamination with pathogens. In the production of fermented dairy products, the risk from the pathogenic bacteria is eliminated during processing (ie, the pasteurization of raw milk, the length of maturation, and the controlled storage temperature) together with the intrinsic properties of cheese such as the composition, moisture content, pH, content of salt, and presence of antimicrobial compounds produced by the starter culture. However, potential temperature abuse of raw milk on the farm, during transportation, and/or prolonged storage at the milk processing plant could promote the growth of psychrotrophic pathogenic bacteria, such as *L. monocytogenes*, to unacceptable levels, if not entirely eliminated by the low treatment temperatures (60–63°C) commonly used in the manufacture of cheese. The prevalence rates of *L. monocytogenes* in retail cheese are higher in soft and semisoft cheese compared with hard cheese (EFSA, 2011). The high pH value of soft cheeses with white molds and a red-smeared surface constitute a suitable environment for the growth of *L. monocytogenes*. The presence of *L. monocytogenes* in cheese is a concern due to potential growth or long-term survival under refrigeration conditions (Kousta et al., 2010; Melo et al., 2015).

In dry fermented sausages, which should reach a pH of 5.3 or less at the end of the fermentation period and the subsequent drying should reduce the a_w to 0.90 or less, the combination of several factors (ie, concentration of NaCl, nitrite and lactic acid, water activity, pH, choice of starter cultures, and smoking) results in the inhibition of growth of most pathogens. However, *L. monocytogenes* and *E. coli* O157:H7 can survive in fermented sausage and thus may not be completely eliminated during the production process. The recognition of dry fermented sausages as a potential food safety hazard has led to regulations to minimize the risk of Shiga-toxin-producing *E. coli*. In the United States, Canada, and Australia, implemented regulations for producers state that production should be validated to ensure that levels of Shiga-toxin-producing *E. coli* are reduced by 5 $\log_{10}$ in United States and Canada or 3 $\log_{10}$ in Australia (Pragalaki et al., 2013).

While initial *L. monocytogenes* levels in contaminated foods are usually low, the ability of *L. monocytogenes* to survive and multiply at low temperatures allows it to reach levels that are high enough to cause human disease; particularly if contaminated foods, which allow *L. monocytogenes* growth, are stored for prolonged times under refrigeration (Kotzekidou, 2013a). According to the European Commission Regulation 2073/2005 & 1441/2007 (EC, 2005,2007), RTE foods other than those intended for infants or special medical purposes (for which zero tolerance has been adopted) are subdivided into those that are able to support the growth of *L. monocytogenes* (ie, products "with pH $>$ 4.4 or $a_w > 0.92$, products with pH $>$ 5.0 and $a_w > 0.94$ and products with a shelf life of more than five days") and into those that are not (ie, products "with pH $\leq$ 4.4 or $a_w \leq 0.92$, products with pH $\leq$ 5.0 and $a_w \leq 0.94$ and products with a shelf-life of less than five days"). For RTE foods that are able to support the growth of *L. monocytogenes*, the regulation demands the absence of the pathogen (in 25 g) "before the food has left the immediate control of the food business operator, who has produced it," but allows up to 100 CFU/g in "products placed on the market during their shelf-life." Regarding RTE foods that are unable to support *L. monocytogenes* growth, the 100 CFU/g limit applies throughout the shelf life. The same criteria have been established for *L. monocytogenes* in Codex Alimentarius in 2009 (Codex Alimentarius Commission, 2009). Regulations in Canada, New Zealand, and Australia also differentiate between RTE food products that can support the growth of *L. monocytogenes* (absent in 25-g sample) and other food products ($<$100 CFU/g).

There is an increased risk of listeriosis when RTE foods are stored at 10°C (Okada et al., 2013). Storage at slight abuse temperatures (eg, 7–10°C) of refrigerated RTE foods that permit *L. monocytogenes* growth increases greatly the risk that *L. monocytogenes*, if present, will reach numbers that could cause human disease (Kotzekidou, 2013b). In the United States, a zero-tolerance policy has been adopted with respect to *Listeria* in RTE foods. However, the US Food and Drug Administration has published a proposal for *L. monocytogenes* in RTE foods not able to support growth of this pathogen, to be permitted up to the level of 100 CFU/g (FDA, 2012).

The activity of the epiphytic microbiota, the structure of the microenvironment where microbial growth occurs, and the nutrient availability constitute stressful factors, which may introduce high variability in the growth potential of foodborne pathogens on RTE foods. On liquid media, the daughter cells drift away from the mother cell, after each bacterial division. Thus, the possibility of "crowding effect" (ie, the growth deceleration due to interactions of closely located cells) is limited. On the contrary, when cells are immobilized on a solid matrix, as in the case of structured foods, cells grow as colonies and, hence, interactions may occur between cells within a colony or between adjacent colonies. The inhibitory interference of the epiphytic microbiota against pathogens may be attributed to the competition for the available nutrients, the production of antimicrobial substances (ie, lactic acid, bacteriocins) and/or the alteration of the physicochemical properties (ie, pH, redox potential) of the microenvironment, in which the growth of pathogens may occur. The product structure (ie, solid vs liquid) also influences the pathogens' growth due to the slower diffusion rate of nutrients or inhibitory substances compared to juices (Manios et al., 2013). The effects of food structure include constraints on the mechanical distribution of water, the chemical redistribution of organic acids, including food preservatives, and physical constraints on mobility of microorganisms (Pragalaki et al., 2013). Food matrix-related characteristics may have an influence not only on the ability of foodborne pathogens like *L. monocytogenes* to survive and grow in food, but also on the potential to tolerate host defenses and cause disease. Data from studies using real food matrices may be advantageous, when compared to those based on liquid media. As they can better reproduce the real milieu encountered by a foodborne pathogen prior to host infection and how gastrointestinal stress is sensed when the bacterial cells are previously exposed to a food environment (Melo et al., 2015).

Edible coatings have been used to improve the quality of fresh-cut produce and to minimize changes due to processing. Edible coatings provide a partial barrier to gas exchange (including water vapor), and help to maintain or improve color, texture, mechanical integrity, volatile flavors, and reduce microbial growth. Thus, edible coatings contribute to extend the shelf-life of fresh-cut fruits by reducing moisture and solute migration, gas exchange, respiration, and oxidative reaction rates, as well as by reducing or even suppressing physiological disorders (Bierhals et al., 2011). The effects of edible coatings depend on the type of material with which they are elaborated, such as polysaccharides, protein, and lipids. The efficiency of coatings is not the same for different products, due to their different chemical composition and moisture. The choice of coating used on fresh-cut products is very critical due to the hydrophilic nature of cut surfaces where some coatings may not adhere. In addition, some coatings may not resist water vapor pressure. Lipids confer important water barrier characteristics, but may give a gummy mouth feel to the product, while hydrophilic polymers have less barrier properties, especially with high relative humidity values (RH above 0.70). In addition, antioxidants, fungicides, and preservatives may be added to the emulsion mix to increase the coating performance, while the addition of minerals or vitamins may enhance the nutritional value of fresh-cut products (Francis et al., 2012). Polyethylene glycol film is an approved material by Food and Drug Administration (US-FDA) with antifouling properties that is possible to be removed by washing with water before consumption (Zhang et al., 2014). The application of ultrathin polyethylene glycol film by employing oxygen plasma may be used to effectively inhibit the attachment of pathogenic bacteria on tomato surfaces (Zhang et al., 2014).

The packaging contributes to product safety and prevents microbial contamination as well as growth of microorganisms. A major packaging function is to prevent the product from mechanical injuries and contamination from the environment. However, when packaging is designed for lowering or preventing the growth of microorganisms, environment and/or packaging conditions should be properly controlled; that is, under modified atmosphere packaging storage temperature should be kept at low levels (Luning et al., 2011). Time temperature indicators that are embedded in the packaging may have a significant role in ensuring that a safe storage temperature is used. Some of the latest packaging technologies, such as antimicrobial films/wraps or "smart packaging," may be effective in reducing the growth of foodborne pathogens (like *Listeria*) and provide essential protection during storage and transportation. Modified atmosphere packaging with $\geq 30\%$ CO_2 is an efficient technique to help perishable chilled foods extend their shelf-life. The preserving effect of modified atmosphere packaging derives mainly from the use of CO_2 gas, which inhibits or retards microbial growth. The effectiveness of modified atmosphere packaging in preserving a product is due to the change in the pattern of food spoilage, and thus other spoilage organisms, such as anaerobic microorganisms, may outcompete the aerobic bacteria, thus extending the shelf-life (Lee et al., 2008).

PERSONNEL HYGIENE

RTE foods prepared by hand are often implicated in foodborne illness outbreaks, as this direct contact may lead to an increased incidence of contamination with potential foodborne pathogens (Colombari et al., 2007). The hands of food handlers are an important vehicle of food cross-contamination, as microorganisms found on hands vary significantly due to poor personal hygiene and contacted surfaces. The most common isolated bacteria are *Bacillus* spp., *Staphylococcus aureus* and coagulase-negative staphylococci, *E. coli*, and some other members of Enterobacteriaceae (Ayçiçek et al., 2004; Shojaei et al., 2006). Human norovirus, which is a leading cause of foodborne gastroenteritis worldwide, is easily transferred via RTE foods that are prepared by infected food handlers. Virus contamination during preparation of RTE foods that are not heated before consumption, such as delicatessen sandwiches, results in a health risk to consumers. In RTE foods, human norovirus is the most important foodborne pathogen in terms of disease outbreaks. It is mainly linked to ill food handlers, with an estimated 58% of foodborne outbreaks being associated with noroviruses in the United States (Sharps et al., 2012). Food handlers can contaminate RTE foods either by introducing the virus onto food via poor hand hygiene, or by cross-contamination while handling contaminated food. One reason for the worldwide importance of noroviruses is their rapid transmission, which is facilitated by factors such as the low infectious dose of 10–100 norovirus particles and their high environmental stability (Verhaelen et al., 2013). Inadequate hand hygiene and gloving seem to play major roles in human norovirus transmission linked to food handlers. Wearing gloves reduces the risk of virus transfer from contaminated hands and can partially, but not completely, protect food from contamination by the food handler. Glove wearing is recommended in the US Food Code (FDA, 2009) and the Codex Alimentarius (2012) for the preparation of RTE foods. Although, isolation rates of bacteria on gloved hands are lower than on bare hands (Ayçiçek et al., 2004), microorganisms do adhere to the surfaces of gloves. Disposable gloves must be changed frequently, otherwise they can be sources of cross-contamination. It is also important that the hands are washed with an antimicrobial product prior to wearing gloves, because microorganisms can rapidly multiply inside the moist and warm environment of the gloves.

Effective hand hygiene, including hand washing with soap and water, is crucial in preventing contamination of food ingredients by noroviruses (Rönnqvist, et al., 2014). Appropriate hand hygiene is probably the most relevant measure to prevent virus contamination of fresh produce by food handlers (Mokhtari and Jaykus, 2009). Washing hands under running water with soap for 20 s and subsequent hand drying with paper towels is recommended (CAC, 2012; EFSA, 2011). Washing hands for 10 s with soap reduces roughly 1 log-unit of norovirus genomes from hands (Liu et al., 2010).

Apart from the actual virus reduction on hands, compliance to good hand hygiene practices determines the successful prevention of norovirus introduction by infected food handlers. Handwashing compliance amongst foodservice operations personnel can be enhanced by educational and training programs and easy access to handwashing facilities (Soon and Baines, 2012). Food handlers may also contaminate food if they are working when continuing to be infectious. The exclusion of food handlers, which shed human norovirus, would be certainly the most efficient prevention measure to reduce the introduction of human norovirus into the food chain by food handlers. However, this approach seems not feasible, because infections may be asymptomatic and food handlers cannot be excluded from work for the whole period in which viruses may be shed (Verhaelen et al., 2013).

Food handlers can spread human norovirus to a large number of food items, making prevention of virus contamination on food handlers' hands crucial for food safety. Nevertheless, complete prevention is unlikely to be achieved in practice, giving the need for reliable and effective intervention measures. Interventions to achieve a 1 log-unit reduction of infectious human norovirus on the food handlers' hands or contaminated product, may lower the public health risk substantially. A 1 log-unit reduction of infectious virus particles on food handlers' hands, achieved by, for example, hand washing, decreases the health risk of consumers. Two widely used types of hand hygiene products are antibacterial liquid soaps and alcohol-based hand sanitizers. Sodium hypochlorite affects the integrity of both capsid protein and viral RNA, while ethanol affects the viral capsid protein, but not the nucleic acid. Ethanol-based hand sanitizers are an effective deterrent to the transmission of enveloped viruses such as influenza virus, hepatitis B virus, and herpes simplex viruses. However, they appear to be less effective for controlling the transmission of human enteric viruses, particularly the epidemiologically important human noroviruses. The relatively poor efficacy of ethanol-based hand sanitizers against norovirus has important implications for infection control in food service establishments, where human norovirus outbreaks are most frequent. Even though water washing, with or without the addition of an antibacterial soap, can remove some noroviruses from fingers, it should not be relied upon to eliminate virus that can be shed in titers of up to 10^{12} genome copies per gram of stool. Clearly, a need remains to develop commercial hand hygiene agents with specific, high-level activity (≥4 log reduction) against human gastrointestinal viruses (Liu et al., 2010).

Dishcloths are commonly used to clean surfaces, kitchen equipment and utensils, crockery and cutlery, and so on, enhancing the potential for cross-contamination between food-related habitats. Foodborne bacteria are also commonly present in kitchens and households, and form an important source of foodborne illness. High bacterial concentrations (up to 7.47 log colony-forming units (CFU)/cm^2 total aerobic bacteria and up to 4 log CFU/cm^2 *E. coli*) have been detected in used dishcloths (Chavatte et al., 2014). Used dishcloths carry the largest load of total coliforms and fecal coliforms compared to other kitchen surfaces. The moist conditions, presence of food residues and storage at room temperature favor the survival and growth of pathogenic bacteria, such as *S. aureus* and *Arcobacter butzleri*. *Arcobacter* is an emerging foodborne pathogen which can cause human infections through the consumption of contaminated food such as poultry products, pork, and beef; drinking water is also a potential contamination route. Free-living protozoa are known vectors for pathogenic bacteria, and the simultaneous presence of both free-living protozoa and bacteria in dishcloths is affected by environmental factors (eg, number of days that the dishcloths had been in use and the use of detergent). Detergents reduce both free-living protozoa and bacterial numbers, being an effective measure of good hygiene in kitchen environments (Chavatte et al., 2014).

CONCLUSIONS

Intrinsic factors, epiphytic microbiota, structure of product, edible coatings, and packaging contribute to the safety of RTE foods. Climate changes affecting ambient temperature, increased use of improperly composted manure, and extreme weather conditions influence the contamination and internalization of enteric pathogens in fresh produce. Temperature control and maintenance of refrigeration are essential to reduce the growth rate and survival of foodborne pathogens in RTE foods. Food preparation surfaces, slicing equipment, dishcloths used for cleaning utensils, may be reservoirs for microbial contamination of RTE foods. Biofilm formation on food contact surfaces should be avoided by washing and disinfection on a regular basis. Glove wearing, appropriate hand washing, and exclusion of personnel showing clinical symptoms contribute to preventing contamination and norovirus transfer by infected food handlers.

REFERENCES

Ayçiçek, H., Aydoğan, H., Küçükkaraaslan, A., Baysallar, M., Başustaoğlu, A.C., 2004. Assessment of the bacterial contamination on hands of hospital food handlers. Food Control 15, 253–259.

Bierhals, V.S., Chiumarelli, M., Hubinger, M.D., 2011. Effect of cassava starch coating on quality and shelf life of fresh-cut pineapple (*Ananas comosus* L. Merril cv "Pérola"). J. Food Sci. 76 (1), E62–E72.

Burlingame, B., Pineiro, M., 2007. The essential balance: risks and benefits in food safety and quality. J. Food Composit. Anal. 20, 139–146.

CAC (Codex Alimentarius Commission). 2003. Code of hygienic practice for fresh fruits and vegetables, CAC/RCP 53–2003.

CAC (Codex Alimentarius Commission), 2012. Guidelines on the application of general principles of food hygiene to the control of viruses in food, CAC/GL 79–2012.

Carrasco, E., Morales-Rueda, A., García-Gimeno, R.M., 2012. Cross-contamination and recontamination by *Salmonella* in foods: a review. Food Res. Int. 45, 545–556.

CDC (Centers for Disease Control and Prevention), Dec 24 2010. *Salmonella* Montevideo infections associated with salami products made with contaminated imported black pepper and red pepper—United States, July 2009–April 2010. Morbidity and Mortality Weekly Report 59 (50), pp. 1647–1650.

Chavatte, N., Baré, J., Lambrecht, E., Van Damme, I., Vaerewijck, M., Sabbe, K., et al., 2014. Co-occurrence of free-living protozoa and foodborne pathogens on dishcloths: implications for food safety. Int. J. Food Microbiol. 191, 89–96.

Codex Alimentarius Commission, 2009. Risk assessment of *Listeria monocytogenes*. In: Ready to Eat Foods: Guidelines. FAO/WHO Microbiological Risk Assesment Series, vol. 4. Food and Agriculture Organisation, Rome.

Colombari, V., Mayer, M.D.B., Laicini, Z.M., Mamizuka, E., Franco, B.D.G.M., Destro, M.T., et al., 2007. Foodborne outbreak caused by *Staphylococcus aureus*: phenotypic and genotypic characterization of strains of food and human sources. J. Food Prot. 70, 489–493.

Doyle, M.P., Erickson, M.C., 2008. Summer meeting 2007—the problems with fresh produce: an overview. J. Appl. Microbiol. 105, 317–330.

EFSA, 2007. Scientific opinion of the panel on biological hazards on a request from the European Commission on request for updating the former SCVPH opinion on *Listeria monocytogenes* risk related to ready-to-eat foods and scientific advice on different levels of *Listeria monocytogenes* in ready-to-eat foods and the related risk for human illness. EFSA J. 599, 1–42.

EFSA, 2011. Scientific opinion, an update on the present knowledge on the occurrence and control of foodborne viruses. EFSA J. 9, 2190–2286.

European Commission, 2004. Regulation (EC) no 852/2004 of the European Parliament and of the Council of 29 April 2004 on the hygiene of foodstuffs. Off. J. Eur. Union L 139, 1–54.

European Commission, 2005. Commission regulation (EC) no. 2073/2005 of 15 November 2005 on microbiological criteria for foodstuffs. Off. J. Eur. Union L 338, 1–26.

European Commission, 2007. Commission regulation (EC) No 1441/2007 of 5 December 2007 amending regulation (EC) No 2073/2005 on microbiological criteria for foodstuffs. Off. J. Eur. Union L 332, 12–29.

FAO/WHO, 2001. Codex Alimentarius-Recommended International Code of Hygiene Practice-General Principles of Food Hygiene (CAC/RCP 1-1969, Rev. 3 (1997), Amended 1999. FAO/WHO, Rome. <http://www.codexalimentarius.net/download/standards/297/CXP_021e.pdf/>.

FDA, 2009. U.S. Food and Drug Administration, Food Code 2009. Available at: <http://www.fda.gov/downloads/Food/FoodSafety/RetailFoodProtection/FoodCode/FoodCode2009/UCM189448.pdf>.

FDA, 2011. Food safety Modernization Act. 111th Congress Public Law 353.

FDA, 2012. CPG Sec. 555.320 *Listeria monocytogenes*. <http://www.fda.gov/ICECI/ComplianceManuals/CompliancePolicyGuidanceManual/ucm136694.htm/>.

Francis, G.A., Gallone, A., Nychas, G.J., Sofos, J.N., Colelli, G., Amodio, M.L., et al., 2012. Factors affecting quality and safety of fresh-cut produce. Crit. Rev. Food Sci. Nutr. 52, 595–610.

Gil, M.I., Selma, M.V., López-Gálvez, F., Allende, A., 2009. Fresh-cut product sanitation and wash water disinfection: problems and solutions. Int. J. Food Microbiol. 134, 37–45.

González, D., Vitas, A.I., Díez-Leturia, M., García-Jalón, I., 2013. *Listeria monocytogenes* and ready-to-eat seafood in Spain: study of prevalence and temperatures at retail. Food Microbiol. 36, 374–378.

Health Protection Agency (HPA), 2010. Shellfish consumption and the risk of norovirus infection. Available at: <http://www.hpa.org.uk/Topics/InfectiousDiseases/InfectionsAZ/Norovirus/oysterconsumptionnorovirus/>.

Ho, A.J., Lappi, V.R., Wiedmann, M., 2007. Longitudinal monitoring of *Listeria monocytogenes* contamination patterns in a farmstead dairy processing facility. J. Dairy Sci. 90, 2517–2524.

ICMSF, 1996. Microorganisms in Foods 5. Characteristics of Microbial Pathogens. Blackie Academic and Professional, London.

James, S.J., James, C., 2010. The food cold-chain and climate change. Food Res. Int. 43, 1944–1956.

Kotzekidou, P., 2013a. Microbiological examination of ready-to-eat foods and ready-to-bake frozen pastries from university canteens. Food Microbiol. 34, 337–343.

Kotzekidou, P., 2013b. Survey of *Listeria monocytogenes, Salmonella* spp. and *Escherichia coli* O157:H7 in raw ingredients and ready-to-eat products by commercial real-time PCR kits. Food Microbiol. 35, 86–91.

Kousta, M., Mataragas, M., Skandamis, P., Drosinos, E.H., 2010. Prevalence and sources of cheese contamination with pathogens at farm and processing levels. Food Control 21, 805–815.

Lee, K.-E., Kim, H.J., An, D.S., Lyu, E.S., Lee, D.S., 2008. Effectiveness of modified atmosphere packaging in preserving a prepared ready-to-eat food. Packag. Technol. Sci. 21, 417–423.

Little, C., Gormley, F., Mook, P., Gillespie, I., Foster, K., Harris, J., et al., 2011. Outbreaks of infection associated with ready-to-eat food. January 2011(ACM/1014). www.food.gov.uk/multimedia/pdfs/committee/acm1014hpa.pdf.

Liu, P., Yuen, Y., Hsiao, H.M., Jaykus, L.A., Moe, C., 2010. Effectiveness of liquid soap and hand sanitizer against Norwalk virus on contaminated hands. Appl. Environ. Microbiol. 76, 394–399.

Luning, P.A., Marcelis, W.J., Rovira, J., van Boekel, M.A.J.S., Uyttendaele, M., Jacxsens, L., 2011. A tool to diagnose context riskiness in view of food safety activities and microbiological safety output. Trends Food Sci. Technol. 22, S67–S79.

Manios, S.G., Konstantinidis, N., Gounadaki, A.S., Skandamis, P.N., 2013. Dynamics of low (1-4 cells) *vs* high populations of *Listeria monocytogenes* and *Salmonella* Typhimurium in fresh-cut salads and their sterile liquid or solidified extracts. Food Control 29, 318–327.

Melo, J., Andrew, P.W., Faleiro, M.L., 2015. *Listeria monocytogenes* in cheese and the dairy environment remains a food safety challenge: the role of stress responses. Food Res. Int. 67, 75–90.

Miettinen, H., Wirtanen, G., 2005. Prevalence and location of *Listeria monocytogenes* in farmed rainbow trout. Int. J. Food Microbiol. 104, 135–143.

Mokhtari, A., Jaykus, L.A., 2009. Quantitative exposure model for the transmission of norovirus in retail food preparation. Int. J. Food Microbiol. 133, 38–47.

Morris, C.E., Monier, J.-M., 2003. The ecological significance of biofilm formation by plant associated bacteria. Annu. Rev. Phytopathol. 41, 429–453.

Nicholson, F.A., Groves, S.J., Chambers, B.J., 2005. Pathogen survival during livestock manure storage and following land application. Bioresour. Technol. 96, 135–143.

Okada, Y., Ohnuki, I., Suzuki, H., Igimi, S., 2013. Growth of *Listeria monocytogenes* in refrigerated ready-to-eat foods in Japan. Food Addit. Contam. Part A 30, 1446–1449.

Olaimat, A.N., Holley, R.A., 2012. Factors influencing the microbial safety of fresh produce: a review. Food Microbiol. 32, 1–19.

Pragalaki, T., Bloukas, J.G., Kotzekidou, P., 2013. Inhibition of *Listeria monocytogenes* and *Escherichia coli* O157:H7 in liquid broth medium and during processing of fermented sausage using autochthonous starter cultures. Meat Sci. 95, 458–464.

Rönnqvist, M., Aho, E., Mikkelä, A., Ranta, J., Tuominen, P., Rättö, M., et al., 2014. Norovirus transmission between hands, gloves, utensils, and fresh produce during simulated food handling. Appl. Environ. Microbiol. 80, 5403–5410.

Sharps, C.P., Kotwal, G., Cannon, J.L., 2012. Human norovirus transfer to stainless steel and small fruits during handling. J. Food Prot. 75, 1437–1446.

Shojaei, H., Shooshtaripoor, J., Amiri, M., 2006. Efficacy of simple hand-washing in reduction of microbial hand contamination of Iranian food handlers. Food Res. Int. 39, 525–529.

Soon, J.M., Baines, R.N., 2012. Food safety training and evaluation of hand washing intention among fresh produce farm workers. Food Control 23, 437–448.

Suslow, T.V., Oria, M.P., Beuchat, L.R., Garrett, E.H., Parish, M.E., Harris, L.J., et al., 2003. Production practices as risk factors in microbial food safety of fresh and fresh cut produce. Compr. Rev. Food Sci. Food Saf. 2, 38–77.

Tessema, A.G., Gelaye, K.A., Chercos, D.H., 2014. Factors affecting food handling practices among food handlers of Dangila town food and drink establishments, North West Ethiopia. BMC Public Health 14, 571.

Tirado, M.C., Clarke, R., Jaykus, L.A., McQuatters-Gollop, A., Frank, J.M., 2010. Climate change and food safety: a review. Food Res. Int. 43, 1745–1765.

Uyttendaele, M., Jaykus, L.-A., Amoah, P., Chiodini, A., Cunliffe, D., Jacxsens, L., et al., 2015. Microbial hazards in irrigation water: standards, norms, and testing to manage use of water in fresh produce primary production. Compr. Rev. Food Sci. Food Saf. 14, 336–356.

Van Doren, J.M., Neil, K.P., Parish, M., Gieraltowski, L., Gould, L.H., Gombas, K.L., 2013. Foodborne illness outbreaks from microbial contaminants in spices, 1973–2010. Food Microbiol. 36, 456–464.

Van Haute, S., Sampers, I., Jacxsens, L., Uyttendaele, M., 2015. Selection criteria for water disinfection techniques in agricultural practices. Crit. Rev. Food Sci. Nutr. 55, 1529–1551.

Verhaelen, K., Bouwknegt, M., Carratala, A., Lodder-Verschoor, F., Diez-Valcarce, M., Rodriguez-Lazaro, D., et al., 2013. Virus transfer proportions between gloved fingertips, soft berries, and lettuce, and associated health risks. Int. J. Food Microbiol. 166, 419–425.

Warriner, K., Namvar, A., 2010. The tricks learnt by human enteric pathogens to persist within the plant environment. Curr. Opin. Biotechnol. 21, 131–136.

World Health Organization (WHO), 1992. WHO surveillance programme for control of foodborne infections and intoxications in Europe. Sixth report (1990–1992). World Health Organization, Geneva.

World Health Organization (WHO), 2008. Microbiological risk assessment series. Microbiological hazards in fresh fruits and vegetables. Available from: <http://www.who.int/foodsafety/publications/micro/MRA_FruitVeges.pdf/> (accessed 14.08.14.).

Zhang, Y., Bi, P., Hiller, J.E., 2010. Climate variations and *Salmonella* infection in Australian subtropical and tropical regions. Sci. Total Environ. 408, 524–530.

Zhang, M., Yang, F., Pasupuleti, S., Kyun, Oh, J., Kohli, N., Lee, I.-S., et al., 2014. Preventing adhesion of *Escherichia coli* O157:H7 and *Salmonella* Typhimurium LT2 on tomato surfaces via ultrathin polyethylene glycol film. Int. J. Food Microbiol. 185, 73–81.

CHAPTER

Foodborne viruses in ready-to-eat foods

4

Leena Maunula and Carl-Henrik von Bonsdorff

Department of Food Hygiene and Environmental Health, Faculty of Veterinary Medicine, University of Helsinki, Helsinki, Finland

INTRODUCTION

The role of viruses in causing foodborne outbreaks after consumption of shellfish, such as oysters, has been recognized for decades. However, only in late 1990s did more data become available regarding viral foodborne outbreaks, in addition to those caused by contaminated shellfish, mainly due to improved detection of noroviruses (NoV). NoV is currently recognized as one of the most common causative agents of foodborne gastroenteritis outbreaks in developed countries. Small and large viral outbreaks emerge depending on the mode of food contamination and distribution (Fig. 4.1).

Lynch et al. (2009) mentioned the growing burden of foodborne outbreaks due to contaminated fresh produce. Besides bacterial contamination, such as *Salmonella* and Enterohaemorrhagic Escherichia coli (EHEC), NoV and hepatitis A virus (HAV) are also mentioned. The authors set five points which especially affect contamination: (1) microbiological quality of water used for produce; (2) protection of food from fecal contamination; (3) appropriate washing and sanitizing conditions of food; (4) cold storage; and (5) good hygiene of food handlers. While viral contamination fails to diminish with cold storage and viruses do not grow in food, the other points are relevant for viruses as well.

GENERAL ASPECTS OF FOODBORNE VIRUSES

Viruses multiply inside host cells and thus the viral burden in the environment or in food slowly decreases temporally. Most enteric viruses are, however, quite resistant to conditions in the environment. NoV have been found to retain infectivity at least for 60 days in ground water (Seitz et al., 2011). In addition, in a study from the United States it is estimated that viruses cause 59% of foodborne illnesses (Scallan et al., 2011).

P. Kotzekidou (Ed): Food Hygiene and Toxicology in Ready-to-Eat Foods.
DOI: http://dx.doi.org/10.1016/B978-0-12-801916-0.00004-2

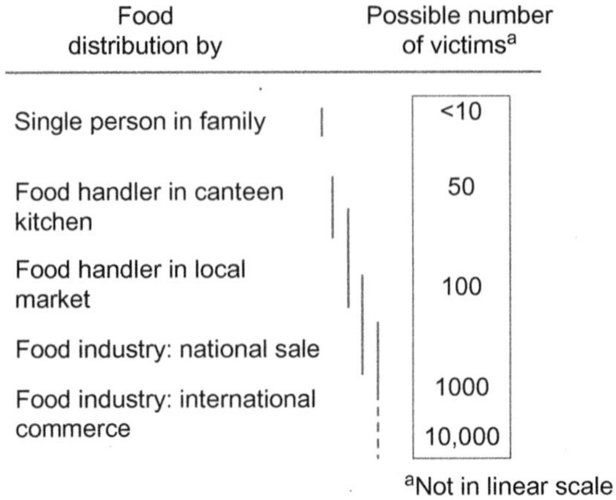

FIGURE 4.1

The effect of the mode of contamination on the potential resulting outbreak.

NOROVIRUS

The general features of human NoV are described in several reviews, such as Rodriguez-Lazaro et al. (2012) and Robilotti et al. (2015). NoV is a single-stranded, positive-sense RNA virus with a genome size of about 7.5 kb. This non-enveloped virus, 23–40 nm in diameter, belongs to the *Caliciviridae* family as a member of the Norovirus genus. The genome consists of three open reading frames that encode for nonstructural proteins (ORF1, includes polymerase), VP1 capsid protein (ORF2), and a phosphorylated VP2 protein (ORF3). NoV can be divided into six genogroups (I–VI), and human pathogens belong to genogroups I, II, and IV. A more detailed division into distinct genotypes reveals about 10 genotypes in GI (GI.1, etc.), over 20 in GII (GII.1, etc.), and less than 10 genotypes in GIII (GIII.1, etc.). The GII.4 is the most commonly detected type and emerging GII.4 variant viruses cause pandemics sequentially. The distribution of the most common genotypes is universal.

Regarding NoV infection, infective dose is low (1–10 virions). NoV infection is characterized by a 1–2-day incubation period after which symptoms appear, typically including diarrhea and vomiting. The infection affects all ages and is usually relatively mild, although increased mortality of hospitalized elderly patients during NoV epidemic peak seasons has been observed. Subclinical infections are common and symptoms last for 1–2 days, rarely 4–5 days, after which the person excretes viruses in feces for a couple of weeks (Atmar et al., 2008). Vomit also contains viruses and virus excretion may even precede symptoms (Atmar et al., 2008). Genome-based diagnostics is employed.

HEPATITIS A VIRUS

HAV is a single-stranded, positive-sense RNA virus, 35 nm in diameter with a genome of about 7.5 kb, in line with it belonging to the *Picornaviridae* family in which it is the only member of a Hepevirus genus (reviewed by Coudray-Meunier et al., 2014). Serologically, all HAV fall into one genus. The genome structure allows, however, a subdivision into distinct genotypes of which there are six (I–VI) primate-specific. Of these, types I, II, and III are human-specific and are further divided into A and B classes (Costa-Mattioli et al., 2002).

HAV infection starts as enteric and is characterized by a 2–6-week incubation period, after which the typical icteric liver infection takes place. The infection in adults is rather serious, with a long recovery period but a mortality of 2.3%. Children are mostly symptomless, which must be considered in dealing with outbreaks.

Globally, HAV is widely prevalent and its abundance is generally correlated to the hygiene (and economic) status of the county or region. A geographic map has been constructed based on HAV seroprevalence (Mohd Hanafiah et al., 2011). Three levels can be discerned: (1) high endemicity (Africa, Middle East, India, Central and South America); (2) moderate endemicity (Eastern Europe, Mediterranean area, Southeast Asia); and (3) low endemicity area (North America, Western Europe, Australia, and Japan). The HAV genotype distribution is largely geographically linked; the most prevalent types are IA and IB (Nainan et al., 2006).

The clinical diagnosis for HAV infection is mostly based on serology. A recent infection is identified by the presence of IgM class antibodies. For epidemiological reasons, a genome-based diagnosis is more desirable, because it allows the identification of the common source and geographical origin of the virus due to the numerous subspecies of the virus (Coudray-Meunier et al., 2014). The use of the same genome sequence, preferentially the VP1–A2 gene junction, would be a prerequisite for identification of multinational outbreaks (Koopmans and Duizer, 2004).

HEPATITIS E VIRUS

HEV is a small, naked, positive-strand RNA virus with a genome of about 7.2 kb in the *Hepeviridae* family. HEV occurs in man, and also in a wide variety of animals including rat, rabbit, camel, deer, wild boar, chicken, and fish (Sridhar et al., 2015). Mammalian HEV are divided into four species (A–D). The group A is further divided into four genotypes (1–4): genotypes 1 and 2 are purely human and 3 and 4 are zoonotic with a variety of hosts, primarily pigs (Smith and Simmonds, 2015; Pavio et al., 2015; van der Poel, 2014). Thus, HEV may occur either as purely human, or as a zoonotic infection.

HEV are known from extended large outbreaks, mostly waterborne, in hyper-endemic regions like India, Bangladesh, China (genotype 1), and Mexico (genotype 2). In a single year (2005), an estimated 20 million cases of HEV occurred worldwide (Rein et al., 2012). The majority of the infections were subclinical or very mild, but icteric hepatitis occurred in 3.4 million cases resulting in 70,000 deaths. A mortality rate of about 20% for pregnant women was noteworthy.

Zoonotic HEV has been found circulating globally in most pig farms (Meng et al., 1997; Kantala et al., 2015) but also in the environment (Diez-Valcarce et al., 2012; Maunula et al., 2013; Iaconelli et al., 2015). Viruses are found in liver and muscle of the infected animal. The environmental spread suggests, however, that the "natural" spread of enteric viruses via sewage to natural waters and subsequently to fresh produce (Brassard et al., 2012; Maunula et al., 2013) and seafood (Crossan et al., 2012) is possible. Antibody surveys seem to suggest that the infections are more common than the registered hepatitis cases would indicate (Baylis et al., 2010). Thus IgG antibodies are in some areas quite common, for example, among veterinarians exposed to swine (Meng et al., 2002; Bouwknegt et al., 2008) without apparent disease.

Stable viruses other than NoV, HAV, and HEV are also excreted in sewage and are thus able to contaminate food by the same transmission pathways (Rodriguez-Lazaro et al., 2012). For several reasons, these viruses, human astrovirus, rotavirus, sapovirus, and adenovirus, are more seldom linked to foodborne outbreaks, however. This may be because their target population is mainly children, thus food handlers are less critical in their transmission.

RISK READY-TO-EAT FOOD ITEMS

Currently, the apparent sources for foodborne NoV and HAV outbreaks can be grouped as follows: shellfish, vegetables, berries, and most often, food handlers implanting the virus on any raw or insufficiently heated foodstuff (Table 4.1). Virus prevalence in food items is not yet well known. The report by Baert et al. (2011) indicates a 6.7–34.5% prevalence of the NoV genome in soft red fruits, and even 28.2–50% in leafy greens. Studies with the adenovirus genome showed that about 3.2% of market frozen raspberries (0.7% fresh berries) were contaminated (Maunula et al., 2013). NoV GII.4 contamination has been reported in 13.6% (15/110) of dried tomatoes (in oil or bulk) of Italian or Turkish origin in Italian markets (Serracca et al., 2012), and also a single finding in Turkish bars and restaurants for tomato and green onion (Yilmaz et al., 2011). In a large study, oysters contained viruses in growing areas with a maximal load of 16,000–18,000 genomic copies/g (Lowther et al., 2012). The prevalence ranged monthly between 45.7% and 100%. A lower prevalence of 1.7% has been reported in Australia (Brake et al., 2014). There seems to be specific binding of genogroup I NoV to shellfish (Le Guyader et al., 2012). In Spain, bivalve species (mussels, clams, and cockles) growing in B region showed contamination, not only for NoV (25.6–32.1%), but also for HAV (10.1%) (Polo et al., 2015). Viral contamination ranged from 100 to 1000 RNA

Table 4.1 Viruses and RTE Food, Categories Follow Loosely the Definition for RTE by the FDA Food Code, 2009[a]

RTE Category	Viral Outbreaks Reported or Virus Detected in RTE		
	NoV	**HAV**	**HEV**
Raw animal food	Yes (shellfish fresh or frozen: oysters, clams, mussels)	Yes (shellfish fresh or frozen: oysters, clams, mussels)	Virus detected in shellfish
Raw fruits and vegetables washed and plant food not washed, cooked, processed, only naturally present rinds, peels, husks or shells removed	Yes (soft berries, leafy greens)	Yes (soft berries, leafy greens, green onions, dates)	Virus detected in berries (not common)
Spices, seasonings, sugar (plant substances)	Possible	Possible	Not known
Bakery items, (bread, cakes, pies, fillings, uncooked icings)	Yes (berry-containing icings, fillings)	Yes (berry-containing icings, fillings)	Not known
Dry fermented sausages, salt-cured or dried meat products	Not known	Not known	Yes (fermented pork liver sausages)
Cooked or frozen animal food	Possible, if handled after cooking	Possible, if handled after cooking	Not known
All potentially hazardous food cooked and cooled	Possible, if handled after cooking	Possible, if handled after cooking	Not known

[a] *http://www.fda.gov/Food/GuidanceRegulation/RetailFoodProtection/FoodCode/ucm186464.htm.*

copies/g of digestive tissue. In the 2007 US market survey, NoV was detected in 3.9–4.4% of live oysters (DePaola et al., 2010). The best-established sources for HEV infections are raw or undercooked deer meat (Kanayama et al., 2015) and sausages made of raw pork liver (Renou et al., 2014; Kanayama et al., 2015). HEV has also been detected as a rare finding in market frozen raspberries (Maunula et al., 2013). More data about foodborne outbreaks related to RTE foods can be found in outbreak descriptions.

PERSISTENCE OF FOODBORNE VIRUSES ON FOOD AND SURFACES

Murine norovirus MNV-1, that can be cultivated, has been used as a surrogate to study the persistence of human NoV. MNV-1 could be detected in lettuce left at room temperature for 4 days with 1-log reduction (Fallahi and Mattison, 2011). The same reduction took 7–15 days on stainless steel disks (Mormann et al., 2015). No reduction was observed when the steel disks were stored at 7°C. HAV (after drying) persisted on the surface of alfalfa seeds over 20 days with <1 log reduction (Wang et al., 2013b). When frozen, HAV and rotaviruses persisted for at least 90 days without loss of infectivity (Butot et al., 2008). HAV retained its infectivity in marinated mussels for 4 weeks (Hewitt and Greening, 2004). According to Verhaelen et al. (2012), MNV-1 persisted at 4°C on strawberries and on raspberries (no reduction after 7 days), but at 21°C there was a difference in time for 1-log reduction (1 vs 3 days, on strawberries and raspberries, respectively). Thus viruses, if entering the food chain, will survive the trip from field to fork quite well and survive on food until served, if consumed raw.

VIRUS TRANSMISSION INTO FOOD SUPPLY CHAIN

Food gets contaminated with viruses, either when the food is in contact with contaminated sewage water or when food is contaminated by a food handler, either during the harvesting/mild processing or during food handling while preparing a meal. In the food chain the critical points are naturally those in which produce is in contact with any water that can be contaminated with viruses and those where humans excreting viruses are handling them. During irrigation of berries and spraying of the plants with pesticides (if diluted in contaminated water; Verhaelen et al., 2013), and during washing of strawberries or leafy greens, viruses can contaminate the produce. For human food handling, collection of the berries into baskets, and manual handling of berries before entering the freezing tunnel enable transmission of viruses from hands to food produce. Rapid freezing of berries may be beneficial for survival of viruses, since there is evidence that frozen berries are more often contaminated than berries sold fresh (Maunula et al., 2013). Even in the vegetable food supply chain, risk assessment studies showed that hand contact contributed more than irrigation or the water used for rinsing of vegetables (Bouwknegt et al., 2015). Several factors, such as regulations in countries, can be expected to affect the proportion of hand contact and water in produce contamination.

Internalization of foodborne viruses into lettuce has been proposed as a route for viral contamination. Urbanucci et al. (2009) showed that NoV surrogate canine Calicivirus RNA could be detected in lettuce leaves, when plants were cultivated hydroponically in Calicivirus-containing water. Infectious Sapovirus as a NoV surrogate was shown to transfer from roots to leaves of romaine lettuce through the xylem system (Esseili et al., 2012a). Dicaprio et al. (2012) also demonstrated internalization of Tulane viruses, MNV-1 and NoV GII.4 via roots and viral dissemination to leaves and shoots. According to Esseili et al. (2012b) NoV Virus-like particles (VLPs) attached to carbohydrates on the lettuce cell wall material.

Microscopy has demonstrated viruses or VLPs attached to lettuce leaf surface, inside cuts, stomata, and along veins (Esseili et al., 2012b; Tian et al., 2011; Wei et al., 2010). The role of this transmission route is, however, regarded as low.

As drought and lack of water has become more common in some regions due to climate change, graywater or treated wastewater is used for irrigation of vegetables. Several Australian researchers have studied the risks for this kind of irrigation in some circumstances using quantitative microbial risk assessment. In the case of household graywater use, Barker et al. (2013) reported that based on frequency of NoV in household graywater, it is more recommendable to use bathroom water than laundry water if graywater irrigation is used for home-produced lettuce. The study of Mok et al. (2014) reveals that upon accumulation of pathogens from previous irrigation events, disinfection treatments should be included in wastewater treatment. Especially, this should be done if the water is used for irrigation of vegetables (lettuce, broccoli, cabbage, Asian vegetables, cucumber).

ROLE OF FOOD HANDLERS IN TRANSMISSION ROUTES OF VIRUSES

In outbreak situations, NoV (GI.6) has been detected on kitchen utensils, such as a knife grip, and the virus was also found on the hands of a staff member preparing RTE food (Boxman et al., 2009). Wearing gloves diminishes virus dissemination, but experiments have shown that gloves might be contaminated from virus-containing hands while donning gloves (Rönnqvist et al., 2014). Viruses are transmitted readily during food preparation, as several studies have shown (Rönnqvist et al., 2014; Shieh et al., 2014; Wang et al., 2013a). During simulated preparation of a cucumber sandwich, virus (genome) was transmitted from MNV-1-contaminated hands or cucumber to the first contact point, but usually no further. Shieh et al. (2014) showed that several tomatoes were contaminated by a tomato slicer when they were sliced after a MuNoV-containing tomato. As hands can contain 1000 genome copies of NoV (Liu et al., 2013), it is a realistic scenario to transfer virus from hands to gloves and further to food during food handling (Rönnqvist et al., 2014).

FOODBORNE OUTBREAKS RELATED TO RTE FOODS

As hundreds of outbreak reports have been published, only a small selection is discussed here. NoV outbreaks caused by shellfish and vegetables (unless caused by food handlers) are mostly left out. From all NoV outbreaks about 14% are of foodborne origin (Verhoef et al., 2015). All NoV genotypes can be transmitted through food, but the genotype profile in foodborne outbreaks differ somewhat from that of person-to-person transmission.

Shellfish is by far the most common source of viral outbreaks, often with a background of sewage contamination and/or undercooking. In a systematic review by Bellou et al. (2013) oysters, that are usually consumed raw, were found to cause 58.4% of food-related outbreaks; of which, NoV caused 83.7% and HAV 12.8%. The hitherto largest HAV outbreak with about 250,000 cases in Shanghai was caused by local-river mussels (Cooksley, 2000). The "industrialized," large-scale farming of bivalves has resulted in national or multinational outbreaks (Guillois-Becel et al., 2009; Shieh et al., 2009; Pinto et al., 2009). Accurate genome-based identification enabled widely spread infections to be connected to imported frozen mussels in Spain (D'Andrea et al., 2015). Despite the fact that each hepatitis infection is notifiable in low-endemic countries, outbreaks remain unrecorded (Tricco et al., 2006) or the infections remain unrelated to food (Petrignani et al., 2014). In an analysis of HEV cases in Germany, Wichmann et al. (2008) found a rather strong epidemiological link to consumption of mollusks. Seafood, primarily oyster, consumption has also been labeled as the cause of a HEV outbreak on a cruise ship when 33 of the 789 passengers were found to possess IgM-class antibodies (Said et al., 2009). A direct link between HEV infection and shellfish consumption was presented by Takahashi and Okamoto (2014).

Although viral contamination is recorded to be sparse in lettuce, and especially soft fruit, in the European food-supply chain (Bouwknegt et al., 2015), outbreaks linked to these products do occur. Imported salads caused two simultaneous but separate HAV outbreaks in Finland, both through central kitchens delivering lunch food (Pebody et al., 1998). Less sensitive vegetables, like onions, may carry the virus over longer times and distances. An extensive outbreak was caused by green onions grown in a defined farm in Mexico and distributed to a number of restaurants in the United States (Wheeler et al., 2005; Amon et al., 2005); of all 601 reported cases of HAV, 100 were hospitalized and 3 victims died. In a subsequent study, the route was established labeling growing onions at the producing farm with biomarkers (fluorescent microspheres and HAV vaccine strain virus). Another product that has caused a multinational spread of HAV is semidried tomatoes canned in oil. The first outbreak was recognized in Australia (Donnan et al., 2012) but subsequently outbreaks also occurred in Europe (Gallot et al., 2011; Petrignani et al., 2010). In this latter outbreak, close to 200 cases were found, but apparently many more remained unrecognized.

In northern Europe, frozen berries have caused multiple foodborne NoV outbreaks, as exemplified by Maunula et al. (2009), since berries are mostly eaten unheated. However, the largest outbreak with 11,000 ill persons occurred in Germany. The source of the virus was strawberries imported from China. Genotyping revealed three NoV genotypes (GI.9, GII.6, and GII.p16/GII.13) in different berry lots, an indication of sewage contamination (Mäde et al., 2013). A quite extensive HAV outbreak was recognized in the United States, where imported strawberries carried the infection to 250 schoolchildren in the states Michigan and Maine. Quite recently two separate outbreaks in Europe spread over several countries (Gillesberg Lassen et al., 2013). In both cases the source of the virus was a mix of frozen berries.

In many cases, it is difficult to resolve whether the cause of an outbreak is the contaminated product or the person preparing the food. Zoonotic HEV makes an exception, as exemplified by an outbreak caused by pork liver sausage (Colson et al., 2010). In a report describing outbreaks caused by various sources from the United States, 15% of foodborne outbreaks linked to melon fruits were caused by NoV (Walsh et al., 2014), and in South Korea, NoV was found emerging, especially linked to pickled vegetables (Moon et al., 2014). In a European multinational HAV outbreak, the common source of infection was the freshly squeezed orange juice in an Egyptian holiday resort hotel (Frank et al., 2007). Here the juice was most probably contaminated during the manufacturing process.

Several examples of food-handler-derived viral RTE outbreaks have been reported. In Austria, a NoV (GII.7) outbreak was linked to handmade salad prepared by a kitchen assistant who fell ill at work (Schmid et al., 2007). Malek et al. (2009) reported a NoV (GII.14) outbreak where a food handler, after recovering from gastroenteritis, used bare hands to slice delicatessen meat that was packaged and frozen for 7–28 days and later served to river rafters: virus was found in the meat packages. In Austria in 2014, kebab that was likely contaminated with NoV (GII.P21) by restaurant workers caused an outbreak among schoolchildren (Lin et al., 2015). HAV outbreaks have also been caused, for example, by RTE products prepared by butchers (Schmid et al., 2009) or bakers (Guzman-Herrador et al., 2014).

METHODS FOR DETECTION OF VIRUSES FROM FOOD

Even a low amount of viruses in food can cause infections in consumers. The detection of viruses (viral genome) from food is challenging (Summa et al., 2012), as long as human enteric viruses, especially NoV, do not readily grow in cell culture. In nucleic acid detection, the infectious nature of the virus remains unknown. An ISO technical specification was published for detection of NoV and HAV in food and feed (ISO/TS 15216, 2013). The methods comprise risk produce, bottled water, shellfish, soft fruit, and leafy greens. Different food matrices are treated with tailored procedures that increase the amount of viruses in proportion to sample volume; for example, by water filtration, enzymatic treatment, or polyethylene glycol precipitation. Then, after nucleic-acid extraction, virus-specific primers, and probes are used for gene amplification in a reverse-transcription-quantitative Polymerase chain reaction (PCR) assay (Loisy et al., 2005; Costafreda et al., 2006). Viability PCR has been tested for leafy greens with some promise (Moreno et al., 2015). In addition to viral detection from food, swabbing is a convenient tool for tracing viruses on environmental surfaces.

PREVENTION OF VIRAL TRANSMISSION THROUGH FOOD

Boiling inactivates NoV, but is not applicable for RTE food, such as for oysters or salads. Washing and mild disinfection treatments can be used to remove viruses from lettuce (Table 4.2). Studies concerning the effect of rinsing and washing show that more or less 1-log reduction in virus load is obtained (Bae et al., 2011; Fraisse et al., 2011). Chlorine or peroxyacetic acid treatment combined with washing further reduces the viral load on lettuce depending on virus, and usually 100 times reduction is achieved (Fraisse et al., 2011; Baert et al., 2009). Predmore and Li (2011) report over 3-log reduction in MNV-1 in berries

Table 4.2 Effect of Washing and Mild Disinfection Treatments for Load of HuNoV, HAV, or Surrogate Viruses on Lettuce

			Log Reduction (Approx.)				
Treatment	**Food**	**Method**	**HuNoV**	**FCV**	**MNV**	**HAV**	**Reference**
Immersion, 2 min and rinsing 30 s, tap water	Iceberg lettuce	RT-qPCR	0.9				Bae et al. (2011)
Washing (immersion, 2 min and rinsing 30 s) + detergent[a]	Iceberg lettuce	RT-qPCR	1.05				Bae et al. (2011)
Washing with tap water, 2 min (with or without ultrasound)	Fresh butter lettuce	Cell culture		0.7	1	0.7	Fraisse et al. (2011)
Washing + chlorine (15 ppm active chlorine)	Fresh butter lettuce	Cell culture		2.9	1.4	1.9	Fraisse et al. (2011)
Washing, PAA (100 ppm), 2 min	Fresh butter lettuce	Cell culture		3.2	2.3	0.7	Fraisse et al. (2011)
Sodium hypochlorite, (200 mg/liter NaOCl)	Shredded iceberg lettuce	Cell culture			1[b]		Baert et al. (2009)
PAA (250 mg/liter)	Shredded iceberg lettuce	Cell culture			1[b]		Baert et al. (2009)
Chlorine (200 ppm) + surfactant (SDS 50 ppm) 2 min	Strawberry, raspberry, lettuce, cabbage	Cell culture			2–3		Predmore and Li (2011)

[a]*A commodity (class I detergent) used to wash fruits and vegetables (South Korea), 0.1% solution.*
[b]*Additional to viral reduction in tap water.*

and leafy greens with chlorine combined with detergent. HAV is generally more resistant to disinfection than NoV surrogates, as reviewed in Li et al. (2012b).

For oysters, depuration or relaying is implemented. Stronger nonthermal treatments for reduction of viruses have been tested, such as high hydrostatic pressure, UV treatment, vaporized H_2O_2 in combination with UV, gamma irradiation, and electron beam irradiation. They could also be suitable for treatment of oysters and soft fruit. Novel techniques, such as cold atmospheric gaseous plasma, ozone, and lactic acid bacteria (antiviral biopreservative) have also been proposed. Some natural compounds, such as essential oils and grapefruit seed extract, have virucidal effect, as reviewed by Li et al. (2013). Grapefruit seed extraction (2 mg/mL) in wash water decreased the amount of MNV-1 in lettuce 1.5–2 log Plaque-forming unit (PFU) (Li et al., 2012a).

For food handlers, hygiene, education, and implementation of hygienic behavior are important elements in the prevention of viral foodborne outbreaks. Grove et al. (2015) showed that handwashing with liquid soap or foam reduced infectious MNV-1 about 3 log, but not totally. Through HAV vaccination of food handlers, the transmission of HAV can be prevented, and vaccine for NoV is under development. In addition to virus-free ingredients and hygienic food handlers, the surface materials of the kitchen environment may reduce the transmission of viruses.

FUTURE PERSPECTIVES

During the past decade, the number of international foodborne outbreaks in industrialized countries has increased. This is partly a result of improved surveillance methods, but also a better coordination of international surveillance. In the United States, the cross-continental outbreak data are gathered under a single instance, the Centers for Disease Control and Prevention (CDC). In Europe, a similar role is carried out by the European Center for Disease Prevention and Control (ECDC). These instances can enforce common practices, especially genome-based diagnostics directed to the same genome sites to be applied. Especially, the risk for extensive HAV outbreaks in poorly immune populations is imminent, as has been pointed out (Wenzel and Allerberger, 2014).

The global ubiquity of NoV, together with the short-lived immunity, makes them first-hand markers for hygiene failures in the food chain. For HEV, the crossing of the species barrier seems to be a limiting factor for extensive outbreaks to take place. The threat would be that some of the HEV genotypes undergo mutations that make humans more susceptible to the infection. Another threat is posed by human HEVs in that international commerce extends its risk product search to the endemic areas, a development that has already started (eg, shellfish, berries).

The international commerce involving RTE food and its ingredients is growing, both regarding volume and item selection. Due to well-developed transportation systems, more sensitive food items, like fresh produce and berries, are imported from developing countries. On the other hand, seasonal workers are commonly brought in from countries of low economy and poor education. The teaching of good hygiene practices must therefore be extended to the production sites, wherever they may be.

The improvement of food-chain safety regarding viruses must include control measures. Since pertinent microbiological hygiene indicators are not valid for viruses, risk assessments encompassing them must be made (Bouwknegt et al., 2015). For the producers, an accurate labeling of the origin of products down to the exact production site and time must become obligatory.

REFERENCES

Amon, J.J., Devasia, R., Xia, G., Nainan, O.V., Hall, S., Lawson, B., et al., 2005. Molecular epidemiology of foodborne hepatitis A outbreaks in the United States, 2003. J. Infect. Dis. 192, 1323–1330.

Atmar, R.L., Opekun, A.R., Gilger, M.A., Estes, M.K., Crawford, S.E., Neill, F.H., et al., 2008. Norwalk virus shedding after experimental human infection. Emerg. Infect. Dis. 14, 1553–1557.

Bae, J.Y., Lee, J.S., Shin, M.H., Lee, S.H., Hwang, I.G., 2011. Effect of wash treatments on reducing human norovirus on iceberg lettuce and perilla leaf. J. Food Prot. 74, 1908–1911.

Baert, L., Vandekinderen, I., Devlieghere, F., Van Coillie, E., Debevere, J., Uyttendaele, M., 2009. Efficacy of sodium hypochlorite and peroxyacetic acid to reduce murine norovirus 1, B40-8, *Listeria monocytogenes*, and *Escherichia coli* O157:H7 on shredded iceberg lettuce and in residual wash water. J. Food Prot. 72, 1047–1054.

Baert, L., Mattison, K., Loisy-Hamon, F., Harlow, J., Martyres, A., Lebeau, B., et al., 2011. Review: norovirus prevalence in Belgian, Canadian and French fresh produce: a threat to human health? Int. J. Food Microbiol. 151, 261–269.

Barker, S.F., O'Toole, J., Sinclair, M.I., Leder, K., Malawaraarachchi, M., Hamilton, A.J., 2013. A probabilistic model of norovirus disease burden associated with greywater irrigation of home-produced lettuce in Melbourne, Australia. Water Res. 47, 1421–1432.

Baylis, S.A., Nick, S., Blumel, J., Micha Nubling, C., 2010. Hepatitis E virus and blood donors in Germany. Vox Sang. 98, 479.

Bellou, M., Kokkinos, P., Vantarakis, A., 2013. Shellfish-borne viral outbreaks: a systematic review. Food Environ. Virol. 5, 13–23.

Bouwknegt, M., Engel, B., Herremans, M.M., Widdowson, M.A., Worm, H.C., Koopmans, M.P., et al., 2008. Bayesian estimation of hepatitis E virus seroprevalence for populations with different exposure levels to swine in The Netherlands. Epidemiol. Infect. 136, 567–576.

Bouwknegt, M., Verhaelen, K., Rzezutka, A., Kozyra, I., Maunula, L., von Bonsdorff, C. H., et al., 2015. Quantitative farm-to-fork risk assessment model for norovirus and hepatitis A virus in European leafy green vegetable and berry fruit supply chains. Int. J. Food Microbiol. 198, 50–58.

Boxman, I., Dijkman, R., Verhoef, L., Maat, A., van Dijk, G., Vennema, H., et al., 2009. Norovirus on swabs taken from hands illustrate route of transmission: a case study. J. Food Prot. 72, 1753–1755.

Brake, F., Ross, T., Holds, G., Kiermeier, A., McLeod, C., 2014. A survey of Australian oysters for the presence of human noroviruses. Food Microbiol. 44, 264–270.

Brassard, J., Gagne, M.J., Genereux, M., Cote, C., 2012. Detection of human food-borne and zoonotic viruses on irrigated, field-grown strawberries. Appl. Environ. Microbiol. 78, 3763–3766.

Butot, S., Putallaz, T., Sanchez, G., 2008. Effects of sanitation, freezing and frozen storage on enteric viruses in berries and herbs. Int. J. Food Microbiol. 126, 30–35.

Colson, P., Borentain, P., Queyriaux, B., Kaba, M., Moal, V., Gallian, P., et al., 2010. Pig liver sausage as a source of hepatitis E virus transmission to humans. J. Infect. Dis. 202, 825–834.

Cooksley, W.G., 2000. What did we learn from the Shanghai hepatitis A epidemic? J. Viral. Hepat. 7 (Suppl. 1), 1–3.

Costafreda, M.I., Bosch, A., Pinto, R.M., 2006. Development, evaluation, and standardization of a real-time TaqMan reverse transcription-PCR assay for quantification of hepatitis A virus in clinical and shellfish samples. Appl. Environ. Microbiol. 72, 3846–3855.

Costa-Mattioli, M., Cristina, J., Romero, H., Perez-Bercof, R., Casane, D., Colina, R., et al., 2002. Molecular evolution of hepatitis A virus: a new classification based on the complete VP1 protein. J. Virol. 76, 9516–9525.

Coudray-Meunier, C., Fraisse, A., Mokhtari, C., Martin-Latil, S., Roque-Afonso, A.M., Perelle, S., 2014. Hepatitis A virus subgenotyping based on RT-qPCR assays. BMC Microbiol. 14, 296.

Crossan, C., Baker, P.J., Craft, J., Takeuchi, Y., Dalton, H.R., Scobie, L., 2012. Hepatitis E virus genotype 3 in shellfish, United Kingdom. Emerg. Infect. Dis. 18, 2085–2087.

D'Andrea, L., Perez-Rodriguez, F.J., de Castellarnau, M., Manzanares, S., Lite, J., Guix, S., et al., 2015. Hepatitis A virus genotype distribution during a decade of universal vaccination of preadolescents. Int. J. Mol. Sci. 16, 6842–6854.

DePaola, A., Jones, J.L., Woods, J., Burkhardt III, W., Calci, K.R., Krantz, J.A., et al., 2010. Bacterial and viral pathogens in live oysters: 2007 United States market survey. Appl. Environ. Microbiol. 76, 2754–2768.

Dicaprio, E., Ma, Y., Purgianto, A., Hughes, J., Li, J., 2012. Internalization and dissemination of human norovirus and animal caliciviruses in hydroponically grown romaine lettuce. Appl. Environ. Microbiol. 78, 6143–6152.

Diez-Valcarce, M., Kokkinos, P., Soderberg, K., Bouwknegt, M., Willems, K., de Roda-Husman, A.M., et al., 2012. Occurrence of human enteric viruses in commercial mussels at retail level in three European countries. Food Environ. Virol. 4, 73–80.

Donnan, E.J., Fielding, J.E., Gregory, J.E., Lalor, K., Rowe, S., Goldsmith, P., et al., 2012. A multistate outbreak of hepatitis A associated with semidried tomatoes in Australia, 2009. Clin. Infect. Dis. 54, 775–781.

Esseili, M.A., Wang, Q., Saif, L.J., 2012a. Binding of human GII.4 norovirus virus-like particles to carbohydrates of romaine lettuce leaf cell wall materials. Appl. Environ. Microbiol. 78, 786–794.

Esseili, M.A., Wang, Q., Zhang, Z., Saif, L.J., 2012b. Internalization of sapovirus, a surrogate for norovirus, in romaine lettuce and the effect of lettuce latex on virus infectivity. Appl. Environ. Microbiol. 78, 6271–6279.

Fallahi, S., Mattison, K., 2011. Evaluation of murine norovirus persistence in environments relevant to food production and processing. J. Food Prot. 74, 1847–1851.

Fraisse, A., Temmam, S., Deboosere, N., Guillier, L., Delobel, A., Maris, P., et al., 2011. Comparison of chlorine and peroxyacetic-based disinfectant to inactivate Feline calicivirus, Murine norovirus and Hepatitis A virus on lettuce. Int. J. Food Microbiol. 151, 98–104.

Frank, C., Walter, J., Muehlen, M., Jansen, A., van Treeck, U., Hauri, A.M., et al., 2007. Major outbreak of hepatitis A associated with orange juice among tourists, Egypt, 2004. Emerg. Infect. Dis. 13, 156–158.

Gallot, C., Grout, L., Roque-Afonso, A.M., Couturier, E., Carrillo-Santisteve, P., Pouey, J., et al., 2011. Hepatitis A associated with semidried tomatoes, France, 2010. Emerg. Infect. Dis. 17, 566–567.

Gillesberg Lassen, S., Soborg, B., Midgley, S.E., Steens, A., Vold, L., Stene-Johansen, K., et al., 2013. Ongoing multi-strain food-borne hepatitis A outbreak with frozen berries as suspected vehicle: four Nordic countries affected, October 2012 to April 2013. Eurosurveillance 18, 20467.

Grove, S.F., Suriyanarayanan, A., Puli, B., Zhao, H., Li, M., Li, D., et al., 2015. Norovirus cross-contamination during preparation of fresh produce. Int. J. Food Microbiol. 198, 43–49.

Guillois-Becel, Y., Couturier, E., Le Saux, J.C., Roque-Afonso, A.M., Le Guyader, F.S., Le Goas, A., et al., 2009. An oyster-associated hepatitis A outbreak in France in 2007. Eurosurveillance 14, 19144.

Guzman-Herrador, B., Jensvoll, L., Einoder-Moreno, M., Lange, H., Myking, S., Nygard, K., et al., 2014. Ongoing hepatitis A outbreak in Europe 2013 to 2014: imported berry mix cake suspected to be the source of infection in Norway. Eurosurveillance 19, 20775.

Hewitt, J., Greening, G.E., 2004. Survival and persistence of norovirus, hepatitis A virus, and feline calicivirus in marinated mussels. J. Food Prot. 67, 1743–1750.

Iaconelli, M., Purpari, G., Libera, S.D., Petricca, S., Guercio, A., Ciccaglione, A.R., et al., 2015. Hepatitis A and E viruses in wastewaters, in river waters, and in Bivalve Molluscs in Italy. Food Environ. Virol. 7, 316–324.

Kanayama, A., Arima, Y., Yamagishi, T., Kinoshita, H., Sunagawa, T., Yahata, Y., et al., 2015. Epidemiology of domestically-acquired hepatitis E virus infection in Japan: assessment of the nationally reported surveillance data, 2007–2013. J. Med. Microbiol. 64, 752–758.

Kantala, T., Heinonen, M., Oristo, S., von Bonsdorff, C.H., Maunula, L., 2015. Hepatitis E virus in young pigs in Finland and characterization of the isolated partial genomic sequences of genotype 3 HEV. Foodborne. Pathog. Dis. 12, 253–260.

Koopmans, M., Duizer, E., 2004. Foodborne viruses: an emerging problem. Int. J. Food Microbiol. 90, 23–41.

Le Guyader, F.S., Atmar, R.L., Le Pendu, J., 2012. Transmission of viruses through shellfish: when specific ligands come into play. Curr. Opin. Virol. 2, 103–110.

Li, D., Baert, L., Zhang, D., Xia, M., Zhong, W., Van Coillie, E., et al., 2012a. Effect of grape seed extract on human norovirus GII.4 and murine norovirus 1 in viral suspensions, on stainless steel discs, and in lettuce wash water. Appl. Environ. Microbiol. 78, 7572–7578.

Li, J., Predmore, A., Divers, E., Lou, F., 2012b. New interventions against human norovirus: progress, opportunities, and challenges. Ann. Rev. Food Sci. Technol. 3, 331–352.

Li, D., Baert, L., Uyttendaele, M., 2013. Inactivation of food-borne viruses using natural biochemical substances. Food Microbiol. 35, 1–9.

Lin, Y.C., Hipfl, E., Lederer, I., Allerberger, F., Schmid, D., 2015. A norovirus GII.P21 outbreak in a boarding school, Austria 2014. Int. J. Infect. Dis. 37, 25–29.

Liu, P., Escudero, B., Jaykus, L.A., Montes, J., Goulter, R.M., Lichtenstein, M., et al., 2013. Laboratory evidence of norwalk virus contamination on the hands of infected individuals. Appl. Environ. Microbiol. 79, 7875–7881.

Loisy, F., Atmar, R.L., Guillon, P., Le Cann, P., Pommepuy, M., Le Guyader, F.S., 2005. Real-time RT-PCR for norovirus screening in shellfish. J. Virol. Methods. 123, 1–7.

Lowther, J.A., Gustar, N.E., Powell, A.L., Hartnell, R.E., Lees, D.N., 2012. Two-year systematic study to assess norovirus contamination in oysters from commercial harvesting areas in the United Kingdom. Appl. Environ. Microbiol. 78, 5812–5817.

Lynch, M.F., Tauxe, R.V., Hedberg, C.W., 2009. The growing burden of foodborne outbreaks due to contaminated fresh produce: risks and opportunities. Epidemiol. Infect. 137, 307–315.

Mäde, D., Trubner, K., Neubert, E., Hohne, M., Johne, R., 2013. Detection and typing of norovirus from frozen strawberries involved in a large-scale gastroenteritis outbreak in Germany. Food Environ. Virol. 5, 162–168.

Malek, M., Barzilay, E., Kramer, A., Camp, B., Jaykus, L.A., Escudero-Abarca, B., et al., 2009. Outbreak of norovirus infection among river rafters associated with packaged delicatessen meat, Grand Canyon, 2005. Clin. Infect. Dis. 48, 31–37.

Maunula, L., Roivainen, M., Keranen, M., Makela, S., Soderberg, K., Summa, M., et al., 2009. Detection of human norovirus from frozen raspberries in a cluster of gastroenteritis outbreaks. Eurosurveillance 14, 731–733.

Maunula, L., Kaupke, A., Vasickova, P., Soderberg, K., Kozyra, I., Lazic, S., et al., 2013. Tracing enteric viruses in the European berry fruit supply chain. Int. J. Food Microbiol. 167, 177–185.

Meng, X.J., Purcell, R.H., Halbur, P.G., Lehman, J.R., Webb, D.M., Tsareva, T.S., et al., 1997. A novel virus in swine is closely related to the human hepatitis E virus. Proc. Natl. Acad. Sci. 94, 9860–9865.

Meng, X.J., Wiseman, B., Elvinger, F., Guenette, D.K., Toth, T.E., Engle, R.E., et al., 2002. Prevalence of antibodies to hepatitis E virus in veterinarians working with swine and in normal blood donors in the United States and other countries. J. Clin. Microbiol. 40, 117–122.

Mohd Hanafiah, K., Jacobsen, K.H., Wiersma, S.T., 2011. Challenges to mapping the health risk of hepatitis A virus infection. Int. J. Health Geogr. 10, 57.

Mok, H.F., Barker, S.F., Hamilton, A.J., 2014. A probabilistic quantitative microbial risk assessment model of norovirus disease burden from wastewater irrigation of vegetables in Shepparton, Australia. Water Res. 54, 347–362.

Moon, S., Sohn, I.W., Hong, Y., Lee, H., Park, J.H., Kwon, G.Y., et al., 2014. Emerging pathogens and vehicles of food- and water-borne disease outbreaks in Korea, 2007–2012. Osong Public Health Res. Perspect. 5, 34–39.

Moreno, L., Aznar, R., Sanchez, G., 2015. Application of viability PCR to discriminate the infectivity of hepatitis A virus in food samples. Int. J. Food Microbiol. 201, 1–6.

Mormann, S., Heissenberg, C., Pfannebecker, J., Becker, B., 2015. Tenacity of human norovirus and the surrogates feline calicivirus and murine norovirus during long-term storage on common nonporous food contact surfaces. J. Food Prot. 78, 224–229.

Nainan, O.V., Xia, G., Vaughan, G., Margolis, H.S., 2006. Diagnosis of hepatitis a virus infection: a molecular approach. Clin. Microbiol. Rev. 19, 63–79.

Pavio, N., Meng, X.J., Doceul, V., 2015. Zoonotic origin of hepatitis E. Curr. Opin. Virol. 10, 34–41.

Pebody, R.G., Leino, T., Ruutu, P., Kinnunen, L., Davidkin, I., Nohynek, H., et al., 1998. Foodborne outbreaks of hepatitis A in a low endemic country: an emerging problem? Epidemiol. Infect. 120, 55–59.

Petrignani, M., Harms, M., Verhoef, L., van Hunen, R., Swaan, C., van Steenbergen, J., et al., 2010. Update: a food-borne outbreak of hepatitis A in the Netherlands related to semi-dried tomatoes in oil, January–February 2010. Eurosurveillance 15, 19572.

Petrignani, M., Verhoef, L., Vennema, H., van Hunen, R., Baas, D., van Steenbergen, J.E., et al., 2014. Underdiagnosis of foodborne hepatitis A, The Netherlands, 2008–2010 (1.). Emerg. Infect. Dis. 20, 596–602.

Pinto, R.M., Costafreda, M.I., Bosch, A., 2009. Risk assessment in shellfish-borne outbreaks of hepatitis A. Appl. Environ. Microbiol. 75, 7350–7355.

Polo, D., Varela, M.F., Romalde, J.L., 2015. Detection and quantification of hepatitis A virus and norovirus in Spanish authorized shellfish harvesting areas. Int. J. Food Microbiol. 193, 43–50.

Predmore, A., Li, J., 2011. Enhanced removal of a human norovirus surrogate from fresh vegetables and fruits by a combination of surfactants and sanitizers. Appl. Environ. Microbiol. 77, 4829–4838.

Rein, D.B., Stevens, G.A., Theaker, J., Wittenborn, J.S., Wiersma, S.T., 2012. The global burden of hepatitis E virus genotypes 1 and 2 in 2005. Hepatology 55, 988–997.

Renou, C., Roque-Afonso, A.M., Pavio, N., 2014. Foodborne transmission of hepatitis E virus from raw pork liver sausage, France. Emerg. Infect. Dis. 20, 1945–1947.

Robilotti, E., Deresinski, S., Pinsky, B.A., 2015. Norovirus. Clin. Microbiol. Rev. 28, 134–164.

Rodriguez-Lazaro, D., Cook, N., Ruggeri, F.M., Sellwood, J., Nasser, A., Nascimento, M.S., et al., 2012. Virus hazards from food, water and other contaminated environments. FEMS Microbiol. Rev. 36, 786–814.

Rönnqvist, M., Aho, E., Mikkelä, A., Ranta, J., Tuominen, P., Rättö, M., et al., 2014. Norovirus transmission between hands, gloves, utensils, and fresh produce during simulated food handling. Appl. Environ. Microbiol. 80, 5403–5410.

Said, B., Ijaz, S., Kafatos, G., Booth, L., Thomas, H.L., Walsh, A., et al., 2009. Hepatitis E outbreak on cruise ship. Emerg. Infect. Dis. 15, 1738–1744.

Scallan, E., Hoekstra, R.M., Angulo, F.J., Tauxe, R.V., Widdowson, M.A., Roy, S.L., et al., 2011. Foodborne illness acquired in the United States--major pathogens. Emerg. Infect. Dis. 17, 7–15.

Schmid, D., Stuger, H.P., Lederer, I., Pichler, A.M., Kainz-Arnfelser, G., Schreier, E., et al., 2007. A foodborne norovirus outbreak due to manually prepared salad, Austria 2006. Infection 35, 232–239.

Schmid, D., Fretz, R., Buchner, G., Konig, C., Perner, H., Sollak, R., et al., 2009. Foodborne outbreak of hepatitis A, November 2007–January 2008, Austria. Eur. J. Clin. Microbiol. Infect. Dis. 28, 385–391.

Seitz, S.R., Leon, J.S., Schwab, K.J., Lyon, G.M., Dowd, M., McDaniels, M., et al., 2011. Norovirus infectivity in humans and persistence in water. Appl. Environ. Microbiol. 77, 6884–6888.

Serracca, L., Rossini, I., Battistini, R., Goria, M., Sant, S., De Montis, G., et al., 2012. Potential risk of norovirus infection due to the consumption of "ready to eat" food. Food Environ. Virol. 4, 89–92.

Shieh, Y.C., Stewart, D.S., Laird, D.T., 2009. Survival of hepatitis A virus in spinach during low temperature storage. J. Food Prot. 72, 2390–2393.

Shieh, Y.C., Tortorello, M.L., Fleischman, G.J., Li, D., Schaffner, D.W., 2014. Tracking and modeling norovirus transmission during mechanical slicing of globe tomatoes. Int. J. Food Microbiol. 180, 13–18.

Smith, D.B., Simmonds, P., 2015. Hepatitis E virus and fulminant hepatitis--a virus or host-specific pathology? Liver Int. 35, 1334–1340.

Sridhar, S., Lau, S.K., Woo, P.C., 2015. Hepatitis E: A disease of reemerging importance. J. Formos. Med. Assoc. 114, 681–690.

Summa, M., von Bonsdorff, C.H., Maunula, L., 2012. Evaluation of four virus recovery methods for detecting noroviruses on fresh lettuce, sliced ham, and frozen raspberries. J. Virol. Methods 183, 154–160.

Takahashi, M., Okamoto, H., 2014. Features of hepatitis E virus infection in humans and animals in Japan. Hepatol. Res. 44, 43–58.

Tian, P., Yang, D., Mandrell, R., 2011. Differences in the binding of human norovirus to and from romaine lettuce and raspberries by water and electrolyzed waters. J. Food Prot. 74, 1364–1369.

Tricco, A.C., Pham, B., Duval, B., De Serres, G., Gilca, V., Vrbova, L., et al., 2006. A review of interventions triggered by hepatitis A infected food-handlers in Canada. BMC Health Serv. Res. 6, 157.

Urbanucci, A., Myrmel, M., Berg, I., von Bonsdorff, C.H., Maunula, L., 2009. Potential internalisation of caliciviruses in lettuce. Int. J. Food Microbiol. 135, 175–178.

van der Poel, W.H., 2014. Food and environmental routes of Hepatitis E virus transmission. Curr. Opin. Virol. 4, 91–96.

Verhaelen, K., Bouwknegt, M., Lodder-Verschoor, F., Rutjes, S.A., de Roda Husman, A. M., 2012. Persistence of human norovirus GII.4 and GI.4, murine norovirus, and human adenovirus on soft berries as compared with PBS at commonly applied storage conditions. Int. J. Food Microbiol. 160, 137–144.

Verhaelen, K., Bouwknegt, M., Rutjes, S.A., de Roda Husman, A.M., 2013. Persistence of human norovirus in reconstituted pesticides--pesticide application as a possible source of viruses in fresh produce chains. Int. J. Food Microbiol. 160, 323–328.

Verhoef, L., Hewitt, J., Barclay, L., Ahmed, S.M., Lake, R., Hall, A.J., et al., 2015. Norovirus genotype profiles associated with foodborne transmission, 1999–2012. Emerg. Infect. Dis. 21, 592–599.

Walsh, K.A., Bennett, S.D., Mahovic, M., Gould, L.H., 2014. Outbreaks associated with cantaloupe, watermelon, and honeydew in the United States, 1973–2011. Foodborne Pathog. Dis. 11, 945–952.

Wang, Q., Erickson, M., Ortega, Y.R., Cannon, J.L., 2013a. The fate of murine norovirus and hepatitis A virus during preparation of fresh produce by cutting and grating. Food Environ. Virol. 5, 52–60.

Wang, Q., Hirneisen, K.A., Markland, S.M., Kniel, K.E., 2013b. Survival of murine norovirus, Tulane virus, and hepatitis A virus on alfalfa seeds and sprouts during storage and germination. Appl. Environ. Microbiol. 79, 7021–7027.

Wei, J., Jin, Y., Sims, T., Kniel, K.E., 2010. Manure- and biosolids-resident murine norovirus 1 attachment to and internalization by Romaine lettuce. Appl. Environ. Microbiol. 76, 578–583.

Wenzel, J.J., Allerberger, F., 2014. Hepatitis A as a foodborne infection. Lancet Infect. Dis. 14, 907–908.

Wheeler, C., Vogt, T.M., Armstrong, G.L., Vaughan, G., Weltman, A., Nainan, O.V., et al., 2005. An outbreak of hepatitis A associated with green onions. N. Engl. J. Med. 353, 890–897.

Wichmann, O., Schimanski, S., Koch, J., Kohler, M., Rothe, C., Plentz, A., et al., 2008. Phylogenetic and case-control study on hepatitis E virus infection in Germany. J. Infect. Dis. 198, 1732–1741.

Yilmaz, A., Bostan, K., Altan, E., Muratoglu, K., Turan, N., Tan, D., et al., 2011. Investigations on the frequency of norovirus contamination of ready-to-eat food items in Istanbul, Turkey, by using real-time reverse transcription PCR. J. Food Prot. 74, 840–843.

CHAPTER

Parasitic protozoa in salad vegetables

5

Lucy J. Robertson

Department of Food Safety and Infection Biology, Norwegian University of Life Sciences, Oslo, Norway

INTRODUCTION

Salad vegetables are particularly suited to transmission of any pathogen that is transmitted as a contaminant; this is not only because they are eaten raw, but also because they are stored cool, eaten when fresh, and, for certain salad vegetables, particularly lettuce, vulnerable to contamination during production and harvesting. In addition, removal of pathogens from salad vegetables, such as lettuce, can be difficult as the surface area is large, and pathogens can become attached, enter stomata, or hide in crevices. Also, unlike other fresh produce, lettuce does not tolerate vigorous washing.

Although it has been long recognized that fresh produce may become contaminated from environmental sources, it has only been relatively recently that foods of nonanimal origin, such as salad vegetables, have been associated recurrently with foodborne illness. In the United States, for example, of all reported foodborne outbreaks with a known food item, produce-associated outbreaks accounted for less than 1% in the 1970s but this elevated to 6% in the 1990s (Sivapalasingam et al., 2004). These data, and others, demonstrate that major health problems can arise from consumption of contaminated prepared salads when hygiene practices fail (Little and Gillespie, 2008).

Those pathogens that can be transmitted via contaminated food can also often be transmitted by contaminated water. As drinking water is a community commodity, outbreaks associated with contaminated water are likely to affect a considerable proportion of the population of any community with a common water source, and are therefore more likely to be identified than outbreaks associated with food, which are likely to affect only clusters or individuals, depending on the food stuff contaminated and its consumption. For example, salad contaminated with an infectious pathogen and eaten in a communal setting, such as a canteen, or at an event like a wedding, is more likely to result in being detected as a vehicle of foodborne infection, than sporadic or cluster infections where an individual or small group of consumers is exposed to the same infectious agent on the same product.

P. Kotzekidou (Ed): Food Hygiene and Toxicology in Ready-to-Eat Foods.
DOI: http://dx.doi.org/10.1016/B978-0-12-801916-0.00005-4

Whilst bacteria have been the initial focus of foodborne outbreaks of disease associated with salad vegetables, with particular focus on *Escherichia coli* O157: H7 (and other enterohemorrhagic *E. coli* serotypes), *Salmonella* spp., and *Listeria monocytogenes* (Mercanoglu Taban and Halkman, 2011), it is clear that other pathogens can also be transmitted by this route. In this chapter the focus is upon parasitic protozoa and their potential for transmission by contaminated salad vegetables. Although these pathogens will not proliferate on salad vegetables, the transmission stages of many protozoa are very robust and unlikely to be inactivated by sanitization regimes used in the fresh produce industry.

PARASITIC PROTOZOA THAT MAY CONTAMINATE SALAD VEGETABLES

Parasitic protozoa that may be transmitted by ingestion of contaminated salad vegetables are largely intestinal protozoa for which the infective stage is excreted in huge quantities in the feces of the definitive host in a robust form (oocyst or cyst), which may contaminate the environment. These parasites include *Balantidium coli*, *Cryptosporidium* spp., *Cyclospora cayetanensis, Entamoeba histolytica*, *Giardia duodenalis*, and *Toxoplasma gondii*. In addition, various species of *Sarcocystis*, such as *Sarcocystis nesbitti*, for which humans may act as accidental intermediate hosts, may theoretically be transmitted by contaminated salad vegetables, and also parasites such as *Blastocystis hominis* or *Cystoisospora belli*. However, these are not considered further here as information on transmission of these protozoa via salad vegetables is lacking. Furthermore, *Trypanosoma cruzi*, a blood parasite for which infection of humans if often vectorborne, can also be transmitted by contaminated produce. However, as most foodborne infections with *T. cruzi* are associated with fruit juices, this parasite is not included here.

The salient biological features of the six selected parasites of relevance, *B. coli*, *Cryptosporidium* spp., *C. cayetanensis, E. histolytica*, *G. duodenalis*, and *T. gondii* are summarized in Table 5.1. Transmission of each of these parasites by contaminated salad vegetables is then considered in turn, by parasite, in the sections below. Section length is determined by the information available for that particular parasite in terms of its association with salad vegetables; although this is a theoretical route of infection for all these parasites, documentation and investigation are not evenly distributed.

It should be noted that a global risk ranking of foodborne parasites that was conducted during 2012 by FAO/WHO listed three of these protozoa (*T. gondii*—number 4; *Cryptosporidium* spp.—number 5; *E. histolytica*—number 6) amongst the most highly ranked six parasites of 24 potentially foodborne parasites (FAO/ WHO, 2014)—although the "top three slots" were taken by cestodes (all of which may also infect humans via contaminated salad vegetables), due to their

Table 5.1 Relevant Biological Features of Protozoan Parasites That Could Be Transmitted by Contaminated Salad Vegetables

Parasite	Phylum	Transmission Stage—Infectious When Excreted	Zoonotic Potential	Clinical Notes	Treatment Options
Balantidium coli	Ciliophora	Cyst—yes	Yes, pigs, wild boar	Usually abdominal symptoms, diarrhea etc. Possible extraintestinal spread	Available
Cryptosporidium spp.	Apicomplexa	Oocyst—yes	Species-dependent—some species zoonotic	Usually abdominal symptoms, watery diarrhea; postinfection sequelae noted	Problematic
Cyclospora cayetanensis	Apicomplexa	Oocyst—no	Minimal	Usually abdominal symptoms, diarrhea, etc.	Available
Entamoeba histolytica	Sarco-mastigophora	Cyst—yes	Minimal	Dysentery, possible extraintestinal spread	Available
Giardia duodenalis	Metamonda	Cyst—yes	Minimal; unclear	Usually abdominal symptoms, fatty diarrhea; postinfection sequelae noted	Available
Toxoplasma gondii	Apicomplexa	Oocyst—no	Yes—cat infections of most relevance for salad vegetables	Usually asymptomatic in immunocompetent, ocular symptoms; problematic in immunocompromised and with congenital transmission	Can be problematic

potentially very severe pathology when humans act as inadvertent intermediate hosts. *Giardia duodenalis*, *C. cayetanensis*, and *B. coli* were ranked at positions 11, 13, and 18, respectively.

BALANTIDIUM COLI

This ciliated protozoan parasite of the large intestine is thought to infect around 1% of the world's population; endemic areas tend to be tropical or semitropical (Theel and Pritt, 2012). Although extraintestinal spread may occur, this is considered to be rare. The other three forms of infection are asymptomatic carriage, chronic infection with intermittent abdominal pain, and fluctuating episodes of diarrhea and constipation, and acute or fulminating balantidiosis, which has sudden, severe onset and symptoms include dysentery and other abdominal symptoms. Mortality rates for acute balantidiosis may be over 25%, with dehydration, intestinal perforation, and hemorrhage being serious complications.

A major reservoir of infection is pigs, and contact with swine is considered a risk factor for infection, with the infection passed directly from pigs to humans or via environmental contamination. Although waterborne balantidiosis has been described, foodborne infection has not been documented, although it is theoretically possible; as transmission is via ingestion of the viable cysts, this can be directly as hand-to-mouth contamination, or by a vehicle such as water or food. Thus, salad vegetables are a logical and suitable vehicle for infection and may be particularly relevant in places where pig manure is used for cultivation of salad crops. In Nigeria, *B. coli* cysts have been reported as contaminants of vegetables for sale in markets (Ogbolu et al., 2009), while another study demonstrated that over 50% of pigs in one region of Nigeria were infected with *B. coli* (Yatswako et al., 2007).

CRYPTOSPORIDIUM SPP.

There are at least 25 species of *Cryptosporidium*, of which several are zoonotic (infectious to both humans and other animals), whilst others are more host-specific. Although *Cryptosporidium parvum* (zoonotic; a relatively common parasite amongst domestic ruminants such as sheep and cattle, particularly young) and *Cryptosporidium hominis* (limited zoonotic potential) are the two species most commonly reported from human infections (Ryan et al., 2014), several other species and genotypes are also important in human infections. Other species associated with human infections include: *Cryptosporidium meleagridis* (type host, turkey), *Cryptosporidium ubiquitum* (type host, cattle), *Cryptosporidium cuniculus* (type host, rabbit), and *Cryptosporidium viatorum* (Ryan et al., 2014). For all species, the sporulated, infective oocysts are excreted in the host feces and, upon ingestion by a susceptible host, excyst, usually in the intestine, releasing sporozoites that then invade epithelial cells. Here, the parasites undergo first asexual multiplication and then sexual replication.

Following fertilization of the macrogamonts by microgametes, oocysts are produced that sporulate in the infected host. *Cryptosporidium* species have a global distribution, and *Cryptosporidium* may be found in between 15% and 25% of children with diarrhea (Checkley et al., 2015). While children in developing countries and immunocompromised individuals are disproportionately affected by cryptosporidiosis, disease incidence is also increasing in industrialized countries largely as a result of outbreaks in recreational water facilities (Shirley et al., 2012). The major symptom of cryptosporidiosis is watery diarrhea and studies have demonstrated that cryptosporidiosis is associated with longer duration of diarrhea and greater childhood morbidity and mortality than are other causes (Checkley et al., 2015). Lack of effective pharmaceutical treatment for the most vulnerable patients (children and the immunocompromised) increases the potential severity of this infection.

There have been many waterborne outbreaks of cryptosporidiosis, with both drinking water and recreational water implicated. Due to the high excretion rate, low infectious dose, small size, and robust characteristics of *Cryptosporidium* oocysts, this infection is particularly well-suited to waterborne infection. These characteristics are also important factors promoting foodborne transmission.

As previously described, waterborne outbreaks are more likely to be identified than foodborne outbreaks, largely due to the population size exposed. Nevertheless, salad vegetables have been identified as a vehicle of infection in various cryptosporidiosis outbreaks (see Table 5.2). It is noticeable that of the nine outbreaks described in Table 5.2, six (two-thirds) of them are reported from Nordic countries. This preponderance has been previously noted and investigated (Robertson and Chalmers, 2013), but the authors concluded that there was no biological or environmental reason that such outbreaks should be more likely to occur in Nordic countries than anywhere else. The authors suggested that superior infrastructure and resources in these countries, enabling more directed investigation and reporting, may result in more outbreaks being identified. Furthermore, it is noteworthy that consumers in these countries seem less concerned about microbiological contamination of foods and generally trust the authorities to provide appropriate information.

Due to the proven transmission route (documented outbreaks) of *Cryptosporidium* by salad vegetables, there has been considerable focus on contamination of fresh produce with this parasite. A Standard ISO method (ISO/DIS 18744) is under preparation (and near to completion) for analyzing "fresh leafy green vegetables" for *Cryptosporidium* (and *G. duodenalis*), and should be available within the near future. It is based largely upon the standard methods for analyzing water for these parasites, but with an elution step in a glycine buffer as the initial step in the procedure. The basis of the method has been validated in interlaboratory trials (Cook et al., 2006). However, because it includes immunomagnetic separation (IMS), as used for water analyses, it is very expensive, probably prohibitively so in some countries, which, ironically, may be the places from which such analyses would be of most value. However, a cheaper method

Table 5.2 Outbreaks of Cryptosporidiosis Associated With Contaminated Salad Vegetables[a]

No. of Cases (Age Group)	Country (Year of Outbreak)	Associated Food	Suspected Source of Contamination	Other Information	References
>300 (mostly adults)	England and Scotland (2012)	Precut mixed leaves from a specific supermarket chain	During production—not by a local food-handler as the outbreak was widely distributed	Convincing foodborne outbreak with a wide geographical distribution making vehicle identification difficult. Of 182 cases, 142 (78%) were *C. parvum* were *gp60* subtype IIaA15G2R1	McKerr et al. (2015)
250 (adults, age range = 24–66 years)	Finland (2012)	Probably imported frisée lettuce	Possibly preharvest contamination in the Netherlands	Convincing foodborne outbreak, spread out through five different outbreaks associated with restaurants in four different cities—in two outbreaks, the same *C. parvum* the same *gp60* subtype, IIdA17G1, was found in patients' stool samples	Åberg et al. (2015)
89 (adults, age range = 24–70 years)	Sweden (2010)	Possibly herbs served as a salad garnish (doubled the probability of becoming ill)	Unknown	Convincing foodborne outbreak, but specific food item less certain. *Cryptosporidium parvum gp60* subtype IIdA24G1 identified	Gherasim et al. (2012)
46 (adults and children at a youth summer camp)	USA (2009)	Sandwich bar ingredients (particularly ham and lettuce, possibly tomatoes or onions)	Possibly livestock at the camp farm (calves, goat kids, piglets)	Convincing foodborne outbreak. Twelve cases lab-confirmed. *Cryptosporidium parvum, gp60* subtype IIaA17G2R1 from seven cases, and most animals. Difficulties in the outbreak investigation limit the strength of the findings, but a food source of infection seems definitive	CDC (2011)

74 (63 children and 11 adults at a mountain farm/wildlife park camp)	Norway (2009)	Possibly peeled carrots	Possibly farm livestock (lambs, goat kids, calves)	Foodborne transmission postulated, but other transmission routes also possible (eg, contaminated water, direct transmission associated with animal handling)	Rimšelienė et al. (2011)
18	Sweden (2008)	Arugula salad	Unknown	Convincing foodborne outbreak. Four subtypes of *C. parvum* associated with outbreak	Insulander et al. (2013)
72 (adults)	Finland (2008)	Lettuce mixture, packed in Sweden and originating from five different European countries	Unknown	Convincing foodborne outbreak. Twelve cases lab-confirmed. No oocysts detected on salad remains. PCR results report *C. parvum*	Pönka et al. (2009)
99 (adults)	Denmark (2005)	Salad bar items (particularly whole carrots, grated carrots, red peppers)	Unknown, but infected food handler (although not someone involved in food preparation) speculated as source of contamination of the buffet	Convincing foodborne outbreak. Twelve cases lab-confirmed. *Cryptosporidium hominis* identified	Ethelberg et al. (2009)
15 (not stated; attendees of a social event)	USA (1995)	Chicken salad	Foodhandler. Food prepared in domestic kitchen of a licenced daycare home	Convincing foodborne outbreak. It is implied that the contamination from the foodhandler may have originated from a child in the foodhandler's care	Anonymous (1996)

[a]*Table modified and updated from a previous publication (Robertson and Chalmers, 2013).*

with a modified IMS procedure has been developed and also subjected to interlaboratory testing, and provided recovery efficiencies comparable to those of the standard method (Utaaker et al., 2015). Surveys of salad vegetables for contamination with *Cryptosporidium* oocysts have been conducted in various areas of Europe using methods similar to the draft ISO method (Robertson and Gjerde, 2001; Di Benedetto et al., 2006; Rzezutka et al., 2010; Amorós et al., 2010; see summary in Robertson and Chalmers, 2013), and contamination with *Cryptosporidium* oocysts has been reported at generally low levels from various salad vegetables, including celery, cabbage, leeks, lettuce, and mung bean sprouts. A study from Canada used nonquantitative molecular methods for detection in a similar survey of salad vegetables, particularly precut salads, and identified *Cryptosporidium* DNA in approximately 6% of samples, with *C. parvum* the species identified (Dixon et al., 2013). This is the species most associated with produce-associated outbreaks (see Table 5.2), and may reflect the potential proximity of vegetable cultivation and livestock production at the farm level of the food-chain.

CYCLOSPORA CAYETANENSIS

This parasite was first identified in the early 1990s, and, although in the same taxonomic phylum as *Cryptosporidium*, the Apicomplexa, differs fundamentally in that the oocysts excreted from an infected host are not immediately infectious. In order to become infectious the parasites require 1–2 weeks at temperatures of around 20°C before they sporulate, such that each oocyst contains two sporocysts, each containing two potentially infective sporozoites. Upon ingestion and digestion of the oocyst and sporocyst, the sporozoites infect the intestinal epithelial cells. Asexual and sexual reproduction occur sequentially, and the unsporulated oocysts that are produced are excreted in the feces. *Cyclospora cayetanensis* was previously considered an infrequent, sporadic parasite, with some association with traveler's diarrhea, but in the mid-1990s several foodborne outbreaks occurred in the United States, with thousands of people infected. The symptoms of cyclosporiasis are generally gastrointestinal, with diarrhea, nausea, and abdominal pain most frequently reported. The distribution of endemic infection does not appear to be global, perhaps due to the requirement for warm environmental temperatures for sporulation, but cases have been reported worldwide. In endemic countries, prevalences from less than 1% to over 10% have been reported from selected populations, particularly children (Shields and Ortega, 2012).

Although most foodborne outbreaks of cyclosporiasis have been associated with berries, particularly raspberries imported to North America from South or Central America, salad vegetables can also act as transmission vehicles and more recent outbreaks appear to be associated with salad vegetables. An overview of outbreaks of cyclosporiasis associated with salad vegetables is provided in Table 5.3. It is interesting that foodborne outbreaks of cyclosporiasis seem to be more common than waterborne outbreaks, an opposite pattern to that seen with

Table 5.3 Outbreaks of Cyclosporiasis Associated With Contaminated Salad Vegetables[a]

No. Cases	Location (Year of Outbreak)	Associated Food	Other Information	References
>300	Maryland, US (1997)	Basil	A number of clusters of cases—food not analyzed for contamination	CDC (1997)
12	Florida, US (1997)	Mesclun salad	Food not analyzed for contamination	Herwaldt (2000)
>60	Missouri, US (1999)	Various salad dishes	Three different clusters of cases associated with salads—for one cluster *Cyclospora* was detected in the implicated salad by molecular methods	Lopez et al. (2001)
34	Baden-Wuerttemberg, Germany (2000)	Side-salads	Four independent clusters; food not analyzed for contamination	Döller et al. (2002)
70	Monterrey, Mexico (2001)	Watercress	Associated with a wedding reception; food not analyzed for contamination	Ayala-Gaytan et al. (2004)
17	British Columbia, Canada (2001)	Thai basil, imported from US	Various small clusters of infection; food analyzed for contamination, but oocysts not detected	Hoang et al. (2005)
56	Medellin, Columbia (2002)	Salad	University outbreak with juice also a potential vehicle; salad analyzed for contamination, but oocysts not detected	Botero-Garces et al. (2006)
96	Pennsylvania, US (2004)	Snow peas in pasta salad	Outbreak occurred at an event at a residential facility; food not analyzed for contamination. Snow peas imported from Guatemala and eaten raw	CDC (2004)
35	Connecticut, US (2004)	Bruschetta with basil and house salad	Restaurant-associated outbreak; food not analyzed for contamination	Anonymous (2005)

(Continued)

Table 5.3 Outbreaks of Cyclosporiasis Associated With Contaminated Salad Vegetables[a] *Continued*

No. Cases	Location (Year of Outbreak)	Associated Food	Other Information	References
20	Quebec, Canada (2005)	Basil	Restaurant-associated outbreak; food testing results inconclusive	Milord et al. (2012)
12	Stockholm, Sweden (2009)	Sugarsnap peas	A number of small clusters; sugarsnap peas imported from Guatemala	Insulander et al. (2010)
>300	Cruise ship from Australia traveling around South-East Asian waters (2010)	Lettuce (fresh produce taken on board)	Lettuce was the food item most frequently eaten by cases, had the strongest univariate association with illness, and was included in the multivariate model; food not analyzed for contamination	Gibbs et al. (2013)
643	25 US states, predominantly Texas (278 cases), Iowa (153), & Nebraska (86) (2013)	Fresh produce, some in salads	Some restaurant-associated cases in Iowa and Nebraska linked to a salad mix containing iceberg lettuce, romaine lettuce, red cabbage, and carrots. In Texas salads containing cilantro, possibly imported from Mexico, seemed more associated with infection	CDC (2013)

[a] *Table modified and updated from a previous publication (Shields and Ortega, 2012).*

Cryptosporidium. Again this is presumably related to the higher temperature requirement for sporulation, although other epidemiological factors may also play a role. While for salad-associated *Cryptosporidium* outbreaks, contamination from animals at the farm level has been implicated, as *C. cayetanensis* only infects humans, the inference is that poor hygiene at any point along the farm-to-fork continuum must be important. It is worth speculating that the initial contamination may actually be of water (from human sewage discharge), but this subsequently contaminates produce during irrigation, application of pesticides, or at other points such as produce washing, and sporulation occurs during this process and/or subsequent to contamination. The implication arises from the wide extent of contamination, suggesting that a single handling event from a person with poor hygiene is unlikely to result in such wide-scale contamination as seen in the 2013 multistate outbreak in the United States in which over 640 people were infected (Table 5.3). Due to the prolonged sporulation requirement, poor hygiene from a food handler in a restaurant is unlikely to be a transmission route.

Although *C. cayetanensis* is clearly recognized as a foodborne parasite, standard methods for examination of fresh produce have yet to be developed. The lack of commercially available monoclonal antibodies against the oocyst wall means that detection using immunofluorescent antibody testing (IFAT) and isolation from food matrices using IMS (as used for standard analysis of fresh produce for *Cryptosporidium* and *Giardia*) are currently not possible. Detection usually relies on molecular-based methods or microscopy (Steele et al., 2003; Lalonde and Gajadhar, 2008) following elution from the produce under investigation (Chandra et al., 2014; Shields et al., 2012), and work has also been done to assess the most suitable DNA extraction approach (Shields et al., 2013). For microscopy-based detection methods, the autofluorescent properties of *Cyclospora* oocyst walls for aiding identification are often used. However, even with relatively sensitive, polymerase chain reaction (PCR)-based analysis, detection efficiencies may be limited when concentrations are low, and may be matrix-specific. In one study, detection of contamination of mesclun lettuce with *Cyclospora* was less sensitive than detection of contamination of raspberries and basil (Steele et al., 2003). A purification step may assist in increasing detection sensitivity; a concentration step using lectin-coated magnetic beads has been developed (Robertson et al., 2000), but is not widely used. This may be because the authors did not report increased recovery efficiencies, rather that microscopy-based analysis was improved due to a reduction in food matrix debris. It would be interesting to investigate whether this purification step improved molecular detection sensitivity.

Some surveys of fresh produce for contamination with *Cyclospora* oocysts have been conducted. Whilst most surveys for *Cryptosporidium* have been conducted in Europe, for *Cyclospora* most studies have been conducted in a range of South American, African, and Asian countries, including Costa Rica, Peru, Egypt, Nigeria, Cambodia, Nepal, and Vietnam (Shields and Ortega, 2012). Generally, results from salad vegetables indicate a broad spread of low-level contamination ($<2->12\%$), while another study on salad vegetables, particularly precut salads, in Canada identified *Cyclospora* DNA in approximately 2% of samples (Dixon et al., 2013).

ENTAMOEBA HISTOLYTICA

Entamoeba histolytica is the only definitive human pathogen in the *Entamoeba* genus, and even this species does not always result in symptomatic disease. This is probably partly a reflection of the wide genetic diversity, with hundreds of unique genotypes, some of which may be more pathogenic than others

(Theel and Pritt, 2012). Ingestion of viable *E. histolytica* cysts by a susceptible host results in cyst excystation, with the resultant trophozoites colonizing the mucosa of the large intestine where they multiply by asexual binary fission. Invasion may occur, resulting in symptoms that include colitis, abdominal pain, and bloody diarrhea, but the factors that precipitate this are not fully elucidated. Encystation occurs in the intestine, and cysts are excreted with the feces. Although most infections are asymptomatic, of those that result in symptomatic infection, a minority (<1%) may progress to fulminating amebic colitis, often associated with colonic perforation and peritonitis, which may be fatal. Extraintestinal infection may also occur, usually resulting in amebic liver abscess.

Estimation of the prevalence of *E. histolytica* infection is difficult. Earlier studies that did not involve molecular characterization almost certainly misidentified the morphologically identical, but nonpathogenic, species *Entamoeba dispar* as *E. histolytica*. Thus, the global prevalence of symptomatic infection is probably rather low (approximately 1%), but, nevertheless, probably tens of millions suffer from symptomatic amebiasis annually, and it has been estimated that complications from invasive infection probably result in over 40,000 deaths every year (Theel and Pritt, 2012).

Although transmission of *E. histolytica* is via cysts in the environment, and therefore contamination of food that subsequently acts as a transmission vehicle is very probable, there are no documented cases or outbreaks in which salad vegetables, or other food types, are identified as such. Various waterborne outbreaks of amebiasis have been reported (see Karanis et al., 2007; Baldursson and Karanis, 2011), indicating the potential for further contamination of salad vegetables, and, anecdotally, unwashed salad is often considered a source of infection. Nevertheless, published outbreaks or cases in which an association with consumption of salad has been demonstrated are not available.

Possibly due to the lack of substantiated foodborne outbreaks or cases of amebiasis, there are no standardized or validated methods for examining salad vegetables for contamination with *E. histolytica* cysts. Many of the studies that have been published on surveys of different fresh produce for contamination do not distinguish between harmless *E. dispar* and potentially pathogenic *E. histolytica*. In the absence of the correct data it is impossible to estimate the risks or determine the requirement for interventions.

GIARDIA DUODENALIS

Giardia duodenalis is a flagellated protozoan in the Metamonda phylum. It is currently considered a species complex, divided into different Assemblages or genotypes that are morphologically very similar, but differ genetically and may have other physiological differences, including host-specificity. Thus, *G. duodenalis* in Assemblages A and B are infectious to humans, and have some, but apparently limited, zoonotic potential (Ryan and Cacciò, 2013), whereas

G. duodenalis in the Assemblages C–H are generally considered noninfectious to humans, but may infect a range of other mammals, with some degree of host-specificity depending on Assemblage. The transmission stage of *Giardia*, the cyst, is infectious at excretion, and, if ingested by a susceptible host, excysts in the small intestine. The rapidly multiplying, noninvasive trophozoites replicate by binary fission and attach onto enterocytes of the mucosal surface of the small intestine by an adhesive disk, or may remain motile in the intestinal lumen. Pathology is related to changes in intestinal architecture, and symptoms include fatty, often explosive diarrhea, abdominal pain and distension, flatulence, and nausea. Long-term sequelae have been noted.

Giardia is considered to be one of the most common intestinal parasites of humans, with a global occurrence, and a prevalence ranging from <10% (in countries with good infrastructure) to up to around 30% or sometimes higher in places where there are particularly challenging circumstances (Feng and Xiao, 2011).

Giardia has been associated with several waterborne outbreaks (Karanis et al., 2007; Baldursson and Karanis, 2011), and therefore has the clear potential to contaminate salad vegetables. However, relatively few outbreaks of foodborne giardiasis have been documented, and these have been associated with a variety of foodstuffs including Christmas pudding, tripe soup, home-canned salmon, and fresh produce (Cook and Lim, 2012). Only two giardiasis outbreaks (with a total of 39 persons infected) have been published in which salad is specifically implicated as the vehicle of infection; in both cases, one of which involved a noodle salad (Petersen et al., 1988) and the other of which involved serving raw sliced vegetables in an office cafeteria (Mintz et al., 1993), a food-handler was indicated as the source of the infection, suggesting that perhaps *Giardia* cysts do not survive for extended periods along the food production line. Nevertheless, a study from Germany indicated daily consumption of green salad as a risk factor for symptomatic giardiasis (Espelage et al., 2010), while a study from Mexico reported that people purchasing vegetables at a city market had higher infection rates than those buying at the village store (Cifuentes et al., 2000). However, this may represent an associated risk, not identified in this analysis, and need not necessarily reflect specific contamination of a particular food item.

Despite the absence of repeated or large outbreaks of foodborne giardiasis, an ISO standard method for detection of *Giardia* cysts on fresh produce has been developed. This is probably largely because there is a similar standard method for analyzing water, and that *Cryptosporidium*, which has caused considerable outbreaks of gastroenteric diseases associated with both water and fresh produce, is analyzed by an almost identical method, such that the analysis can be conducted concurrently with analysis for *Giardia*.

Surveys of salad vegetables for contamination with *Giardia* oocysts have been conducted in various areas of Europe using methods similar to the draft ISO method (Robertson and Gjerde, 2001; Amorós et al., 2010), and contamination at generally low levels has been reported from a number of salad vegetables including cabbage, dill, lettuce, radish sprouts, and mung bean sprouts. A study from Canada used nonquantitative molecular methods for detection in a similar survey of salad vegetables, particularly precut salads, and identified *Giardia* DNA in approximately 2% of samples (Dixon et al., 2013). Nevertheless, whether microscopy-based or molecular tools are used for identification of contamination, neither provides information on the viability or infectivity of the *Giardia* cysts or DNA detected, and given the relatively few outbreaks of food-borne giardiasis found in the literature, it might be assumed that this infection route is relatively infrequent.

TOXOPLASMA GONDII

Toxoplasma gondii is the only species in the *Toxoplasma* genus, within the phylum Apicomplexa, and is perhaps the most widespread protozoan parasite affecting humans; it has been estimated that between 1 and 2 billion of the world's population is currently infected. *Toxoplasma gondii* infects not only humans, but virtually all warm-blooded animals. All mammals and birds appear to be susceptible to infection. The lifecycle is somewhat complicated, with felids, such as the cat, as the definitive host where sexual reproduction occurs, resulting in the excretion of oocysts, whereas asexual reproduction occurs in all hosts. Following asexual multiplication of the parasites in the cells of the lamnia propria of the intestine of the intermediate hosts, the cells rupture, releasing the parasites into the surrounding tissues resulting in systemic infection. The circulating parasites multiply further by asexual reproduction in cells throughout the body, but eventually form tissue cysts. Oral infection can occur either by ingestion of tissue cysts in meat or tissues, or by ingestion of oocysts that have sporulated in the environment following excretion from the cat definitive host subsequent to sexual reproduction. With respect to transmission via salad vegetables, the relevant infectious stages are the oocysts that have been excreted from cats.

Intrauterine transmission can also occur from an infected pregnant mother to her unborn fetus; this is an important cause of abortion, particularly in sheep, but can also result in young, or babies in humans, being born with congenital problems. Although infection with *Toxoplasma* via the oral route (undercooked meat or from oocysts in the environment) is often asymptomatic in the immunocompetent, it may be associated with mild to moderate "flu-like" symptoms and ocular problems. Symptom severity seems to be associated with the strain of *Toxoplasma*, with strains from South America being particularly virulent. In the immunosuppressed, the most important manifestation is encephalitis and this can be fatal.

Due to the multiplicity of infection routes and the fact that infection is often asymptomatic, the role of salad vegetables as a vehicle of transmission of this parasite has probably not been properly acknowledged. Nevertheless, waterborne outbreaks of toxoplasmosis have been recognized, including relatively large outbreaks in British Colombia, Canada, in Santa Isabel do Ivaí, Brazil, and in

Coimbatore, India (Karanis et al., 2007; Baldursson and Karanis, 2011; Bowie et al., 1997; Balasundaram et al., 2010; Vaudaux et al., 2010), and have been responsible for considerable morbidity, particularly with respect to ocular problems. Furthermore, subsequent congenital infection has been reported in association with these outbreaks due to pregnant women becoming infected. The waterborne outbreaks indicate that oocyst transmission from ingestion of contaminated items is entirely possible; in support of this, a study that investigated sporozoites, indicating oocyst infection rather than meat cyst infection, suggested that human infections with *Toxoplasma* in North America may be frequently caused by ingestion of oocysts, probably as contaminants of either food or water (Boyer et al., 2011). Indeed, an outbreak of acute toxoplasmosis that occurred in an industrial plant in the state of São Paulo, Brazil, was shown by a case-control study to be associated with ingestion of green vegetables, but not with meat or water (Ekman et al., 2012). Furthermore, a number of studies have investigated risk factors for seropositive status for *Toxoplasma*, and several have pinpointed consumption of raw vegetables as being a significant risk factor (eg, Mahmoudvand et al., 2015; Daryani et al., 2014; Alvarado-Esquivel et al., 2013; Fakhfakh et al., 2013; Jafari et al., 2012).

As with *Cyclospora*, an absence of commercially available monoclonal antibodies against *Toxoplasma* oocyst walls means that methods to analyze fresh produce for contamination based on those used for *Cryptosporidium* and *Giardia* are not possible. Nevertheless, one study investigated contamination of fruits and vegetables in Poland, using a flocculation method followed by a molecular method for detection of *Toxoplasma* DNA and approximately 10% of samples were found to be positive (Lass et al., 2012). Given that *Toxoplasma* oocysts are just as robust as *Cryptosporidium* oocysts, that an infected felid excretes many millions of oocysts, and that the survey from Poland indicates a high contamination rate of fresh produce it is perhaps surprising that more cases of toxoplasmosis associated with salad vegetables have not been reported. A possible explanation is that toxoplasmosis tends not to be an acute infection, indeed the majority of cases are usually asymptomatic, and when symptoms arise it may be difficult to link back to consumption of salad vegetables, especially given the multiplicity of potential routes of infection.

CONCLUSIONS

Although various protozoan parasites may be transmitted via contaminated salad vegetables, our knowledge on the extent of this transmission route is relatively limited, despite the widespread prevalence of these parasites in the human population. Although parasites will not proliferate on salad vegetables, the robust transmission stages mean that they are unlikely to be inactivated by standard sanitizing regimes. Indeed, the reason for the lack of outbreaks, or even cases, of protozoan parasite infection associated with salad vegetables is more likely to reflect the relatively long incubation periods between infection and symptoms that mean that the association between the illness and the contaminated product is often never made. The intention of this chapter is to provide an overview of those protozoan parasites that may be transmitted via salad vegetables, giving brief insights into the protozoa and the diseases that they cause, but also a consideration of the evidence that we would currently have regarding the potential of salad vegetables to act as transmission vehicles of infection.

REFERENCES

Åberg, R., Sjöman, M., Hemminki, K., Pirnes, A., Räsänen, S., Kalanti, A., et al., 2015. *Cryptosporidium parvum* caused a large outbreak linked to frisée salad in Finland, 2012. Zoonoses. Public Health (in Press). http://dx.doi.org/10.1111/zph.12190.

Alvarado-Esquivel, C., Campillo-Ruiz, F., Liesenfeld, O., 2013. Seroepidemiology of infection with *Toxoplasma gondii* in migrant agricultural workers living in poverty in Durango, Mexico. Parasit. Vectors. 6, 113.

Amorós, I., Alonso, J.L., Cuesta, G., 2010. *Cryptosporidium* oocysts and *Giardia* cysts on salad products irrigated with contaminated water. J. Food. Prot. 73, 1138–1140.

Anonymous, 1996. Foodborne outbreak of diarrhoea illness associated with Cryptosporidium parvum, Minnesota, 1995. Morb. Mortal. Wkly. Rep. 45, 783–784.

Anonymous, 2005. Restaurant-associated outbreak of cyclosporiasis, New Haven County, June 2005. Department of Public Health, Connecticut Department of Public Health. 25.

Ayala-Gaytan, J.J., Diaz-Olachea, C., Riojas-Montalvo, P., Palacios-Martinez, C., 2004. Cyclosporidiosis: clinical and diagnostic of an epidemic outbreak. Rev. Gastroenterol Méx. 69, 226–229.

Balasundaram, M.B., Andavar, R., Palaniswamy, M., Venkatapathy, N., 2010. Outbreak of acquired ocular toxoplasmosis involving 248 patients. Arch. Ophthalmol. 128 (1), 28–32.

Baldursson, S., Karanis, P., 2011. Waterborne transmission of protozoan parasites: review of worldwide outbreaks—an update 2004–2010. Water. Res. 45, 6603–6614.

Botero-Garces, J., Montoya, M.N., Barguil, J.I., Castano-Gonsalez, A., 2006. An outbreak of *Cyclospora cayetanensis* in Medellín, Colombia. Rev. Salud Publica 8, 258–268.

Bowie, W.R., King, A.S., Werker, D.H., Isaac-Renton, J.L., Bell, A., Eng, S.B., et al., 1997. Outbreak of toxoplasmosis associated with municipal drinking water. The BC Toxoplasma Investigation Team. Lancet 350 (9072), 173–177.

Boyer, K., Hill, D., Mui, E., Wroblewski, K., Karrison, T., Dubey, J.P., et al., 2011. Toxoplasmosis Study Group. Unrecognized ingestion of *Toxoplasma gondii* oocysts leads to congenital toxoplasmosis and causes epidemics in North America. Clin. Infect. Dis. 53 (11), 1081–1089.

Caddick, M.X., 1994. Nitrogen metabolite repression. In: Martinelli, S.D., Kinghorn, J.R. (Eds.), Aspergillus: 50 Years on, Progress in Industrial Microbiology, vol. 29. Elsevier Science, Amsterdam, pp. 323–353.

CDC, 1997. Outbreak of cyclosporiasis—Northern Virginia-Washington, DC-Baltimore, MD, Metropolitan Area, (1997). Morb. Mortal. Wkly Rep. 46, 689–691.

CDC, 2004. Outbreak of cyclosporiasis associated with snow peas—Pennsylvania, (2004). Morb. Mortal. Wkly Rep. 53, 876–878.

CDC, 2011. Cryptosporidiosis outbreak at a summer camp—North Carolina. Morb. Mortal. Wkly Rep. 60, 918–922.

CDC, 2013. Outbreaks of cyclosporiasis—United States, June–August 2013. Morb. Mortal. Wkly Rep. 62 (43), 862.

Chandra, V., Torres, M., Ortega, Y.R., 2014. Efficacy of wash solutions in recovering *Cyclospora cayetanensis*, *Cryptosporidium parvum*, and *Toxoplasma gondii* from basil. J. Food. Prot. 77 (8), 1348–1354.

Checkley, W., White Jr., A.C., Jaganath, D., Arrowood, M.J., Chalmers, R.M., Chen Jr., X. M., et al., 2015. A review of the global burden, novel diagnostics, therapeutics, and vaccine targets for *Cryptosporidium*. Lancet Infect. Dis. 15 (1), 85–94.

Cifuentes, E., Gomez, M., Blumenthal, U., Tellez-Rojo, M.M., Romieu, I., Ruiz-Palacios, G., et al., 2000. Risk factors for *Giardia intestinalis* infection in agricultural villages practicing wastewater irrigation in Mexico. Am. J. Trop. Med. Hyg. 62 (3), 388–392.

Cook, N., Lim, Y.A.L., 2012. Section 1, Chapter 4: *Giardia duodenalis*. In: Robertson, L.J., Smith, H.V. (Eds.), Foodborne Protozoan Parasites. Nova Publishers, New York, NY, ISBN: 978-1-61470-008-1.

Cook, N., Paton, C.A., Wilkinson, N., Nichols, R.A., Barker, K., Smith, H.V., 2006. Towards standard methods for the detection of *Cryptosporidium parvum* on lettuce and raspberries. Part 2: validation. Int. J. Food. Microbiol. 109 (3), 222–228.

Daryani, A., Sarvi, S., Aarabi, M., Mizani, A., Ahmadpour, E., Shokri, A., et al., 2014. Seroprevalence of *Toxoplasma gondii* in the Iranian general population: a systematic review and meta-analysis. Acta. Trop. 137, 185–194.

Di Benedetto, M.A., Di Piazza, F., Oliveri, R., Cerame, G., Valenti, R., Firenze, A., 2006. Development of a technique for recovering *Giardia* cysts and *Cryptosporidium* oocysts from fresh vegetables. Ann. Ig. 18 (2), 101–107.

Dixon, B., Parrington, L., Cook, A., Pollari, F., Farber, J., 2013. Detection of *Cyclospora*, *Cryptosporidium*, and *Giardia* in ready-to-eat packaged leafy greens in Ontario, Canada. J. Food. Prot. 76 (2), 307–313.

Döller, P.C., Dietrich, K., Filipp, N., Brockmann, S., Dreweck, C., Vonthein, S., et al., 2002. Cyclosporiasis outbreak in Germany associated with the consumption of salad. Emerg. Infect. Dis. 8, 992–994.

Ekman, C.C., Chiossi, M.F., Meireles, L.R., Andrade Júnior, H.F., Figueiredo, W.M., Marciano, M.A., et al., 2012. Case-control study of an outbreak of acute toxoplasmosis in an industrial plant in the state of São Paulo, Brazil. Rev. Inst. Med. Trop. São Paulo 4 (5), 239–244.

Espelage, W., An der Heiden, M., Stark, K., Alpers, K., 2010. Characteristics and risk factors for symptomatic *Giardia lamblia* infections in Germany. BMC Public Health 10, 41.

Ethelberg, S., Lisby, M., Vestergaard, L.S., Enemark, H.L., Olsen, K.E., Stensvold, C.R., et al., 2009. A foodborne outbreak of *Cryptosporidium hominis* infection. Epidemiol. Infect. 137, 348–536.

Fakhfakh, N., Kallel, K., Ennigro, S., Kaouech, E., Belhadj, S., Chaker, E., 2013. Risk factors for *Toxoplasma gondii* and immune status of pregnant women: cause and effect? La Tunisie Medicale 91 (3), 188–190.

FAO/WHO (Food and Agriculture Organization of the United Nations/World Health Organization), 2014. Multicriteria-based ranking for risk management of foodborne parasites. Microbial Risk Assessment Series. Published by FAO and WHO. Available at: <http://www.fao.org/3/a-i3649e.pdf>.

Feng, Y., Xiao, L., 2011. Zoonotic potential and molecular epidemiology of *Giardia* species and giardiasis. Clin. Microbiol. Rev. 24 (1), 110–140.

Gherasim, A., Lebbad, M., Insulander, M., Decraene, V., Kling, A., Hjertqvist, M., et al., 2012. Two geographically separated food-borne outbreaks in Sweden linked by an unusual *Cryptosporidium parvum* subtype, October 2010. Eurosurveillance 17, 46.

Gibbs, R.A., Nanyonjo, R., Pingault, N.M., Combs, B.G., Mazzucchelli, T., Armstrong, P., et al., 2013. An outbreak of *Cyclospora* infection on a cruise ship. Epidemiol. Infect. 141 (3), 508–516.

Herwaldt, B.L., 2000. *Cyclospora cayetanensis*: a review, focusing on the outbreaks of cyclosporiasis in the 1990s. Emerg. Infect. Dis. 31, 1040–1057.

Hoang, L., Fyee, M., Ong, C., Harb, J., Champagne, S., Dixon, B., et al., 2005. Outbreak of cyclosporiasis in British Columbia associated with imported Thai basil. Epidemiol. Infect. 131, 23–27.

Insulander, M., Svenungsson, B., Lebbad, M., Karlsson, L., de Jong, B., 2010. A foodborne outbreak of *Cyclospora* infection in Stockholm, Sweden. Foodborne Pathog. Dis. 7, 1585–1587.

Insulander, M., Silverlås, C., Lebbad, M., Karlsson, L., Mattsson, J.G., Svenungsson, B., 2013. Molecular epidemiology and clinical manifestations of human cryptosporidiosis in Sweden. Epidemiol. Infect. 141 (5), 1009–1020.

Jafari, R., Sadaghian, M., Safari, M., 2012. Seroprevalence of *Toxoplasma gondii* infection and related risk factors in Tabriz City, Iran, 2008. J. Res. Health Sci. 12 (2), 119–121.

Karanis, P., Kourenti, C., Smith, H., 2007. Waterborne transmission of protozoan parasites: a worldwide review of outbreaks and lessons learnt. J. Water Health 5 (1), 1–38.

Lalonde, L.F., Gajadhar, A.A., 2008. Highly sensitive and specific PCR assay for reliable detection of *Cyclospora cayetanensis* oocysts. Appl. Environ. Microbiol. 74 (14), 4354–4358.

Lass, A., Pietkiewicz, H., Szostakowska, B., Myjak, P., 2012. The first detection of *Toxoplasma gondii* DNA in environmental fruits and vegetables samples. Eur. J. Clin. Microbiol. Infect. Dis. 31 (6), 1101–1108.

Little, C.L., Gillespie, I.A., 2008. Prepared salads and public health. J. Appl. Microbiol. 105 (6), 1729–1743.

Lopez, A.S., Dodson, D.R., Arrowood, M.J., Orlandi Jr., P.A., da Silva, A.J., Bier, J.W., et al., 2001. Outbreak of cyclosporiasis associated with basil in Missouri in 1999. Clin. Infect. Dis. 32, 1010–1017.

Mahmoudvand, H., Saedi Dezaki, E., Soleimani, S., Baneshi, M.R., Kheirandish, F., Ezatpourm, B., et al., 2015. Seroprevalence and risk factors of *Toxoplasma gondii* infection among healthy blood donors in south-east of Iran. Parasite Immunol. 37 (7), 362–367.

McKerr, C., Adak, G.K., Nichols, G., Gorton, R., Chalmers, R.M., Kafatos, G., et al., 2015. An outbreak of *Cryptosporidium parvum* across England & Scotland associated with consumption of fresh pre-cut salad leaves, May 2012. PLoS ONE 10 (5), e0125955.

Mercanoglu Taban, B., Halkman, A.K., 2011. Do leafy green vegetables and their ready-to-eat [RTE] salads carry a risk of foodborne pathogens?. Anaerobe 17 (6), 286–287.

Milord, F., Lampron-Goulet, E., St-Amour, M., Levac, E., Ramsay, D., 2012. *Cyclospora cayetanensis*: a description of clinical aspects of an outbreak in Quebec, Canada. Epidemiol. Infect. 140 (4), 626–632.

Mintz, E.D., Hudson-Wragg, M., Mshar, P., Cartter, M.L., Hadler, J.L., 1993. Foodborne giardiasis in a corporate office setting. J. Infect. Dis. 167 (1), 250–253.

Ogbolu, D.O., Alli, O.A., Ogunleye, V.F., Olusoga-Ogbolu, F.F., Olaosun, I., 2009. The presence of intestinal parasites in selected vegetables from open markets in southwestern Nigeria. Afr. J. Med. Med. Sci. 38 (4), 319–324.

Petersen, L.R., Cartter, M.L., Hadler, J.L., 1988. A food-borne outbreak of *Giardia lamblia*. J. Infect. Dis. 157 (4), 846–848.

Pönka, A., Kotilainen, H., Rimhanen-Finne, R., Hokkanen, P., Hänninen, M.L., Kaarna, A., et al., 2009. A foodborne outbreak due to *Cryptosporidium parvum* in Helsinki, November 2008. Eurosurveillance 14, 28.

Rimšelienė, G., Vold, L., Robertson, L., Nelke, C., Søli, K., Johansen, Ø.H., et al., 2011. An outbreak of gastroenteritis among schoolchildren staying in a wildlife reserve: thorough investigation reveals Norway's largest cryptosporidiosis outbreak. Scand. J. Public Health 39, 287–295.

Robertson, L.J., Chalmers, R.M., 2013. Foodborne cryptosporidiosis: is there really more in Nordic countries? Trends. Parasitol. 29 (1), 3–9.

Robertson, L.J., Gjerde, B., 2001. Occurrence of parasites on fruits and vegetables in Norway. J. Food. Prot. 64, 1793–1798.

Robertson, L.J., Gjerde, B., Campbell, A.T., 2000. Isolation of *Cyclospora* oocysts from fruits and vegetables using lectin-coated paramagnetic beads. J. Food. Prot. 63, 1410–1414.

Ryan, U., Cacciò, S.M., 2013. Zoonotic potential of *Giardia*. Int. J. Parasitol. 43 (12–13), 943–956.

Ryan, U., Fayer, R., Xiao, L., 2014. *Cryptosporidium* species in humans and animals: current understanding and research needs. Parasitology 141 (13), 1667–1685.

Rzezutka, A., Nichols, R.A., Connelly, L., Kaupke, A., Kozyra, I., Cook, N., et al., 2010. *Cryptosporidium* oocysts on fresh produce from areas of high livestock production in Poland. Int. J. Food. Microbiol. 139 (1–2), 96–101.

Shields, J.M., Ortega, Y., 2012. Section 1, Chapter 3: *Cyclospora cayetanensis*. In: Robertson, L.J., Smith, H.V. (Eds.), Foodborne Protozoan Parasites. Nova Publishers, New York, NY, ISBN: 978-1-61470-008-1.

Shields, J.M., Lee, M.M., Murphy, H.R., 2012. Use of a common laboratory glassware detergent improves recovery of *Cryptosporidium parvum* and *Cyclospora cayetanensis* from lettuce, herbs and raspberries. Int. J. Food. Microbiol. 153 (1–2), 123–128.

Shields, J.M., Joo, J., Kim, R., Murphy, H.R., 2013. Assessment of three commercial DNA extraction kits and a laboratory-developed method for detecting *Cryptosporidium* and *Cyclospora* in raspberry wash, basil wash and pesto. J. Microbiol. Methods 92 (1), 51–58.

Shirley, D.A., Moonah, S.N., Kotloff, K.L., 2012. Burden of disease from cryptosporidiosis. Curr. Opin. Infect. Dis. 25 (5), 555–563.

Sivapalasingam, S., Friedman, C.R., Cohen, L., Tauxe, R.V., 2004. Fresh produce: a growing cause of outbreaks of foodborne illness in the United States, 1973 through 1997. J. Food. Prot. 67 (10), 2342–2353.

Steele, M., Unger, S., Odumeru, J., 2003. Sensitivity of PCR detection of *Cyclospora cayetanensis* in raspberries, basil, and mesclun lettuce. J. Microbiol. Methods 54 (2), 277–280.

Theel, E., Pritt, B.S., 2012. Section 1, Chapter 1: *Balantidium coli* and *Entamoeba histolytica*. In: Robertson, L.J., Smith, H.V. (Eds.), Foodborne Protozoan Parasites. Nova Publishers, New York, NY, ISBN: 978-1-61470-008-1.

Utaaker, K.S., Huang, Q., Robertson, L.J., 2015. A reduced-cost approach for analyzing fresh produce for contamination with *Cryptosporidium* oocysts and/or *Giardia* cysts. Food Res. Int. (in Press). http://dx.doi.org/10.1016/j.foodres.2015.05.010.

Vaudaux, J.D., Muccioli, C., James, E.R., Silveira, C., Magargal, S.L., Jung Jr., C., et al., 2010. Identification of an atypical strain of *Toxoplasma gondii* as the cause of a waterborne outbreak of toxoplasmosis in Santa Isabel do Ivai, Brazil. J. Infect. Dis. 202 (8), 1226–1233.

Yatswako, S., Faleke, O.O., Gulumbe, M.L., Daneji, A.I., 2007. *Cryptosporidium* oocysts and *Balantidium coli* cysts in pigs reared semi-intensively in Zuru, Nigeria. Pak. J. Biol. Sci. 10 (19), 3435–3439.

CHAPTER

Foodborne pathogens in ready-to-eat peanut butter-containing products

6

Elizabeth M. Grasso-Kelley

Department of Food Science and Nutrition, Illinois Institute of Technology, Chicago, IL, United States

INTRODUCTION

Peanut butter and peanut butter-containing products are widely consumed ready-to-eat commodities. From 2006 to 2012, worldwide production of peanuts ranged from 31 to 36 million metric tons per year, with a 7% increase in peanut consumption from 2011 to 2012 (INC, 2013). In 2012, the top producers of peanuts were China, India, the United States, and Nigeria with 16, 5, 3, and 1.6 million metric tons, respectively, accounting for 73% of global production (INC, 2013). In the United States the total value of peanuts produced was over $1.1 billion in 2014 (USDA, 2014).

Besides being consumed as a food itself, the most common product produced from peanuts is peanut butter. Peanut butter produced in the United States contains a minimum of 90% peanuts, a maximum of 10% added ingredients including oil, sweeteners, and salt to enhance flavor or stabilizers to prevent oil separation, and the final product shall not contain greater than 55% fat (21 CFR 164.150) (Woodroof, 1983).

During the manufacture of commercial peanut butter, peanuts are first roasted at 160°C for 40–60 min to develop flavors (Woodroof, 1983). Roasted peanuts are then air-blast cooled on conveyors to stop the roasting process and maintain quality (Woodroof, 1983). Following cooling, the peanuts are dry blanched (138°C, 25 min) and brushed to remove the skin, kernel, dust, molds, and other foreign material present on the exterior (Woodroof, 1983). Processed peanuts are then ground and packaged at 48°C. Peanut butter is available in ready-to-eat packaging or in institutional containers for use in bakery and confectionery products (Woodroof, 1983). Peanut butter can be found in 94% of households in the United States with Americans consuming on average over 1.5 billion pounds of peanut butter and peanut products each year (Board, 2015).

P. Kotzekidou (Ed): Food Hygiene and Toxicology in Ready-to-Eat Foods.
DOI: http://dx.doi.org/10.1016/B978-0-12-801916-0.00006-6

OUTBREAKS

Low-moisture foods, including peanut butter and peanut butter-containing products, have been acknowledged as foods whose consumption may lead to foodborne illness. Previously this category of foods was thought to be safe as the low water activity does not permit the growth of foodborne microorganisms. The water activity of peanut butter is dependent on formulation but a peanut butter with 15% moisture will have a water activity ~0.70 (FDA, 1984). Research has shown that while pathogenic foodborne microorganisms are unable to grow in these foods, they are able to persist in the products at both refrigerated and ambient conditions for a long period of time (Grasso et al., 2010; Kenney and Beuchat, 2004; Park et al., 2008).

The viability of even a small number of *Salmonella* cells present in peanut butter can lead to illness. As little as 10–100 cells in high-fat foods can cause infections (Bell and Kyriakides, 2002). The high lipid content of peanut butter (~55%) may protect the microorganisms and allow their transport into the small intestine where they can multiply, colonize, and produce enterotoxin which may cause inflammation and diarrhea (Bell and Kyriakides, 2002).

Since the mid-1990s there have been six recorded outbreaks associated with commercial peanut butter, peanut-based products, and peanut-flavored savory snacks (Table 6.1).

The first outbreak occurred in 1994–95, and was an international outbreak ($n = 71$) that was epidemiologically and microbiologically linked to a peanut-flavored ready-to-eat kosher savory snack, comprised of corn grits, hydrogenated soybean oil, antioxidants, encapsulated vitamins, salt, and a peanut butter coating, which was produced in Israel (Killalea et al., 1996; Shohat et al., 1996; Threlfall et al., 1996). The international outbreak involving *Salmonella* Agona Phage Type 15 affected individuals in the United States, the United Kingdom, and Israel. In the United States, England, and Wales 27 cases were identified, 26 of which were children that consumed a peanut-flavored ready-to-eat kosher savory snack imported from Israel (Killalea et al., 1996). The contaminated snacks were manufactured on at least separate dates during a 4-month period. Microbiological testing found 83% (44/53) of the snack packets manufactured on November 6, 1994 were found to be contaminated with *S.* Agona PT 15 (Killalea et al., 1996). Additional enumeration of eight of the contaminated snack packets found estimated counts ranging from 2 to 8 organisms per 25-g packet (Killalea et al., 1996). *S.* Agona was not isolated from any of the environmental swabs taken from the production line, samples of ingredients, and stool samples from workers, but *Salmonella enteritidis* was isolated from a reusable plastic bag in which the snack was stored for 48 h prior to packaging (Shohat et al., 1996). Officials from the Israel Ministry of Health's Food Service examined all potential points of contamination, but were unable to locate the specific source (Shohat et al., 1996). The cause of the contamination remains unknown.

Table 6.1 Worldwide Outbreaks Related to *Salmonella*-Contaminated Ready-to-Eat Peanut Products

Year(s)	Product(s)	*Salmonella* Serovar(s)	Location(s)	Number Affected	Reference(s)
1994–95	Peanut-covered savory snack	*S.* Agona PT 15	Israel, United Kingdom, and United States	71	Killalea et al. (1996) Shohat et al. (1996) Threlfall et al. (1996)
1996	Peanut butter	*S.* Mbandaka	Australia	15	Scheil et al. (2008) Ng et al. (1996) Oliver (1996)
2001	Peanuts	*S.* Stanley and *S.* Newport	Australia, Canada, and United Kingdom	109 97 *S.* Stanley 12 *S. Newport*	Kirk et al. (2004)
2006–07	Peanut butter	*S.* Tennessee	United States	715	CDC (2007) Sheth et al. (2011)
2008–09	Peanut butter Peanut butter products	*S.* Typhimurium	United States and Canada	714	CDC (2009a) Cavallaro et al. (2011)
2012	Peanut butter and other nut butter products	*S.* Bredeney	United States	42	CDC (2014a) FDA (2012)

Between April and June 1996, 15 confirmed cases of salmonellosis were attributed to consumption of store or generic-brand peanut butter in South Australia (Scheil et al., 2008). Following the outbreak, trace-back investigations conducted by the Communicable Disease Control Branch of South Australia found *Salmonella* Mbandaka in six unopened containers of peanut butter and two opened containers removed from consumer residences (Scheil et al., 2008). Further analysis also isolated *Salmonella* Senftenberg from the peanut butter (Oliver, 1996). Roasted peanuts purchased from a separate state, used for the manufacture of peanut butter, were found to be the point of contamination that led to this outbreak (Scheil et al., 2008).

An increase in *Salmonella* Stanley infections in Australia in September 2001 led to a multicontinent investigation dealing with *Salmonella*-contaminated dry flavored Asian-style peanuts imported from Asia. Molecular-based investigations in Australia, Canada, and the United Kingdom (England, Wales, and Scotland) linked 55, 42, and eight *Salmonella* cases, respectively (Kirk et al., 2004). Of those cases, 95 cases were from *Salmonella* Stanley and 12 from *Salmonella* Newport. Unopened packages ($n = 142$) of dry flavored, garlic-flavored, and roasted peanuts were tested and found to be contaminated with *Salmonella* Stanley, *S.* Newport, *S.* Kottbus, *S.* Lexington, and one unidentified *Salmonella* serovar at contamination levels ranging from <0.03 to 2 organisms/g of in-shell peanuts (Kirk et al., 2004). Despite inquires with the country of origin, it was impossible to determine the source of contamination.

The first recorded outbreak of *Salmonella*-contaminated peanut butter in the United States occurred throughout 2006–07. Illnesses were reported as early as December 2006 and continued throughout March 2007 with a total of 715 cases of *Salmonella* Tennessee in 48 states associated with consumption of store-brand and one name-brand peanut butter manufactured in a Sylvester, GA, processing facility (CDC, 2007). US Food and Drug Administration investigations of plant operations were limited as the company halted operations prior to inspection. The company was not able to provide actual roasting temperatures of peanuts during normal operations and it was found that some roasted peanuts were held in uncovered bins after heat treatment (Sheth et al., 2011). While company records did not show any *Salmonella*-positive samples resulting from finished product and environmental testing between 2005 and 2006, records indicate that *Salmonella* had been isolated from peanut butter jars in October 2004 (Sheth et al., 2011). Samples from unopened jars and environmental samples linked the outbreak strain, *S.* Tennessee, with finished product and two positive environmental samples from the roasting room (CDC, 2007). Following this outbreak the company pled guilty to shipping contaminated peanut butter and paid $11.2 million in fines. Economic investigations noted that the company spent ~$78 million on recalls (ie, notifying customers, implementing a toll-free hotline, and collecting and destroying product), $15–20 million on plant renovations, and may have lost $55 million in sales (McGrath, 2009).

In 2008–09 a *Salmonella* Typhimurium outbreak occurred in the United States associated with peanut butter and peanut-butter-containing products. This outbreak was responsible for 714 illnesses, 166 hospitalizations and 9 recorded deaths throughout 46 states (CDC, 2009a; FDA, 2009b). The manufacturer produced peanut butter products as well as being an ingredient supplier for many institutions, food service providers, food manufacturers, and distributors. Peanut butter and peanut paste are commonly used ingredients in many products, including cookies, crackers, cereal, candy, ice cream, pet treats, and other foods (CDC, 2009a). A joint investigation by the US FDA and Georgia Department of Agriculture at the Blakely, GA, facility found multiple possible sources of contamination, including evidence of rain and other water leakage into the roasted peanut storage areas, possible cross-contamination practices between raw and roasted peanuts, and uncertainty of peanut roasting temperatures (Cavallaro et al., 2011). Concurrent investigations at a facility in Texas found several sanitation problems, including unsealed air-handling systems, rain leakage into storage areas, and rodent-accessible entryways (Cavallaro et al., 2011). Company records indicated that *Salmonella* was previously isolated from peanut paste and other products. Upon receiving negative results from reanalysis, products were shipped into commerce. Microbial investigations resulting from the outbreak found 171/315 *Salmonella*-positive food and environmental samples including open peanut butter, unopened peanut butter, and peanut butter crackers (Cavallaro et al., 2011). In 2014 a federal jury found three individuals guilty of criminal charges associated with this outbreak (Flynn, 2014). The owner was found guilty of conspiracy and other charges, the peanut broker of conspiracy for the shipment of tainted products and cover-up of laboratory results, and the former quality control manager of obstruction (Flynn, 2014).

A 2012 *Salmonella* Bredeney outbreak associated with peanut and other nut butter products produced in a Portales, NM, facility was associated with 42 cases, including 12 hospitalizations, in 20 states (CDC, 2014a). On November 26, 2012 the US FDA suspended the food facility registration, utilizing this power for the first time (CDC, 2014a). By suspending registration, the company was prohibited from introducing food into inter- and intrastate commerce. Subsequent investigations by US FDA inspectors found samples of *Salmonella* in 28 different locations throughout the plant, in 13 samples taken from nut butters and in 1 sample from raw peanuts (FDA, 2012). Inspectors found improper handling of the products, unclean equipment, and uncovered trailers of peanuts exposed to rain and birds (FDA, 2012). It was noted that employees did not have access to handwashing sinks and the company failed to clean production and packaging equipment between runs (FDA, 2012). Record analysis documented at least 9 and up to 13 separate *Salmonella* serovars in peanut butter products the company produced and distributed (FDA, 2012).

In addition to the aforementioned outbreaks resulting from *Salmonella*-contaminated peanut butter products, outbreaks have resulted from *Salmonella*-contaminated almonds (CDC, 2004; Isaacs et al., 2005; Ledet Muller et al., 2007), cashew products (CDC, 2014b), pine nuts (CDC, 2011b), and pistachios (CDC, 2009b). *Escherichia coli* contaminated in-shell hazelnuts (CDC, 2011a) and walnuts (CFIA, 2011).

Table 6.2 Peanut-Related Recalls in Ready-to-Eat Foods in the United States (FDA, 2014)

Year	Product(s)	Pathogen	Notes
2014	Peanut butter and cranberry crunch nutrition bars	*Salmonella*	
2014	Nut butters	*Salmonella*	Distributed across the United States, Canada, Hong Kong, United Arab Emirates, and Dominican Republic
2014	Peanut butter	*Listeria monocytogenes*	
2014	Peanut butter	*Salmonella*	Occurred in Philippines
2012	Peanuts, peanut butter, candy, nutrition bars, ice cream, snacks, snack bars, peanut butter fudge, bistro boxes, gelato, spring rolls, peanut butter power chews, crackers, brownies, cookies, granola bars, bird seed pet treats, and dog food	*Salmonella* Bredeney	Over 57 separate recalls for products associated with 2012 outbreak
2011	Peanut butter	*Salmonella*	
2010	Peanut butter	*Listeria monocytogenes*	
2009	Peanut butter and peanut-based ingredients	*Salmonella* Typhimurium	Over 3900 recalls from 350 producers associated with 2008–09 outbreak

RECALLS

Thorough environmental and finished product testing conducted by manufacturers and the federal government have identified potentially contaminated peanut butter products introduced into commerce. When this occurs, product recalls are conducted to remove the offending products from the market. A variety of ready-to-eat peanut butter and peanut butter-containing products contaminated with *Listeria monocytogenes* or *Salmonella* serovars have been recalled in the United States (Table 6.2) (FDA, 2014). In addition to peanut-butter-related recalls, numerous other nut-related products have been recalled due to pathogenic contamination (FDA, 2014).

PREVALENCE OF PATHOGENS

To aid in determination of appropriate control measures, it is important to assess the incoming risk of pathogens on raw materials. Peanuts, a major ingredient in peanut butter can become contaminated during growth, harvest, transportation, or storage (Mattick et al., 2006). Multiple studies have been conducted in the United States to evaluate the prevalence and counts of *Salmonella* and enterohemorrhagic *E. coli* in raw peanuts. One study that investigated 10,162 raw, shelled runner peanuts found three samples positive for Shiga toxin-producing *E. coli* and 68 samples positive for *Salmonella* (Miksch et al., 2013). Upon enumeration of positive *Salmonella* samples, the average contamination was 1.04 MPN/350-g sample (Miksch et al., 2013). A separate study tested 944 raw, shelled peanut samples for *Salmonella* and found a prevalence of 2.33% with samples ranging from <0.03 to 2.4 MPN/g (Calhoun et al., 2013). A study completed in Australia involving 195 samples of peanuts found 100% of the peanuts to be acceptable with <3 CFU/g *E. coli* and no *Salmonella* detected in a 25-g sample (NSWFA, 2012).

In addition to microbial pathogens, peanuts and peanut-containing products may contain mycotoxins, such as aflatoxins. Studies of peanuts in south-western Nigeria detected aflatoxins in 43/48 samples. Approximately 25% exceeded the 20 ng/g limit for aflatoxin B_1 while 83% and 79% of all samples contained AFB_1 and total aflatoxins above the European Union limits of 2 and 4 ng/g, respectively (Afolabi et al., 2014). Survey studies conducted in Bulawayo, Zimbabwe, on peanut and peanut butter samples found 3/18 and 10/11 samples were contaminated by aflatoxins (Mupunga et al., 2014).

CROSS-CONTAMINATION

Salmonella may contaminate the production environment through introduction of contaminated raw product or through other sources such as rodents, insects, contaminated personnel and equipment, incomplete plant maintenance, or insufficient sanitation (CDC, 2007; Grasso et al., 2010).

Laboratory-based studies investigated the effect of contamination level (high, medium and low; ~9, 6, and 3 log CFU/g, respectively) on the transfer of *Salmonella* Oranienburg and Tennessee from inoculated nut butters to four food-contact surfaces (Delrin, stainless steel, polyurethane, and polyethylene) representing conveyors, machinery, etc. (Zheng et al., 2012). Microbial analysis conducted after physically removing all visible nut butter determined that *S.* Oranienberg and Tennessee followed similar trends for microbial transfer in regards to inoculum level and food-contact surface. Transfer rates, defined as the percent *Salmonella* on coupons remaining following a visual clean (37.8–47.8%), were not statistically different ($p > 0.05$) for the four food contact surfaces. While the amount of *Salmonella* transferred to the food contact surface was dependent on inoculation level, 0.78 ± 0.09, 1.30 ± 0.30, and 2.84 ± 0.21 log CFU/10 cm^2 for low, medium, and high inoculum level, respectively, the transfer rate was ~40% regardless of original inoculum level (Zheng et al., 2012). Further analysis investigating the transfer of *Salmonella* from the previously contaminated visibly clean surface to uninoculated nut butter found cross-contamination to be a risk.

SURVIVAL STUDIES

Laboratory studies have shown that pathogens inoculated into peanuts, peanut butter, and peanut spread could survive for as long as 52 weeks. Brar et al. (2015) inoculated raw peanuts with *Salmonella*, *E. coli*, and *L. monocytogenes* at 3 log CFU/g. Peanuts were stored for 1 year at −24°C, 4°C, and 22°C. Pathogen levels were reduced over time, but were still detected >0.3 log CFU/g for all conditions (Brar et al., 2015). Grasso et al. (2010) investigated the reduction of *Salmonella* Typhimurium over a 28-day storage period at 20°C. Minimal reductions (1.23 log CFU/g) with the greatest decrease over the first 3–6 days were seen, with over 5.5 log CFU/g remaining in inoculated peanut butter (Grasso et al., 2010). Keller et al. (2012) found similar results regarding the viability of *Salmonella* Tennessee and *S.* Oranienburg in peanut butter; regardless of propagation method (plate vs broth grown cells), with the largest decrease in populations within the first 48 h at 25°C. They found significantly more survival in peanut butter ($p < 0.001$) when grown on solid media with <1-log CFU/g loss over 2 weeks as opposed to a 1- to 2- log CFU/g loss when grow in liquid culture (Keller et al., 2012). Park et al. (2008) inoculated a three-serovar cocktail of *S.* Tennessee in five commercial peanut butters stored at refrigerated and ambient conditions (4°C and 22°C, respectively). Populations decreased 0.34–1.28-log CFU/g at ambient conditions and 0.15–0.65-log CFU/g at refrigerated conditions over 14 days, from original levels ~6–7 log CFU/g (Park et al., 2008). Burnett et al. (2000) inoculated a five-serotype mixture of *Salmonella* (5.68 log CFU/g) into peanut products (a_w 0.2–0.33) and found that at both 5°C and 21°C the cocktail was still detectable in all products, except the natural peanut butter, after 24 weeks of storage. Kataoka et al. (2014) investigated the survival of heat-stressed *S.* Tennessee, *S.* Typhimurium, and *Enterococcus faecium* (~5–6 log CFU/g) in four model peanut paste formulations with a combination of two water activities (a_w 0.3 and 0.6) and two fat levels (47% and 56%) over 12 months at 20°C. Microbial reductions following storage were 1.3–2.4-, 1.8–2.8-, and 1.1–2.1-log CFU/g for *S.* Tennessee, *S.* Typhimurium, and *E. faecium*, respectively. Fat level did not significantly affect survival, but greater survivability was observed in peanut pastes with lower water activities for all microorganisms (Kataoka et al., 2014). Kenney and Beuchat (2004) investigated the survival of *L. monocytogenes* inoculated into peanut butter at two water activities. During storage at 20°C, *L. monocytogenes* decreased from 4.42 log CFU/g by 0.62-log CFU/g in 24 weeks at a_w 0.33 and 0.48-log CFU/g in 8 weeks at a_w 0.65. *L. monocytogenes* inoculated into chocolate peanut butter spread at 3.37 log CFU/g decreased 0.90-log CFU/g at a_w 0.33 and 0.95-log CFU/g at a_w 0.65 in 16 and 4 weeks at 20°C, respectively (Kenney and Beuchat, 2004). Similar results are seen for almonds (Kimber et al., 2012; Uesugi and Harris, 2006), pecans (Beuchat and Mann, 2010), walnuts (Blessington et al., 2012, 2013), and pistachios (Kimber et al., 2012).

INTERVENTION STRATEGIES

In response to outbreaks of *Salmonella* related to nuts and nut products, the FDA issued guidance recommending 5-log reduction in *Salmonella* for peanuts (FDA, 2009a). To comply with such regulations and guidelines, food processors must develop processes with a kill step(s) capable of delivering the appropriate reduction in *Salmonella*. For a similar commodity, Monte Carlo simulations assessing the risk of salmonellosis from consumption of raw almonds in the United States suggest the risk can be reduced from 78% to less than 1% by using a process that provides a 5-log reduction (Danyluk et al., 2006).

THERMAL INACTIVATION

Previous outbreaks related to ready-to-eat peanut-based products have shown the importance of *Salmonella* reduction during manufacture. Sanders and Calhoun (2014) investigated the effect of laboratory-scale oil and dry roasting on survival of *Salmonella* inoculated onto raw peanuts at various time and temperature processing conditions. Peanuts were dry roasted at five temperatures between 129 and 163°C and oil roasted at 120–160°C for three different time durations. Dry roasting inoculated peanuts at 154°C provided ~4.6-log and 5.4-log CFU/g reduction at 10 and 15 min, respectively. Oil roasting inoculated peanuts for 1.5 min at 150°C resulted in >6-log CFU/g reductions (Sanders and Calhoun, 2014). Studies conducted using *E. faecium* found the surrogate provided more conservative results than *Salmonella* (Sanders and Calhoun, 2014). Poirier et al. (2014) scaled-up investigations and found a 5-log CFU/g reduction of *E. faecium* at 149°C, 163°C, and 177°C after 21, 15, and 11 min, respectively in an industrial peanut roaster.

As previous research has shown minimal reductions in *Salmonella* levels in contaminated ready-to-eat peanut butter and peanut products over time, thermal inactivation has been investigated as a potential intervention strategy. Ma et al. (2009) investigated the rates of thermal inactivation of *Salmonella* Tennessee, *S.* Enteritidis, *S.* Typhimurium, and *S.* Heidelberg inoculated into peanut butter at 71°C, 77°C, 83°C, and 90°C. They found rapid death in the first 20 min followed by lower death rates thereafter. Their work showed a nonlinear destruction of the microorganisms that required approximately 120 min at 90°C for a 7-log CFU/g decrease, with the first 30 min accounting for an almost 5-log CFU/g reduction. The outbreak strains they tested were more thermally tolerant than other *Salmonella* tested. Thermal treatments at 90°C for less than 30 min were not able to achieve a 5-log reduction (Ma et al., 2009). Shachar and Yaron (2006) found similar nonlinear destruction of *Salmonella* Agona, *S.* Typhimurium, and *S.* Enteritidis in peanut butter as Ma et al. (2009). Surviving *Salmonella* populations did not change significantly after 20 min of exposure to the tested temperatures and remained viable even at 90°C for 50 min (Shachar and Yaron, 2006). Li et al. (2014) investigated the thermal survival of a six serovar cocktail of *Salmonella* inoculated into regular, reduced fat, reduced sugar, and omega 3 formulated peanut butters. They found *Salmonella* in reduced-fat peanut butter spread exhibited significantly higher thermal resistance than other peanut butters. The omega 3 formulation provided the least protective effect (Li et al., 2014). He et al. (2013) found peanut butters with increased water activities (a_w 0.2, 0.4, 0.6, and 0.8) significantly reduced the heat resistance of desiccation stressed *S.* Tennessee K4643, *S.* Enteritidis BSS-1045, and *S.* Typhimurium LT2. Differences in thermal resistance were less notable when treated at 126°C. Thermally treated (90°C) peanut butter at a_w 0.2 resulted in <3-log CFU *S.* Tennessee reduction at 20 min (He et al., 2013). Keller et al. (2012) investigated the effects of inoculum preparation (plate-grown and broth-grown cells) of *S.* Tennessee and *S.* Oranienburg immediately after inoculation, and after 1 and 2 weeks postinoculation. They found the rate of thermal inactivation showed no significant dependence on the length of incubation on the rate of survival at 85°C. Both serovars tested showed similar thermal resistance in peanut butter regardless of growth conditions. An analysis of covariance showed that plate-grown cells had a more linear reduction curve and *D*-value could be calculated using standard first-order kinetics. $D_{85°C}$ of *S.* Tennessee and *S.* Oranienburg were 11.95 ± 1.55 and 12.83 ± 2.35 min, respectively (Keller et al., 2012).

Studies have found that microorganisms are much more heat resistant in low-water activity environments than at levels which support growth. While thermal resistance increases with decreasing water activity, it is difficult to predict the extent of this increase, and it does vary with the type of solute present (Beuchat et al., 2011). Research has shown that pathogenic *Salmonella* serovars are not completely inactivated by heat typical in the processing of peanut butter and thus can be present in the consumer product (He et al., 2013; Keeny and Beuchat, 2004; Keller et al., 2012; Ma et al., 2009; Shachar and Yaron, 2006).

NONTHERMAL INACTIVATION

Thermal inactivation has been shown to provide minimal reductions in *Salmonella* via typical processing conditions (He et al., 2013; Kenney and Beuchat, 2004; Keller et al., 2012; Ma et al., 2009; Shachar and Yaron, 2006). Nonthermal intervention strategies, including high-pressure processing, irradiation, and chemicals, have been investigated to aid in the destruction of *Salmonella* in peanut butter and other ready-to-eat peanut products while maintaining product quality.

Grasso et al. (2010) investigated the effect of high-pressure treatment (600 MPa, 5 min, 45°C) on *Salmonella* Typhimurium-inoculated creamy peanut butter and peanut-based model systems. High-pressure processing (HPP) led to minimal (<1 log CFU/g) reductions of *S.* Typhimurium inoculated into commercial creamy and natural peanut butters. Results for a 25% peanut protein in peanut oil mixture exhibited similar results to that of commercial peanut butters tested. The 2%, 5%, and 10% protein peanut flour/oil mixtures had a

1.7-, 1.6-, and 1.0-log reduction, respectively. When levels of protein were reduced, the survivability of *Salmonella* also decreased, implying that peanut proteins may aid in the resistance (Grasso et al., 2010). D'Souza et al. (2014) found similar results for HPP-treated creamy peanut butter with modified composition. There was no difference for 10%, 15%, or 25% added moisture, which corresponded to a_w 0.67, 0.79, and 0.87, respectively, for samples treated at 600 MPa for 18 min. A steady and statistically significant increase in log reduction was seen as added moisture was increased from 50% to 90% (a_w 0.93–0.96) (D'Souza et al., 2014). D'Souza et al. (2014) also investigated the effects of nisin at 2.5, 5, 12.5, and 25 ppm alone and in combination with HPP on *Salmonella*-inoculated peanut butter. Nisin alone had no significant antimicrobial effect. A combination of nisin and HPP showed a maximum reduction similar to that of HPP alone (D'Souza et al., 2014).

Matak et al. (2010) and Hvizdak et al. (2010) studied the effects of electron beam irradiation on the reduction and recovery of *S.* Typhimurium and *S.* Tennessee in peanut butter. Hvizdak et al. (2010) found >4-log CFU/g reduction of *Salmonella* at 3.0 kGy for 10-g samples treated at 1 mm thickness. Matak et al. (2010) found that *Salmonella* will survive in peanut butter when exposed to sublethal doses of irradiation. Ban and Kang (2014) studied the effects of gamma (cobalt 60) irradiation on inactivating a three-serovar cocktail of *S.* Typhimurium in peanut butter. Samples (5 g) were spread into a thin layer and exposed to 0–3 kGy. Treated samples were held for up to 14 days. Maximum reductions of 1.9-, 2.8-, and 4-log reductions were seen at 1, 2, and 3 kGy, respectively. *Salmonella* reached undetectable levels in samples treated at 3 kGy and held at 25°C for 14 days (Ban and Kang, 2014).

Since nonthermal treatments do not appear to improve the safety of contaminated peanut butter, it is essential to identify strategies that prevent pathogen contamination of peanut butter products.

PREVENTIVE CONTROLS

Implementation of evidence-based preventive controls and sanitation procedures are required to ensure product safety as *Salmonella* can survive in low-moisture foods, such as peanut butter and other peanut-based products. A hazard analysis should be conducted to determine risk factors and potential points and sources of contamination (Hung et al., 2015).

Suitable precautions should be taken to protect the nuts from contamination by domestic animals, rodents, birds, insects, mites, and other arthropods during handling and storage. Peanuts should be moved to suitable storage as soon as possible after harvesting or drying. Where nuts are likely to become infested with insects during or after harvesting, suitable treatment such as fumigation or application of a pesticide spray should be carried out as a preventive measure. Nuts should be stored in covered containers, buildings, or under covering (Codex, 1979).

Dry clean procedures should be utilized to avoid wet spots in which microorganisms can propagate and contaminate contacted product. Even though water may not be used directly on equipment, spray and elevated humidity from continuous use can increase moisture in organic matter trapped in crevices in equipment, such as conveyors, to the point where microorganisms can proliferate (Codex, 1979). Grasso et al. (2014) evaluated the efficacy of a peanut butter push-through clean of a section of previously contaminated peanut butter processing equipment at both the laboratory and pilot scale. In laboratory-scale studies, 6- and 7-log CFU/g *Salmonella* reduction was found for stainless steel and silicone contact surfaces, respectively, but push-through processes did not completely eliminate *Salmonella* from within the system. In pilot-scale studies *Salmonella* was detected in the peanut butter remaining in the piping (6.0 log CFU/g) and in the exit valve (5.2 log CFU/g) following the push-through. Physically cleaned inner surfaces of piping, appearing "visibly clean," had 0.14–0.17 log CFU/cm^2 *Salmonella*, indicating that bacteria were transferred to the stainless steel surface of the piping during the push-through procedure (Grasso et al., 2013). Grasso et al. (2015) investigated the cleaning and sanitation of *Salmonella*-contaminated pilot-scale peanut butter-processing equipment. A two-step hot-oil (90°C, 2 h) circulation followed by a 60% isopropanol (1 h) with or without added quaternary ammonium compounds reduced *Salmonella* levels below detection at $<$0.16 log CFU/cm^2.

If pathogenic contamination does occur within a peanut-processing facility the contaminated product should be reprocessed to eliminate the pathogen or destroyed. If product has been released into commerce, a recall should be initiated. Prior to production, mitigation strategies to reduce the risk of recontamination events should be established. These may include (but are not limited to) implementation of a preventive controls plan including new supplier verification activities, procedures to reduce the risk of cross-contamination, as well as a revised environmental testing program. Additionally, a validated cleaning and sanitation procedure should be conducted, facilities repaired, and new equipment installed (as needed).

CONCLUSIONS

Historically, low-moisture ready-to-eat foods, such as peanut butter and peanut butter-containing products, were thought to be safe as the low water activity levels of these products do not allow pathogens to grow. Worldwide salmonellosis outbreaks associated with peanuts, peanut butter, and peanut butter-containing products have brought the attention to this category of ready-to-eat foods. While research has shown pathogens are unable to grow, studies have indicated that they are able to persist in these products at both refrigerated and ambient conditions for long periods of time. Thermal inactivation strategies are able to eliminate contamination on raw peanuts, but are unable to eliminate pathogens in processed peanut butter products. Evaluation of nonthermal intervention strategies, including high-pressure processing, irradiation, and chemical additives, provide minimal reductions of *Salmonella* and other microorganisms. As elimination of pathogens in processed peanut butter products has not been achieved to date, it is imperative to implement preventive control measures to mitigate the risk of illness to consumers.

REFERENCES

Afolabi, C.G., Ezekiel, C.N., Kehinde, I.A., Olaolu, A.W., Ogunsanya, O.M., 2014. Contamination of groundnut in South-Western Nigeria by aflatoxigenic fungi and aflatoxins in relation to processing. J. Phytopathol. 163, 279–286.

Ban, G.-H., Kang, D.-H., 2014. Effects of gamma irradiation for inactivating *Salmonella* Typhimurium in peanut butter product during storage. Int. J. Food. Microbiol. 171, 48–53.

Bell, C., Kyriakides, A., 2002. *Salmonella*: A Practical Approach to the Organism and Its Control in Foods. Iowa State University Press, Ames, IA.

Beuchat, L.R., Mann, D.A., 2010. Factors affecting infiltration and survival of *Salmonella* in in-shell pecans and nutmeats. J. Food Prot. 73, 1257–1268.

Beuchat, L.R., Komitopoulou, E., Betts, R., Beckers, H., Bourdichon, F., Joosten, H., et al., 2011. Persistence and survival of pathogens in dry foods and dry food processing environments. International Life Sciences Institute Europe Report Series. <http://www.ilsi.org/Eurpope/Documents/Persistence%20and%20survival%20report> (accessed September 2012).

Blessington, T., Mitcham, E.J., Harris, L.J., 2012. Survival of *Salmonella enterica, Escherichia coli* O157:H7, and *Listeria monocytogenes* on inoculated walnut kernels during storage. J. Food. Prot. 75, 245–254.

Blessington, T., Theofel, C.G., Mitcham, E.J., Harris, L.J., 2013. Survival of foodborne pathogens on inshell walnuts. Int. J. Food Microbiol. 166, 341–348.

Brar, P.K., Proano, L.G., Friedrich, L.M., Harris, L.J., Danyluk, M.D., 2015. Survival of *Salmonella, Escherichia coli O*157:H7, and *Listeria monocytogenes* on raw peanut and pecan kernels stored at −24, 4, and 22°C. J. Food Prot. 78, 323–332.

Burnett, S.L., Gehm, E.R., Weissinger, W.R., Beuchat, L.R., 2000. Survival of *Salmonella* in peanut butter and peanut butter spread. J. Appl. Microbiol. 89, 472–477.

Calhoun, S., Post, L., Warren, B., Thompson, S., Bontempo, A.R., 2013. Prevalence and concentration of *Salmonella* on raw shelled peanuts in the United States. J. Food Prot. 76, 575–579.

Cavallaro, E., Date, K., Medus, C., Meyer, S., Miller, B., Kim, C., et al., 2011. *Salmonella* Typhimurium infections associated with peanut products. N. Engl J. Med. 365, 601–610.

CDC, 2004. Outbreak of *Salmonella* serotype Enteritidis infections associated with raw almonds—United States and Canada, 2003–2004. Morb. Mortal. Wkly. Rep. 53, 484–487.

CDC, 2007. Multistate outbreak of *Salmonella* serotype Tennessee infections associated with peanut butter—United States, 2006–2007. Morb. Mortal. Wkly. Rep. 56, 521–524.

CDC, 2009a. Multistate outbreak of *Salmonella* infections associated with peanut butter and peanut butter containing products—United States, 2008–2009. Morb. Mortal. Wkly. Rep. 58, 1–6.

CDC, 2009b. Multistate outbreak of *Salmonella* infections linked to pistachio nuts (final update). <http://www.cdc.gov/salmonella/pistachios/index.html> (accessed March 2014).

CDC, 2011a. Investigation update: multistate outbreak of *E. coli* O157:H7 infections associated with in-shell hazelnuts. <http://www.cdc.gov/ecoli/2011/hazelnuts0157/index.html> (accessed March 2015).

CDC, 2011b. Multistate outbreak of human *Salmonella* Enteritidis infections linked to Turkish pine nuts. <http://www.cdc.gov/salmonella/pinenuts-enteriditis/index.html> (accessed March 2014).

CDC, 2014a. Multistate outbreak of *Salmonella* Bredeney infections linked to peanut butter manufactured by Sunland, Inc. (final update). <http://www.cdc.gov/salmonella/bredeney-09-12/index.html?s_cid-tw_cdc1615> (accessed March 2015).

CDC, 2014b. Multistate outbreak of *Salmonella* Stanley infections linked to raw cashew cheese (final update). <http://www.cdc.gov/salmonella/stanley-01-14/index.html> (accessed March 2015).

CFIA, 2011. Certain bulk and prepackaged bulk raw shelled walnuts may contain *E. coli* O157:H7 bacteria. <http://www.inspection.gc.ca/english/corpaffr/recarapp/2011/20110404e.shtml> (accessed July 2015).

Codex, 1979. Recommended international code of hygienic practice for groundnuts (peanuts). FAO/WHO Joint Publications. <www.codexalimentarius.net/input/download/standards/267/CXP_022e.pdf> (accessed November 2011).

Danyluk, M.D., Harris, L.J., Schaffner, D.W., 2006. Monte Carlo simulations assessing the risk of salmonellosis from consumption of almonds. J. Food. Prot. 69, 1594–1599.

D'Souza, T., Karwe, M., Schaffner, D.W., 2014. Effect of high hydrostatic pressure on *Salmonella* inoculated into creamy peanut butter with modified composition. J. Food. Prot. 77, 1664–1668.

FDA, 1984. Inspection technical guides: water activity (aw) in food. <http://www.fda.gov/ICECI/Inspections/InspectionGuides/InspectionTechnicalGuides/ucm072916.htm> (accessed June 2015).

FDA, 2009a. Guidance for industry: measures to address the risk for contamination by *Salmonella* species in food containing a peanut-derived product as an ingredient, <http://www.fda.gov/Food/GuidanceComplianceRegulatoryInformation/GuidanceDocuments/ProduceandPlanProducts/ucm115386.htm> (accessed March 2011).

FDA, 2009b. Peanut butter and other peanut containing products recall list. <http://www.accessdata.fda.gov/scripts/peanutbutterrecall/index.cfm> (accessed December 2009).

FDA, 2012. Inspection report: Sunland, Inc., In: Department of Health and Human Services (Ed.), FDA. <http://www.fda.gov/downloads/AboutFDA/CentersOffices/OfficeofGlobalRegulatoryOperationsandPolicy/ORA/ORAElectronicReadingRoom/UCM324793.pdf> (accessed November 2012).

FDA, 2014. Recalls, market withdrawals, and safety alerts. <http://www.fda.gov/Safety/Recalls/default.htm> (accessed July 2015).

Flynn, D., 2014. Jury verdicts: guilty, guilty, and guilty in PCA criminal trial. Food Safety News http://www.foodsafetynews.com/2014/09/guilty-guilty-and-guilty-in-pca-criminal-trial-in-albany-ga/#.VavtPU3bLZ4 (accessed July 2015).

Grasso, E.M., Somerville, J.A., Balasubramaniam, V.M., Lee, K., 2010. Minimal effects of high-pressure treatment on *Salmonella enterica* serovar Typhimurim inoculated into peanut butter and peanut products. J. Food Sci. 75, E522–E526.

Grasso, E.M., Keller, S.E., Anderson, N.M., Grove, S.F., 2013. Push-through sanitation of peanut butter processing equipment. International Association for Food Protection, Annual Meeting, Charlotte, NC.

Grasso, E.M., Chen, X., Keller, S.E., Anderson, N.M., 2014. Push-through sanitation of lab-scale peanut butter pumping equipment. Institute of Food Technologists, Annual Meeting, New Orleans, LA.

Grasso, E.M., Grove, S.F., Halik, L.A., Arritt, F., Keller, S.E., 2015. Cleaning and sanitation of *Salmonella*-contaminated peanut butter processing equipment. Food Microbiol. 46, 100–106.

He, Y., Li, Y., Salazar, J.K., Yang, J., Tortorello, M.L., Zhang, W., 2013. Increased water activity reduces the thermal resistance of *Salmonella* Enterica in peanut butter. Appl. Environ. Microbiol. 79, 4763–4767.

Hung, Y.-T., Liu, C.-T., Peng, I.-C., Hsu, C., Yu, R.-C., 2015. The implementation of a hazard analysis and critical control point management system in a peanut butter ice cream plant. J. Food Drug Anal. http://www.sciencedirect.com/science/article/pii/S1021949815000319 (accessed July 2015).

Hvizdak, A.L., Beamer, S., Jackzynski, J., Matak, K.E., 2010. Use of electron beam radiation for the reduction of *Salmonella enterica* serovars Typhimurium and Tennessee in peanut butter. J. Food Prot. 73, 353–357.

INC, 2013. INC global statistical review reflects the continued increase of the global consumption of nuts and dried fruits. International Nut and Dried Fruit Council. <http://www.nutfruit.org/en/inc-global-statistical-review-reflects-the-continued-increase-of-the-global-consumption-of-nuts-and-dried-fruits_70983> (accessed July 2015).

Isaacs, S., Aramini, J., Ciebin, B., Farrar, J.A., Ahmed, R., Middleton, D., et al., 2005. An international outbreak of salmonellosis associated with raw almonds contaminated with a rare phage type of *Salmonella* Enteritidis. J. Food Prot. 68, 191–198.

Kataoka, A., Enache, E., Black, D.G., Elliott, P.H., Napier, C.D., Podolak, R., et al., 2014. Survival of *Salmonella* Tennessee, *Salmonella* Typhimurium DT104, and *Enterococcus faecium* in peanut paste formulations at two different levels of water activity and fat. J. Food Prot. 77, 1252–1259.

Keller, S.E., Grasso, E.M., Halik, L.A., Fleischman, G.J., Chirtel, S.J., Grove, S.F., 2012. The effect of growth on the thermal resistance and survival of *Salmonella* Tennessee and Oranienburg in peanut butter using a new thin-layer thermal death time device. J. Food Prot. 75, 1125–1130.

Kenney, S.J., Beuchat, L.R., 2004. Survival, growth, and thermal resistance of *Listeria monocytogenes* in products containing peanut and chocolate. J. Food Prot. 67, 2205–2211.

Killalea, D., Ward, L.R., Roberts, D., 1996. International epidemiological and microbiological study of outbreak of *Salmonella* Agona infection from a ready-to-eat savoury snack. I: England and Wales and the United States. BMJ 313, 1105–1107.

Kimber, M.A., Kaur, H., Wang, L., Danyluk, M.D., Harris, L.J., 2012. Survival of *Salmonella, Escherichia coli* O157:H7, and *Listeria monocytogenes* on inoculated almonds and pistachios stored at −19, 4, and 24°C. J. Food Prot. 75, 1394–1403.

Kirk, M.D., Little, C.L., Lem, M., Fyfe, M., Genobile, D., Tan, A., et al., 2004. An outbreak due to peanuts in their shell caused by *Salmonella enterica* serotypes Stanley and Newport—sharing molecular information to solve international outbreaks. Epidemiol. Infect. 132, 571–577.

Ledet Muller, L., Hjertqvist, M., Payne, L., Pettersson, H., Olsson, A., Plym Forshell, L., et al., 2007. Cluster of *Salmonella* Enteritidis in Sweden, 2005–2006-suspected source: almonds. Eurosurveillance 12, 153–155.

Li, C., Huang, L., Chen, J., 2014. Comparative study of thermal inactivation kinetics of *Salmonella* spp. In peanut butter and peanut butter spread. Food Control 45, 143–149.

Ma, L., Guodong, A., Gerner-Smidt, P., Vijaya, M., Ezeoke, I., Doyle, M.P., 2009. Thermal inactivation of *Salmonella* in peanut butter. J. Food Prot. 8, 1596–1601.

Matak, K.E., Hvizdak, A.L., Beamer, S., Jackzynski, J., 2010. Recovery of *Salmonella enterica* serovars Typhimurium and Tennessee in peanut butter after electron beam exposure. J. Food Sci. 75, M462–M467.

Mattick, K.L., Jorgensen, F., Legan, J.D., Lappin-Scott, H.M., Humphrey, T.J., 2006. Habituation of *Salmonella* spp. at reduced water activity and its effect on heat tolerance. Appl. Environ. Microbiol. 66, 4921–4925.

McGrath, J., 2009. 10 costly food recalls. <http://money.howstuffworks.com/10-food-recalls.htm#page=6> (accessed July 2015).

Miksch, R.R., Leek, J., Myoda, S., Nguyen, T., Tenney, K., Svidenko, V., et al., 2013. Prevalence and counts of *Salmonella* and enterohemorrhagic *Escherichia coli* in raw, shelled runner peanuts. J. Food Prot. 76, 1668–1675.

Mupunga, I., Lebelo, S.L., Mngqawa, P., Rheeder, J.P., Katerere, D.R., 2014. Natural occurrence of aflatoxins in peanuts and peanut butter from Mulawayo, Zimbabwe. J. Food Prot. 77, 1814–1818.

National Peanut Board, 2015. Consumption facts. <http://nationalpeanutboard.org/the-facts/fun-facts/> (accessed July 2015).

Ng, S., Rouch, G., Dedman, R., Harries, B., Boyden, A., McLennan, L., et al., 1996. Human salmonellosis and peanut butter. Commun. Dis. Intell. 20, 326.

NSWFA, 2012. Report on the prevalence of *Salmonella* and *E. coli* in ready to eat nuts and nut products sold in Australia. <http://www.foodauthority.nsw.gov.au/_Documents/science/national_nut_survey.pdf> (accessed July 2015).

Oliver, G., 1996. The *Salmonella* Mbandaka outbreak—an Australian overview. Commun. Dis. Intell. 20, 326.

Park, E.-J., Oh, S.-W., Kang, D.-H., 2008. Fate of *Salmonella* Tennessee in peanut butter at 4 and 22°C. J. Food Sci. 73, M82–M86.

Poirier, D., Sanders, T.H., Davis, J.P., 2014. *Salmonella* surrogate reduction using industrial peanut dry roasting parameters. Peanut Sci. 41, 72–84.

Sanders, T.H., Calhoun, R.S., 2014. Effect of oil and dry roasting on peanuts at various temperatures and times on survival of *Salmonella* and *Enterococcus faecium*. Peanut Sci. 41, 65–71.

Scheil, W., Cameron, S., Dalton, C., Murry, C., Wilson, D., 2008. A South Australian *Salmonella* Mbandaka outbreak investigation using a database to set controls. Aust. N. Z. J. Public Health 22, 536–539.

Shachar, D., Yaron, S., 2006. Heat tolerance of *Salmonella enterica* serovars Agona, Enteritidis, and Typhimurium in peanut butter. J. Food Prot. 69, 2687–2691.

Sheth, A.N., Hoekstra, M., Patel, N., Ewald, G., Lord, C., Clarke, C., et al., 2011. A national outbreak of *Salmonella* serotype Tennessee infections from contaminated peanut butter: a new food vehicle for salmonellosis in the United States. Clin. Infect. Dis. 53, 356–362.

Shohat, T., Green, M.S., Marom, D., 1996. International epidemiological and microbiological study of outbreak of *Salmonella* Agona infection from a ready-to-eat savoury snack. II: Israel. BMJ 313, 1107–1109.

Threlfall, E.J., Hampton, M.D., Ward, L.R., Rowe, B., 1996. Application of pulsed-field gel electrophoresis to an international outbreak of *Salmonella* Agona. Emerg. Infect. Dis. 2, 130–132.

Uesugi, A.R., Harris, L.J., 2006. Growth of *Salmonella* Enteritidis phage type 30 in almond hull and shell slurries and survival in drying almond hulls. J. Food Prot. 69, 712–718.

USDA, 2014. Survey results pertaining to the value of peanuts. <http://quickstats.nass.usda.gov/results/E3084E61-E415-3978-9882-689D1F26A3DF> (accessed July 2015).

Woodroof, J.G., 1983. Peanuts: Production, Processing, Products, third ed. AVI Publishing Company, Inc, Westport, CT.

Zheng, Y., Grasso, E.M., Keller, S.E., 2012. Impact of inoculum level on the transfer of *Salmonella* serovars from contaminated almond butter to food-contact materials. Institute of Food Technologists, Annual Meeting, Las Vegas, NV.

CHAPTER

7 Handling of hamburgers and cooking practices

Daniel A. Unruh[1], Justin J. Kastner[2], Jacob R. Jenott[1] and Sara E. Gragg[1]

[1]Department of Animal Sciences and Industry, Kansas State University, Olathe, KS, United States [2]Department of Diagnostic Medicine and Pathobiology, Kansas State University, Manhattan, KS, United States

INTRODUCTION

Fletcher Davis is credited with making the first hamburger in the late 19th century, and consumption by the American public has been commonplace ever since (Grimes, 1998). In fact, the popularity of the hamburger is staggering: an estimated 50 billion hamburgers are consumed annually in the United States (Rolfes, 2012). This comprises 60% of total beef production in the United States and equates to over 16 billion pounds (Crandall et al., 2013). Quick-service restaurants (more commonly known as fast-food establishments) have played a vital role in the growing popularity of hamburgers, particularly when considering convenience and price. It is estimated that fast-food restaurants account for more than 50% of hamburgers consumed, with another 30% consumed in the home (a number that has been increasing) (Ralston et al., 2002). A smaller percentage of hamburgers are consumed at "sit-down" restaurants, cafeterias, residential dining locations, and convenience stores (Ralston et al., 2002; Røssvoll et al., 2014). Additionally, hamburger is a component of many commonly consumed dishes, including tacos, spaghetti, meatballs, breakfast items, ethnic dishes, and more.

The following chapter sheds light on the processes involved in hamburger production throughout the supply chain, as well as the importance and shared responsibility, from animal harvest to consumer practices, of safe handling. This includes current industry practices used to protect consumers, recommendations for at-home preparation, and potential public health implications of improper handling at any point in the supply chain. Additionally, government regulations to ensure hamburger safety, as well as current research into pathogen reduction in ground beef, will be discussed.

P. Kotzekidou (Ed): Food Hygiene and Toxicology in Ready-to-Eat Foods.
DOI: http://dx.doi.org/10.1016/B978-0-12-801916-0.00007-8

HAMBURGER: CHARACTERISTICS AND COOKERY

HAMBURGER AND GROUND BEEF

The *Code of Federal Regulations* (CFR) establishes a clear definition for hamburger. However, before that definition can be understood, the term for ground beef must be established. Thus, Title 9 of the CFR § 319.15 (a) defines ground beef (or chopped beef) as consisting of "chopped fresh and/or frozen beef with or without seasoning and without the addition of beef fat as such, shall not contain more than 30% fat, and shall not contain added water, phosphates, binders, or extenders" (Code of Federal Regulations, 2015b). Hamburger is subsequently defined as consisting of "chopped fresh and/or frozen beef with or without the addition of beef fat as such and/or seasoning, shall not contain more than 30% fat, and shall not contain added water, phosphates, binders, or extenders" (Code of Federal Regulations, 2015b). The definitions as such are similar and, in fact, Title 7 of the CFR § 65.155 states that ground beef, as defined, "includes products defined by the term 'hamburger'"(Code of Federal Regulations, 2012a). Careful reading of the text clarifies that the main difference between hamburger and ground beef, from a law and definition perspective, is that beef fat can be added to hamburger but not ground beef (USDA-FSIS, 2013a). If either product contains more than 30% fat, it is misbranded per federal regulations (USDA-FSIS, 2013a).

For simplicity, and keeping these definitions in mind, the terms ground beef and hamburger are used interchangeably in this chapter. Both hamburger and ground beef require the same hygienic practices, and thus are approached as one product.

GROUND BEEF PRODUCTION

Categories of ground beef products

It is important to distinguish the different categories of ground beef products that can be purchased in retail outlets. Products labeled as ground chuck, ground round, or ground sirloin are ground beef products that contain only skeletal muscle and skeletal trimmings that originate from the muscle primal specified on the label. As an example, a ground round product contains only skeletal muscle and skeletal trimmings from the round primal. A product labeled simply as ground beef is produced by combining skeletal muscle, skeletal trimmings, head meat trimmings and up to 25% of cheek meat. In addition to the components allowable in a ground beef product, products designated as hamburger can also contain added beef fat (Anonymous, 2014). Regular or coarse grind ground beef and hamburger can be purchased at varying ratios of lean to fat (eg, 80% lean/20% fat), which provides an array of products with differing price points that are suitable for a variety of applications and personal preferences (Anonymous, 2014). Unlike water, seasonings, phosphates, and other additives (eg, extenders or binders) are allowable in ground beef and hamburger products (Anonymous, 2014; USDA-FSIS, 2013a).

Nutrient and chemical composition of ground beef

What exactly is hamburger composed of? Work conducted by Prayson et al. (2008) (an analysis of fast-food hamburgers from eight different restaurants) demonstrated that hamburgers are primarily water (49%) and that meat content ranged from 2.1% to 14.8%. The "meat" portion of the samples was comprised of skeletal tissue, connective tissue, blood vessels, peripheral nerves, adipose tissue, bone, cartilage, and plant material (Prayson et al., 2008). Ground beef is nutrient-dense, accounting for high percent daily values of total fat, saturated fat, cholesterol, protein, and vitamins B12 and B6 (USDA-ARS, 2015). Sufficient levels of various minerals, such as phosphorus and potassium, are also bioavailable in ground beef (USDA-ARS, 2015).

Lean finely textured beef

With ground beef products (including hamburgers) being the most popular category of beef production in the United States, food technologists have innovatively worked to enhance not only the safety but also the leanness of these products. Amongst an array of technologies introduced during the last decade and a half, lean finely textured beef (LFTB) stands out as a manufactured beef product that has provided both (1) an exceptionally lean component in blended ground beef (and other meat products such as lunch meats and sausages) and (2) an unfortunate source of controversy that stems from the omnipresent tension between food-production innovation and consumer perceptions of risk (Greene, 2012). Pioneered by Beef Products, Inc. (BPI), LFTB is the product of a process involving the dissolution of residual beef (typically, 50% lean beef trimmings) obtained from carcasses processed in United States Department of Agriculture (USDA)-inspected processing facilities, separating the fat and connective tissue, and reconstituting the resultant—and exceptionally lean (94–97% lean beef)—product for incorporation into ground beef products including hamburgers. Significantly, BPI also pioneered an accompanying USDA-approved microbial-intervention process whereby food-grade ammonium gas interacts with water in the beef product to yield ammonium hydroxide and a concomitant increase in pH, thereby facilitating pathogen reduction. As explained in a case study authored by the International Food Information Council, a high-profile media story erupted in 2012 around concerns related to LFTB and, in particular, the ammonia-related aspects of BPI's technologies; this story progressed into a story of "an animated graphic explaining how the meat is collected, treated with ammonia, liquefied, pressed into brick form, and distributed to grocery stores" (Benson, 2014). A number of policy and private responses ensued. These included the USDA announcing that it would allow school districts participating in the National School Lunch Program to choose whether or not to purchase products containing LFTB, and grocery store chains such as Food Lion and Kroger announcing the cessation of sales of beef containing LFTB. Negatively referred to as "pink slime," LFTB suffered a consumer-perception blow, but the USDA—including the Under Secretary for the USDA's Food Safety and

Inspection Service (FSIS)—publicly insisted on its safety (Hagen, 2012). Furthermore, notable food safety activists such as Nancy Donley (of STOP Foodborne Illness) lamented the unfair and misinformation-filled "pink slime" media stories. However, in an effort to promote transparency, some firms have instituted some labeling schemes to acknowledge the presence of LTFB in their products, but such labeling was embraced for marketing, not safety-assurance, purposes.

HANDLING AND COOKING OF HAMBURGER

Storage

Upon purchase, ground beef should be refrigerated or frozen as soon as possible. This practice both preserves freshness and, importantly, slows the growth of any bacteria present in the beef. Fresh ground beef can be stored for 1–3 days at a temperature below 40°F (4.4°C), with an optimum temperature of 28°F (−2.2°C). If vacuum packaged, fresh ground beef can be stored under these conditions for up to 14 days, depending on the supplier. Frozen ground beef should be stored at, or below, 0°F (−17.8°C) for up to 90 days (Anonymous, 2014). If properly held under these conditions, frozen ground beef is considered safe indefinitely; however, the quality will degrade throughout storage (USDA-FSIS, 2013a). Following cooking, ground beef can be refrigerated for 2–3 days below 40°F (4.4°C) and frozen up to 90 days at 0°F (−17.8°C) or below (Anonymous, 2014). If the ground beef is to be used soon, it is appropriate to refrigerate or freeze it in the original packaging. If the product will be stored in the freezer for extended periods of time it should be wrapped in aluminum foil, heavy-duty plastic wrap, freezer paper, or plastic freezer bags prior to freezing (USDA-FSIS, 2013a).

Frozen ground beef can be thawed safely in the refrigerator and should be cooked or refrozen within 1–2 days (Anonymous, 2014; USDA-FSIS, 2013a). It is also appropriate to use a microwave oven to defrost frozen ground beef; however, the ground beef should be cooked immediately, as portions of the product may have begun to cook while defrosting. Submerging frozen ground beef in cold water can also be a safe defrosting method, if the meat is placed in a waterproof plastic bag and the water is replaced every 30 min. Ground beef thawed in this manner should be cooked immediately. Ground beef defrosted in the microwave oven or submerged in cold water should never be refrozen, unless it has been cooked prior to freezing (USDA-FSIS, 2013a).

Following storage and thawing guidelines is important for ensuring quality as well as safety of ground beef products. Both spoilage and pathogenic microorganisms may be present in ground beef and can rapidly multiply between 40°F and 140°F (4.4°C and 60°C), which is known as the temperature "danger zone." Growth of spoilage microorganisms can degrade product quality, while pathogenic microbial growth poses a risk of foodborne illness (USDA-FSIS, 2013a). Before cooking, consumers may notice ground beef packaging containing a blood-like liquid remaining after taking the meat out. This liquid is known as "purge" and is a result of cellular breakage and moisture loss from the ground beef. It is completely normal and often becomes more pronounced as temperature increases or the longer the product sits in the package (USDA-FSIS, 2011).

Handling raw hamburger

Microorganisms present in raw ground beef and purge can cross-contaminate surfaces, hands, and equipment upon contact. When manipulating ground beef, it is important to be mindful of this risk in order to prevent cross-contamination of foods that have been safely cooked or foods that will be consumed raw (eg, salad ingredients). Raw ground beef packaging materials should never be reused. Hands should be thoroughly washed with soap and warm water for at least 20 seconds after handling raw ground beef. Utensils and surfaces that have come into contact with the raw product should (1) be thoroughly washed in soap and (2) be sanitized in a bleach solution (1 tablespoon of unscented, liquid chlorine bleach mixed with 1 gallon (3.8 L) of water). Unless washed and sanitized as described, cooked hamburgers should never be handled with utensils, or placed on the same platter, that were used for the raw hamburgers (USDA-FSIS, 2013a).

Cooking hamburger

The importance of cooking hamburger correctly must be underscored as it is the last form of protection consumers have to prevent foodborne illness. The actual methodology for cooking a hamburger is largely up to consumer discretion and preference, taking into account the size of patties and whether the ground beef will be cooked as a patty or as a component in another dish. Research has demonstrated that a higher-fat ground beef product (70% lean:30% fat) needs a longer cook time compared to a leaner counterpart due to less efficient heat conduction (Luchansky et al., 2013a). According to a 2014 survey, the most popular method for cooking hamburger was pan frying, followed by grilling (Røssvoll et al., 2014). Less common methods include broiling, baking, and steaming.

Indicators of doneness

Color is not an effective measure of doneness (Hunt et al., 1999; Lyon et al., 2000). Studies have demonstrated that premature browning, less pink coloration, and an increase in cooked texture can occur in hamburger patties that are only cooked to 135°F (57.2°C) and 150°F (65.6°C), far below the recommended target temperature of 160°F (71.1°C) (Lyon et al., 2000; USDA-FSIS, 2013a). While consumers may prefer a slight pink hue in the center of the burger or argue that undercooking enhances flavor, the only true indicator of doneness is temperature. Failing to reach the required temperature allows the potential survival and regrowth of pathogenic bacteria, explored later in this chapter.

Previously cooked hamburgers

The main concern when reheating previously cooked hamburgers is avoiding excessive exposure to the aforementioned "danger zone," in which bacterial replication can easily occur (USDA-FSIS, 2013a). Hamburgers (as in patty form) must be reheated to 165°F (73.9°C). Hamburger in a sauce (eg, a meat sauce for spaghetti), soup, or gravy should be brought to a rolling boil (USDA-FSIS, 2012a). Hamburger reheated in a microwave should be done to ensure even heating: the food should be covered and rotated as necessary, and cold spots should be avoided (USDA-FSIS, 2012a).

PUBLIC HEALTH IMPLICATIONS OF POOR HAMBURGER HYGIENE

Neglecting the proper cooking of hamburger leads to an increased likelihood of pathogenic bacteria in the food at the point of consumption, which can result in serious implications for public health. This section examines the potential consequences of improper hamburger hygiene, including pathogens that can lead to illness, methods to increase hamburger hygiene at the processing facility, and current regulations that aim to increase hamburger hygiene along the supply chain.

COMMON PATHOGENS FOUND IN GROUND BEEF

Escherichia coli

Escherichia coli (*E. coli*) is one of the most commonly mentioned organisms when discussing pathogens in beef. Although not as common as other foodborne pathogens, numerous life-altering illnesses and deaths have been attributed to *E. coli* in beef. Diarrheagenic *E. coli* is classified in six categories, with one category, enterohemorrhagic *E. coli* (EHEC) causing the most severe illnesses. The Shiga-toxin-producing *E. coli* (STEC) include EHEC, as EHEC produce Shiga toxins (eg, Stx-1, Stx-2 or both); however, only those STEC that lead to bloody diarrhea are categorized as EHEC (Montville et al., 2012). *Escherichia coli* O157:H7, the predominantly recognized STEC, is estimated to cause 63,153 illnesses annually, and non-O157 STEC (serogroups O26, O45, O103, O111, O121, and O145) cause 112,752 illnesses annually (Scallan et al., 2011). Non-O157 STEC and *E. coli* O157:H7 cause the same disease and have similar illnesses; however, rates of hospitalization and death are greater for O157 than non-O157 (Scallan et al., 2011).

For a majority of those infected with STEC, infection will cause mild gastrointestinal discomfort, including non-bloody diarrhea. However, for susceptible populations and the immunocompromised, a STEC infection may lead to hemorrhagic colitis (severe pain and bloody diarrhea) (Murray et al., 2013) or hemolytic uremic syndrome, a sequelae leading to kidney failure and possibly death (Gyles, 2007). STEC have a low infectious dose; disease can occur from ingestion of as few as 10 cells (Moore, 2004).

E. coli, including STEC, are nonpathogenic colonizers of the bovine gastrointestinal tract (Gyles, 2007; Wells et al., 1991). Isolates of STEC have been recovered from cattle and humans, with the majority of the human isolates identified as being of cattle origin (Beutin et al., 1993, 2004; Gyles, 2007). It is estimated that a beef source was implicated in 55% of O157 outbreaks and 50% of non-O157 STEC outbreaks that occurred between 1998 and 2008 (Painter et al., 2013).

Title 21 of United States Code (USC) mandates that food be kept free from adulteration. Specifically, 21 U.S.C. § 601(m)(1) states: "The term 'adulterated' shall apply to any carcass, or part thereof, meat or meat food product . . . if it has, bears or contains any poisonous or deleterious substance which may render it injurious to health. . . ." Furthermore, § 601(m)(4) states that a product is adulterated "if it has been prepared, packed, or held under insanitary conditions whereby it may have become contaminated with filth, or whereby it may have been rendered injurious to health" (United States Code, 2006). In the *Code of Federal Regulations* (9 CFR § 300.2(b)(1)), the USDA-FSIS is granted authority to implement the *Federal Meat Inspection Act* (FMIA) as amended in 21 U.S.C. 601 *et seq.* (Code of Federal Regulations, 2015a). In 1994, the FSIS exercised that right, declaring *E. coli* O157:H7 an adulterant in ground beef *per se*, and extended adulteration status to all non-intact raw beef product and intact raw beef product intended for use in raw, non-intact beef product (USDA-FSIS, 1999a). Adulterant status was again expanded in 2012, when the FSIS declared that routine verification testing for six non-O157 STEC (O26, O45, O103, O111, O121, and O145) would commence for raw beef manufacturing trimmings (USDA-FSIS, 2012b).

Salmonella spp.

According to foodborne illness estimates, *Salmonella* is the second leading foodborne illness agent in the United States. An estimated 1 million illnesses from non-typhoidal *Salmonella* occur in the United States annually. Of these illnesses, it is estimated that over 19,000 hospitalizations and nearly 400 deaths result (CDC, 2011a).

Cattle have been identified as an important reservoir for *Salmonella* and ground beef can serve as a source of human exposure (Bosilevac et al., 2009). In 2013, USDA-FSIS analyzed 17,161 samples of ground beef and reported a *Salmonella* prevalence of 1.6% (USDA-FSIS, 2014). Bosilevac and colleagues reported a 4.2% prevalence of *Salmonella* for the 4136 ground beef samples analyzed between July 2005 and June 2007 (Bosilevac et al., 2009). According to the Centers for Disease Control and Prevention (CDC), salmonellosis outbreaks have been associated with ground beef products on multiple occasions (CDC, 2011b, 2012, 2013). Cattle hides are recognized as the primary source of carcass contamination with *Salmonella* (Arthur et al., 2008; Barkocy-Gallagher et al., 2001, 2003; Brichta-Harhay et al., 2008). Following hide removal, but prior to the application of postharvest interventions, *Salmonella* carcass contamination can vary, with one study reporting 3–24.9% contamination among fed cattle carcasses over the course of 1 year (Barkocy-Gallagher et al., 2003). However, postharvest antimicrobial interventions applied during slaughter are highly effective at controlling *Salmonella* on the carcass (Barkocy-Gallagher et al., 2003). A study that evaluated *Salmonella* contamination on carcasses harvested at two geographically distinct commercial beef operations reported prevalence to be less than 1% (Rivera-Betancourt et al., 2004) following antimicrobial application. Despite the efficacy of postharvest antimicrobial interventions, Bosilevac et al. (2009) reported *Salmonella* prevalence to be 4.2% in ground beef. This discrepancy suggests that carcass contamination may not be the only source of *Salmonella* in ground beef.

Recently, a number of studies have identified the potential for *Salmonella* carriage in cattle lymph nodes to serve as a vehicle for ground beef contamination. Initial studies investigated cattle mesenteric lymph nodes for *Salmonella* carriage. However, these lymph nodes are discarded during evisceration and do not serve as a vehicle for ground beef contamination (Arthur et al., 2008; Brichta-Harhay et al., 2012). Conversely, cattle peripheral lymph nodes (ie, subiliac) are located within the fatty tissues that are often destined for further processing into ground beef; thus, these lymph nodes can provide a direct source of contamination when harboring *Salmonella* (Arthur et al., 2008; Brichta-Harhay et al., 2012; Gragg et al., 2013a,b; Haneklaus et al., 2012; Li et al., 2015).

Gragg et al. (2013a) analyzed 3,327 subiliac lymph nodes from fed and cull cattle in the United States. These lymph nodes were collected from several regions of the country across several seasons in a 12-month period. An arithmetic mean *Salmonella* prevalence of 7.5% was reported. The data also indicated that *Salmonella* prevalence in cattle lymph nodes is significantly affected by animal type, season, and region.

Salmonella cells encased within cattle lymph nodes are protected from the effects of antimicrobial interventions applied to the carcass surface (Gragg et al., 2013a; Li et al., 2015). This, combined with the discrepancy between *Salmonella* contamination frequency on the carcass and in ground beef, has recently resulted in a great deal of attention directed towards cattle lymph nodes as a plausible source of ground beef contamination (Li et al., 2015).

Other pathogens of concern

In addition to STEC and *Salmonella* spp., ground beef can harbor other foodborne pathogens. A recent study isolated *Listeria monocytogenes*, a serious pathogen that can cause meningitis, septicemia, and spontaneous abortion, from 3% of analyzed ground beef samples ($n = 1750$) (Lecuit, 2005; Samadpour et al., 2006). Other common foodborne pathogens, such as *Campylobacter jejuni*, *Staphylococcus aureus*, and *Clostridium perfringens* (*C. perfringens*) can contaminate ground beef. A 2010 study analyzed 1,109 retail ground beef samples and detected *Campylobacter* spp. in 6.31% of samples (Vipham et al., 2010). *Staphylococcus aureus* can be found at a much higher rate (30% of samples) while *C. perfringens* has been isolated in even greater proportions (50%) (Kelman et al., 2011; Ladiges et al., 1974). All of the aforementioned organisms are predominately environmental and likely originate from the animal, particularly its digestive tract or hide. *Staphylococcus aureus* is also commonly harbored on human skin and nasal passages, which may contribute to its elevated prevalence from cross-contamination by employees (Waters et al., 2011).

HAMBURGER AND PATHOGEN REDUCTION

The most effective way to destroy pathogens is through heat (also known as thermal inactivation). When subjected to heat, microorganism death is orderly and mostly logarithmic when held at a constant temperature (Stumbo, 1973). Numerous studies have been performed demonstrating the decimal reduction time (*D*-value) of *E. coli* in ground beef, highlighting the effects of fat level, freezing, and the similarity in behavior of O157 and non-O157 STEC (Line et al., 1991; Luchansky et al., 2013b; Stringer et al., 2000; Unruh, 2014).

An increasingly common practice in American households is to "make" ground beef by grinding whole beef primal cuts in-home. The impression that the resulting ground beef is safer, or has lower pathogen incidence, is a common misconception held by consumers. There is logic in the assumption since primal cuts can be cooked safely to a lower temperature (145°F or 62.8°C). However, as soon as the grinding process begins, pathogens present on the outside are mixed throughout the entire product. Thus, all ground beef, ground in-home or not, must be cooked to 160°F (71.1°C). The USDA warns against home-grinding and cautions that the process is potentially more dangerous than retail ground beef (USDA-FSIS, 2013a). Since commercial beef plants routinely sample for *E. coli* and follow sanitation standard operating procedures (SSOPs), the likelihood for pathogen presence is lower than in lesser-regulated primal cuts (USDA-FSIS, 2013a). Whole beef subprimal cuts are not usually sampled for *E. coli*, as bacteria present on the surface are presumed to be inactivated upon cooking (although, it is estimated that only 0.02% of subprimal cuts harbor STEC) (USDA-FSIS, 2002).

Of course, interventions can, and must, occur before the product reaches a restaurant or home. Work by Nutsch (1998) and Fung et al. (2001) summarizes these in-plant options. Interventions may include proper handling of carcasses (eg, minimizing contact of the hide with the skinned carcass, performing environmental sanitation, and ensuring rapid carcass chilling), water washing, hot water washing, chlorine decontamination, use of organic acid treatments (eg, acetic acid, lactic acid, peroxyacetic acid), trimming to remove filth and contaminants, steam pasteurization, and steam vacuuming (Fung et al., 2001; Nutsch, 1998).

REGULATIONS AND RESEARCH

A brief history of US regulations regarding hygienic foods

The history of legislative and regulatory attempts to ensure the safety and hygiene of ground beef products have their heritage in the century-old 1906 US FMIA. A key factor in the passage of the act—which was the first major US law to require inspection of meat products—was the publication of Upton Sinclair's exposé, "The Jungle." First in 1904, as part of a weekly serial and later as the book *The Jungle*, the socialist activist Sinclair decried working-condition and food-safety atrocities in Chicago beef packing plants. Sinclair's writing provided political room for Congress to pass, and President Theodore Roosevelt to sign, food safety legislation (including the FMIA, but also the *Pure Food and Drug Act*) in 1906. Since 1906, primary policies (eg, legislative acts) as well as secondary policies (eg, regulations) have encouraged the development of new technological and business-management approaches to ensuring meat safety, including the safety of ground beef products (Hobbs et al., 2000). Following the watershed events associated with the hamburger-implicated *E. coli* O157:H7 outbreak of 1993 in the Pacific Northwest (involving Jack-in-the-Box), the USDA's FSIS redoubled its efforts to ensure the safety of ground beef products (Golan et al., 2004). In July 1996, the FSIS mandated that meat processors (including ground beef processors) adopt Hazard Analysis Critical Control Point (HACCP) systems in their plants. The regulation, called "Pathogen Reduction; Hazard Analysis and Critical Control Point (HACCP) Systems," was expressly designed to encourage food-hygiene-related technological innovation. The regulatory announcement, published in the US Federal Register, explicitly acknowledged stated the following:

> *HACCP, combined with the other measures required by this rulemaking, will substantially improve the ability of meat and poultry establishments and FSIS to target and systematically prevent and reduce food safety hazards and. . .to continuously improve food safety as science and technology improve.*
>
> **USDA-FSIS (1996)**

Simultaneous to the development and issuance of the "Pathogen Reduction" regulation, the FSIS convened a number of beef safety technology conferences involving scientists and businesses (USDA-FSIS, 1996). Indeed, the 1996 regulation influenced the design and use of new technologies for beef safety—including, steam pasteurization of beef carcasses, the use of organic acids, and other interventions (Fung et al., 2001).

Meanwhile, nation-state governments including, but not limited to, the US government have acted to ensure the safety of beef products, and some of these enactments leverage common safety- and quality-related business practices; for example, the European Union requires that food businesses provide "use-by" dates on labels for highly perishable foods. "Safe handling" instructions have, for years, also appeared on the labels of ground beef products in grocery stores.

Other, regulatory actions regarding hamburger safety include requirements that ground-beef processors remove from supply chains beef tissue (eg, central nervous system tissue) that might expose consumers to the causative agent of bovine spongiform encephalopathy (USDA-FSIS, 2004).

Performance standards

Performance standards exist for raw and finished ground beef products. As aforementioned, the USDA-FSIS has a "zero-tolerance" policy for the presence of *E. coli* O157:H7, O26, O45, O103, O111, O121, and O145 in non-intact raw beef products, which includes ground beef (USDA-FSIS, 2012b). The performance standard for *Salmonella* in raw ground beef states that a maximum of five ground beef samples ($n = 53$; 7.5%) can be positive for the pathogen in a sampling event (USDA-FSIS, 2013b).

Processed meat products, some of which likely include hamburger, must undergo certain performance standards established by the USDA-FSIS. In finished meat products, lethality performance standards for ready-to-eat products require a 6.5-log reduction of *Salmonella* throughout the finished product (USDA-FSIS, 1999b) (also codified at 9 CFR § 301, 317, 318, 320, and 381). Similarly, methods to control *C. perfringens* and *Clostridium botulinum* are also included in the performance standards. These stabilization standards require that a product's maximum internal temperature not remain between 130°F (54.4°C) and 80°F (26.7°C) for more than 1.5 hours nor between 80°F (26.7°C) and 40°F (4.4°C) for more than 5 hours (USDA-FSIS, 1999b). Colloquially, the lethality performance standards are known as "Appendix A" and stabilization performance standards are known as "Appendix B."

For producers of uncured meat patties (a common route for ground beef usage), further guidelines have been established in 9 CFR § 318.23. Namely, patties that are heat-processed must undergo permitted heat-processing temperature and time combinations. Patties that are partially cooked or "char-marked" patties must bear labeling warning customers that the product is uncooked and must reach 160°F. Stabilization protocols similar to those in Appendix B are also spelled out in the regulation (Code of Federal Regulations, 2012b).

Hazard Analysis and Critical Control Points

This chapter previously discussed one of the government's answers to foodborne illness associated with ground beef—HACCP. It is possible that ground beef may become contaminated with biological, chemical or physical hazards prior to consumption by the consumer. A proactive, prevention-based food safety program, HACCP aims to ensure farm-to-fork food safety by reducing (or eliminating) these hazards to acceptable levels. More specifically, the objectives of HACCP are to (1) make the product as safe as possible and (2) prove the product was produced as safely as possible. The implementation of HACCP is mandated by the Food and Drug Administration (FDA) and the USDA for domestic and imported seafood, juices, meat, and poultry (Aramouni and Boyle, 2010).

An effective HACCP plan is not a standalone food safety program; rather, HACCP is built upon preexisting Good Manufacturing Practices (GMPs) and SSOPs currently in place at a facility. Before a HACCP plan can be developed, a HACCP team consisting of line workers, sanitation personnel, quality assurance personnel, marketing and communication personnel, and an external food safety expert must be assembled (Aramouni and Boyle, 2010).

FDA Food Code for restaurants

The FDA Food Code is a set of uniform provisions disseminated to retail and foodservice establishments (and available electronically to the public) to address food safety issues and ensure public health. It is described as a "model code," with the most current edition published in 2013, and is based upon recommendations from the biennial meeting of the Conference for Food Protection in 2012. The Food Code provides numerous important facts and references for handlers of hamburger, including proper thermometer selection and use, heating protocols, cooling protocols, methods for holding partially cooked food for later cooking, and when product must be disposed of. The Code also outlines process deviation correction measures (eg, what employees should do if hamburgers have not attained proper temperature for the fully prescribed time) (FDA, 2013).

Further, the Food Code stipulates messages that restauranteurs must include on menus if they are serving animal products that are raw, undercooked, or not otherwise processed to eliminate pathogens, including hamburger. This is comprised of both a disclosure and a reminder. The disclosure can describe the raw nature of the product, or identify animal-derived foods by asterisking them to a footnote stating the items are served raw or undercooked. The reminder involves asterisking the animal-derived foods to one of three footnotes:

> *(1) Cooked meat and poultry such as beef, pork, ham, lamb, chicken, turkey, duck. Regarding the safety of these items, written information is available upon request;*
>
> *(2) Consuming raw or undercooked meats, poultry, seafood, shellfish, or eggs may increase your risk of foodborne illness; or*
>
> *(3) Consuming raw or undercooked meats, poultry, seafood, shellfish, or eggs may increase your risk of foodborne illness, especially if you have certain medical conditions.*
>
> **FDA (2013)**

REFERENCES

Anonymous, 2014. Ground beef fact sheet, Beef University. Beef Foodservice. Online. <http://www.beefu.org/p/beef-cuts/clip/ground-beef-fact-sheet>.

Aramouni, F., Boyle, E. 2010. Principles of HACCP Course Notes. Kansas State University, Manhattan, KS.

Arthur, T.M., Brichta-Harhay, D.M., Bosilevac, J.M., Guerini, M.N., Kalchayanand, N., Wells, J.E., et al., 2008. Prevalence and characterization of *Salmonella* in bovine lymph nodes potentially destined for use in ground beef. J. Food Prot. 71, 1685–1688.

Barkocy-Gallagher, G.A., Arthur, T.M., Siragusa, G.R., Keen, J.E., Elder, R.O., Laegreid, W.W., et al., 2001. Genotypic analyses of *Escherichia coli* O157:H7 and O157 nonmotile isolates recovered from beef cattle and carcasses at processing plants in the Midwestern states of the United States. Appl. Environ. Microbiol. 67, 3810–3818.

Barkocy-Gallagher, G.A., Arthur, T.M., Rivera-Betancourt, M., Nou, X., Shackelford, S.D., Wheeler, T.L., et al., 2003. Seasonal prevalence of Shiga toxin-producing *Escherichia coli*, including O157:H7 and non-O157 serotypes, and *Salmonella* in commercial beef processing plants. J. Food Prot. 66, 1978–1986.

Benson, A., 2014. International Food Information Council, with Files from Deborah Sellnow and Tim Sellow, 2014. Case Study: When "Lean Finely Textured Beef" Is portrayed as "Pink Slime". International Food Information Council, Washington, DC.

Beutin, L., Geier, D., Steinruck, H., Zimmerman, S., Scheutz, F., 1993. Prevalence and some properties of verotoxin (Shiga-like toxin)-producing *Escherichia coli* isolated from human patients in Germany. Med. Microbiol. Immunol. 183, 13–21.

Beutin, L., Krause, G., Zimmerman, S., Kaulfuss, S., Gleier, K., 2004. Characterization of Shiga toxin-producing *Escherichia coli* strains isolated from human patients in Germany over a 3-year period. J. Clin. Microbiol. 42, 1099–1108.

Bosilevac, J.M., Guerini, M.N., Kalchayanand, N., Koohmaraie, M., 2009. Prevalence and characterization of *Salmonella* in commercial ground beef in the United States. Appl. Environ. Microbiol. 75, 1892–1900.

Brichta-Harhay, D.M., Guerini, M.N., Arthur, T.M., Bosilevac, J.M., Kalchayanand, N., Shackelford, S.D., et al., 2008. *Salmonella* and *Escherichia coli* O157:H7 contamination on hides and carcasses of cull cattle presented for slaughter in the United States: an evaluation of prevalence and bacterial loads by immunomagnetic separation and direct plating methods. Appl. Environ. Microbiol. 74, 6289–6297.

Brichta-Harhay, D.M., Arthur, T.M., Bosilevac, J.M., Kalchayanand, N., Schmidt, J.W., Wang, R., et al., 2012. Microbiological analysis of bovine lymph nodes for the detection of *Salmonella enterica*. J. Food Prot. 75, 854–858.

CDC, 2011a. CDC estimates of foodborne illness in the United States. Online. <http://www.cdc.gov/foodborneburden/2011-foodborne-estimates.html>.

CDC, 2011b. Multistate outbreaks of *Salmonella* Typhimurium infections linked to ground beef. Online. <http://www.cdc.gov/salmonella/typhimurium-groundbeef/index.html>.

CDC, 2012. Multistate outbreaks of *Salmonella* Enteritidis infections linked to ground beef. Online. <http://www.cdc.gov/salmonella/enteritidis-07-12/>.

CDC, 2013. Multistate outbreaks of *Salmonella* Typhimurium infections linked to ground beef. 2015. Online. <http://www.cdc.gov/salmonella/typhimurium-01-13/>.

Code of Federal Regulations, 2012a. Ground beef, 7 CFR § 65.155.

Code of Federal Regulations, 2012b. Heat-processing and stabilization requirements for uncured meat patties, 9 CFR § 300.2.

Code of Federal Regulations, 2015a. FSIS responsibilities, 9 CFR § 300.2.

Code of Federal Regulations, 2015b. Miscellaneous beef products, 9 CFR § 319.15.

Crandall, P.G., O'Bryan, C.A., Babu, D., Jarvis, N., Davis, M.L., Buser, M., et al., 2013. Whole-chain traceability, is it possible to trace your hamburger to a particular steer, a U. S. perspective. Meat Sci. 95, 137–144.

FDA, 2013. Food Code. Online. <http://www.fda.gov/downloads/Food/GuidanceRegulation/RetailFoodProtection/FoodCode/UCM374510.pdf>.

Fung, D.Y.C., Hajmeer, M.N., Kastner, C.L., Kastner, J.J., Marsden, J.L., Penner, K.P., et al., 2001. Meat safety. In: Hui, Y.H., Nip, W.-K., Rogers, R., Young, O.A. (Eds.), Meat Science and Applications. Marcel Dekker, Inc, New York, NY, pp. 171–205. (Chapter 8).

Golan, E., Roberts, T., Salay, E., Caswell, J., Ollinger, M., Moore, D., 2004. Food safety innovation in the United States: evidence from the meat industryIn: USDA (Ed.), Agricultural Economic Report. United States Department of Agriculture, Washington, DC.

Gragg, S.E., Loneragan, G.H., Brashears, M.M., Arthur, T.M., Bosilevac, J.M., Kalchayanand, N., et al., 2013a. Cross-sectional study examining *Salmonella enterica* carriage in subiliac lymph nodes of cull and feedlot cattle at harvest. Foodborne Pathog. Dis. 10, 368–374.

Gragg, S.E., Loneragan, G.H., Nightingale, K.K., Brichta-Harhay, D.M., Ruiz, H., Elder, J.R., et al., 2013b. Substantial within-animal diversity of *Salmonella* isolates from lymph nodes, feces, and hides of cattle at slaughter. Appl. Environ. Microbiol. 79, 4744–4750.

Greene, J.L., 2012. Lean Finely Textured Beef: The "Pink Slime" Controversy. Congressional Research Service, Washington, DC.

Grimes, W., 1998. The pride of Texas on a bun. The New York Times. The New York Times Company, New York, NY, p. F6.

Gyles, C.L., 2007. Shiga toxin-producing *Escherichia coli*: an overview. J. Anim. Sci. 85 (Electronic Suppl.), E45–E62.

Hagen, E., 2012. Setting the Record Straight on Beef. USDA Food Safety and Inspection Service, Washington, DC, Online. <http://blogs.usda.gov/2012/03/22/setting-the-record-straight-on-beef/>.

Haneklaus, A.N., Harris, K.B., Griffin, D.B., Edrington, T.S., Lucia, L.M., Savell, J.W., 2012. *Salmonella* prevalence in bovine lymph nodes differs among feedyards. J. Food Prot. 75, 1131–1133.

Hobbs, J.E., Fearne, A., Spriggs, J., 2000. Incentive structures for food safety and quality assurance: an international comparison. In: Mini-Symposium on the "Economics of Food Safety," International Agricultural Economics Association Conference, Berlin.

Hunt, M.C., Sørheim, O., Slinde, E., 1999. Color and heat denaturation of myoglobin forms in ground beef. J. Food Sci. 64, 847–851.

Kelman, A., Soong, Y., Dupuy, N., Shafer, D., Richbourg, W., Johnson, K., et al., 2011. Antimicrobial susceptibility of *Staphylococcus aureus* from retail ground meats. J. Food Prot. 74, 1625–1629.

Ladiges, W.C., Foster, J.F., Ganz, W.M., 1974. Incidence and viability of *Clostridium perfringens* in ground beef. J. Milk Food Technol. 12, 593–635.

Lecuit, M., 2005. Understanding how *Listeria monocytogenes* targets and crosses heat barriers. Clin. Microbiol. Infect. 11, 430–436.

Li, M., Malladi, S., Hurd, H.S., Goldsmith, T.J., Brichta-Harhay, D.M., Loneragan, G.H., 2015. *Salmonella* spp. in lymph nodes of fed and cull cattle: relative assessment of risk to ground beef. Food Control 50, 423–434.

Line, J.E., Fair Jr., J.R., Moran, A.B., Martin, L.M., Lechowich, R.V., Carosella, J.M., et al., 1991. Lethality of heat to *Escherichia coli* O157:H7: D-value and z-value determinations in ground beef. J. Food Prot. 54, 762–766.

Luchansky, J.B., Porto-Fett, A.C.S., Shoyer, B.A., Phillips, J., Chen, V., Eblen, D.R., et al., 2013a. Fate of Shiga toxin-producing O157:H7 and non-O157:H7 *Escherichia coli* cells within refrigerated, frozen, or frozen then thawed ground beef patties cooked on a commercial open-flame gas or a clamshell electric grill. J. Food Prot. 76, 1500–1512.

Luchansky, J.B., Porto-Fett, A.C.S., Shoyer, B.A., Phillips, J., Eblen, D., Evans, P., et al., 2013b. Thermal inactivation of a single strain each of serotype O26:H11, O45:H2, O103:H2, O104:H4, O111:H-, O121:H19, O145:NM, and O157:H7 cells of Shiga toxin-producing *Escherichia coli* in wafers of ground beef. J. Food Prot. 76, 1434–1437.

Lyon, B.G., Berry, B.W., Soderberg, D., Clinch, N., 2000. Visual color and doneness indicators and the incidence of premature brown color in beef patties cooked to four end point temperatures. J. Food Prot. 63, 1389–1398.

Montville, T.J., Matthews, K.R., Kniel, K.E., 2012. Enterohemorrhagic *Escherichia coli*, Food Microbiology: An Introduction, third ed. ASM Press, Washington, DC.

Moore, J.E., 2004. Gastrointestinal outbreaks associated with fermented meats. Meat Sci. 67, 565–598.

Murray, P.R., Rosenthal, K.S., Pfaller, M.A., 2013. Enterobacteriaceae, Medical Microbiology, seventh ed. Elsevier-Saunders, Philadelphia, PA.

Nutsch, A.L., 1998. Bacterial Decontamination of Meat Surfaces Through the Application of a Steam Pasteurization Process, Ph.D. Dissertation, Animal Sciences and Industry. Kansas State University, Manhattan, KS.

Painter, J.A., Hoekstra, R.M., Ayers, T., Tauxe, R.V., Braden, C.R., Angulo, F.J., et al., 2013. Attribution of foodborne illnesses, hospitalizations, and deaths to food commodities by using outbreak data, United States, 1998–2008. Emerg. Infect. Dis. 19, 407–415.

Prayson, B., McMahon, J.T., Prayson, R.A., 2008. Fast food hamburgers: what are we really eating? Ann. Diagn. Pathol. 12, 406–409.

Ralston, K., Brent, C.P., Starke, Y., Riggins, T., Lin, C.T.J., 2002. Consumer Food Safety Behavior: A Case Study in Hamburger Cooking and Ordering. United States Department of Agriculture Economic Research Service, Washington, DC.

Rivera-Betancourt, M., Shackelford, S.D., Arthur, T.M., Westmoreland, K.E., Bellinger, G., Rossman, M., et al., 2004. Prevalence of *Escherichia coli* O157:H7, *Listeria monocytogenes*, and *Salmonella* in two geographically distant commercial beef processing plants in the United States. J. Food Prot. 67, 295–302.

Rolfes, E., 2012. The hidden costs of hamburgers. PBS Newshour. Online. <http://www.pbs.org/newshour/rundown/the-hidden-costs-of-hamburgers/>.

Røssvoll, E., Sørheim, O., Heir, E., Møretrø, T., Olsen, N.V., Langsrud, S., 2014. Consumer preferences, internal color and reduction of shigatoxigenic *Escherichia coli* in cooked hamburgers. Meat Sci. 96, 695–703.

Samadpour, M., Barbour, M.W., Nguyen, T., Cao, T.M., Buck, F., Depavia, G.A., et al., 2006. Incidence of enterohemorrhagic *Escherichia coli*, *Escherichia coli* O157, *Salmonella*, and *Listeria monocytogenes* in retail fresh ground beef, sprouts, and mushrooms. J. Food Prot. 69, 441–443.

Scallan, E., Hoekstra, R.M., Angulo, F.J., Tauxe, R.V., Widdowson, M.-A., Roy, S.L., et al., 2011. Foodborne illness acquired in the United States—major pathogens. Emerg. Infect. Dis. 17, 7–14.

Stringer, S.C., George, S.M., Peck, M.W., 2000. Thermal inactivation of *Escherichia coli* O157:H7. J. Appl. Microbiol. 88, 79S–89S.

Stumbo, C.R., 1973. Thermobacteriology in Food Processing, second ed. Academic Press, Orlando, FL.

United States Code, 2006. Title 21—Food and Drugs. (Chapter 12). Section 601.

Unruh, D.A., 2014. Destruction of Non-O157 Shiga Toxin-Producing Escherichia coli (STECs) by Heat and High Pressure, M.S. Thesis, Food Science and Technology. University of Nebraska-Lincoln, Lincoln, NE.

USDA-ARS, 2015. USDA National Nutrient Database for Standard Reference, Release 27 (revised). Online. <http://ndb.nal.usda.gov/ndb/beef/show>.

USDA-FSIS, 1996. Pathogen reduction; Hazard Analysis and Critical Control Point (HACCP) Systems. Fed. Regist. 61, 38806–38989.

USDA-FSIS, 1999a. Beef products contaminated with *Escherichia coli* O157:H7. Fed. Regist. 64, 2803–2805.

USDA-FSIS, 1999b. Performance standards for the production of certain meat and poultry products. Fed. Regist. 64, 732–749.

USDA-FSIS, 2002. Comparative risk assessment for intact (non-tenderized) and non-intact (tenderized) beef: technical report. Online. <http://www.fsis.usda.gov/shared/PDF/Beef_Risk_Assess_Report_Mar2002.pdf>.

USDA-FSIS, 2004. Meat produced by advanced meat/bone separation machinery and meat recovery (AMR) systems: interim final rule and request for comment. Fed. Regist. 69, 1874–1885.

USDA-FSIS, 2011. Water in meat and poultry. Online. <http://www.fsis.usda.gov/wps/wcm/connect/42a903e2-451d-40ea-897a-22dc74ef6e1c/Water_in_Meats.pdf?MOD=AJPERES>.

USDA-FSIS, 2012a. Leftovers and food safety. Online. <http://www.fsis.usda.gov/shared/PDF/Leftovers_and_Food_Safety.pdf>.

USDA-FSIS, 2012b. Shiga toxin-producing *Escherichia coli* in certain raw beef products. Fed. Regist. 77, 31975–31981.

USDA-FSIS, 2013a. Ground beef and food safety. Online. <http://www.fsis.usda.gov/wps/portal/fsis/topics/food-safety-education/get-answers/food-safety-fact-sheets/meat-preparation/ground-beef-and-food-safety/CT_Index>.

USDA-FSIS, 2013b. *Salmonella* and *Campylobacter* verification program for raw meat and poultry products. FSIS Directive. 10,250.1.

USDA-FSIS, 2014. Progress report on *Salmonella* and *Campylobacter* testing of raw meat and poultry products, CY 1998–2013. Online. <http://www.fsis.usda.gov/wps/wcm/connect/c33b69fe-7041-4f50-9dd0-d098f11d1f13/Beef_from_Farm_to_Table.pdf?MOD=AJPERES>.

Vipham, J., Brooks, C., Miller, M., Brashears, M., Echeverry, A., 2010. A Survey of *Campylobacter* in beef cuts at retail, Beef Research. Online. <http://www.beefresearch.org/CMDocs/BeefResearch/Safety_Project_Summaries/FY09_A_survey_of_Campylobacter_in_beef_cuts.pdf>.

Waters, A.E., Contente-Cuomo, T., Buchhagen, J., Liu, C.M., Watson, L., Pearce, K., et al., 2011. Multidrug-resistant *Staphylococcus aureus* in US Meat and Poultry. Clin. Infect. Dis. 52, 1–4.

Wells, J.G., Shipman, L.D., Greene, K.D., Sowers, E.G., Green, J.H., Cameron, D.N., et al., 1991. Isolation of *Escherichia coli* O157:H7 and other Shiga-like-toxin-producing *E. coli* from dairy cattle. J. Clin. Microbiol. 29, 985–989.

CHAPTER

8

Salmonella and *Listeria monocytogenes* in ready-to-eat leafy vegetables

Rafael Djalma Chaves[1], Rafael Chacon Ruiz Martinez[1], Ana Carolina Bortolossi Rezende[1], Monyca Dias Rocha[1], Juan M. Oteiza[2] and Anderson de Souza Sant'Ana[1]

[1]Department of Food Science, Faculty of Food Engineering, University of Campinas Rua Monteiro Lobato, Campinas, Sao Paulo, Brazil [2]Center for Research and Technical Assistance to Industry, Centenario, Neuquén, Argentina

INTRODUCTION

Leafy vegetables are important components of the diet, and their regular consumption in adequate amounts can prevent the onset of cardiovascular diseases and some types of cancers (Taban and Halkman, 2011). Leafy vegetables are included in the group of ready-to-eat (RTE) foods, which means they can be consumed directly without the need for cooking or using any other measurement for effective elimination or reduction to an acceptable level of the microorganisms of concern (Commission Regulation 2073, 2005). RTE products are widely commercialized, primarily due to their ease of use (Ha and Kang, 2015).

Data from the European Food Safety Authority (EFSA, 2011, 2012, 2013, 2014, 2015) have shown a significant increasing trend of cases of listeriosis between 2009 and 2013 (Table 8.1), probably due to a number of factors, such as improvement of control programs and health surveillance (Harris et al., 2003), changes in agricultural practices, and increased international trade demand (Cabedo et al., 2008). Furthermore, the increase of populations at risk, that is, elderly and immunocompromised patients, may also be considered as a cause of the rise in reported cases (Garrido et al., 2012; González et al., 2013). In contrast, cases of salmonellosis in European countries have been systematically decreasing in recent years, and the adoption of sanitary inspection has contributed to this significant reduction (EFSA, 2015) (Table 8.1).

According to the EFSA, the main foods implicated in outbreaks are RTE fish and meat products (EFSA, 2015), although fruits and vegetables are considered vehicles for the transmission of these diseases. In recent years, an increasing

P. Kotzekidou (Ed): Food Hygiene and Toxicology in Ready-to-Eat Foods.
DOI: http://dx.doi.org/10.1016/B978-0-12-801916-0.00008-X

Table 8.1 Occurrence of Listeriosis and Salmonellosis in the European Union in the Period 2009–2013

Year of Occurrence	Confirmed Cases	Deaths	References
Listeriosis			
2009	1645	~270	EFSA (2011)
2010	1601	~272	EFSA (2012)
2011	1476	134	EFSA (2013)
2012	1658	198	EFSA (2014)
2013	1763	191	EFSA (2015)
Salmonellosis			
2009	108,614	~90	EFSA (2011)
2010	99,020	62	EFSA (2012)
2011	95,548	56	EFSA (2013)
2012	91,034	61	EFSA (2014)
2013	82,694	59	EFSA (2015)

number of diseases due to transmission of pathogens by vegetables and RTE salads has been observed, most likely due to improvements in microbiological methods for monitoring programs and changes in food processing, together with the increasing number of immunocompromised consumers (Beuchat, 2002; Harris et al., 2003; De Giusti et al., 2014). Although human pathogens are not part of the regular microbial population of fruits and vegetables, the occurrence and importance of these pathogens in fresh products should be considered, given the aspects mentioned above (Suslow, 2002). This scenario raises concerns for consumers and governments, as well as for food-producing industries (Sant'Ana et al., 2014).

In minimally processed vegetables (MPVs), natural barriers including peel and skin are removed during cutting, which reduces their own natural defense against bacteria and allows the multiplication of microorganisms (Fatemi et al., 2006). Therefore, MPVs are more susceptible to pathogen proliferation in comparison to intact fresh products, including *Salmonella*, *Listeria monocytogenes*, *Shigella*, pathogenic *Escherichia coli*, among others (Olaimat and Holley, 2012; Tomas-Callejas et al., 2012; Taban and Halkman, 2011).

According to the Centers for Disease Control and Prevention (CDC, 2013) in 2013, in the United States, 818 foodborne disease outbreaks were reported, resulting in 13,360 infected people and 16 deaths. *Salmonella* was the second most common agent of outbreaks, affecting 3553 people (34% of the reported cases). Although outbreaks caused by *Listeria* are rare, hospitalization and death rates are higher when compared to other agents (91% and 13%, respectively; CDC, 2015).

As reported by Taban and Halkman (2011), both in Europe and in the United States, recent data revealed an increased association of leafy vegetables, especially lettuce and spinach and their RTE salads, with foodborne disease outbreaks.

This may be related to the use of untreated irrigation water, improper use of organic fertilizers and pesticides, in addition to the possibility of contact of these vegetables with sources of contamination during harvesting, handling, processing, and packaging.

SALMONELLA

Salmonella spp. comprises a group of Gram-negative, facultative anaerobic bacteria belonging to the family *Enterobacteriaceae*. Although members of this genus are mobile due to the presence of peritrichous flagella, there are nonmobile variants such as *Salmonella* Pullorum and *Salmonella* Gallinarum. According to the analyses of the differences in sequences of the 16 S ribosomal RNA 2463 serotypes of *Salmonella* are known, which can be grouped into two species: *Salmonella enterica* (2443 serotypes) and *Salmonella bongori* (20 serotypes). This classification is used by collaborating centers of the World Health Organization (WHO) and CDC, among others. According to this taxonomic scheme, *S. enterica* is divided into six subspecies. *S. enterica* subsp. *enterica* I is largely isolated from warm-blooded animals and corresponds to about 99% of clinical isolates (Li et al., 2013). In humans, *Salmonella* infection can be very severe, causing four different clinical manifestations, including gastroenteritis, bacteremia, enteric fever (typhoid fever, related to serotypes *Typhi* and *Paratyphi*), and an asymptomatic carrier state (Ryan and Ray, 2004).

Salmonella spp. can survive in the environment for long periods, and their primary hosts are animals, including cattle, pigs, poultry, birds, and insects, which are capable of excreting the microorganism in their feces for weeks or months (Newell et al., 2010; Pui et al., 2011). This microorganism is able to adapt to temperature, pH, and water activity conditions different from those usually required for growth, which means a major challenge to food safety. Furthermore, *Salmonella* spp. is not capable of forming endospores, is relatively heat-sensitive, and can be destroyed at 60°C for 15–20 min ($D_{62.8°C} = 0.06$ min) (Garrity et al., 2005; Tindall et al., 2005; Forsythe, 2013).

Although meat, poultry, eggs, and dairy products have been considered the main food vehicles for *Salmonella* spp., in recent years, fruit and vegetable products have become important sources of this pathogen (Jay et al., 2005; Bouchrif et al., 2009; Doyle and Buchanan, 2012). Several factors have contributed to this situation, including major changes in agricultural and processing practices, the increased per capita consumption of RTE vegetables, and the increase in the international trade of these products (Lynch et al., 2009; Taban and Halkman, 2011). Moreover, the growing popularity of precut and prepackaged foods introduces new public health concerns since the damaged plant tissue can release nutrients that favor bacterial growth (Warriner et al., 2009; Francis et al., 2012; Doyle and Buchanan, 2012; Leff and Fierer, 2013).

Contamination of fruits and vegetables by *Salmonella* may occur due to the penetration of the microorganism through cuts in plant tissues, uptake of the pathogen by root systems, and surface contamination of flowering plants and subsequent entrapment of the pathogen during embryogenesis of the fruit or vegetable. In addition, the edible part of the plant can be contaminated during cutting or pressing of fresh fruits to obtain the juice (Goldberg et al., 2011; Kroupitski et al., 2011; Doyle and Buchanan, 2012). Table 8.2 presents some data on the incidence of *Salmonella* in leafy vegetables.

Considering that *Salmonella* is able to adhere to plant surfaces and be internalized, the use of chemical sanitizers may not be effective for the inactivation of the pathogen in contaminated foods. Thus, appropriate control measurements to avoid contamination in farms or during processing and packaging stages, such as irradiation, can increase the microbiological safety of these food (Lynch et al., 2009).

According to the EFSA, the presence of *Salmonella* spp. in RTE leafy vegetables puts these foods in first place in terms of risk of probability of occurrence of outbreaks (McLauchlin, 2014). The following topics will address the incidence of *Salmonella* from other sources, such as irrigation water, soil, and manure. Several *Salmonella* outbreaks related to the consumption of leafy vegetables have been reported in the literature (Table 8.3).

LISTERIA MONOCYTOGENES

Listeria monocytogenes is a Gram-positive, facultative anaerobic, nonsporeforming, and facultative intracellular bacterium. After ingestion, the microorganism can cross the intestinal epithelium and spread to deeper tissues through bloodstream and lymph. The liver and spleen are the primary target organs for bacteria proliferation, leading to abscess formation. An important characteristic of *L. monocytogenes* is the ability to cross two major epithelial barriers, blood–brain and placental, which can lead to encephalitis or infection of the fetus and eventually abortion, death of newborn, and neonatal meningitis (Stavru et al., 2011). In humans, the incidence of listeriosis, with clinical symptoms ranging from gastroenteritis to the severe forms described above, is relatively low, partly due to strict food control (Allerberger and Wagner, 2010). However, the disease has a high rate of mortality among immunocompromised patients, infants, and fetuses, which can exceed 30% (Watson, 2009).

Listeria monocytogenes is a major concern for agribusiness. Due to its ubiquitous characteristics, the microorganism can contaminate processing plants, survive for longer periods in the environment (soil, plant, and water), multiply at low temperatures (2–4°C), and form biofilms (Ryser and Buchanan, 2013; Vivant et al., 2013). Additionally, *L. monocytogenes* can multiply in a wide pH range (4.3–9.6) and also in the presence of 10–12% sodium chloride (Low and Donachie, 1997; Ryser and Buchanan, 2013).

Listeria monocytogenes isolates are often serotyped to provide data for public health surveillance service, as well as to assist in outbreak investigations. There are 13 known *L. monocytogenes* serotypes, three of which (1/2a, 1/2b, and 4b)

Table 8.2 Incidence of *Salmonella* in Leafy Vegetables in the Last Two Decades

Vegetable	Year	Country	Serotype	Samples Analyzed/ Positive Results (Incidence)	References
Watercress RTE[a]	2004	Brazil	*Salmonella* spp.	13/1 (7.7%)	Fröder et al. (2007)
	2004–2005	Mexico	*Salmonella* spp.	100/7 (7.0%)	Quiroz-Santiago et al. (2009)
	2007	Turkey	*Salmonella* spp.	20/1 (5.0%)	Aytac et al. (2010)
	2011	Brazil	*Salmonella* spp.	172/50 (29.1%)	Maistro et al. (2012)
Lettuce	2004–2005	Mexico	*Salmonella* spp.	200/10 (5.0%)	Quiroz-Santiago et al. (2009)
	2006–2007	The Netherlands	*S*. Typhimurim, *S*. Montevideo	565/2 (0.4%)	Wijnands et al. (2014)
Lettuce RTE	2004	Brazil	*Salmonella* spp.	41/1 (2.4%)	Fröder et al. (2007)
	2007	Turkey	*Salmonella* spp.	12/2 (12.7%)	Aytac et al. (2010)
	2008	Brazil	*S*. Typhimurium	152/1 (0.7%)	Sant'Ana et al. (2011)
Chicory	2004	Brazil	*Salmonella* spp.	12/1 (8.3%)	Fröder et al. (2007)
Cabbage	2007	Turkey	*Salmonella* spp.	16/1 (6.25%)	Aytac et al. (2010)
Endive	2006–2007	The Netherlands	*S*. Typhimurium	370/3 (0.8%)	Wijnands et al. (2014)
Spinach	2004–2005	Mexico	*Salmonella* spp.	100/7 (7.0%)	Quiroz-Santiago et al. (2009)
Basil RTE	2007	Turkey	*Salmonella* spp.	8/1 (12.5%)	Aytac et al. (2010)
Arugula RTE	2007	Turkey	*Salmonella* spp.	20/4 (20.0%)	Aytac et al. (2010)
	2008	Brazil	*S. enterica enterica* O:47:z4,z23	19/1 (5.3%)	Sant'Ana et al. (2011)
Parsley RTE	2004–2005	Mexico	*Salmonella* spp.	100/12 (12%)	Quiroz-Santiago et al. (2009)
	2007	Turkey	*Salmonella* spp.	19/3 (15.8%)	Aytac et al. (2010)

[a]*RTE, ready-to-eat.*

Table 8.3 Outbreaks of Salmonellosis Related to the Consumption of Leafy Vegetables

Year	Country	Food	Serotype	Confirmed Cases	Deaths	References
2000	England, Wales, the Netherlands, Scotland, Germany, and Iceland	Lettuce	Typhimurium PT 204b	>392	0	Crook et al. (2003)
2001	England and Wales	Lettuce and leafy vegetable salad	Newport PT33	19	0	Sagoo et al. (2003)
2004	Northern Ireland	Lettuce	Newport	372	0	Gillespie (2004)
2005	Finland	American lettuce	Typhimurium var. Copenhagen DT 104b	60	0	Takkinen et al. (2005)
2007	England, Wales, Scotland, Denmark, the Netherlands, and US	Basil	Senftenberg	74	0	Pezzoli et al. (2008)
2008	Finland	Lettuce	Newport, Reading	107	2	Lienemann et al. (2011)
2009	US[a]	American lettuce	Typhimurium	145	0	CDC (2015)
2010	US	Lettuce	Infantis	21	0	CDC (2015)
2011	US	Lettuce	Hartford	5	0	CDC (2015)
2012	US	Lettuce	Utah	8	0	CDC (2015)
2012	US	Romaine lettuce	Newport	15	0	CDC (2015)
2013	US	Vegetable unidentified	Braenderup	75	0	CDC (2015)

[a]*US, United States.*

represent more than 95% of isolates obtained from infected humans (Kathariou, 2002). Serotype 1/2a corresponds to more than 50% of *L. monocytogenes* isolates from food and environmental samples, while most of the outbreaks of listeriosis in humans are related to strains belonging to serotype 4b (Pan et al., 2009).

The FDA (United States Food and Drug Administration), in collaboration with the USDA (United States Department of Agriculture Food Safety and Inspection Service), and the CDC developed a ranking to predict the potential relative risk of acquiring listeriosis by eating different foods, divided into 23 categories, considering three population groups: perinatal (16 weeks after fertilization to 30 days after birth), elderly (over 60 years), and middle age (under 60 years). According to this assessment, RTE meat products, including deli meat and sausages, presented the greatest risk. The consumption of vegetables, in turn, is associated with low risk of infection (Anonymous, 2003). In fact, *L. monocytogenes* growth rate on plants is typically lower when compared to that observed in meat and dairy products (Chen et al., 2006). However, the presence of the pathogen in these foods is a major concern as it constitutes a threat to public health (Ajayeoba et al., 2015); furthermore, the detection of *L. monocytogenes* in leafy vegetables is relatively common (Table 8.4).

Although *Listeria* species are commonly isolated from plants, mainly due to their ubiquitous nature, there are few reported cases of leafy vegetables implicated in foodborne diseases outbreaks caused by this microorganism. However, it is worth emphasizing that many foodborne diseases are sporadic and not taken into account in outbreak studies (Gandhi and Chikindas, 2007). One of them occurred in 1979 and involved the consumption of celery, lettuce, and tomato, affecting patients from eight hospitals in Boston (Ho et al., 1986). Allerberger and Guggenbichler (1989) have reported that the consumption of organic vegetables was the cause of a small listeriosis outbreak, affecting nine newborns and their mothers, and two adults in Austria in 1986. Another listeriosis outbreak also occurred in Texas (United States) in 2010 and was related to the consumption of celery, leading to four deaths (FDA, 2010). In March 2015, a recall of approximately 74,000 organic spinach products possibly contaminated by *L. monocytogenes* was carried out, coming from a US company. No hospitalizations or deaths have been recorded until now (FDA, 2015). The incidence of *L. monocytogenes* in environmental sources such as irrigation water, soil, and manure is discussed in the following topics.

SALMONELLA AND *L. MONOCYTOGENES* SOURCES AND CONTAMINATION PATHWAYS IN THE FIELD

According to Brackett (1994), both preharvest and postharvest steps (transportation, processing, packaging, distribution, and sale) of leafy vegetables have potential for contamination and/or multiplication of pathogenic microorganisms (Fig. 8.1).

Table 8.4 *L. monocytogenes* Isolated from Leafy Vegetables

Vegetable	Year	Country	Samples Analyzed/Positive Results (Incidence)	References
Watercress RTE[a]	2001	Brazil	50/2 (4.0%)	Porto and Eiroa (2001)
	2007	Turkey	20/1 (5.0%)	Aytac et al. (2010)
	2008	Brazil	18/1 (5.6%)	Sant'Ana et al. (2012)
	2011	Brazil	172/1 (0.6%)	Maistro et al. (2012)
Lettuce	2001	Brazil	100/4 (4.0%)	Porto and Eiroa (2001)
	2001	Spain	10/1 (10.0%)	Soriano et al. (2001)
	2002	Norway	10/1 (10.0%)	Johannessen et al. (2002)
Lettuce RTE	2001	Spain	152/3 (2.0%)	Soriano et al. (2001)
	2008	Brazil	30/1 (3.3%)	Sant'Ana et al. (2012)
Cabbage RTE	2009	Brazil	111/1 (0.9%)	Oliveira et al. (2010)
Endive	2006–2007	The Netherlands	13/3 (23.1%)	Wijnands et al. (2014)
Escarole	2008	Brazil	172/1 (0.6%)	Sant'Ana et al. (2012)
	2011	Brazil	13/1 (7.7%)	Maistro et al. (2012)
Spinach RTE	2000–2005	Chile	181/1 (0.6%)	Cordano and Jacquet (2009)
	2004	Brazil	11/1 (9.1%)	Fröder et al. (2007)
	2008	Brazil	334/19 (6%)	Sant'Ana et al. (2012)
	2014	United States	8/3 (37.5%)	Weller et al. (2015a,b)
Basil RTE	2007	Turkey	214/2 (0.9%)	Aytac et al. (2010)
Mixed greens RTE	2008	Brazil	19/1 (5.3%)	Sant'Ana et al. (2012)
Cabbage RTE	2009–2010	Croatia	50/2 (4.0%)	Kovasevic et al. (2013)
Rocket	2013	Greek	19/2 (10.5%)	Hadjilouka et al. (2014)
Parsley	2001	Brazil	22/1 (4.5%)	Porto and Eiroa (2001)
	2007	Turkey	19/2 (10.5%)	Aytac et al. (2010)
	2009	Brazil	22/1 (4.5%)	Oliveira et al. (2010)

[a]*RTE, ready-to-eat.*

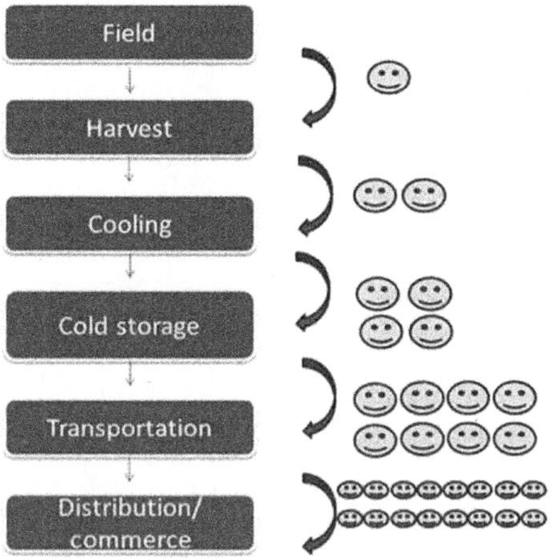

FIGURE 8.1

Schematic presentation of bacterial multiplication throughout the production chain.

The prerequisites to avoid contamination are mainly hygiene and sanitary conditions in the processing industry and cold-chain management from production to distribution of vegetables (Matthews et al., 2014).

The adoption of good manufacturing practices (GMPs) and hazard analysis and critical control points (HACCP), commonly used in food-processing industries could be applied on farms to minimize contamination (Soon et al., 2010; Gil et al, 2015). However, small farms typically have limited resources when compared to large producers, thus HACCP application may be unviable in the production chain.

Luber et al. (2011) have suggested five points to be controlled during all production stages of RTE products, as shown in Fig. 8.2.

Leafy vegetables are extremely susceptible to contamination originating from the soil, vegetation, and irrigation water due to their growing conditions (FAO/WHO, 2008), and constitute potential sources of *L. monocytogenes* and *Salmonella* spp. contamination in the primary food chain (Ivanek et al., 2006). Fecal contamination in soil and manure or fertilizer used by farmers are also proven as contamination sources. Tables 8.5 and 8.6 provide, respectively, a number of studies that evaluated the incidence of *L. monocytogenes* and *Salmonella* spp. in different farm/environmental sources.

Once present in the soil, *Listeria* and *Salmonella* can remain viable over relatively long periods. Strawn et al. (2013a,b) evaluated the presence of pathogens in 253 growing fields in the United States. Among them, *Salmonella* spp. and *L. monocytogenes* were recovered from 6.1% to 17.5% of fields, respectively. Through the analysis of soil fertilized with manure and contaminated water, *Salmonella* spp. was found 161 days after cultivation of lettuce, and 231 days after cultivation of salsa. These results highlight the importance of manure treatment to prevent vegetable contamination (Islam et al., 2004).

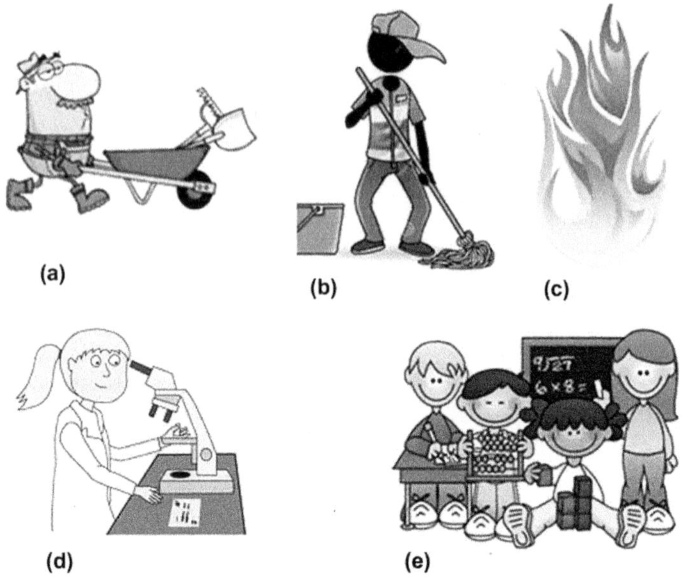

FIGURE 8.2

Schematic representation of controlling measures at all production stages of RTE products. (A) Good practices during primary production; (B) cleaning and disinfection of processing equipment and facilities; (C) thermal and nonthermal process to reduce viable cells; (D) microbiological testing to validate cleaning and disinfection process; and (E) education of all stakeholders.

Images retrieved from public websites.

The irrigation water quality and treated manure are determining factors for the possible presence of pathogens in vegetables (Benjamim et al., 2013; Cooley et al., 2014). In developing countries, the rate of water samples contaminated by *Salmonella* spp. can reach 35% (Ndiaye et al., 2011). According to Koseki et al. (2011), the contamination by *Salmonella* and *L. monocytogenes* via the roots is significantly greater than the direct infection on the leaves. Water from natural sources such as rain can also favor pathogen development and multiplication (Simental and Urtaza, 2008; Cevallos-Cevallos et al., 2012; Weller et al., 2015a,b). The contact of vegetables with contaminated water, that is, irrigation water, increases the risk of bacterial adherence to the surfaces and the subsequent internalization and proliferation of the pathogen during handling, storage, and distribution steps (Matthews et al., 2014). Climate and seasonal changes, as well as geographical differences such as latitude and longitude are also factors responsible for contamination of vegetables by pathogens (Matthews et al., 2014; Liu et al., 2013).

Table 8.5 Occurrence of *L. monocytogenes* in Different Environmental Samples

Sample	Samples Analyzed/Positive Results (Incidence)	References
Soil	504/120 (23.8%)	Nightingale et al. (2004)
	298/57 (19.1%)	Fox et al. (2009)
	21/5 (23.9%)	Jeyaletchumi et al. (2011)
	53/9 (17.0%)	Locatelli et al. (2013)
	526/51 (9.7%)	Strawn et al. (2013a,b)
	200/10 (5.0%)	Soni et al. (2014)
	467/28 (6.0%)	Linke et al. (2014)
	1000/55 (5.5%)	Szymczak et al. (2014)
Water	314/75 (23.9%)	Lyautey et al. (2007)
	30/10 (33.3%)	Mawak et al. (2009)
	105/15 (14.3%)	Mohammed et al. (2010)
	36/19 (52.6%)	Ijabadeniyi et al. (2011)
	78/27 (34.6%)	Ndiaye et al. (2011)
	74/22 (29.7%)	Strawn et al. (2013a,b)
	33/10 (30.0%)	Weller et al. (2015a,b)
Animal feces/manure	508/100 (19.7%)	Nightingale et al. (2004)
	825/255 (30.9%)	Ho et al. (2007)
	343/93 (27.1%)	Esteban et al. (2009)
	115/32 (27.8%)	Klein et al. (2010)
	5/1 (20.0%)	Bonetta et al. (2011)
	250/1 (0.4%)	Foerster et al. (2012)
	1738/105 (6.0%)	Hasegawa et al. (2014)

CONTROL OF *SALMONELLA* AND *L. MONOCYTOGENES* CONTAMINATION IN RTE VEGETABLES

The best way to reduce the chances of occurrence of foodborne diseases associated with vegetables by pathogens is to prevent contamination. However, as this is not always effective, disinfection of MPVs is crucial to prevent these illnesses.

The use of heat treatment to reduce the microbial load or maintain the microbiological stability has long been used in the food industry (Bermudez-Aguirre and Corradini, 2012). However, heat inactivation may also compromise food structure by altering sensory characteristics and destroying nutrients (Bermudez-Aguirre and Corradini, 2012). Alternative treatments that have gained prominence include the use of chemical agents, high pressure, plasma, ultraviolet (UV) light, and irradiation (Pereira and Vicente, 2010) (Table 8.7).

Table 8.6 Occurrence of *Salmonella* spp. in Different Environmental Samples

Sample	Samples Analyzed/Positive Results (Incidence)	References
Soil	48/11 (22.9%)	Rodriguez et al. (2006)
	72/34 (47.8%) mean of four sampling periods	Pangloli et al. (2008)
	617/16 (2.6%)	Gorski et al. (2011)
	120/1 (0.8%)	Micallef et al. (2012)
	39/4 (10.3%)	Jay-Russell et al. (2013)
Water	635/179 (28.2%)	Luo et al. (2015)
	1405/996 (65%)	Cooley et al. (2014)
	96/6 (6%)	Benjamim et al. (2013)
	78/27 (35%)	Ndiaye et al. (2011)
	252/18 (7.1%)	Gorski et al. (2011)
	1429/89 (6.2%)	Johnson et al. (2003)
Animal feces/manure	380/6 (1.6%)	Holzel and Bauer (2008)
	359/113 (31.5%)	Farzan et al. (2010)
	20/9 (45%)	Jay-Russell et al. (2013)
	115/6 (5.2%)	Klein et al. (2010)
	795/17 (2.14%)	Gorski et al. (2011)

Table 8.7 Nonthermal Methods to Reduce Microorganisms in Vegetables

Method	Reduction (log)	Advantages	Disadvantages	References
Atmospheric pressure plasma	3.0	Quick treatments (30 s)	Changing the plant structure in treatments over 60 s	Baier et al. (2014)
High hydrostatic pressure	3.0–4.0	Homogeneity of treatment, shelf-life similar to pasteurization	Ineffective against spores	Myers et al. (2013)
Atmospheric cold plasma	5.0	Biofilm removal	Ineffective in internalized bacteria	Ziuzina et al. (2014)
Ultraviolet	1.4	Low cost; preserved nutritional quality	Possible change in color	Birmpa et al. (2013)
Ultrasound	2.0	Low cost; preserved nutritional quality	Possible change in color	Birmpa et al. (2013)
Organic acids (citric acid)	No significant reduction after 15 min exposure	No changes on product quality	Effective only in combination with other treatments	Bermudez-Aguirre and Barbosa-Canovas (2013)

The use of GRAS (generally recognized as safe) additives, such as organic acids (Park et al., 2011; Sagong et al., 2011) and essential oils (Barbosa et al., 2016), can also be considered alternatives to control the multiplication of pathogenic microorganisms.

The FDA approves irradiation of leafy vegetables at doses up to 1 kGy (Gomes et al., 2011). UV-C radiation is most commonly used for pathogen inactivation studies in plants (Kim et al., 2013; Raybaudi-Massilia et al., 2013; Mukhopadhyay et al., 2014). However, the continued use of UV-C radiation can cause changes in product attributes, especially with regard to color (Bermudez-Aguirre and Barbosa-Canovas, 2013). Gamma radiation is also capable of reducing *Salmonella* and *L. monocytogenes* populations, without significantly affecting sensory attributes of fresh-cut minimally processed spinach (Rezende et al., 2014).

The use of ultrasound techniques for reducing microbial load in different agricultural products has been successfully described (Bilek and Turantas, 2013; Birmpa et al., 2013; De São José et al., 2014.). Studies have shown that the synergistic effect of ultrasound combined with organic acids improves microbial inactivation, resulting in further reduction of 0.8−1 log cycle, when compared to single treatments (Sagong et al., 2011).

MPVs are usually stored under a modified atmosphere (MAP), using combinations of gases such as oxygen, carbon dioxide, and nitrogen, which aims to change the atmosphere of the interior of the package and, therefore, maintain the product's quality and increase its shelf-life (Rojas-Graü et al., 2009). MAP is commonly used for inhibiting oxidative browning of lettuces and controlling microbial growth (Horev et al., 2012), although it may promote the proliferation of pathogenic, facultative anaerobic bacteria, such as *L. monocytogenes* and *E. coli* O157: H7 (Gil and Allende, 2012). Despite the antimicrobial effect generally occurring by addition of CO_2 (Bennik et al., 1995), pathogenic microorganisms such as *Salmonella* can tolerate high CO_2 concentrations and persist in the product (Hintlian and Hotchkiss, 1987). Sant'Ana et al. (2013) reported that *Salmonella* spp. and *L. monocytogenes* were able to survive in samples of lettuce and cabbage packaged under MAP (5% O_2, 15% CO_2, and 80% N_2) for 6 days stored at 7°C and 15°C.

Sanitizers are widely used for food disinfection, especially concerning RTE vegetables, in which the application of heat is not feasible. In addition, consumers should be aware of the risk due to pathogens and the importance of good washing and disinfection practices to reduce the risk of foodborne disease outbreaks (Domenech et al., 2013). However, the most important factor in disinfection efficiency is the antimicrobial product itself (Nastou et al., 2012). Different foods require specific treatments, and the same method cannot be extrapolated to different vegetables (Nastou et al., 2012).

Antimicrobial agents such as chlorine, chlorine dioxide, ozone, and peracetic acid may be added to the water of the processing tanks, aiming to reduce the total microbial load, including any human pathogens that may be present. However, the antimicrobial activity of these chemicals against bacteria attached to the

Table 8.8 Sanitizing Agents Used in the Food Industry

Agent (Compound Based on)	Unit	Product	Microorganism	Reduction (log CFU/mL)	References
Quaternary ammonium	0.0125%	Parsley	*S. Typhimurium*	0.7	Shirron et al. (2009)
Peracetic acid	80 mg/L	Spinach	*Salmonella* spp.	0.8	Neal et al. (2012)
	80 ppm	Romaine lettuce	*L. monocytogenes*	1.0	Beuchat et al. (2004)
ClO_2 (gas, generated by sachet after 3 h)	8.7 mg/20 L	Lettuce	*Salmonella* spp.	5.4	Lee et al. (2004)
	8.7 mg/20 L	Lettuce	*L. monocytogenes*	5.4	Lee et al. (2004)
Cl_2	200 μg/mL	Parsley	*Salmonella* spp.	5.0	Lang et al. (2004)
	200 μg/mL	Parsley	*L. monocytogenes*	3.4	Lang et al. (2004)

surfaces of vegetables is restricted to a population reduction of 90–99% (Sapers, 2002). By definition, an effective sanitizer must reduce the contamination level at 99.999% (Schmidt, 1997).

Although chlorine has been widely used in processing plants, its use within the limits allowed by the FDA is generally not sufficient to eliminate important pathogens such as *Salmonella* (Zhang and Farber, 1996). Alternative sanitizers, such as chlorine dioxide and peracetic acid, appear as an option for reducing the microbial load on vegetables (Beuchat, 1998), and the combination of different agents may be an effective barrier against microorganisms. A study on the effect of peracetic acid, neutral electrolyzed water, and UV-C radiation combined with storage at low temperatures have extended the shelf-life of broccoli for 14 days (Martinez-Hernandez et al., 2015).

Table 8.8 provides the main sanitizing agents used in the food industry and the reductions of microbial loads obtained.

Shredded lettuce samples were artificially inoculated with *L. monocytogenes* before and after washing for 3 min in water containing 100 mg/L of total chlorine at 4°C and 47°C, and stored at 1°C and 10°C in oxygen-permeable containers. The washing step performed with cold water (4°C) impaired the survival of the microorganism independent of the storage temperature applied. However, washing using heated chlorinated water favored microbial growth in the samples stored at 10°C (Delaquis et al., 2002). This study demonstrated that temperature abuse conditions allowed *L. monocytogenes* multiplication in RTE, which represents remarkable health risks to the consumers, especially pregnant women and immunocompromised individuals.

The temperature control of leafy vegetables must be constant throughout the production cycle, once temperature changes may lead to pathogen multiplication, especially of psychrotrophic strains, such as *L. monocytogenes*. Thus, regardless of the control method, the cold chain must be maintained until the final consumer.

Considering that *Salmonella* and *L. monocytogenes* are generally inactivated by the appropriate sanitizer recommended by the manufacturer (Barker et al., 2003), their persistence can be due to greater tolerance to sanitizers, higher resistance to different sublethal methods, or higher adhesion ability to surfaces by a particular strain (Carpentier and Cerf, 2011).

Persistency describes how long a pathogen can survive in the environment, whether in a human host or food matrices. Enteric pathogens, including *E. coli* and *Salmonella*, can survive in the surface of fruits or vegetables, usually in micro-holes or other physical damage for a time greater than the product's shelf-life (Olaimat and Holley, 2012). Once attached, bacteria removal using the washing procedure alone becomes very difficult (Warning and Datta, 2013) due to the biofilm formation, among other circumstances not well defined yet (Yaron and Romling, 2014).

Oliveira et al. (2007) suggested that bacteria adhesion is strongly dependent on inherent characteristics of the strain, including, for example, the production of polysaccharides. Well-developed biofilms formed in a plant by pioneer bacterial cells, for example, *Pseudomonas fluorescens*, provide a safe environment for the proliferation of *L. monocytogenes* (Carmichael et al., 1999).

The disinfection process using chlorine compounds and other antimicrobial agents can rarely reduce more than one logarithmic cycle of strongly adherent biofilms. Moreover, enteric pathogens are less susceptible to the action of chlorine-based sanitizers, allowing the remaining cells to multiply and remain in the food (Shirron et al., 2009).

Apart from the characteristics mentioned above, *Salmonella* spp. and *L. monocytogenes* have the ability to penetrate plant tissues. Confocal microscopy of thale cress leaves dipped in *L. monocytogenes* inoculum (8.8 log CFU/g) showed the presence of the pathogen within the stomata, as well as deeper intracellular spaces of leaf tissue (Milillo et al., 2008). Studies using the scanning electron microscopy technique have also confirmed the internalization capability of *S. enterica* in spinach leaf stomata (Kroupitski et al., 2009).

Besides the ability of a pathogen to resist an inactivation process and survive in the environment, cross-contamination also brings concern to producers. If not carried out safely, routine procedures used during processing stages, such as washing, transportation, and storage can provide favorable conditions for the multiplication of pathogenic and spoilage microorganisms (Fig. 8.3).

Washing is one of the most important steps during processing, since in addition to removing the dirt, it also favors the reduction of the microbial load (Haute et al., 2015). Water from a contaminated source can, therefore, contaminate hydro refrigerators, discharge tanks, or troughs, leading to cross-contamination when recycled without further treatment (Sapers, 2002).

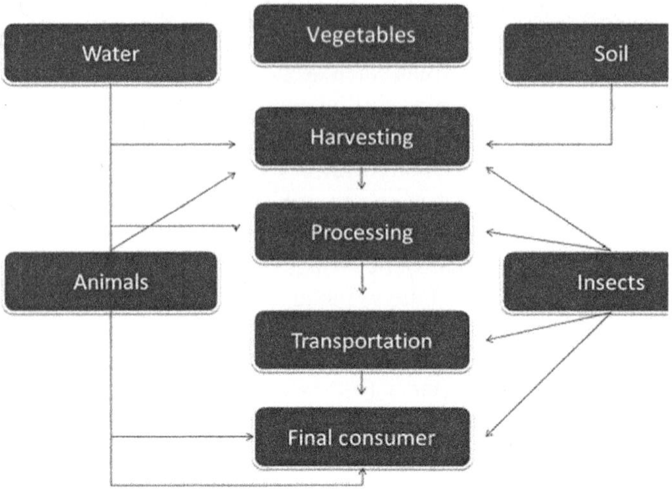

FIGURE 8.3

Schematic representation of the possible pathways for cross-contamination of vegetables from farm to consumers.

Cross-contamination can also occur during processing when the equipment in contact with potentially contaminated products is not regularly cleaned and sanitized. The multiplication of the remaining microorganisms after sanitization of leafy vegetables can occur, depending on the time and storage temperature conditions used, or due to accumulated moisture into the package due to the respiration process. At the market, microbial population levels can compromise the product quality by reducing its shelf-life, as well as pose a risk to consumer health (Beuchat, 1996).

Utensils used in both domestic and industrial kitchens can become sources of such contamination. Vegetables are among the products implicated in foodborne disease outbreaks, and *S. enterica* is one of the most commonly found pathogens. Studies have shown that contaminated knives can act as carrier agents and transfer up to 3 log cycles of the microorganism to lettuce leaves (Ravishankar et al., 2010).

The application of GMP in domestic kitchens and the implementation of a rigid plan of HACCP in the industry are prophylactic measures that really minimize the occurrence of foodborne disease outbreaks (Duff et al., 2003; Verhoeff-Bakkenes et al., 2008; Kennedy et al., 2011).

CONCLUSION

Despite the efforts of researchers and government agencies to control or reduce the number of foodborne diseases, outbreaks have still been reported around the world. Considering the wide diversity of potentially pathogenic microorganisms colonizing these food products, preventing their contamination is a critical step for safeguarding public health.

REFERENCES

Ajayeoba, T.A., Atanda, O.O., Obadina, A.O., Bankole, M.O., Adelowo, O.O., 2015. The incidence and distribution of *Listeria monocytogenes* in ready-to-eat vegetables in South-Western Nigeria. Food Sci. Nutr. Available from: http://dx.doi.org/10.1002/fsn3.263.

Allerberger, F., Guggenbichler, J.P., 1989. Listeriosis in Austria: a report of an outbreak in 1986. Acta Microbiologica Hungrica 36, 149–152.

Allerberger, F., Wagner, M., 2010. Listeriosis: a resurgent foodborne infection. Clin. Microbiol. Infect. 16, 16–23.

Anonymous, 2003. Interpretive summary: quantitative assessment of relative risk to public health from foodborne *Listeria monocytogenes* among selected categories of ready-to-eat foods. Center for Food Safety and Applied Nutrition, Food and Drug Administration, US Department of Health and Human Services, Food Safety and Inspection Service, US Department of Agriculture. Available from: <http://www.fda.gov/downloads/Food/FoodScienceResearch/UCM197330.pdf> (accessed 01.06.15.).

Aytac, S.A., Ben, U., Cengiz, C., Taban, B.M., 2010. Evaluation of *Salmonella* and *Listeria monocytogenes* contamination on leafy green vegetables. J. Food Agricult. Environ. 8, 275–279.

Baier, M., Gorgen, M., Ehlbeck, J., Knorr, D., Herppich, W.B., Schluter, O., 2014. Non-thermal atmospheric pressure plasma: screening for gentle process conditions and anti-bacterial efficiency on perishable fresh produce. Innov. Food Sci. Emerg. Technol. 22, 147–157.

Barbosa, I.M., Medeiros, J.A.C., Oliveira, K.A.R., Gomes-Neto, N.J., Tavares, J.F., Magnani, M., et al., 2016. Efficacy of the combined application of oregano and rosemary essential oils for the control of *Escherichia coli, Listeria monocytogenes* and *Salmonella* Enteritidis in leafy vegetables. Food Contr. 59, 468–477.

Barker, J., Naeeni, M., Bloomfield, S.F., 2003. The effects of cleaning and disinfection in reducing *Salmonella* contamination in a laboratory model kitchen. J. Appl. Microbiol. 95, 1351–1360.

Benjamim, L., Atwill, E.R., Jay-Russell, M., Cooley, M., Carychao, D., 2013. Occurrence of generic *Escherichia coli, E. coli* O157 and *Salmonella* spp. in water and sediment from leafy green produce farms and streams on the Central California coast. Int. J. Food Microbiol. 165, 65–76.

Bennik, M.H.J., Smid, E.J., Rombouts, F.M., Gorris, L.G.M., 1995. Growth of psychrotrophic foodborne pathogens in a solid surface model system under the influence of carbon dioxide and oxygen. Food Microbiol. 12, 509–519.

Bermudez-Aguirre, D., Barbosa-Canovas, G.V., 2013. Disinfection of selected vegetables under nonthermal treatments: chlorine, acid nitric, ultraviolet light and ozone. Food Contr. 29, 82–90.

Bermudez-Aguirre, D., Corradini, M.G., 2012. Inactivation kinetics of *Salmonella* spp. under thermal and emerging treatments: a review. Food Res. Int. 45, 700–712.

Beuchat, L.R., 1996. *Listeria monocytogenes*: incidence on vegetables. Food Contr. 7, 223–228.

Beuchat, L.R., 1998. Food Safety Issues: Surface Decontamination of Fruits and Vegetables Eaten Raw: A Review. Food Safety Unit. World Health Organization (WHO), Geneva, Switzerland, p. 42.

Beuchat, L.R., 2002. Ecological factors influencing survival and growth of human pathogens on raw fruits and vegetables. Microbes Infect. 4, 413–423.

Beuchat, L.R., Adler, B.A., Lang, M.M., 2004. Efficacy of chlorine and a peroxyacetic acid sanitizer in killing *Listeria monocytogenes* on iceberg lettuce and romaine lettuce using simulated commercial processing conditions. J. Food Protect. 67, 1238–1242.

Bilek, S.E., Turantas, F., 2013. Decontamination efficiency of high power ultrasound in the fruit and vegetable industry, a review. Int. J. Food Microbiol. 166, 155–162.

Birmpa, A., Sfika, V., Vantarakis, A., 2013. Ultraviolet light and ultrasound as non-thermal treatments for the inactivation of microorganisms in fresh ready-to-eat foods. Int. J. Food Microbiol. 167, 96–102.

Bonetta, S., Ferretti, E., Bonetta, S., Fezia, G., Carraro, E., 2011. Microbiological contamination of digested products from anaerobic co-digestion of bovine manure and agricultural by-products. Lett. Appl. Microbiol. 53, 552–557.

Bouchrif, B., Paglietti, B., Murgia, M., Piana, A., Cohen, N., Ennaji, M., et al., 2009. Prevalence and antibiotic-resistance of *Salmonella* isolated from food in Morocco. J. Infect. Develop. Countr. 28, 35–40.

Brackett, R.E., 1994. Microbiological spoilage and pathogens in minimally processed fruits and vegetables. In: Wiley, R.C. (Ed.), Minimally Processed Refrigerated (MRP) Fruits and Vegetables. Van Nostrand Reinhold, New York, pp. 269–312.

Cabedo, L., Barrot, L.P., Canelles, A.T., 2008. Prevalence of *Listeria monocytogenes* and *Salmonella* in ready to eat food in Catalonia, Spain. J. Food Protect. 71, 855–859.

Carmichael, I., Harper, I.S., Coventry, M.J., Taylor, P.W.J., Wan, J., Hickey, M.W., 1999. Bacterial colonization and biofilm development on minimally processed vegetables. J. Appl. Microbiol. Symp. Suppl. 85, 45–51.

Carpentier, B., Cerf, O., 2011. Review—persistence of *Listeria monocyotgenes* in food industry equipment and premises. Int. J. Food Microbiol. 145, 1–8.

Centers for Disease Control and Prevention—CDC. Foodborne Outbreak Online Database (FOOD). 2015. Available from: <http://wwwn.cdc.gov/foodborneoutbreaks/Default.aspx> (accessed 14.08.15.).

Centers for Disease Control and Prevention—CDC. New CDC Data on Foodborne Disease Outbreaks. Surveillance for Foodborne Disease Outbreaks. United States, 2013: Annual Report. Available from: <http://www.cdc.gov/foodsafety/pdfs/foodborne-disease-outbreaks-annual-report-2013-508c.pdf> (accessed 13.08.15.).

Cevallos-Cevallos, J.M., Danyluk, M.D., Gu, G., Vallad, G.E., van Bruggen, A.H., 2012. Dispersal of *Salmonella* Typhimurium by rain splash onto tomato plants. J. Food Protect. 75, 472–479.

Chen, Y., Ross, W.H., Gray, M.J., Wiedmann, M., Whiting, R.C., Scott, V.N., 2006. Attributing risk to *Listeria monocytogenes* subgroups: dose response in relation to genetic lineages. J. Food Protect. 69, 335–344.

Commission Regulation (EC) No 2073/2005 of 15 November 2005 on microbiological criteria for foodstuffs. Official Journal of the European Union, L338/1–L338/26, 2005. Available from: <http://eur- ex.europa.eu/LexUriServ/LexUriServ.do?uri=OJ:L:2005:338:0001:0026:EN:PDF> (accessed 11.06.15.).

Cooley, M.B., Quinones, B., Oryang, D., Mandrell, R.E., Gorski, L., 2014. Prevalence of shiga toxin producing *Escherichia coli, Salmonella enterica, and Listeria monocytogenes* at public access watershed sites in a California Central Coast agricultural region. Front. Cellular Infect. Microbiol. 4, 1–13.

Cordano, A.M., Jacquet, C., 2009. *Listeria monocytogenes* isolated from vegetable salads sold at supermarkets in Santiago, Chile: prevalence and strain characterization. Int. J. Food Microbiol. 132, 176–179.

Crook, P.D., Aguilera, J.F., Threlfall, E.J., O'Brien, S.J., Sigmundsdóttir, G., Wilson, D., et al., 2003. A European outbreak of *Salmonella enterica* serotype Typhimurium definitive phage type 204b in 2000. Clin. Microbiol. Infect. 9, 839–845.

De Giusti, M., Solimini, A.G., Cottarelli, A., De Vito, C., Aurigemma, C., Tufi, D., et al., 2014. Temporal pattern of microbial indicators of ready-to-eat salads during shelf life. Annali dell Instituto Superiore di Sanitá 50, 90–95.

Delaquis, P., Stewart, S., Cazaux, S., Toivonen, P., 2002. Survival and growth of *Listeria monocytogenes* and *E. coli* O157:H7 in ready-to-eat iceberg lettuce washed in warm chlorinated water. J. Food Protect. 65, 459–464.

De São José, J.F.B., Andrade, N.J., Ramos, A.M., Vanetti, M.C.D., Stringheta, P.C., Chaves, J.B.P., 2014. Decontamination by ultrasound application in fresh fruits and vegetables. Food Contr. 45, 36–50.

Domenech, E., Botella, S., Ferrus, M.A., Escriche, I., 2013. The role of the consumer in the reduction of *Listeria monocytogenes* in lettuces by washing at home. Food Contr. 29, 98–102.

Doyle, M.P., Buchanan, R.L., 2012. Food Microbiology: Fundamentals and Frontiers, fourth ed ASM Press, Washington, DC.

Duff, S.B., Scott, E.A., Mafilios, M.S., Todd, E.C., Krilov, L.R., Geddes, A.M., et al., 2003. Cost-effectiveness of a targeted disinfection program in household kitchens to prevent foodborne illnesses in the United States, Canada, and the United Kingdom. J. Food Prot. 66, 2103–2115.

EFSA (European Food Safety Authority), 2011. The European Union summary report on trends and sources of zoonoses, zoonotic agents and food-borne outbreaks in 2009. EFSA J. 9 (3), 2090, 378 pp.

EFSA (European Food Safety Authority), 2012. The European Union summary report on trends and sources of zoonoses, zoonotic agents and food-borne outbreaks in 2010. EFSA J. 10 (3), 2597, 442 pp.

EFSA (European Food Safety Authority), 2013. The European Union summary report on trends and sources of zoonoses, zoonotic agents and food-borne outbreaks in 2011. EFSA J. 11 (4), 3129, 250 pp.

EFSA (European Food Safety Authority), 2014. The European Union summary report on trends and sources of zoonoses, zoonotic agents and food-borne outbreaks in 2012. EFSA J. 12 (2), 3547, 312 pp.

EFSA (European Food Safety Authority), 2015. The European Union summary report on trends and sources of zoonoses, zoonotic agents and food-borne outbreaks in 2013. EFSA J. 13 (1), 3991, 162 pp.

Esteban, J.I., Oporto, B., Aduriz, G., Juste, R.A., Hurtado, A., 2009. Faecal shedding and strain diversity of *Listeria monocytogenes* in healthy ruminants and swine in Northern Spain. BMC Veterin. Res. 5, 2.

FAO/WHO (Food and Agriculture Organization of the United Nations/World Health Organization), 2008. Microbiological hazards in fresh leafy vegetables and herbs. Meeting Report, Microbiological Risk Assessment Series, n 14. Rome, 151 pp.

Farzan, A., Friendship, R.M., Cook, A., Pollari, F., 2010. Occurrence of *Salmonella, Campylobacter, Yersinia enterocolitica, Escherichia coli* O157 and *Listeria monocytogenes* in Swine. Zoonoses Publ. Health 57, 388–396.

Fatemi, P., LaBorde, L.F., Patton, J., Sapers, G.M., Annous, B.A., Knabel, S.J., 2006. Influence of punctures, cuts, and surface morphologies of golden delicious apples on penetration and growth of *Escherichia coli* O157:H7. J. Food Prot. 69, 267–275.

FDA (US Food and Drug Administration), 2010. DSHS (Texas Dept. of Health Services) orders Sangar produce to close, recall products. Recall: State press release. Available from. <http://www.fda.gov/Safety/Recalls/ucm230709.htm> (accessed 20.05.15.).

FDA (US Food and Drug Administration), 2015. Amy's kitchen recalls various products because of possible health risk. Available from: <http://www.fda.gov/Safety/Recalls/ucm439397.htm> (accessed 14.08.15.).

Foerster, C., Vidal, L., Troncoso, M., Figueroa, G., 2012. Characterization of *Listeria monocytogenes* isolates from cattle and ground beef by pulsed-field gel electrophoresis. Revista Argentina de Microbiología 44, 195–200.

Forsythe, S.J., 2013. Microbiologia da segurança dos alimentos, second ed. Artmed, Porto Alegre.

Fox, E., O'Mahony, T., Clancy, M., Dempsey, R., O'Brien, M., Jordan, K., 2009. *Listeria monocytogenes* in the Irish dairy farm environment. J. Food Prot. 72, 1450–1456.

Francis, G.A., Gallone, A., Nychas, G.J., Sofos, J.N., Colelli, G., Amodio, M.L., et al., 2012. Factors affecting quality and safety of fresh-cut produce. Crit. Rev. Food Sci. Nutr. 52, 595–610.

Fröder, H., Martins, C.G., De Souza, K.L., Landgraf, M., Franco, B.D., Destro, M.T., 2007. Minimally processed vegetable salads: microbial quality evaluation. J. Food Prot. 70, 1277–1280.

Gandhi, M., Chikindas, M.L., 2007. *Listeria*: a foodborne pathogen that knows how to survive. Int. J. Food Microbiol. 113, 1–15.

Garrido, V., Vitas, A.I., García-Jalón, I., Torroba, L., Navascués, A., 2012. Epidemiology of invasive listeriosis in clinical cases in Navarra (Spain): comparison between 1995–2005 and 2006–2011. In: Romano, A., Giordano, C.F. (Eds.), *Listeria* Infections: Epidemiology, Pathogenesis and Treatment. Nova Science Publishers, New York City, NYC, pp. 141–154.

Garrity, G.M., Brenner, D.J., Krieg, N.R., Staley, J.T., 2005. Bergey's Manual of Systematic Bacteriology, second ed. Springer, New York.

Gil, M.I., Allende, A., 2012. Minimal processing. In: Gómez-López, V.M. (Ed.), Decontamination of Fresh and Minimally Processed Produce. Wiley-Blackwell, Garsington Road, Oxford, pp. 105–120.

Gil, M.I., Selma, M.V., Suslow, T., Jacxsens, L., Uyttendaele, M., Allende, A., 2015. Pre- and postharvest preventive measures and intervention strategies to control microbial food safety hazards of fresh leafy vegetables. Crit. Rev. Food Sci. Nutr. 55, 453–468.

Gillespie, I., 2004. Outbreak of *Salmonella* Newport infection associated with lettuce in UK. Euro surveillance 8, pii = 2562.

Goldberg, D., Kroupitski, Y., Belausov, E., Pinto, R., Sela, S., 2011. *Salmonella* Typhimurium internalization is variable in leafy vegetables and fresh herbs. Int. J. Food Microbiol. 145, 250–257.

Gomes, C., Moreira, R.G., Castell-Perez, E., 2011. Radiosensitization of *Salmonella* spp. and *Listeria* spp. in ready to eat baby spinach leaves. J. Food Sci. 76, 141–148.

González, D., Vitas, A.I., Díez-Leturia, M., García-Jálon, I., 2013. *Listeria monocytogenes* and ready-to-eat seafood in Spain: study of prevalence and temperatures at retail. Food Microbiol. 36, 374–378.

Gorski, L., Parker, T.C., Liang, A., Cooley, M.B., Jay-Russell, M.T., Gordus, A.G., et al., 2011. Prevalence, distribution, and diversity of *Salmonella enterica* in a major produce region of California. Appl. Environ. Microbiol. 77, 2734–2748.

Ha, J.W., Kang, D.H., 2015. Enhanced inactivation of food borne pathogens in ready-to-eat sliced ham by near-infrared heating combined with UV-C irradiation and mechanism of the synergistic bactericidal action. Appl. Environ. Microbiol. 81, 2–8.

Hadjilouka, A., Andritsos, N.D., Paramithiotis, S., Mataragas, M., Drosinos, E.H., 2014. *Listeria monocytogenes* serotype prevalence and biodiversity in diverse food products. J. Food Prot. 77, 2115–2120.

Harris, L.J., Farber, J.N., Beuchat, L.R., Parish, M.E., Suslow, T.V., Garret, E.H., et al., 2003. Outbreaks associated with fresh produce: incidence, growth, and survival of pathogens in fresh and fresh-cut produce. Comprehensive Rev. Food Sci. Food Safety 2, 78–141.

Hasegawa, M., Iwabuchi, E., Yamamoto, S., Muramatsu, M., Takashima, I., Hirai, K., 2014. Prevalence and characteristics of *Listeria monocytogenes* in feces of black beef cattle reared in three geographically distant areas in Japan. Foodborne Pathogens Dis. 11, 96–103.

Haute, S.V., Tryland, I., Veys, A., Sampers, I., 2015. Wash water disinfection of a full-scale leafy vegetables washing process with hydrogen peroxide and the use of a commercial metal ion mixture to improve disinfection efficiency. Food Contr. 50, 173–183.

Hintlian, C.B., Hotchkiss, J.H., 1987. Comparative growth of spoilage and pathogenic organisms on modified atmosphere-packaged cooked beef. J. Food Prot. 50, 218–223.

Ho, A.J., Ivanek, R., Gröhn, Y.T., Nightingale, K.K., Wiedmann, M., 2007. *Listeria monocytogenes* fecal shedding in dairy cattle shows high levels of day-to-day variation and includes outbreaks and sporadic cases of shedding of specific *L. monocytogenes* subtypes. Prevent. Veter. Med. 80, 287–305.

Ho, J.L., Shands, K.N., Friedland, G., Eckind, P., Fraser, D.W., 1986. An outbreak of type 4b *Listeria monocytogenes* infection involving patients from eight Boston hospitals. Arch. Int. Med. 146, 520–524.

Holzel, C., Bauer, J., 2008. *Salmonella* spp. in Bavarian liquid pig manure: occurrence and relevance for the distribution of antibiotic resistance. Zoonoses Publ. Health 55, 133–138.

Horev, B., Sela, S., Vinokur, Y., Gorbatsevich, E., Pinto, R., Rodov, V., 2012. The effects of active and passive modified atmosphere packaging on the survival of *Salmonella enterica* serotype Typhimurium on washed romaine lettuce leaves. Food Res. Int. 45, 1129–1132.

Ijabadeniyi, O.A., Debusho, L.K., Vanderlinde, M., Buys, E.M., 2011. Irrigation water as a potential preharvest source of bacterial contamination of vegetables. J. Food Safety 31, 452–461.

Islam, M., Morgan, J., Doyle, M.P., Phatak, S.C., Millner, P., Jiang, X., 2004. Persistence of *Salmonella enterica* serovar Typhimurium on lettuce and parsley and in soils on which they were grown in fields treated with contaminated manure composts or irrigation water. Foodborne Pathogens Dis. 1, 27–35.

Ivanek, R., Gröhn, Y.T., Wiedmann, M., 2006. *Listeria monocytogenes* in multiple habitats and host populations: review of available data for mathematical modeling. Foodborne Pathogens Dis. 3, 319–336.

Jay, J.M., Loessner, M.J., Golden, D.A., 2005. Modern Food Microbiology. Springer, US, New York City, 790 p.

Jay-Russell, M.T., Madigan, J.E., Bengson, Y., Madigan, S., Hake, A.F., Foley, J.E., et al., 2013. *Salmonella* Oranienburg isolated from horses, wild turkeys and an edible home garden fertilized with raw horse manure. Zoonoses Publ. Health 61, 64–71.

Jeyaletchumi, P., Tunung, R., Selina, P.M., Chai, L.C., Radu, S., Farinazleen, M.G., et al., 2011. Evaluation of *Listeria* spp. and *Listeria monocytogenes* in selected vegetable farms. J. Tropical Agric. Food Sci. 39, 255–266.

Johannessen, G.S., Loncarevic, S., Krusel, H., 2002. Bacteriological analysis of fresh produce in Norway. Int. J. Food Microbiol. 77, 199–204.

Johnson, J.Y.M., Thomas, J.E., Graham, T.A., Townshend, I., Byrne, J., Selinger, L.B., et al., 2003. Prevalence of *Escherichia coli* O157:H7 and *Salmonella* spp. in surface waters of southern Alberta and its relation to manure sources. Can. J. Microbiol. 49, 326–335.

Kathariou, S., 2002. *Listeria monocytogenes* virulence and pathogenicity, a food safety perspective. J. Food Prot. 65, 1811–1829.

Kennedy, J., Nolan, A., Gibney, S., Obrien, S., Mcmahon, M.A.S., Mckenzie, K., et al., 2011. Determinants of cross-contamination during home food preparation. Br. Food J. 113, 280–297.

Kim, Y.H., Jeong, S.G., Back, K.H., Park, K.H., Chung, M.S., Kang, D.H., 2013. Effect of various conditions on inactivation of *Escherichia coli* O157:H7, *Salmonella* Typhimurium, and *Listeria monocytogenes* in fresh-cut lettuce using ultraviolet radiation. Int. J. Food Microbiol. 166, 349–355.

Klein, M., Brown, L., Tucker, R.W., Ashbolt, N.J., Stuetz, R.M., Roser, D.J., 2010. Diversity and abundance of zoonotic pathogens and indicators in manures of feedlot cattle in Australia. Appl. Environ. Microbiol. 76, 6947–6950.

Koseki, S., Mizuno, Y., Yamamoto, K., 2011. Comparison of two possible routes of pathogen contamination of spinach leaves in a hydroponic cultivation system. J. Food Prot. 74, 1536–1542.

Kovasevic, M., Burazin, J., Pavlovic, H., Kopjar, M., Pilizota, V., 2013. Prevalence and level of *Listeria monocytogenes* and other *Listeria* sp. in ready-to-eat minimally processed and refrigerated vegetables. World J. Microbiol. Biotechnol. 29, 707–712.

Kroupitski, Y., Golberg, D., Belausov, E., Pinto, R., Swartzberg, D., Granot, D., et al., 2009. Internalization of *Salmonella enterica* in leaves is induced by light and involves chemotaxis and penetration through open stomata. Appl. Environ. Microbiol. 75, 6076–6086.

Kroupitski, Y., Pinto, R., Belausov, E., Sela, S., 2011. Distribution of Salmonella typhimurium in romaine lettuce leaves. Food Microbiology 28, 990–997.

Lang, M.M., Harris, L.J., Beuchat, L.R., 2004. Survival and recovery of *Escherichia coli* O157:H7, *Salmonella,* and *Listeria monocytogenes* on lettuce and parsley as affected by method of inoculation, time between inoculation and analysis, and treatment with chlorinated water. J. Food Prot. 67, 1092–1103.

Lee, S.Y., Costello, M., Kang, D.H., 2004. Efficacy of chlorine dioxide gas as a sanitizer of lettuce leaves. J. Food Prot. 67, 1371–1376.

Leff, J.W., Fierer, N., 2013. Bacterial communities associated with the surfaces of fresh fruits and vegetables. PLos One 8, e59310.

Li, H., Wang, H., D'Aoust, J.Y., Maurer, J., 2013. *Salmonella* species. In: Doyle, M.P., Buchanan, R.L. (Eds.), Food Microbiology: Fundamentals and Frontiers. ASM Press, Washington, DC, pp. 225–261.

Lienemann, T., Niskanen, T., Guedes, S., Siitonen, A., Kuusi, M., Rimhanen-Finne, R., 2011. Iceberg lettuce as suggested source of a nationwide outbreak caused by two *Salmonella* serotypes, Newport and Reading, in Finland in 2008. J. Food Prot. 74, 1035–1040.

Linke, K., Rückerl, I., Brugger, K., Karpiskova, R., Walland, J., Muri-Klinger, S., et al., 2014. Reservoirs of *Listeria* species in three environmental ecosystems. Appl. Environ. Microbiol. 80, 5583–5592.

Liu, C., Hofstra, N., Franz, E., 2013. Impacts of climate change on the microbial safety of pre-harvest leafy green vegetables as indicated by *Escherichia coli* O157 and *Salmonella* spp. Int. J. Food Microbiol. 163, 119–128.

Locatelli, A., Depret, G., Jolivet, C., Henry, S., Dequiedt, S., Piveteau, P., et al., 2013. Nation-wide study of the occurrence of *Listeria monocytogenes* in French soils using culture-based and molecular detection methods. J. Microbiol. Meth. 93, 242–250.

Low, J.C., Donachie, W., 1997. A review of *Listeria monocytogenes* and listeriosis. Veter. J. 153, 9–29.

Luber, P., Crerar, S., Dufour, C., Farber, J., Datta, A., Todd, E.C.D., 2011. Controlling *Listeria monocytogenes* in ready-to-eat foods: working towards global scientific consensus and harmonization—recommendations for improved prevention and control. Food Contr. 22, 1535–1549.

Luo, Z., Gu, G., Ginn, A., Giurcanu, M.C., Adams, P., Vellidis, G., et al., 2015. Distribution and characterization of *Salmonella enterica* isolates from irrigation ponds in the Southeastern U.S.A. Appl. Environ. Microbiol. Available from: http://dx.doi.org/10.1128/AEM.04086-14.

Lyautey, E., Lapen, D.R., Wilkes, G., McCleary, K., Pagotto, F., Tyler, K., et al., 2007. Distribution and characteristics of *Listeria monocytogenes* isolates from surface waters of the South Nation River Watershed, Ontario, Canada. Appl. Environ. Microbiol. 73, 5401–5410.

Lynch, M.F., Tauxe, R.V., Hedberg, C.W., 2009. The growing burden of foodborne outbreaks due to contaminated fresh produce: risks and opportunities. Epidemiol. Infect. 137, 307–315.

Maistro, L.C., Miya, N.T.N., Sant'Ana, A.S., Pereira, J.L., 2012. Microbiological quality and safety of minimally processed vegetables marketed in Campinas, SP—Brazil, as assessed by traditional and alternative methods. Food Contr. 28, 258–264.

Martinez-Hernandez, G.B., Navarro-Rico, J., Gomez, P.A., Oton, M., Artes, F., Artes-Hernandez, F., 2015. Combined sustainable sanitising treatments to reduce *Escherichia coli* and *Salmonella enteritidis* growth on fresh cut kalian hybrid broccoli. Food Contr. 47, 312–317.

Matthews, K.R., Sapers, G.M., Gerba, C.P., 2014. The Produce Contamination Problem, Second Edition: Causes and Solutions (Food Science and Technology). Academic Press, Whaltham.

Mawak, J.D., Dashen, N.M., Idolor, A.J., Chukwu, O.O.C., 2009. Occurrence of *Listeria monocytogenes* in irrigation water and vegetables at Jos, Plateau State, Nigeria. Int. J. Trop. Agric. Food Syst. 3, 279–282.

McLauchlin, J., 2014. Risk of *Salmonella* and Norovirus in leafy greens eaten raw as salads. Veg-i-trade, closing event: taking the safety of fresh produce to the next level, Brussels, 11–12 June 2014. Public Health England, UK.

Micallef, S.A., Goldstein, R.E.R., George, A., Kleinfelter, L., Boyer, M.S., McLaughlin, C. R., et al., 2012. Occurrence and antibiotic resistance of multiple *Salmonella* serotypes recovered from water, sediment and soil on mid-Atlantic tomato farms. Environ. Res. 114, 31–39.

Milillo, S.R., Badamo, J.M., Boor, K.J., Wiedmann, M., 2008. Growth and persistence of *Listeria monocytogenes* isolates on the plant model *Arabidopsis thaliana*. Food Microbiol. 25, 698–704.

Mohammed, H.O., Atwill, E., Dunbar, L., Ward, T., McDonough, P., Gonzalez, R., et al., 2010. The risk of *Listeria monocytogenes* infection in beef cattle operations. J. Appl. Microbiol. 108, 349–356.

Mukhopadhyay, S., Ukuku, D.O., Juneja, V., Fan, X., 2014. Effects of UV-C treatment on inactivation of *Salmonella enterica* and *Escherichia coli* O157:H7 on grape tomato surface and stem scars, microbial loads, and quality. Food Contr. 44, 110–117.

Myers, K., Cannon, J., Montoya, D., Dickson, J., Lonergan, S., Sebranek, J., 2013. Effects of high hydrostatic pressure and varying concentrations of sodium nitrite from traditional and vegetable-based sources on the growth of *Listeria monocytogenes* on ready-to-eat (RTE) sliced ham. Meat Sci. 94, 69–76.

Nastou, A., Rhoades, J., Smirniotis, P., Makri, I., Kontominas, M., Likotrafiti, E., 2012. Efficacy of household washing treatments for the control of *Listeria monocytogenes* on salad vegetables. Int. J. Food Microbiol. 159, 247–253.

Ndiaye, M.L., Dieng, Y., Niang, S., Pfeifer, H.R., Tonolla, M., Peduzzi, R., 2011. Effect of irrigation water on the incidence of *Salmonella* spp. on lettuces produced by urban agriculture and sold on the markets in Dakar, Senegal. Afr. J. Microbiol. Res. 5, 2885–2890.

Neal, J.A., Marquez-Gonzalez, M., Cabrera-Diaz, E., Lucia, L.M., O'Bryan, C.A., Crandall, P.G., et al., 2012. Comparison of multiple chemical sanitizers for reducing *Salmonella* and *Escherichia coli* O157:H7 on spinach (*Spinacia oleracea*) leaves. Food Res. Int. 45, 1123–1128.

Newell, D.G., Koopmans, M., Verhoef, L., Duizer, E., Aidara-Kane, A., Sprong, H., et al., 2010. Food-borne diseases—the challenges of 20 years ago still persist while new ones continue to emerge. Int. J. Food Microbiol. 139, 3–15.

Nightingale, K.K., Schukken, Y.H., Nightingale, C.R., Fortes, E.D., Ho, A.J., Her, Z., et al., 2004. Ecology and transmission of *Listeria monocytogenes* infecting ruminants and in the farm environment. Appl. Environ. Microbiol. 70, 4458–4467.

Olaimat, A.N., Holley, R.A., 2012. Factors influencing the microbial safety of fresh produce: a review. Food Microbiol. 32, 1–19.

Oliveira, K., Oliveira, T., Teixeira, P., Azeredo, J., Oliveira, R., 2007. Adhesion of *Salmonella* Enteritidis to stainless steel surfaces. Braz. J. Microbiol. 38, 318–323.

Oliveira, M.A., Ribeiro, E.G.A., Bergamini, A.M.M., De Martinis, E.C.P., 2010. Quantification of *Listeria monocytogenes* in minimally processed leafy vegetables using a combined method based on enrichment and 16S rRNA real-time PCR. Food Microbiol. 27, 19–23.

Pan, Y., Breidt, F., Kathariou, S., 2009. Competition of *Listeria monocytogenes* serotype 1/2a and 4b strains in mixed-culture biofilms. Appl. Environ. Microbiol. 75, 5846–5852.

Pangloli, P., Dje, Y., Ahmed, O., Doane, C.A., Oliver, S.P., Draughon, F.A., 2008. Seasonal incidence and molecular characterization of *Salmonella* from dairy cows, calves, and farm environment. Foodborne Pathogens Dis. 5, 87–96.

Park, S.H., Choi, M.R., Park, J.W., Park, K.H., Chung, M.S., Ryu, S., et al., 2011. Use of organic acids to inactivate *Escherichia coli* O157:H7, *Salmonella* Typhimurium, and *Listeria monocytogenes* on organic fresh apples and lettuce. J. Food Sci. 76, 293–298.

Pereira, R.N., Vicente, A.A., 2010. Environmental impact of novel thermal and nonthermal technologies in food processing. Food Res. Int. 43, 1936–1943.

Pezzoli, L., Elson, R., Little, C.L., Yip, H., Fisher, I., Yishai, R., et al., 2008. Packed with *Salmonella*—investigation of an international outbreak of *Salmonella* Senftenberg infection linked to contamination of prepacked basil in 2007. Foodborne Pathogens Dis. 5, 661–668.

Porto, E., Eiroa, M.N.U., 2001. Occurrence of *Listeria monocytogenes* in vegetables. Dairy Food Environ. Sanit. 21, 282–286.

Pui, C.F., Wong, W.C., Chai, L.C., Tunung, R., Jeyaletchumi, P., Moor Hidayah, M.S., et al., 2011. *Salmonella*: a foodborne pathogen. Int. Food Res. J. 18, 465–473.

Quiroz-Santiago, C., Rodas-Suárez, O.R., Vázquez, Q., Fernández, C.R., Quiñones-Ramírez, F.J., Vázquez-Salinas, C, E.I., 2009. Prevalence of *Salmonella* in vegetables from Mexico. J. Food Prot. 72, 1279–1282.

Ravishankar, S., Zhu, L., Jaroni, D., 2010. Assessing the cross contamination and transfer rates of *Salmonella enterica* from chicken to lettuce under different food-handling scenarios. Food Microbiol. 27, 791–794.

Raybaudi-Massilia, R., Calderon-Gabaldon, M.I., Mosqueda-Melgar, J., Tapia, M.S., 2013. Inactivation of *Salmonella enterica* ser. Poona and *Listeria monocytogenes* on fresh cut "maradol" red papaya (*carica papaya* l) treated with UV-C light and malic acid. J. Consumer Prot. Food Safety 8, 37–44.

Rezende, A.C.B., Igarashi, M.C., Destro, M.T., Franco, B.D.G.M., Landgraf, M., 2014. Effect of gamma radiation on the reduction of *Salmonella* strains, *Listeria monocytogenes,* and shiga toxin producing *Escherichia coli* and sensory evaluation of minimally processed spinach (*tetragonia exapnsa)*. J. Food Prot. 77, 1768–1772.

Rodriguez, A., Pangloli, P., Richards, H.A., Mount, J.R., Draughon, F.A., 2006. Prevalence of *Salmonella* in diverse environmental farm samples. J. Food Prot. 11, 2568–2808.

Rojas-Graü, M.A., Oms-Oliu, G., Soliva-Fortuny, R., Martn-Belloso, O., 2009. The use of packaging techniques to maintain freshness in fresh-cut fruits and vegetables: a review. Int. J. Food Sci. Technol. 44, 875–889.

Ryan, K.J., Ray, C.G., 2004. Sherris Medical Microbiology: An Introduction to Infectious Disease, fourth ed. McGraw-Hill, New York.

Ryser, E.T., Buchanan, R.L., 2013. Listeria monocytogenes. In: Doyle, M.P., Buchanan, R.L. (Eds.), Food Microbiology: Fundamentals and Frontiers. ASM Press, Washington, DC, pp. 503–545.

Sagong, H.G., Lee, S.Y., Chang, P.S., Heu, S., Ryu, S., Choi, Y.J., et al., 2011. Combined effect of ultrasound and organic acids to reduce *Escherichia coli* O157:H7, *Salmonella* Typhimurium, and *Listeria monocytogenes* on organic fresh lettuce. Int. J. Food Microbiol. 145, 287–292.

Sagoo, S.K., Little, C.L., Ward, L., Gillespie, I.A., Mitchell, R.T., 2003. Microbiological study of ready-to-eat salad vegetables from retail establishments uncovers a national outbreak of salmonellosis. J. Food Prot. 66, 403–409.

Sant'Ana, A.S., Landgraf, M., Destro, M.T., Franco, B.D.G.M., 2011. Prevalence and counts of *Salmonella* spp. in minimally processed vegetables in São Paulo, Brazil. Food Microbiol. 28, 1235–1237.

Sant'Ana, A.S., Igarashi, M.C., Landgraf, M., Destro, M.T., Franco, B.D.G.M., 2012. Prevalence, populations and pheno- and genotypic characteristics of *Listeria monocytogenes* isolated from ready-to-eat vegetables marketed in São Paulo, Brazil. Int. J. Food Microbiol. 155, 1–9.

Sant'Ana, A.S., Landgraf, M., Destro, M.T., Franco, B.D.G.M., 2013. Growth potential of *Salmonella* and *Listeria monocytogenes* in ready-to-eat lettuce and collard greens packaged under modified atmosphere and in perforated film. J. Food Prot. 76, 888–891.

Sant'Ana, A.S., Franco, B.D.G.M., Schaffner, D.W., 2014. Risk of infection with *Salmonella* and *Listeria monocytogenes* due to consumption of ready-to-eat leafy vegetables in Brazil. Food Contr. 42, 1–8.

Sapers, G.M., 2002. Washing and sanitizing raw materials for minimally processed fruit and vegetable products. In: Novak, J.S., Sapers, G.M., Juneja, V.K. (Eds.), Microbial Safety of Minimally Processed Foods. CRC Press, Boca Raton, FL, pp. 221–253.

Schmidt, R.H., 1997. Basic Elements of Equipment Cleaning and Sanitizing in Food Processing and Handling Operations. University of Florida Institute of Food and Agricultural Sciences Cooperative Extension Services. Available from: <http://www.scribd.com/doc/141323473/Basic-Elements-of-Equipment-Cleaning-and-Sanitizing-in-Food-Processing-and-HandlingOperations#scribd> (accessed 15.08.15.).

Shirron, N., Kisluk, G., Zelikovich, Y., Eivin, I., Shimoni, E., Yaron, S., 2009. A comparative study assaying commonly used sanitizers for antimicrobial activity against indicator bacteria and a *Salmonella* Typhimurium strain on fresh produce. J. Food Prot. 72, 2413–2417.

Simental, L., Urtaza, J.M., 2008. Climate patterns governing the presence and permanence of salmonellae in coastal areas of Bahia de Todos Santos, Mexico. Appl. Environ. Microbiol. 74, 5918–5924.

Soni, D.K., Singh, M., Singh, D.V., Dubey, S.K., 2014. Virulence and genotypic characterization of *Listeria monocytogenes* isolated from vegetable and soil samples. BMC Microbiol. 14, 241.

Soon, J.M., Manning, L., Davies, W.P., Baines, R., 2010. Fresh produce-associated outbreaks: a call for HACCP on farms?. Br. Food J. 114, 553–597.

Soriano, J.M., Rico, H., Moltó, J.C., Mañes, J., 2001. *Listeria* species in raw and ready-to-eat foods from restaurants. J. Food Prot., Des Moines 64, 551–553.

Stavru, F., Archambaud, C., Cossart, P., 2011. Cell biology and immunology of *Listeria monocytogenes* infections: novel insights. Immunol. Rev. 240, 160–184.

Strawn, L., Gröhn, Y.T., Warchocki, S., Worobo, R.W., Bihn, E.A., Wiedmann, M., 2013a. Risk factors associated with *Salmonella* and *Listeria monocytogenes* contamination of produce fields. Appl. Environ. Microbiol. 79, 7618–7627.

Strawn, L.K., Fortes, E.D., Bihn, E.A., Nightingale, K.K., Gröhn, Y.T., Worobo, R.W., et al., 2013b. Landscape and meteorological factors affecting prevalence of three food-borne pathogens in fruit and vegetable farms. Appl. Environ. Microbiol. 79, 588–600.

Suslow, T.V., 2002. Production practices affecting the potential for persistent contamination of plants by microbial foodborne pathogens. In: Lindow, S.E., Hecht-Poinar, E.I., Elliott, V.J. (Eds.), Phyllosphere Microbiology. APS Press, St. Paul, MN, pp. 241–256.

Szymczak, B., Szymczak, M., Sawicki, W., Dąbrowski, W., 2014. Anthropogenic impact on the presence of *L. monocytogenes* in soil, fruits, and vegetables. Folia Microbiol. 59, 23–29.

Taban, B.M., Halkman, A.K., 2011. Do leafy vegetables and their ready-to-eat [RTE] salads carry a risk of foodborne pathogens? Anaerobe 17, 286–287.

Takkinen, J., Nakari, U.M., Johansson, T., Niskanen, T., Siitonen, A., Kuusi, M., 2005. A nationwide outbreak of multiresistant *Salmonella* Typhimurium in Finland due to contaminated lettuce from Spain, May 2005. Euro Surv. 10, E050630.1.

Tindall, B.J., Grimont, P.A., Garrity, G.M., Euzeby, J.P., 2005. Nomenclature and taxonomy of the genus *Salmonella*. Int. J. Syst. Evol. Evol. Microbiol. 55, 521–524.

Tomas-Callejas, A., Lopez-Galvez, F., Sbodio, A., Artes, F., Artes-Hernandez, F., Suslow, T.V., 2012. Chlorine dioxide and chlorine effectiveness to prevent *Escherichia coli* O157:H7 and *Salmonella* cross-contamination on fresh-cut red chard. Food Contr. 23, 325–332.

Verhoeff-Bakkenes, L., Beumer, R.R., de Jonge, R., Van Leusden, R.M., de Jong, A.E.I., 2008. Quantification of *Campylobacter jejuni* cross-contamination via hands, cutlery, and cutting board during preparation of a chicken fruit salad. J. Food Prot. 71, 1018–1022.

Vivant, A.-L., Garmyn, D., Piveteau, P., 2013. *Listeria monocytogenes*, a down-to-earth pathogen. Front. Cellular Infect. Microbiol. 3, 1–10.

Warning, A., Datta, A.K., 2013. Interdisciplinary engineering approaches to study how pathogenic bacteria interact with fresh produce. J. Food Eng. 114, 426–448.

Warriner, K., Huber, A., Namyar, A., Fan, W., Dunfield, K., 2009. Recent advances in the microbial safety of fresh fruits and vegetables. Adv. Food Nutr. Res. 57, 155–208.

Watson, R., 2009. Listeriosis remains a cause for concern in Europe. Br. Med. J. 338, b319.

Weller, D., Wiedmann, M., Strawn, L.K., 2015a. Spatial and temporal factors associated with an increased prevalence of *Listeria monocytogenes* in spinach fields in New York State. Appl. Environ. Microbiol. 81, 6059–6069.

Weller, D., Wiedmann, M., Strawn, L.K., 2015b. Irrigation is significantly associated with an increased prevalence of *Listeria monocytogenes* in produce production environments in New York State. J. Food Prot. 78, 1132–1141.

Wijnands, L.M., Delfgou-Van Asch, E.H.M., Beerepoot-Mensink, M.E., Van Der Meij-Florijn, A., Fitz-James, I., Van Leusden, F.M., et al., 2014. Prevalence and concentrations of bacterial pathogens in raw produce and minimally processed packaged salads produced in and for the Netherlands. J. Food Prot. 77, 388–394.

Yaron, S., Romling, U., 2014. Biofilm formation by enteric pathogens and its role in plant colonization and persistence. Microb. Biotechnol. 7, 496–516.

Zhang, S., Farber, J.M., 1996. The effects of various disinfectants against *Listeria monocytogenes* on fresh-cut vegetables. Food Microbiol. 13, 311–321.

Ziuzina, D., Patil, S., Cullen, P.J., Keener, K.M., Bourke, P., 2014. Atmospheric cold plasma inactivation of *Escherichia coli, Salmonella enterica* serovar Typhimurium and *Listeria monocytogenes* inoculated on fresh produce. Food Microbiol. 42, 109–116.

CHAPTER

Foodborne pathogenic bacteria in fresh-cut vegetables and fruits

9

Hülya Ölmez

TÜBİTAK MRC Material Institute, Gebze, Kocaeli, Turkey

INTRODUCTION

Driven mainly by the increasing consumer demands for fresh and healthier convenience foods, the fresh produce market has grown dramatically over the last two decades. Following the increase in consumption, on the other hand, the foodborne outbreaks associated with fresh produce have also shown a raising trend globally (CSPI, 2014; EFSA, 2011a). The proportion of fresh-produce-associated foodborne illnesses increased from 12% in the 1990s to 24% in the 2010s (Painter et al., 2013; Sivapalasingam et al., 2004). Three main factors may add to the increase in the number of reported illnesses. First, the globalization of food industry has definitely resulted in the spreading of foodborne outbreaks between countries and even continents, resulting in a higher number of people being affected by the same contaminated produce. Second, the changes in the agricultural and processing practices throughout the food chain which aimed at increasing the supply may have enhanced the risk for transmission of pathogens and cross-contamination. Third, the improvements in the methods of microbial detection and identification, as well as in surveillance techniques, have also contributed to the increase in the number of reported illnesses.

Fresh produce was reported to be one of the main vehicles of foodborne outbreaks in Europe, accounting for 8% of all the foodborne outbreaks in 2011 (EFSA, 2011a). In the United States, the produce category was linked to the largest number of foodborne illnesses associated with foodborne outbreaks between the periods of 1998–2007 and 2002–2011, constituting 23% and 24% of all illnesses, respectively (CSPI, 2011). The produce category surpassed the poultry, beef, and seafood categories, accounting for 667 outbreaks and 23,748 illnesses, most of which were associated with vegetables (10,806) (CSPI, 2014). The most commonly identified causative agents were *Salmonella* spp., norovirus, and pathogenic *Escherichia coli* (Painter et al., 2013). The main reason for having fresh-cut fruits and vegetables as one of the major vehicles of foodborne

P. Kotzekidou (Ed): Food Hygiene and Toxicology in Ready-to-Eat Foods.
DOI: http://dx.doi.org/10.1016/B978-0-12-801916-0.00009-1

outbreaks associated with foodborne pathogens is that they are often consumed raw after minimal processing. There is no kill step in the processing which can assure the elimination of the pathogenic bacteria associated with fresh produce. Once contaminated in the field, it is not possible to completely eliminate the pathogenic bacteria from the fresh-cut produce. Therefore, identification of the main sources of contamination and transmission routes of pathogenic bacteria to fresh produce is a key for managing the food safety risks associated with fresh produce.

SOURCES AND ROUTES OF PATHOGEN CONTAMINATION IN FRESH-CUT PRODUCE

Fresh produce may be contaminated with pathogenic bacteria via different routes at different points throughout the preharvest and harvest/postharvest stages (Fig. 9.1). Current contamination sources in the field include manure, irrigation water, soil, fecal contamination, contaminated seeds, geographical location, and climate change. The postharvest steps including the harvesting itself, precooling, processing operations, washing and sanitizing, storage packaging, and transfer

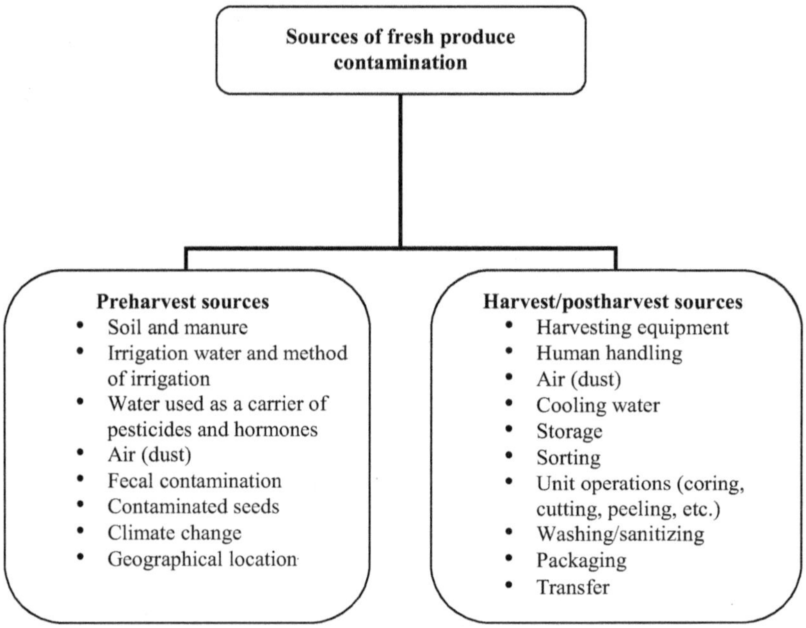

FIGURE 9.1

Main sources of fresh produce contamination.

may lead to pathogen contamination if not controlled properly according to the good manufacturing practices.

Irrigation water and the method of irrigation have been identified as one of the main sources of contamination (Park et al., 2012). The use of contaminated surface water or untreated sewage water results in high numbers of fecal coliforms and leads to the transmission of pathogenic bacteria to the fresh produce. Due to the direct deposition of the contaminated water onto the edible parts of the produce, spray irrigation systems pose a higher risk of pathogenic bacteria contamination compared to other irrigation methods such as surface irrigation and drip irrigation (Aruscavage et al., 2006; Fonseca et al., 2011; Van der Linden et al., 2013). Studies showed that *E. coli* O157:H7 can persist for 2–20 days on spray-irrigated lettuce leaves (Erickson et al., 2014; Solomon et al., 2003). Moreover, the repeated spray irrigation may lead to the internalization of *E. coli* O157:H7 in leafy vegetables such as spinach and parsley (Erickson et al., 2014).

Animal manure used as a soil amendment also represents a source of pathogenic bacteria contamination to fresh produce. The pathogenic bacteria including *E. coli* O157:H7, *Salmonella*, *Listeria monocytogenes*, and *Campylobacter jejuni* can survive for 45–175 days in soil depending on the soil characteristics (salinity, total nitrogen, etc.), moisture level, and temperature (Holley et al., 2006; Lang and Smith, 2007; Nicholson et al., 2005). Moreover, even with very low levels of contamination (10^0–10^1 CFU/g), the pathogenic bacteria are also able to survive for long periods in animal manure, depending on the temperature (Beuchat, 1999). The rate of survival is higher at lower temperatures. *E. coli* O157:H7 can survive in bovine feces for 70 days at 5°C, but only 49 days at 22°C (Semenov et al., 2007). Therefore, manure that has not been properly treated when used as a soil amendment may serve as a source of pathogen transmission to the fresh produce.

As the soil type, moisture level, and temperature are critical factors for the survival of enteric pathogens in soil and manure, the geographical location and the climate change also appear to be important preharvest factors for fresh fruit and vegetable contamination. Moreover, higher intensity of precipitation and rain may increase the potential of pathogenic bacteria contamination to fresh produce by enhancing the survival of the pathogens in soil and manure (Cevallos-Cevallos et al., 2012; Warriner et al., 2009).

Harvesting and postharvest practices can serve as a source of contamination and may enhance both the survival and growth of the pathogenic bacteria on fresh fruits and vegetables. Contaminated harvesting equipment, containers and poor hygiene of farm workers are important factors that influence the cross-contamination of fresh produce (Buchholz et al., 2012; Matthews, 2013; Todd et al., 2008).

The postharvest unit operations (coring, cutting, slicing, peeling, shredding) and even the washing/sanitizing steps cause tissue damage and render the produce more prone to bacterial contamination. The destruction of tissue and cells results in the release of nutrient reach exudates which facilitates both the attachment and the rapid proliferation of pathogenic bacteria. The promotion of the rapid growth of *E. coli* O157:H7 as a result of tissue damage in lettuce has been shown by Brandl (2008). Cut leaf surfaces represent a specific target for the attachment of *Salmonella* cells (Kroupitski et al., 2009b). The use of inadequately sanitized water for cooling or washing/sanitizing operations can represent a source of cross-contamination and can serve as the primary vehicle for the spread of pathogenic bacteria during fresh produce processing (Pérez-Rodríguez et al., 2014).

FOODBORNE OUTBREAKS ASSOCIATED WITH PATHOGENIC BACTERIA IN FRESH-CUT FRUITS AND VEGETABLES

Summaries of the foodborne outbreaks associated with pathogenic bacteria in fresh fruits and vegetables in Europe (2007–2011) and in the United States (2006–2015) are given in Tables 9.1 and 9.2, respectively. In both regions, most of the fresh-produce-associated outbreaks were linked with *Salmonella* spp., followed by pathogenic *E. coli.* In fact, in all the fresh-produce-associated foodborne outbreaks viruses rank first followed by *Salmonella* and *E. coli* O157:H7 as the causative agent for both regions (CDC, 2015; EFSA, 2013). The other pathogenic bacteria identified as the causative agent in fresh-produce-associated outbreaks in the given periods include *Shigella* spp., *Yersinia* spp., *Bacillus* spp., *Clostridium perfringens*, and *Staphylococcus aureus*. Sprouted seeds and leafy greens were the main food vehicles in most of the fresh-produce-associated outbreaks. The use of contaminated seeds was determined to be the major cause of sprout-associated outbreaks (EFSA, 2011a; Emberland et al., 2007; Werner et al., 2007).

In the reported periods, *Salmonella* was the most common bacterial agent, accounting for 71% of the outbreaks associated with bacteria in Europe, and 60% in the United States (Tables 9.1 and 9.2). Moreover, 86% of the human cases in the selected multistate outbreaks in the United States and 31% of the human cases in Europe were linked to *Salmonella* contamination. The multistate outbreak associated with *Salmonella* Saintpaul contaminated cantaloupes (Munnoch et al., 2009) and *Salmonella* spp. contaminated alfalfa sprouts (Compton et al., 2008) in Australia are other examples of *Salmonella* spp. linked fresh produce outbreaks with more than 200 human cases. A wide variety of fresh produce has been associated with *Salmonella*-linked outbreaks, sprouts, leafy greens, cucumbers, and melons being the major ones. The higher ratio of *Salmonella*-associated fresh produce outbreaks is partly linked to the ability of *Salmonella* to internalize within the plant and proliferate to high numbers under favorable conditions (Golberg et al., 2011; Schikora et al., 2008).

Pathogenic *E. coli* was the second major causative agent of the bacteria-associated outbreaks mainly linked to the consumption of contaminated sprouts and leafy greens for the last decade (Tables 9.1 and 9.2). *E. coli* O157:H7 was

Table 9.1 Summary of Strong Evidence Foodborne Outbreaks Associated with Pathogenic Bacteria in Fresh Fruits and Vegetables in Europe (2007–2011)

Food Vehicle	Causative Agent	Number of Outbreaks	Human Cases	Hospitalizations/ Deaths
Leafy greens	*Salmonella* spp.	7	438	29/0
Sprouted seeds	*Salmonella* spp.	11	521	76/1
Watermelon	*Salmonella* Newport	1	17	8/0
Tomatoes	*Salmonella* Strathcona	1	43	0/0
Baby spinach	*Salmonella* Java and others	2	189	0/0
Lettuce	*Salmonella* spp. (*S.* Enteritidis, *S.* Newport, *S.* Napoli, *S.* Java)	5	249	31/0
Bean sprouts	*Salmonella* spp. (*S.* Weltevreden, *S.* Bareilly, *S.* Kottbus)	4	275	35/1
Alfalfa sprouts	*Salmonella* spp. (*S.* Weltevreden, *S.* Stanley, *S.* Bovismorbificans)	4	114	3/0
Mung bean sprouts	*S.* Newport	2	126	35/0
Fenugreek seed sprouts	*E. coli* O104:H4	1	3830	2381/54
Spices and herbs	*Bacillus* spp.	7	343	0/0
Lettuce	*Bacillus cereus*	1	2	0/0
Fresh basil	*S. sonnei*	1	46	4/0
Carrots	*Shigella sonnei*	1	145	5/0
Carrots	*Yersinia pseudotuberculosis*	1	50	10/0
Mixed fresh herbs	*Clostridium perfringens*	1	2	NR/NR
Bean sprouts	*S. aureus*	1	42	NR/NR

Data obtained from EFSA (2013).

the most prevalent strain mostly associated with leafy greens, especially lettuce, and sprouts. Lettuce was the food vehicle most commonly linked to *E. coli*-associated outbreaks in the United States, causing eight of the 46 total *E. coli* outbreaks between 2004 and 2012 (Callejón et al., 2015). The high percentage of enterohemorrhagic *E. coli* (EHEC) and *Salmonella*-associated outbreaks in the United States during the last decade can partly be explained by the increased

Table 9.2 Summary of Strong Evidence Foodborne Outbreaks Associated with Pathogenic Bacteria in Fresh Fruits and Vegetables in Selected Multistate Outbreaks in the United States (2006–2015)

Food Vehicle	Causative Agent	Number of Outbreaks	Human Cases	Hospitalizations/ Deaths
Cucumber	*Salmonella* spp. (*S.* Newport, *S.* Saintpaul *S. Poona)*	3	1196	230/1
Alfalfa sprouts	*Salmonella* spp. (*S.* Saintpaul *S.* Newport, *S.* serotype I 4, *S.* Enteritidis)	4	444	50/0
Bean sprouts	*Salmonella* Enteritidis	1	115	28/0
Tomatoes	*Salmonella* Typhimurium	1	183	22/0
Jalapeno and serrano peppers	*Salmonella* Saintpaul	1	1442	286/2
Mangoes	*Salmonella* Braenderup	1	127	33/0
Cantaloupes	*Salmonella* spp. (*S.* Typhimurium *S.* Litchfield)	2	212	110/3
Papayas	*Salmonella* Agona	1	106	10/0
Frozen mamey fruit pulp	*Salmonella* Typhi	1	9	7/0
Ready-to-eat salads	*E. coli* O157:H7	1	33	7/0
Spinach and spring mix blend	*E. coli* O157:H7	1	33	13/0
Raw clover sprouts	*E. coli* O121, *E. coli* O26	2	48	16/0
Romaine lettuce	*E. coli* O157:H7, *E. coli* O145	2	84	45/0
Fresh spinach	*E. coli* O157:H7	1	199	102/3
Mung bean sprout	*L. monocytogenes*	1	5	5/2
Caramel apples	*L. monocytogenes*	1	35	34/7
Cantaloupes	*L. monocytogenes*	1	147	NR/33

Data obtained from CDC (2015).

prevalence of these pathogens in leafy green vegetables between 2007 and 2013 in California, where more than 70% of the US total leafy green vegetable production is done (Karp et al., 2015). The EHEC prevalence in leafy greens in California was found to increase from <0.1% of samples in 2007 to 2.5% in 2013. The worst fresh-produce-associated outbreak case during the last decade was the large outbreak of hemolytic-uremic syndrome (HUS) caused by Shiga-toxin-producing *E. coli* O104:H4 centered in Germany in 2011, due to the consumption of contaminated fenugreek seed sprouts (Grad et al., 2012). As listed in Table 9.1, infecting nearly 4000 persons, mainly in Germany, it caused more than 2300 cases of hospitalization, more than 900 cases of HUS, and resulted in 54 deaths (EFSA, 2011a,b; Grad et al., 2012). In fact, this outbreak is a clear example of how global food trade can result in the spreading of a contaminated product between countries resulting in high numbers of people being affected.

Listeria outbreaks linked to fresh produce are not as frequent as *Salmonella*- and *E. coli*-associated outbreaks and mainly affects vulnerable groups such as pregnant women, the elderly, infants, and immunocompromised individuals. Although rare, it is an important cause of human illness, especially in the United States. In fact, although *L. monocytogenes* was identified as the causative agent in only three multistate outbreaks in the United States during the last decade, it was responsible for the highest death rate in fresh-produce-associated outbreaks (Table 9.2). The multistate outbreak of listeriosis linked to whole cantaloupes affected 147 persons in 28 states and resulted in 30 deaths and one miscarriage (Laksanalamai et al., 2012).

In general, *Bacillus* spp., *Shigella* spp., *Yersinia* spp., *Clostridium* spp., and *S. aureus* were associated with higher numbers of fresh-produce-linked outbreaks in Europe than in the United States (Tables 9.1 and 9.2) (Callejón et al., 2015). *Bacillus* spp. were linked to eight outbreaks, in seven of which spices and herbs were the food vehicle and only one was linked to lettuce. The rest of these pathogenic bacteria were associated only with one or two specific outbreak cases with much lower reported human cases.

ATTACHMENT, INTERNALIZATION, AND BIOFILM DEVELOPMENT BY FOODBORNE PATHOGENIC BACTERIA ON FRESH-CUT FRUIT AND VEGETABLES

The increase in the numbers of *Salmonella*- and *E. coli* O157:H7-associated outbreaks involving fresh produce during the last decade increased the interest in highlighting the mechanisms of interactions of these pathogens with fresh produce. Many studies revealed that both *Salmonella* and *E. coli* O157:H7 can attach to the surface and internalize within the tissue of a variety of fresh produce. The extent of attachment and internalization depend on many factors including the route of contamination and internalization, the nature of the epidermis, tissue pH, the growth phase of the plant, as well as extrinsic factors imposed by the environment of the plant (Beuchat, 2002; Erickson et al., 2010a; Golberg et al., 2011; Zheng et al., 2013). Both endophytic and epiphytic colonization have been reported for foodborne pathogenic bacteria in fresh produce. The mechanisms of attachment for epiphytic colonization mainly include curli, fimbriae, flagella, and biofilm formation (Kroupitski et al., 2013; Lee et al., 2015; Neal et al., 2012; Rossez et al., 2014; Salazar et al., 2013). Fig. 9.2 summarizes the molecular mechanisms facilitating the attachment and colonization of leafy green vegetables by enteric bacterial pathogens (Martínez-Vaz et al., 2014). The first step in the contamination of fresh produce, that is, the initial attachment of pathogens to the pyhllosphere or rhizosphere tissue, includes cell surface structures (curli, fimbriae, and flagella), followed by biofilm formation.

Functional genomic studies revealed that the genes involved in cell surface structures, virulence, motility, and biofilm formation mediate the plant–bacteria interactions. The attachment of *E. coli* O157:H7 to leafy greens involves multiple mechanisms influenced by the leaf properties, damage to the leaf, and water hardness (Lee et al., 2015). Compared with cellulose-deficient cells, cellulose producing *E. coli* O157:H7 attach more to lettuce surface, but not to spinach. The level of colanic-acid-producing cells changes depending on the water hardness, being greater in higher water hardness (Lee et al., 2015). The genes involved in curli production (*csgA* and *csgB*) and those involved in biofilm

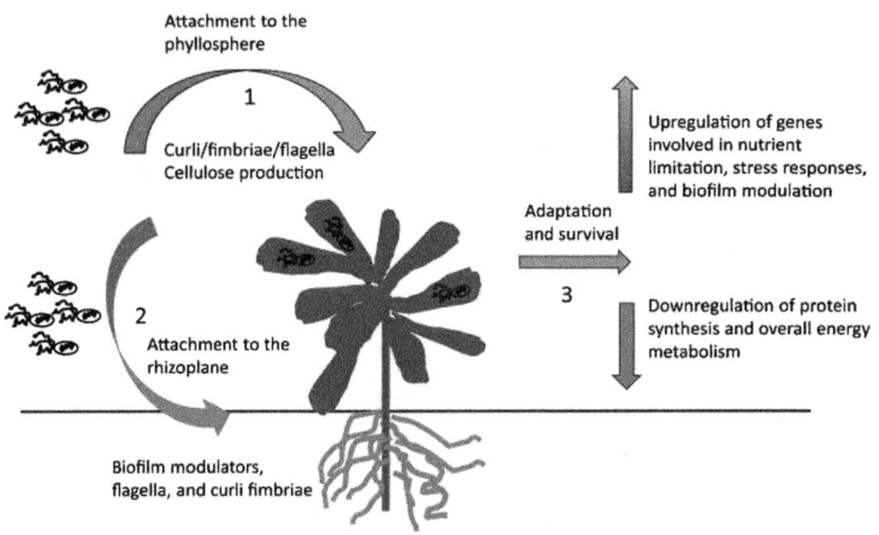

FIGURE 9.2

Molecular mechanisms mediating the colonization of leafy greens by human pathogens (Martínez-Vaz et al., 2014).

formation (*bbsA* and *ybiM*) are upregulated in *E. coli* O157:H7 upon interaction with intact lettuce leaf surfaces (Fink et al., 2012). The expression of the virulance factor O-antigen in *E. coli* O157:H7 enhances attachment to iceberg lettuce, possibly by altering the hydrophobicity and cell surface charge (Boyer et al., 2011). Curli producing *E. coli* O157:H7 attach more strongly to spinach and lettuce leaf surfaces (Macarisin et al., 2012; Patel et al., 2011). On the other hand, cellulose, a constituent of extracellular matrix, was found to be dispensable for the attachment of *E. coli* O157:H7 to spinach (Macarisin et al., 2012). Similarly, the deletion of the curli expressing *agfBA* gene was more effective on the transfer and survival of *S.* Typhimurium on parsley, compared to the deletion of cellulose expressing *bcsA* gene (Lapidot and Yaron, 2009). Moreover, the putative stress regulatory gene, *ycfR*, and the *sirA* gene involved in biofilm formation in *S.* Enterica are essential for attachment to intact spinach leaf and grape tomatoes, and survival during postharvest minimal processing (Salazar et al., 2013).

Several studies showed that enteric pathogens interact with salad leaves by mechanisms that frequently involve flagella. Flagellum is the major adhesin linking enterotoxigenic *E. coli* (Shaw et al., 2011) and EHEC (Xicohtencatl-Cortes et al., 2009) to lettuce, basil, and spinach leaves. Moreover, in enteroaggregative *E. coli*, the binding to the guard cell of stomata is flagella-mediated, whereas AFF (aggregative adherence fimbriae) pilus mediates the binding to the epidermis (Berger et al., 2009b). In fact, *Salmonella* serovars use strain-specific mechanisms to attach to salad leaves. Unlike *S.* Typhimurium, the attachment of *S.* Seftenberg to basil, lettuce, rocket, and spinach leaves is mainly mediated via flagella (Berger et al., 2009a).

The occurrence of naturally formed microbial biofilms on fresh produce (alfalfa, clover, broccoli, sunflower sprouts, lettuce) have been demonstrated in previous studies (Fig. 9.3) (Fett, 2000; Olmez and Temur, 2010; Rayner et al., 2004).

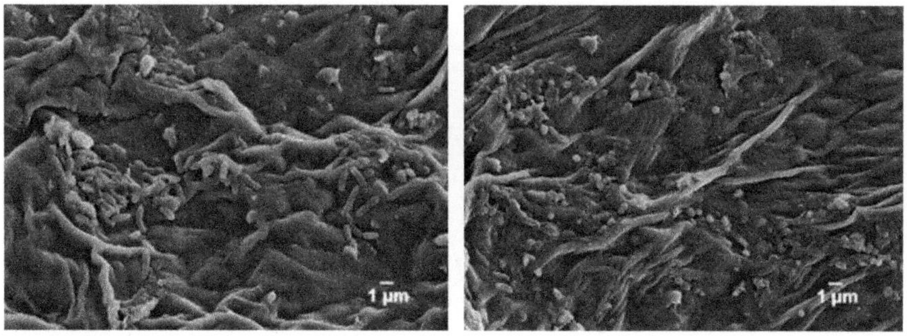

FIGURE 9.3

Scanning electron micrograph images of natural biofilms of mixed microbial populations on the surface of raw lettuce leaves stored overnight below 10°C after harvest (Olmez and Temur, 2010).

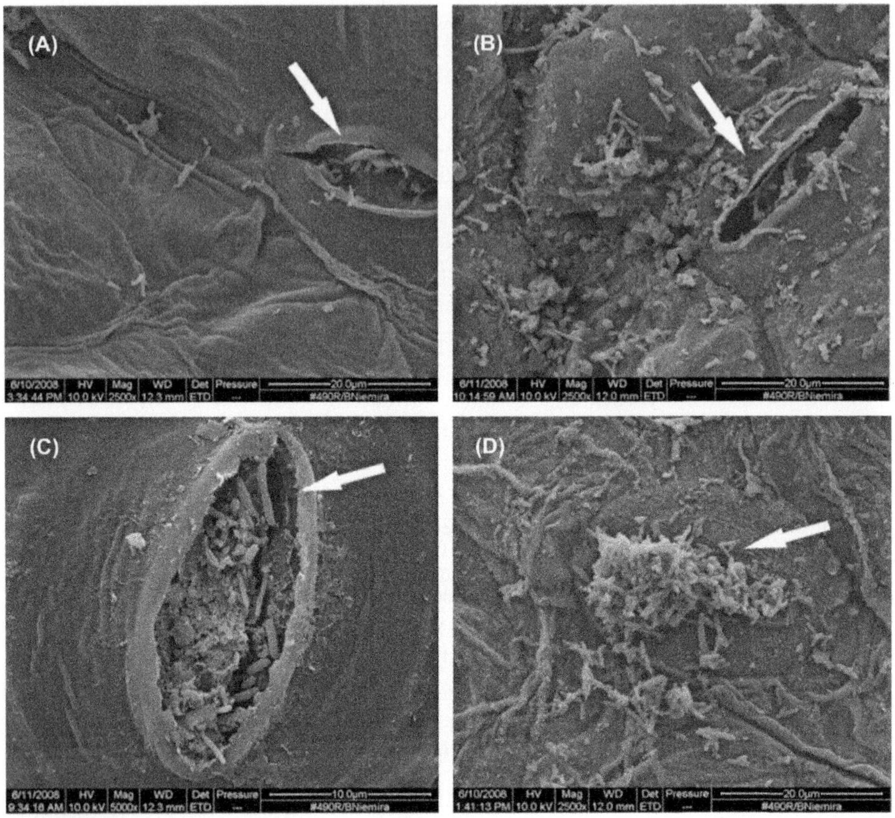

FIGURE 9.4

Scanning electron micrographs of stomata on baby spinach leaves inoculated with *E. coli* O157:H7 and stored at 4°C for 0 h (A), 24 h (B), 48 h (C), or 72 h (D). Intrusion and development of cell populations peripheral to and within the stomatal wells are indicated by arrows (Niemira and Cooke, 2010).

Artificially inoculated foodborne pathogenic bacteria have also been shown to form biofilms on fresh produce surfaces (Fig. 9.4) (Annous et al., 2005; Lapidot et al., 2006; Niemira and Cooke, 2010; Olmez and Temur, 2010). The biofilm-forming ability of bacterial pathogens is highly correlated with the attachment strength and persistence on the fresh produce surface (Kroupitski et al., 2009b; Patel et al., 2013). Solomon et al. (2005) found that curli- and cellulose-deficient *S.* Enterica isolates are low biofilm formers, supporting the correlation between attachment strength and biofilm formation.

The endophytic colonization occurs through natural openings (stomata), damaged tissue of rhizosphere or phyllosphere, and chemotaxis to metabolites found in plant exudates or within the plant (Fig. 9.4) (Bartz et al., 2015; Critzer and Doyle, 2010; Erickson et al., 2010a,b; Golberg et al., 2011; Klerks et al., 2007; Kroupitski et al., 2009a; Neal et al., 2012). Different mechanisms are responsible for the internalization of pathogens within the plant tissue during the pre- and postharvest stages. At the preharvest level, the serovar–cultivar interaction has a significant effect on the type of colonization, whether it will be epiphytic or endophytic, as well as on the level of colonization. Chemotaxis studies revealed that *S. enterica* serovars actively move toward root exudates of a specific lettuce cultivar (Tamburo), as the genes involved in chemotaxis and virulence are activated by the root exudates of this cultivar (Klerks et al., 2007). They also reported that among the five *Salmonella enterica* serovars only *S. enterica* serovar Dublin was able to colonize endophytically. The method, duration, and direction of irrigation have an important effect on the degree of internalization and persistence. Internalization of *E. coli* O157:H7 accompanied with higher persistence occurred in lettuce leaves spray-irrigated on the abaxial side compared to the leaves irrigated on the adaxial sides (Erickson et al., 2010b). At the postharvest level, washing, vacuum cooling, and hydrocooling processes can present an opportunity for the internalization of pathogenic bacteria at various infiltration sites (stomata, stem scar, cut edges, etc.) (Li et al., 2008). Infiltration is enhanced when using water cooler than the produce, applying long immersion periods, using surfactants, and with the presence of cut edges and damaged tissue (Beuchat, 2002; Bordini et al., 2007; Eblen et al., 2004; Erickson, 2012; Fatemi et al., 2006; Janes et al., 2005). The incidence of *Salmonella* internalization varies highly depending on the type of leafy greens, being highest in iceberg lettuce and arugula leaves, moderate in basil and lettuce, and lowest in parsley and tomato leaves (Golberg et al., 2011).

CHALLENGES IN FRESH-CUT FRUIT AND VEGETABLE SAFETY

Pathogenic bacteria associated with fresh-cut fruits and vegetables are able to proliferate to high numbers once deposited onto the injured parts of fruits and vegetables. *E. coli* O157:H7 was shown to grow 11-fold over 4 h of incubation after inoculation onto cut lettuce stems (Brandl, 2008). Harvesting and postharvest operation may cause significant damage to the tissue of fruits and vegetables, which can promote the growth of pathogenic bacteria. Moreover, microarray-based whole-genome transcriptional profiling studies revealed that genes involved in *E. coli* O157:H7 attachment, virulence, DNA repair, resistance to oxidative stress, and resistance to antimicrobials are upregulated when the cells are exposed to the exudates of the damaged plant tissue (Kyle et al., 2010). The enhancement of DNA repair ability, resistance to oxidative stress and antimicrobials also increase the ability of the enteric pathogens to evade the current sanitization treatments, as the fresh-cut industry mainly relies on the use of oxidative agents for the sanitization of produce and wash waters (Martínez-Vaz et al., 2014).

This may be one of the main reasons for increasing numbers of enteric pathogenic bacteria-associated outbreaks involving fresh-cut produce and represents a critical challenge for the fresh produce industry. Attachment of bacterial cells to inaccessible sites, like stomata and cavities of the leaf, are also likely responsible for the inefficacy of sanitizing treatments and represents another challenge for the fresh-cut industry since none of the current methods is able to reach subsurface cells. Therefore, there is a need for the development of produce and wash water sanitization methods able to inactivate internalized bacteria and which are based on inactivation mechanisms that can overcome the issue of increased resistance to oxidative stress and improved DNA repair ability.

REFERENCES

Annous, B.A., Solomon, E.B., Cooke, P.H., Burke, A., 2005. Biofilm formation by *Salmonella* spp. on cantaloupe melons. J. Food Safety 25, 276–287.

Aruscavage, D., Lee, K., Miller, S., LeJeune, J.T., 2006. Interactions affecting the proliferation and control of human pathogens on edible plants. J. Food. Sci. 71, R89–R99.

Bartz, J.A., Yuk, H.-G., Mahovic, M.J., Warren, B.R., Sreedharan, A., Schneider, K.R., 2015. Internalization of *Salmonella enterica* by tomato fruit. Food Contr. 55, 141–150.

Berger, C.N., Shaw, R.K., Brown, D.J., Mather, H., Clare, S., Dougan, G., et al., 2009a. Interaction of *Salmonella enterica* with basil and other salad leaves. ISME J. 3, 261–265.

Berger, C.N., Shaw, R.K., Ruiz-Perez, F., Nataro, J.P., Henderson, I.R., Pallen, M.J., et al., 2009b. Interaction of enteroaggregative Escherichia coli with salad leaves. Environ. Microbiol. Rep. 1, 234–239.

Beuchat, L.R., 1999. Survival of enterohemorrhagic *Escherichia coli* O157: H7 in bovine feces applied to lettuce and the effectiveness of chlorinated water as a disinfectant. J. Food. Prot. 62, 845–849.

Beuchat, L.R., 2002. Ecological factors influencing survival and growth of human pathogens on raw fruits and vegetables. Microbes Infect. 4, 413–423.

Bordini, M.E.B., Ristori, C.A., Jakabi, M., Gelli, D.S., 2007. Incidence, internalization and behavior of *Salmonella* in mangoes, var. Tommy Atkins. Food Contr. 18, 1002–1007.

Boyer, R.R., Sumner, S.S., Williams, R.C., Kniel, K.E., McKinney, J.M., 2011. Role of O-antigen on the *Escherichia coli* O157:H7 cells hydrophobicity, charge and ability to attach to lettuce. Int. J. Food. Microbiol. 147, 228–232.

Brandl, M., 2008. Plant lesions promote the rapid multiplication of *Escherichia coli* O157: H7 on postharvest lettuce. Appl. Environ. Microbiol. 74, 5285–5289.

Buchholz, A.L., Davidson, G.R., Marks, B.P., Todd, E.C., Ryser, E.T., 2012. Quantitative transfer of *Escherichia coli* O157: H7 to equipment during small-scale production of fresh-cut leafy greens. J. Food. Prot. 75, 1184–1197.

Callejón, R.M., Rodríguez-Naranjo, M.I., Ubeda, C., Hornedo-Ortega, R., Garcia-Parrilla, M.C., Troncoso, A.M., 2015. Reported foodborne outbreaks due to fresh produce in the United States and European Union: trends and causes. Foodborne Pathog. Dis. 12, 32–38.

CDC, 2015. Center for Disease Control and Prevention (CDC) online database. <http://www.cdc.gov/foodsafety/outbreaks/multistate-outbreaks/outbreaks-list.html> (accessed 25.11.2015).

Cevallos-Cevallos, J.M., Danyluk, M.D., Gu, G., Vallad, G.E., van Bruggen, A.H., 2012. Dispersal of *Salmonella* Typhimurium by rain splash onto tomato plants. J. Food. Prot. 75, 472–479.

Compton, J.A., Baney, J.A., Donaldson, S.C., Houser, B.A., San Julian, G.J., Yahner, R.H., et al., 2008. *Salmonella* infections in the common raccoon (*Procyon lotor*) in western Pennsylvania. J. Clin. Microbiol. 46, 3084–3086.

Critzer, F.J., Doyle, M.P., 2010. Microbial ecology of foodborne pathogens associated with produce. Curr. Opin. Biotechnol. 21, 125–130.

CSPI, 2011. A 10-Year Review of State Outbreak Reporting, 1st ed. <www.cspinet.org>. Center for Science in the Public Interest, Washington, DC 2005 (accessed 25/11/2015).

CSPI, 2014. A review of foodborne illness in America from 2002–2011. Center for Science in the Public Interest <www.cspinet.org>, pp. 1–17 (accessed 25/11/2015).

Eblen, B.S., Walderhaug, M.O., Edelson-Mammel, S., Chirtel, S.J., De Jesus, A., Merker, R.I., et al., 2004. Potential for internalization, growth, and survival of *Salmonella* and *Escherichia coli* O157: H7 in oranges. J. Food. Prot. 67, 1578–1584.

EFSA, 2011a. Scientific Report of EFSA and ECDC. The European Union Summary Report on Trends and Sources of Zoonoses, Zoonotic Agents and Food-borne Outbreaks in 2011. EFSA J.3129.

EFSA, 2011b. Tracing seeds, in particular fenugreek (*Trigonella foenum-graecum*) seeds, in relation to the Shiga toxin-producing *E. coli* (STEC) O104:H4 2011 outbreaks in Germany and France.

EFSA, 2013. FSA Panel on Biological Hazards (BIOHAZ) Panel; Scientific Opinion on the risk posed by pathogens in food of non-animal origin. Part 1 (outbreak data analysis and risk ranking of food/pathogen combinations). [138 pp.] <http://dx.doi.org/10.2903/j.efsa.2013.3025>.

Emberland, K., Ethelberg, S., Kuusi, M., Vold, L., Jensvoll, L., Lindstedt, B., et al., 2007. Outbreak of *Salmonella* Weltevreden infections in Norway. Euro. Surveill. 12, 4.

Erickson, M.C., 2012. Internalization of fresh produce by foodborne pathogens. Annual Rev. Food Sci. Technol. 3, 283–310.

Erickson, M.C., Liao, J., Payton, A.S., Riley, D.G., Webb, C.C., Davey, L.E., et al., 2010a. Preharvest internalization of *Escherichia coli* O157: H7 into lettuce leaves, as affected by insect and physical damage. J. Food. Prot. 73, 1809–1816.

Erickson, M.C., Webb, C.C., Davey, L.E., Payton, A.S., Flitcroft, I.D., Doyle, M.P., 2014. Internalization and fate of *Escherichia coli* O157: H7 in leafy green phyllosphere tissue using various spray conditions. J. Food. Prot. 77, 713–721.

Erickson, M.C., Webb, C.C., Diaz-Perez, J.C., Phatak, S.C., Silvoy, J.J., Davey, L., et al., 2010b. Surface and internalized *Escherichia coli* O157: H7 on field-grown spinach and lettuce treated with spray-contaminated irrigation water. J. Food. Prot. 73, 1023–1029.

Fatemi, P., LaBorde, L.F., Patton, J., Sapers, G.M., Annous, B., Knabel, S.J., 2006. Influence of punctures, cuts, and surface morphologies of Golden Delicious apples on penetration and growth of *Escherichia coli* O157: H7. J. Food. Prot. 69, 267–275.

Fett, W.F., 2000. Naturally occurring biofilms on alfalfa and other types of sprouts. J. Food. Prot. 63, 625–632.

Fink, R.C., Black, E.P., Hou, Z., Sugawara, M., Sadowsky, M.J., Diez-Gonzalez, F., 2012. Transcriptional responses of *Escherichia coli* K-12 and O157:H7 associated with lettuce leaves. Appl. Environ. Microbiol. 78, 1752–1764.

Fonseca, J., Fallon, S., Sanchez, C., Nolte, K., 2011. *Escherichia coli* survival in lettuce fields following its introduction through different irrigation systems. J. Appl. Microbiol. 110, 893–902.

Golberg, D., Kroupitski, Y., Belausov, E., Pinto, R., Sela, S., 2011. *Salmonella* Typhimurium internalization is variable in leafy vegetables and fresh herbs. Int. J. Food. Microbiol. 145, 250–257.

Grad, Y.H., Lipsitch, M., Feldgarden, M., Arachchi, H.M., Cerqueira, G.C., FitzGerald, M., et al., 2012. Genomic epidemiology of the *Escherichia coli* O104: H4 outbreaks in Europe, 2011. Proc. Natl. Academy Sci. 109, 3065–3070.

Holley, R.A., Arrus, K.M., Ominski, K.H., Tenuta, M., Blank, G., 2006. Survival in manure-treated soils during simulated seasonal temperature exposure. J. Environ. Qual. 35, 1170–1180.

Janes, M.E., Kim, K., Johnson, M., 2005. Transmission electron microscopy study of enterohemorrhagic *Escherichia coli* O157: H7 in apple tissue. J. Food. Prot. 68, 216–224.

Karp, D.S., Gennet, S., Kilonzo, C., Partyka, M., Chaumont, N., Atwill, E.R., et al., 2015. Comanaging fresh produce for nature conservation and food safety. Proc. Natl. Academy Sci. 112, 11126–11131.

Klerks, M.M., Franz, E., van Gent-Pelzer, M., Zijlstra, C., van Bruggen, A.H., 2007. Differential interaction of *Salmonella enterica* serovars with lettuce cultivars and plant-microbe factors influencing the colonization efficiency. ISME J. 1, 620–631.

Kroupitski, Y., Brandi, M.T., Pinto, R., Belausov, E., Tamir-Ariel, D., Burdman, S., et al., 2013. Identification of *Salmonella enterica* genes with a role in persistence on lettuce leaves during cold storage by recombinase-based in vivo expression technology. Phytopathology 103, 362–372.

Kroupitski, Y., Golberg, D., Belausov, E., Pinto, R., Swartzberg, D., Granot, D., et al., 2009a. Internalization of *Salmonella enterica* in leaves is induced by light and involves chemotaxis and penetration through open stomata. Appl. Environ. Microbiol. 75, 6076–6086.

Kroupitski, Y., Pinto, R., Brandl, M., Belausov, E., Sela, S., 2009b. Interactions of *Salmonella enterica* with lettuce leaves. J. Appl. Microbiol. 106, 1876–1885.

Kyle, J.L., Parker, C.T., Goudeau, D., Brandl, M.T., 2010. Transcriptome analysis of *Escherichia coli* O157: H7 exposed to lysates of lettuce leaves. Appl. Environ. Microbiol. 76, 1375–1387.

Laksanalamai, P., Joseph, L.A., Silk, B.J., Burall, L.S., Tarr, C.L., Gerner-Smidt, P., et al., 2012. Genomic characterization of *Listeria monocytogenes* strains involved in a multistate listeriosis outbreak associated with cantaloupe in US. PLoS One 7, e42448.

Lang, N., Smith, S., 2007. Influence of soil type, moisture content and biosolids application on the fate of *Escherichia coli* in agricultural soil under controlled laboratory conditions. J. Appl. Microbiol. 103, 2122–2131.

Lapidot, A., Romling, U., Yaron, S., 2006. Biofilm formation and the survival of *Salmonella* Typhimurium on parsley. Int. J. Food. Microbiol. 109, 229–233.

Lapidot, A., Yaron, S., 2009. Transfer of *Salmonella enterica* Serovar Typhimurium from contaminated irrigation water to parsley is dependent on curli and cellulose, the biofilm matrix components. J. Food. Prot. 72, 618–623.

Lee, C.-C., Chen, J., Frank, J.F., 2015. Role of cellulose and colanic acid in attachment of shiga toxin–producing *Escherichia coli* to lettuce and spinach in different water hardness environments. J. Food. Prot. 78, 1461–1466.

Li, H., Tajkarimi, M., Osburn, B.I., 2008. Impact of vacuum cooling on *Escherichia coli* O157: H7 infiltration into lettuce tissue. Appl. Environ. Microbiol. 74, 3138–3142.

Macarisin, D., Patel, J., Bauchan, G., Giron, J.A., Sharma, V.K., 2012. Role of curli and cellulose expression in adherence of *Escherichia coli* O157:H7 to spinach leaves. Foodborne Pathog. Dis. 9, 160–167.

Martínez-Vaz, B.M., Fink, R.C., Diez-Gonzalez, F., Sadowsky, M.J., 2014. Enteric pathogen–plant interactions: molecular connections leading to colonization and growth and implications for food safety. Microbes Environ. 29, 123.

Matthews, K.R., 2013. Sources of enteric pathogen contamination of fruits and vegetables: future directions of research. Stewart Postharvest Review 9, 1–5.

Munnoch, S., Ward, K., Sheridan, S., Fitzsimmons, G., Shadbolt, C., Piispanen, J., et al., 2009. A multi-state outbreak of *Salmonella* Saintpaul in Australia associated with cantaloupe consumption. Epidemiol. Infect. 137, 367–374.

Neal, J., Cabrera-Diaz, E., Castillo, A., 2012. Attachment of *Escherichia coli* O157: H7 and *Salmonella* on spinach (*Spinacia oleracea*) using confocal microscopy. Agric. Food Anal. Bacteriol. 2, 275–279.

Nicholson, F.A., Groves, S.J., Chambers, B.J., 2005. Pathogen survival during livestock manure storage and following land application. Bioresour. Technol. 96, 135–143.

Niemira, B.A., Cooke, P.H., 2010. *Escherichia coli* O157:H7 biofilm formation on romaine lettuce and spinach leaf surfaces reduces efficacy of irradiation and sodium hypochlorite washes. J. Food. Sci. 75, M270–M277.

Olmez, H., Temur, S.D., 2010. Effects of different sanitizing treatments on biofilms and attachment of *Escherichia coli* and *Listeria monocytogenes* on green leaf lettuce. LWT—Food Sci. Technol. 43, 964–970.

Painter, J.A., Hoekstra, R.M., Ayers, T., Tauxe, R.V., Braden, C.R., Angulo, F.J., et al., 2013. Attribution of foodborne illnesses, hospitalizations, and deaths to food commodities by using outbreak data, United States, 1998–2008. Emerg. Infect. Dis. 19, 407–415.

Park, S., Szonyi, B., Gautam, R., Nightingale, K., Anciso, J., Ivanek, R., 2012. Risk factors for microbial contamination in fruits and vegetables at the preharvest level: a systematic review. J. Food. Prot. 75, 2055–2081.

Patel, J., Sharma, M., Ravishakar, S., 2011. Effect of curli expression and hydrophobicity of *Escherichia coli* O157:H7 on attachment to fresh produce surfaces. J. Appl. Microbiol. 110, 737–745.

Patel, J., Singh, M., Macarisin, D., Sharma, M., Shelton, D., 2013. Differences in biofilm formation of produce and poultry *Salmonella enterica* isolates and their persistence on spinach plants. Food Microbiol. 36, 388–394.

Pérez-Rodríguez, F., Saiz-Abajo, M.J., Garcia-Gimeno, R.M., Moreno, A., González, D., Vitas, A.I., 2014. Quantitative assessment of the *Salmonella* distribution on fresh-cut leafy vegetables due to cross-contamination occurred in an industrial process simulated at laboratory scale. Int. J. Food. Microbiol. 184, 86–91.

Rayner, J., Veeh, R., Flood, J., 2004. Prevalence of microbial biofilms on selected fresh produce and household surfaces. Int. J. Food. Microbiol. 95, 29–39.

Rossez, Y., Holmes, A., Wolfson, E.B., Gally, D.L., Mahajan, A., Pedersen, H.L., et al., 2014. Flagella interact with ionic plant lipids to mediate adherence of pathogenic *Escherichia coli* to fresh produce plants. Environ. Microbiol. 16, 2181–2195.

Salazar, J.K., Deng, K., Tortorello, M.L., Brandl, M.T., Wang, H., Zhang, W., 2013. Genes ycfR, sirA and yigG contribute to the surface attachment of *Salmonella enterica* Typhimurium and Saintpaul to fresh produce. PLoS One. 8, e57272.

Schikora, A., Carreri, A., Charpentier, E., Hirt, H., 2008. The dark side of the salad: *Salmonella* Typhimurium overcomes the innate immune response of *Arabidopsis thaliana* and shows an endopathogenic lifestyle. PLoS One. 3, e2279.

Semenov, A.V., Van Bruggen, A.H., Van Overbeek, L., Termorshuizen, A.J., Semenov, A. M., 2007. Influence of temperature fluctuations on *Escherichia coli* O157: H7 and *Salmonella enterica* serovar Typhimurium in cow manure. FEMS. Microbiol. Ecol. 60, 419–428.

Shaw, R.K., Berger, C.N., Pallen, M.J., Sjoeling, A., Frankel, G., 2011. Flagella mediate attachment of enterotoxigenic Escherichia coli to fresh salad leaves. Environ. Microbiol. Rep. 3, 112–117.

Sivapalasingam, S., Friedman, C.R., Cohen, L., Tauxe, R.V., 2004. Fresh produce: a growing cause of outbreaks of foodborne illness in the United States, 1973 through 1997. J. Food. Prot. 67, 2342–2353.

Solomon, E.B., Niemira, B.A., Sapers, G.M., Annous, B.A., 2005. Biofilm formation, cellulose production, and curli biosynthesis by *Salmonella* originating from produce, animal, and clinical sources. J. Food. Prot. 68, 906–912.

Solomon, E.B., Pang, H.-J., Matthews, K.R., 2003. Persistence of *Escherichia coli* O157: H7 on lettuce plants following spray irrigation with contaminated water. J. Food. Prot. 66, 2198–2202.

Todd, E.C., Greig, J.D., Bartleson, C.A., Michaels, B.S., 2008. Outbreaks where food workers have been implicated in the spread of foodborne disease. Part 5. Sources of contamination and pathogen excretion from infected persons. J. Food. Prot. 71, 2582–2595.

Van der Linden, I., Cottyn, B., Uyttendaele, M., Vlaemynck, G., Heyndrickx, M., Maes, M., 2013. Survival of enteric pathogens during butterhead lettuce growth: crop stage, leaf age, and irrigation. Foodborne. Pathog. Dis. 10, 485–491.

Warriner, K., Huber, A., Namvar, A., Fan, W., Dunfield, K., 2009. Recent advances in the microbial safety of fresh fruits and vegetables. Adv. Food. Nutr. Res. 57, 155–208.

Werner, S., Boman, K., Einemo, I., Erntell, M., De Jong, B., Lindqvist, A., et al., 2007. Outbreak of *Salmonella* Stanley in Sweden associated with alfalfa sprouts, July–August 2007. Euro. Surveill. 12 (42), pii. 3291.

Xicohtencatl-Cortes, J., Sánchez Chacón, E., Saldaña, Z., Freer, E., Girón, J.A., 2009. Interaction of *Escherichia coli* O157: H7 with leafy green produce. J. Food. Prot. 72, 1531–1537.

Zheng, J., Allard, S., Reynolds, S., Millner, P., Arce, G., Blodgett, R.J., et al., 2013. Colonization and internalization of *Salmonella enterica* in tomato plants. Appl. Environ. Microbiol. 79, 2494–2502.

CHAPTER

Stress adaptation of *Listeria monocytogenes* in acidic ready-to-eat products

10

Ranju Paudyal and Kimon Andreas G. Karatzas

Department of Food & Nutritional Sciences, School of Chemistry, Food & Pharmacy, University of Reading, Reading, United Kingdom

READY-TO-EAT FOODS

Ready-to-eat (RTE) foods are a large, heterogeneous category of foods that include any food (including beverages) that is normally consumed in its raw state, or those handled, processed, mixed, cooked or otherwise prepared into a form that is normally consumed without further processing (Codex Alimentarius Commission, 2004). According to another definition, RTE foods are those that go through some kind of processing and can be consumed without any further treatment such as thorough heating. These foods do not require further preparation prior to consumption, except washing, thawing, or moderate reheating (Farber et al., 1996). Consumption of these products has increased dramatically over recent decades and this trend is expected to continue, as they are designed to fit into busy lifestyles by requiring minimal preparation time.

Despite their advantages, RTE foods are associated with increased concerns over their safety. Improper handling of these foods can result in foodborne illness. Food poisoning can occur even through the presence of a few cells of pathogenic microorganisms, being able to affect individual consumers (cases) or large numbers of them (outbreaks). The vast majority of outbreaks are normally caused by RTE foods resulting in illness or even death. Foodborne pathogens commonly associated with these foods are *Salmonella*, *Staphylococcus aureus*, and *Escherichia coli* O157:H7. However, one of the most prominent is *Listeria monocytogenes*, which is the focus of this chapter and is known as the deadliest foodborne pathogen because it causes the most food-related deaths (http://ecdc.europa.eu/en/publications/Publications/EU-summary-report-zoonoses-food-borne-outbreaks-2012.pdf). The majority of these deaths as well as other cases of listeriosis occur following

P. Kotzekidou (Ed): Food Hygiene and Toxicology in Ready-to-Eat Foods.
DOI: http://dx.doi.org/10.1016/B978-0-12-801916-0.00010-8

consumption of RTE foods. A significant property of RTE foods is their ability to support or inhibit growth of such pathogens and as such they are divided into two categories: high- and low-risk RTE foods.

HIGH-RISK FOODS

High-risk foods are RTE foods that can support growth of pathogenic microorganisms which are also more likely to be implicated in food poisoning cases. They are usually high in protein content, require strict temperature control and protection from contamination and include:

1. Cooked meat and poultry such as beef, pork, ham, lamb, chicken, turkey, duck.
2. Cooked meat products such as meat pies and pasties, paté, meat stock and gravy, cook-chill meals.
3. Dairy products such as milk, cream, artificial cream, custards, products containing unpasteurized milk, ripened soft and molded cheeses.
4. Egg products such as cooked eggs, quiche and products containing uncooked or lightly cooked eggs, mayonnaise, mousse, homemade ice cream.
5. Shellfish and other seafoods such as mussels, cockles, cooked prawns, raw oysters.
6. Farinaceous dishes including cooked rice, pasta, couscous.

LOW-RISK FOODS

Low-risk foods are ambient-stable, such as bread, biscuits, cereals, crisps, and cakes (not cream cakes). Such foods are unlikely to be implicated in food poisoning and include:

1. Foods that have been preserved, for example, smoked or salted fish.
2. Dry goods, those that contain minimal amounts of moisture, such as bread, flour, biscuits.
3. Acidic foods, for example, pickled foods, vinegar, and fruit.
4. Fermented products such as salami, pepperoni.
5. Foods with high sugar/fat content, for example, jam or chocolate.
6. Canned food, whilst unopened.

(https://www2.warwick.ac.uk/services/healthsafetywellbeing/guidance/foodhygieneandsafety/highandlowriskfoods).

ACIDIC FOODS

According to US Food and Drug Administration (FDA) CFR 2, acidic foods are those that have a pH of 4.6 or below. "Low-acid foods" are foods, other than alcoholic beverages, with a final equilibrium pH greater than 4.6 and a water activity (a_w) greater than 0.85. "Acidified foods" are low-acid foods to which acid(s) or acid food(s) are added. These products might be called "pickles" or "pickled" and include beans, cucumbers, cabbage, artichokes, cauliflower, puddings, peppers, tropical fruits, fish, vinegar, fruits, cheeses, and other fermented products such as tempeh, miso, sauerkraut, kefir, kombucha, and kimchi. Although most bacteria are unable to survive in acid-containing foods, bacteria like *Salmonella*, *E. coli*, *Clostridium botulinum*, *Clostridium perfringens*, and *Listeria monocytogenes* can survive in some of them.

LISTERIA MONOCYTOGENES

As mentioned above *L. monocytogenes* is one of the most important foodborne pathogens associated with RTE foods. It is a Gram-positive, rod-shaped, facultative anaerobic, bacterium that can be isolated from soil, water, animal feed, feces, and tissues of infected invertebrates and vertebrate animals, including humans (Cooper and Walker, 1998). It is catalase-positive, oxidase-negative, and does not produce any spores and capsules (Bubert et al., 1997; Swartz et al., 1991). The genus *Listeria* is known to contain 10 recognized species, which are *L. monocytogenes*, *L. ivanovii*, *L. innocua*, *L. welshimeri*, *L. seeligeri*, *L. grayi* (McLauchlin and Rees, 2009), *L. marthii* (Graves et al., 2010), *L. weihenstephanensis* (Halter et al., 2012), *L. rocourtiiae* (Leclercq et al., 2010), *L. fleischmannii* (Bertsch et al., 2013). However, lately new species such as *L. floridensis*, *L. aquatica*, *L. cornellensis*, *L. riparia*, and *L. grandensis* have been added to the list (den Bakker et al., 2014). From all these species, *L. monocytogenes* is the only human pathogen, whereas *L. ivanovii* is an animal pathogen mainly affecting cattle and sheep.

Listeria monocytogenes affects humans through consumption of contaminated food and can result in the development of listeriosis. Listeriosis mainly affects immunocompromised individuals, the elderly, pregnant women, and newborn infants, and is characterized by mortality rates as high as 30–40% (Siegman-Igra et al., 2002). Following the consumption of contaminated food, *L. monocytogenes* can survive in the stomach due to its increased acid resistance and in the duodenum due its increased bile tolerance (Begley et al., 2005). Once in the gut, anaerobic conditions trigger the production of internalins such as InlAB (Jydegaard-Axelsen et al., 2004) that facilitate the invasion of *L. monocytogenes* into the intestinal epithelial cells initiating the intracellular stage of the infection which involves phagocytosis, escape from the phagocytic vacuole, and spread from cell to cell to different parts of the body leading to listeriosis (Vázquez-Boland et al., 2001). Symptoms can vary depending on the organs affected by the infection, although it is possible that the intestinal stage of infection is asymptomatic.

LISTERIA MONOCYTOGENES IN RTE FOODS

Despite its very low annual incidence (0.26 cases per 100,000 population in the United States; http://www.cdc.gov/listeria/statistics.html), listeriosis is responsible for 30% of all food pathogen-related deaths in the United States, making it the single most deadly food-related disease and therefore a serious public health concern.

This is a result of the high mortality rate (30%) of this pathogen. The incidence, mortality rates, and other related statistical figures are similar in the EU. Numerous listeriosis outbreaks have been associated with raw products, unpasteurized milk, cheese, cantaloupes, cold cuts, hot dogs, poultry, etc. The first known outbreak of foodborne listeriosis occurred in patients of Boston hospital in 1979, following consumption of contaminated vegetables that were prepared within the hospital. Another outbreak in Canada (1981) occurred after consumption of coleslaw contaminated with *L. monocytogenes*, and for the first time its foodborne transmission was demonstrated based on the similarity of the isolates that caused the outbreak and those found in the coleslaw. The foods associated with outbreaks have varied from the early 1980s to the mid-1990s, from vegetable products to dairy products. In recent years, a multistate outbreak of listeriosis with 21% mortality has been reported due to consumption of contaminated RTE meat and poultry products (Donnelly, 2001).

One of the key attributes that allows *L. monocytogenes* to cause disease is its ability to grow at refrigeration temperatures (Wemekamp-Kamphuis et al., 2002), where most RTE foods are stored. Furthermore, this microorganism is ubiquitous in the environment, for example, soil and water (Sauders and Wiedmann, 2007), and it can easily find its way into processed foods such as RTE foods prior to or during packaging. The presence of *L. monocytogenes* is commonly confirmed on equipment and food-processing environments where RTE foods are prepared (Sauders and Wiedmann, 2007). The presence of this pathogen in RTE foods, apart from the obvious human cost associated with disease, loss of life, or even permanent health damage, can also have major economic implications since food products that do not comply with the regulations should be destroyed and removed from the market. The microbiological criteria in the EU define that RTE foods that are unable to support growth of *L. monocytogenes* should contain less than 100 CFU/g when they are put on the market. According to footnote 8 of Annex I of Regulation (EC) No 2073/2005, products directly included in this group are these with pH $\leq$ 4.4 and $a_w \leq 0.92$ or pH ≤ 5.0 and $a_w \leq 0.94$. Furthermore, products with a shelf-life of less than 5 days, frozen products and other products based on the scientific justification are also included in this category (European Commission, 2008). However, foods for infants or dietary foods for special medical purposes (eg, foods intended for consumption by patients in hospitals) should contain no cells of *L. monocytogenes* in 25 g due to the high vulnerability of these specific consumer target groups to listeriosis (European Commission, 2008).

Countries may differ in their regulatory approach to the presence of *L. monocytogenes* in RTE foods. For example, in the United States there is a "zero tolerance policy" on the presence of *L. monocytogenes* in all RTE foods, and the pathogen should be absent in 25 g of product (Malley et al., 2015). However various scientists and organizations (including the FDA) based on the ubiquitous presence of *L. monocytogenes* in the environment and the difficulty in its elimination, have proposed to loosen up these controls, allowing a maximum limit of 100/g in frozen and refrigerated RTE foods that do not support the growth of *Listeria* (Warriner et al., 2009; FDA, 2004).

Apart from legislations that enforce specific limits in various groups of foods, there are also recommendations intended to specific consumer groups such as pregnant women, the elderly (above 65 years of age), HIV patients, or people who have received chemotherapy or transplantation (NHS Choices, 2011). These recommendations indicate foods that should be avoided, because they pose a higher risk of listeriosis. For example, pregnant women are 14 times more likely to develop listeriosis than the normal population. These foods include prepacked sandwiches, pâtés, butter, soft cheeses (such as Brie or Camembert), or others with a similar rind, soft blue cheeses, cooked sliced meats, smoked salmon, and others. One of the problems with many of these products is that they are not used in one go and they are reused and reopened several times. Due to that, they might be contaminated during use or can be subject to temperature abuse which can both contribute to growth of *L. monocytogenes* as it is able to grow at refrigeration temperatures.

LISTERIA MONOCYTOGENES IN ACIDIC RTE FOODS

Annually around 0.3–7.5 cases of listeriosis per million inhabitants are being reported in European countries (Swaminathan and Gerner-Smidt, 2007). Outbreaks of listeriosis have been known to occur in acidic foods. Coleslaw has been implicated in several listeriosis outbreaks such as one that occurred in 1981 affecting 41 consumers of which 17 died. The cabbage was the main source of the outbreak, and was grown in fields fertilized by manure contaminated with *L. monocytogenes* (Schlech et al., 1983). Dairy products are also commonly associated with listeriosis outbreaks such as one that occurred in California in 1985 caused by cheese with a mortality rate of 34% (Linnan, 1988). Out of 142 cases, 48 were immunocompromised or elderly or had chronic illness. Another outbreak that involved a vacherin-type soft cheese resulted in 122 cases, with half associated with perinatal cases resulting in 31 deaths (Bannerman and Bille, 1988). Listeriosis has also been associated with deli meats which have an acidic pH as a preservative factor. An example of an outbreak involving such products is the one that occurred in 2002 that affected nine states in the United States. It was caused by turkey deli meat and resulted in 44 cases and 8 deaths (Gottlieb et al., 2006). These few examples of outbreaks due to different types of acidic foods show that *L. monocytogenes* is able to grow and even survive under acidic conditions with serious implications for consumers.

ACID RESISTANCE OF *L. MONOCYTOGENES*

As seen from the different categories of RTE foods depending on their ability to support growth of *L. monocytogenes*, the pH factor plays a major role in the inhibition of this microorganism in foods. Therefore, acid resistance is one of the attributes that allow *L. monocytogenes* to grow and survive in various foods where acids are present. Furthermore, it allows it to pass through the stomach and localize in the intestine where it can cause disease. In addition, acid resistance can allow *L. monocytogenes* to survive disinfection regimes using acidic disinfectants for the cleaning of fresh produce or food preparation environments or surfaces. The importance of acid resistance for *L. monocytogenes* and its ability to cause disease have led to research in this area which has revealed various mechanisms of acid resistance. These mechanisms mainly facilitate the maintenance of high intracellular pH during exposure to acidic conditions (Ryan et al., 2008).

The most important determinants of acid resistance in *L. monocytogenes* seem to be amino acid decarboxylase systems that can very rapidly raise the intracellular pH. One of the most important acid resistance systems in *L. monocytogenes* is the glutamate decarboxylase (GAD) system. This is known to promote the passage of *L. monocytogenes* through the stomach, enabling it to reach the intestine, where it can invade the intestinal epithelial cells and initiate the intracellular cycle leading to listeriosis (Cotter et al., 2001). *L. monocytogenes* also possesses an arginine deiminase (ADI) pathway and an agmatine deiminase (AgDI) system which contribute to the maintenance of intracellular pH (Ryan et al., 2008). There are also many other mechanisms contributing to acid resistance in this organism.

THE GAD SYSTEM

The main element of the GAD system is the GAD enzyme that catalyzes the conversion of glutamate to γ-aminobutyric acid (GABA). The GAD system is present in a wide variety of organisms from all Kingdoms of life. Interestingly, although the core reaction described above is universal, organisms from each one of the Kingdoms uses the system for completely different purposes and functions. In animals the GAD system is an integral part of the brain and the nervous system where glutamate and GABA are major excitatory or inhibitory neurotransmitters. In plants the GAD system is activated in response to various stimuli and under a variety of stresses. However, in microorganisms (fungi, archaea, bacteria) the GAD system specifically plays a role in acid resistance (Cotter et al., 2001). This system comprises up to three decarboxylases, accompanied by one or two antiporters. However, there is major interspecies variability in the architecture of the system. *L. monocytogenes* normally possess three decarboxylases, namely GadD1, GadD2, and GadD3, and two antiporters, GadT1 and GadT2, but serotype 4 and several 1/2 strains do not possess the *gadD1T1* operon (Cotter et al., 2005; Karatzas et al., 2012). The antiporter imports glutamate from the extracellular to the intracellular environment and converts it to GABA and CO_2, and this reaction consumes a proton resulting in an increase in the intracellular pH (Figs 10.1 and 10.2). This process is

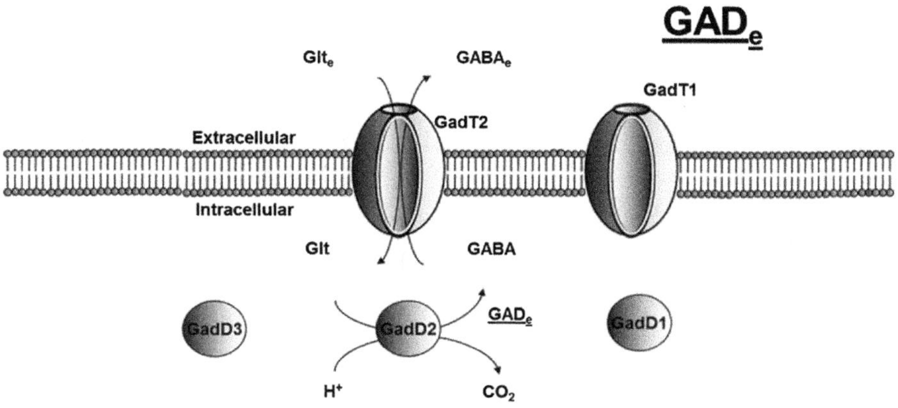

FIGURE 10.1

Description of the current model for the function of the GAD_e system under severe acidic conditions (pH < 4.5). The GadT2 antiporter imports extracellular Glt, which is decarboxylated by GadD2 to GABA with the consumption of a proton (H^+). GABA is then exported by the GadT2 with the simultaneous import of Glt. The above-described process is carried out by the GAD_e, which is depicted by bold lines.

(Karatzas et al., 2012).

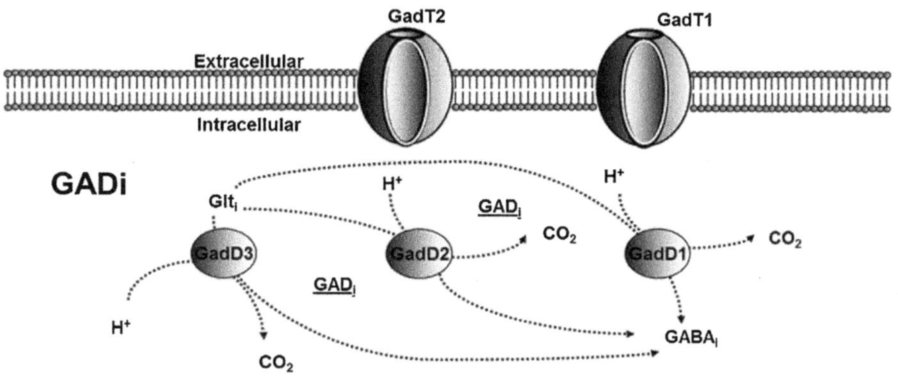

FIGURE 10.2

Description of the current model for the function of the GAD_i system under severe acid conditions (pH < 4.5). Intracellular Glt is decarboxylated by GadD3 and GadD2, resulting in the accumulation of $GABA_i$. The latter process is carried out by GAD_i, which is depicted by dotted lines. The contribution of GadD1 and GadT1 in both the GAD_e and GAD_i processes is minor according to the results presented here. GadD3 has previously been suggested by various authors to be a Glt decarboxylase, but further work is required to prove this.

(Karatzas et al., 2012).

repeated with a high frequency and it is able to maintain a neutral internal pH (Waterman and Small, 1998). GABA is exported via GadT in exchange for a further glutamate molecule. In this process a proton is removed from the intracellular environment to the extracellular environment. The existence of the intracellular GAD system (GAD_i), which has been described by Karatzas et al. (2010, 2012), converts intracellular glutamate to GABA and significantly contributes to acid resistance. GAD_i works independently even in the absence of an antiporter and is able to accumulate high levels of $GABA_i$. GAD_i utilizes intracellular pools of glutamate and it comprises mainly the GadD3 decarboxylase which works to convert glutamate to $GABA_i$ and CO_2. The accumulated $GABA_i$ remains in the intracellular milieu as the antiporter-based GAD system cannot affect the glutamate–GABA balance since the imported glutamate is exported in the form of GABA. This accumulated GABA can be metabolized by ArgD and Lmo0913 via the GABA shunt pathway (Feehily et al., 2013). Function of GAD system is variable between strains of *L. monocytogenes* since strains such as 10403S and LO28 utilize both GAD_i and GAD_e (Karatzas et al., 2010), whereas strains such as EGD-e utilize only GAD_i (Feehily et al., 2013). The GAD system assists the bacteria in certain foods to grow and survive under mild and acidic conditions (Cotter et al., 2005) with GadD1 playing a major role in growth at mild acidic conditions (pH 5.1; Cotter et al., 2005) while GadD2 plays a role in survival at extreme acidic conditions.

The regulation of the GAD system is complex and variable. For example, all regulatory elements in the Gram-negative bacterium *E. coli* (*gadE*, *gadY*, *gadX*, *gadW*) are absent in the Gram-positive *L. monocytogenes.* In *L. monocytogenes* upregulation of *gadT2D2* and *gadD3* but not *gadT1D1* is SigB-dependant during entry to the stationary phase (Wemekamp-Kamphuis et al., 2004). However, this only occurs in rich media and when stationary-phase cells of *L. monocytogenes* grown in defined medium were exposed to acidic conditions, only *gadD1T1* and *gadD3* were upregulated but not *gadT2D2* operon, which is the most important (Karatzas et al., 2010). Apart from acidic conditions, hypoxia has been shown to upregulate the GAD system in both *E. coli* (Blankenhorn et al., 1999) and *L. monocytogenes* (Jydegaard-Axelsen et al., 2004). The GAD system functionality is variable between different bacteria and it probably depends on the environment in which they grow. Here is a clear example of the GAD system being functional in minimal media in *E. coli* (Castanie-Cornet and Foster, 2001), and in *L. monocytogenes* the GAD system in not functional even in the presence of glutamate (Karatzas et al., 2010). There is great variability in the expression and function of the GAD system between various strains of *L. monocytogenes* that seems to be linked to different environmental factors, such as various strains might use it in one medium but not use it in another and vice versa (Karatzas et al., 2012). Further research is required to identify the signals and factors that affect its expression and function as this would allow us to manipulate the acid resistance of *L. monocytogenes* and achieve maximal antilisterial effects.

ARGININE DEIMINASE SYSTEM

This system provides a source of ATP derived from the catabolism of arginine to ornithine, CO_2, and NH_3 in a variety of organisms, including members of the genus *Pseudomonas* and many streptococci (Abdelal, 1979). A number of bacterial genera are able to catabolize arginine via the ADI pathway which is associated with an enhanced ability to survive under acidic extracellular conditions (Cunin et al., 1986). There are three different enzymes, namely ArcA, ArcB, and ArcC, that participate in the ADI-mediated conversion of arginine to ornithine, carbon dioxide, and ammonia (Cunin et al., 1986). Extracellular arginine is imported via antiporter ArcD and converted to citrulline and ammonia by ArcA and then further converted to ornithine and carbamolyphosphate by ArcB. Subsequently, ornithine is exported by the antiporter ArcD in exchange for a further molecule of arginine in an energy-independent manner. The ammonia produced as a by-product results in an increase in the intracellular pH, hence protecting the cell from potentially lethal effects of acidic extracellular environments (Fig. 10.3).

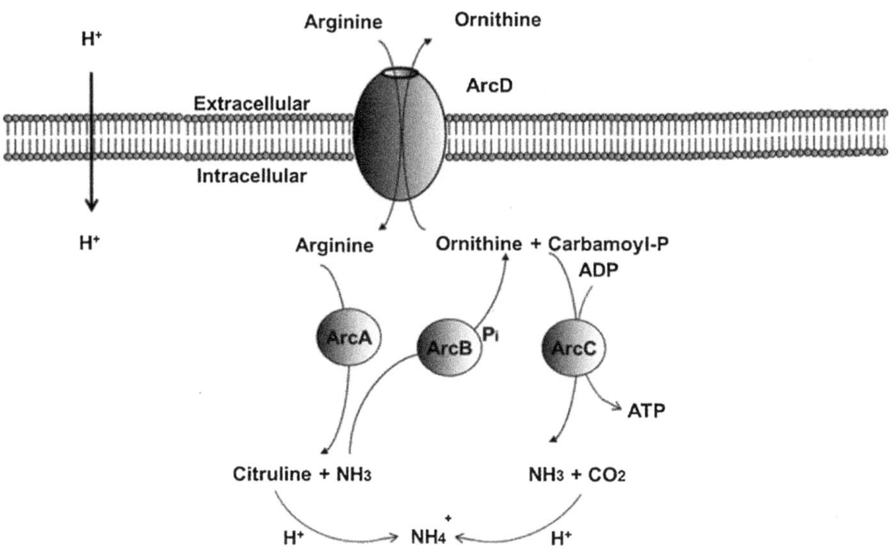

FIGURE 10.3

ADI deiminase pathway.

ADI systems are highly regulated in most of the organisms (Casiano-Colon and Marquis, 1988). Arrangement of the gene and regulation and biological role of ADS differ among species (Galimand et al., 1991). ADI permits growth under anaerobic conditions in some of bacteria including *Pseudomonas aeruginosa* and *Bacillus licheniformis* (Maghnouj et al., 1998). Some bacteria, such as *Streptococcus pyogenes* and oral streptococci, are protected by ADS under acidic conditions by the production of ammonia (Casiano-Colon and Marquis, 1988). A study showed that during infection of Caco-2 human colonic cells, *arcB* gene of *L. monocytogenes* strain EGD was upregulated (Joseph et al., 2006). When *L. monocytogenes* is exposed to low pH and low oxygen tension, there is upregulation of the ADI system (*arcA-D*; Ryan et al., 2009). Apart from a role in acid tolerance in vitro, *L. monocytogenes* ArcA also participates in gastric survival and virulence in mice (Cheng et al., 2013b). ArcA, encoded by *lmo0043*, is the key enzyme in the ADI pathway which initiates the first reaction. Ryan et al. (2009) reported molecular characteristics of the ADI system and its in vitro contribution to acid tolerance in *L. monocytogenes*. Growth of *L. monocytogenes* in mild acidic conditions is impaired and there is a reduction of survival of *L. monocytogenes* in human gastric flow due to deletion of *arcA* (Cheng et al., 2013b). This suggests that *arcA* is involved in listerial resistance to acid stress. As such, it plays an essential role in the survival of *L. monocytogenes* in acidic RTE foods or even in the stomach. Due to that deletion, ArcA reduces the virulence of *L. monocytogenes* in mice (Cheng et al., 2013b).

The ADI system is mediated by the interplay between the virulence regulator, PrfA, dedicated regulator, ArgR, and the stress sigma factor, SigB (Ryan et al., 2009). PrfA gene is a transcriptional activator of *L. monocytogenes* determinants of pathogenicity, the major regulator of virulence genes in this organism. Most of the virulence factors involved in internalization, vacuolar escape, and cell-to-cell spread are coregulated by regulation of the major virulence activator, PrfA (Vázquez-Boland et al., 2001). Other genes include bile salt hydrolase (*bsh*) playing a role in virulence contain binding sites for SigB and PrfA in their promoter regions (Rauch and Goebel, 2005; Zhang et al., 2011). The loss of SigB and PrfA results in decreased transcription of *arcA* and *argR* genes, which indicates their role in regulation of expression of ADI genes (Ryan et al., 2009).

OTHER FACTORS INVOLVED IN ACID RESISTANCE OF *L. MONOCYTOGENES*

AGMATINE DEIMINASE SYSTEM

Listeria monocytogenes also possesses an AgDI system that converts agmatine, *N*-carbamoylputrescine, and NH_3 and participates in acid resistance. This bacterium possesses two putative AgDI (*aguA1* and *aguA2*) which are upregulated at pH 5 (Cheng et al., 2013a). However, it has been proven that *aguA1*, but not *aguA2*, encodes a functional AgDI which enhances survival of *L. monocytogenes* in gastric fluid and in the mouse stomach. It is suspected to also play a role in survival or growth in acidic foods.

SIGB AND ACID RESISTANCE

As environmental conditions can change rapidly, microorganisms have also adapted mechanisms that allow them to fine-tune various mechanisms rapidly and simultaneously to ensure survival. This is usually achieved by proteins such as Sigma factors whose main task is to regulate the expression of many genes simultaneously. The "Stress" sigma factor SigB in *L. monocytogenes* and other Gram-positive microorganisms, such as *Bacillus subtilis*, mainly upregulates stress genes resulting in increased stress resistance. SigB was first recognized in *B. subtilis* as a factor that controls the transcription of genes which is induced at the end of the exponential phase (Wong et al., 1984). SigB plays an important role in the stress responses of several Gram-positive bacteria. After the exposure of cells to stress, there is activation of SigB, which in turn transcribes SigB-regulated genes, the SigB regulon (van Schaik and Abee, 2005). Apart from stress, SigB also has a role in the regulation of virulence of *L. monocytogenes* (van Schaik and Abee, 2005). SigB is the main stress gene regulator that contributes to multiple stress resistance by affecting the expression of more than 150 genes. These genes are involved in resistance against oxidative stress, acid, heat, salt, bile acids, etc. (Völker et al., 1999). The role of general stress transcription factor SigB on acid resistance has been studied (Wideman et al., 1998) concluding that expression of SigB-dependent proteins has an important role in acid resistance and ATR in *L. monocytogenes*. SigB is known to affect the expression of genes of the GAD system which contribute to acid resistance (Wemekamp-Kamphuis et al., 2004). Furthermore, the expression of other genes playing a role in acid resistance, such as *lmo0913* (a putative succinate semialdehyde dehydrogenase), *lmo0796* (hypothetical protein), and *lmo2391* (hypothetical protein), have been found to be regulated by SigB (Abram et al., 2008). According to van Schaik and Abee (2005) the preservation of food and food safety could be upgraded successfully by targeting SigB and trying to inhibit the system regulated by SigB.

OTHER MECHANISMS OF ACID RESISTANCE

Previously a *thiT* (*lmo1429*) mutant in *L. monocytogenes* has been found to be highly sensitive to acidic conditions. ThiT plays a role in thiamine uptake and similarly to the *thiT* mutant, cells depleted of thiamine were very sensitive under acidic conditions. The mechanism behind this phenotype is probably related to acetoin synthesis, which is a proton-consuming process where thiamine plays a role as a cofactor for acetolactate synthase (Madeo et al., 2012).

There is also an important role of a ferritin-like protein (*fri*) in the acid tolerance of *L. monocytogenes*. The mutant *fri* strain of *L. monocytogenes* under osmotic stress showed delayed growth and under acidic stress it showed impaired growth, which indicates that the ferritin-like protein plays an important role in acid tolerance (Milecka et al., 2015). The authors suggest that under acidic conditions the *fri* mutant is more susceptible to DNA strand breaks than the wild-type.

The SOS response is a conserved pathway which is activated under certain conditions and is regulated by the repressor LexA and the activator RecA. RecA is an important factor in the DNA repair and recombination process (Cox, 2007) and the activator of the SOS response. The SOS response is regulated by the RecA activator and LeXA repressor (Cox, 2007). The SOS response of *L. monocytogenes* contributes to survival upon exposure to a range of stress, including acid stress, contributing to its persistence in the environment (van der Veen et al., 2010). However, further work is required to identify the molecular mechanisms behind the involvement of these pathways in acid resistance.

CONCLUSIONS

The persistence of microorganisms in the food-processing environment is a major concern for the food industries due to posing a threat to consumers. There is increasing demand for RTE foods with minimal food processing, which is increasing the risk of food poisoning and even contributing to increased incidence. Listeriosis is a major challenging issue to food safety in food-processing environments as it is a significantly stress-resistant organism. This significant stress resistance and more specifically its acid resistance mean it has been very difficult to establish effective control and implementation in food safety. *Listeria monocytogenes* is a major foodborne pathogen posing a health risk to a number of individuals, especially immunocompromised individuals including the elderly, pregnant, neonates, etc. Various mechanisms discussed here are responsible for the bacterium combating different types of stresses, such as acidic stress, which are found in acidic RTE foods. Although there are a number of mechanisms/systems involved in acid resistance, the GAD system is probably the most important. Further work is essential to increase our understanding of the acid resistance mechanisms which could allow us to manipulate their effectiveness and responsiveness and therefore enhance the acid sensitivity of *L. monocytogenes*. The results of this work could be applied in food products or in disinfection regimes ensuring the safety of food products.

REFERENCES

Abdelal, A.T., 1979. Arginine catabolism by microorganisms. Annu. Rev. Microbiol. 33, 139–168.

Abram, F., Starr, E., Karatzas, K.A.G., Matlawska-Wasowska, K., Boyd, A., Wiedmann, M., et al., 2008. Identification of components of the sigma B regulon in *Listeria monocytogenes* that contribute to acid and salt tolerance. Appl. Environ. Microbiol. 74 (22), 6848–6858.

Bannerman, E.S., Bille, J., 1988. A new selective medium for isolating Listeria spp. From heavily contaminated material. Appl. Environ. Microbiol. 54, 165–167.

Begley, M., Sleator, R.D., Gahan, C.G., Hill, C., 2005. Contribution of the three bile-associated loci, *bsh*, *pva*, and *btlB*, to gastrointestinal persistence and bile tolerance of *Listeria monocytogenes*. Infect. Immun. 73, 894–904.

Bertsch, D., Rau, J., Eugster, M.R., Haug, M.C., Lawson, P.A., Lacroix, C., et al., 2013. *Listeria fleischmannii* sp. nov., isolated from cheese. Int. J. Syst. Evol. Microbiol. 63, 526–532.

Blankenhorn, D., Phillips, J., Slonczewski, J.L., 1999. Acid- and base-induced proteins during aerobic and anaerobic growth of *Escherichia coli* revealed by two-dimensional gel electrophoresis. J. Bacteriol. 181, 2209–2216.

Bubert, A., Riebe, J., Schnitzler, N., Schönberg, A., Goebel, W., Schubert, P., 1997. Isolation of catalase-negative *Listeria monocytogenes* strains from listeriosis patients and their rapid identification by anti-p60 antibodies and/or PCR. J. Clin. Microbiol. 35, 179–183.

Casiano-Colon, A., Marquis, R.E., 1988. Role of the arginine deiminase system in protecting oral bacteria and an enzymatic basis for acid tolerance. Appl. Environ. Microbiol. 54, 1318–1324.

Castanie-Cornet, M.P., Foster, J.W., 2001. *Escherichia coli* acid resistance: cAMP receptor protein and a 20 bp cis-acting sequence control pH and stationary phase expression of the gadA and gadBC glutamate decarboxylase genes. Microbiology 147, 709–715.

Cheng, C., Chen, J., Fang, C., Xia, Y., Shan, Y., Liu, Y., et al., 2013a. *Listeria monocytogenes aguA1*, but not *aguA2*, encodes a functional agmatine deiminase: biochemical characterization of its catalytic properties and roles in acid tolerance. J. Biol. Chem. 288 (37), 26606–26615.

Cheng, C., Chen, J., Shan, Y., Fang, C., Liu, Y., Xia, Y., et al., 2013b. *Listeria monocytogenes* ArcA contributes to acid tolerance. J. Med. Microbiol. 62 (6), 813–821.

Codex Alimentarius Commission, 2004. Risk assessment of *Listeria monocytogenes*. Ready-to-Eat Foods: Guidelines. FAO/WHO Microbiological Risk Assessment Series, No. 4. Food and Agriculture Organization, Rome.

Cooper, J., Walker, R.D., 1998. Listeriosis. Vet. Clin. North Am. Food Anim. Pract. 14, 113–125.

Cotter, P.D., Gahan, C.G., Hill, C., 2001. A glutamate decarboxylase system protects *Listeria monocytogenes* in gastric fluid. Mol. Microbiol. 40, 465–475.

Cotter, P.D., Ryan, S., Gahan, C.G.M., Hill, C., 2005. Presence of GadD1 glutamate decarboxylase in selected *Listeria monocytogenes* strains is associated with an ability to grow at low pH presence of GadD1 glutamate decarboxylase in selected *Listeria monocytogenes* strains is associated with an ability to grow. Appl. Environ. Microbiol. 71, 2832–2839.

Cox, M.M., 2007. Regulation of bacterial RecA protein function. Crit. Rev. Biochem. Mol. Biol. 42, 41–63.

Cunin, R., Glansdorff, N., Piérard, A., Stalon, V., 1986. Biosynthesis and metabolism of arginine in bacteria. Microbiol. Rev. 50, 314–352.

den Bakker, H.C., Warchocki, S., Wright, E.M., Allred, A.F., Ahlstrom, C., Manuel, C.S., et al., 2014. *Listeria floridensis* sp. nov., *Listeria aquatica* sp. nov., *Listeria cornellensis* sp. nov., *Listeria riparia* sp. nov. and *Listeria grandensis* sp. nov., from agricultural and natural environments. Int. J. Syst. Evol. Microbiol. 64, 1882–1889.

Donnelly, C.W., 2001. *Listeria monocytogenes*: a continuing challenge. Nutr. Rev. 59, 183–194.

European Commission, 2008. Guidance document on *Listeria monocytogenes* shelf-life studies for ready-to-eat foods, under Regulation (EC) No 2073/2005 of 15 November 2005 on microbiological criteria for foodstuffs. DG SANCO, Commission of the European Communities. http://ec.europa.eu/food/food/biosafety/salmonella/docs/guidoc_listeria_monocytogenes_en.pdf. (EC) No 2073/2005 of 15 November 2005.

Farber, J.M., Ross, W.H., Harwig, J., 1996. Health risk assessment of *Listeria monocytogenes* in Canada. Int. J. Food. Microbiol. 30, 145–156.

FDA (Food and Drug Administration), 2004. Juice HACCP hazards and controls guidance. <http://www.fda.gov/Food/GuidanceRegulation/GuidanceDocumentsRegulatoryInformation/Juice/ucm072557.htm> (accessed on 09.11.2015).

Feehily, C., O'Byrne, C.P., Karatzas, K.A.G., 2013. Functional γ-aminobutyrate shunt in *Listeria monocytogenes*: role in acid tolerance and succinate biosynthesis. Appl. Environ. Microbiol. 79, 74–80.

Galimand, M., Gamper, M., Zimmermann, A., Haas, D., 1991. Positive FNR-like control of anaerobic arginine degradation and nitrate respiration in *Pseudomonas aeruginosa*. J. Bacteriol. 173, 1598–1606.

Gardan, R., Duché, O., Leroy-Sétrin, S., Labadie, J., 2003. Role of ctc from *Listeria monocytogenes* in osmotolerance. Appl. Environ. Microbiol. 69, 154–161.

Gottlieb, S.L., Newbern, E.C., Griffin, P.M., Graves, L.M., Hoekstra, R.M., Baker, N.L., et al., 2006. Multistate outbreak of listeriosis linked to turkey deli meat and subsequent changes in US regulatory policy. Clin. Infect. Dis.29–36.

Graves, L.M., Helsel, L.O., Steigerwalt, A.G., Morey, R.E., Daneshvar, M.I., Roof, S.E., et al., 2010. *Listeria marthii* sp. nov., isolated from the natural environment, Finger Lakes National Forest. Int. J. Syst. Evol. Microbiol. 60, 1280–1288.

Halter, E.L., Neuhaus, K., Scherer, S., 2012. *Listeria weihenstephanensis* sp. nov., isolated from the water plant *Lemna trisulca* of a German fresh water pond. Int. J. Syst. Evol. Microbiol. 63, 641–647.

Joseph, B., Przybilla, K., Stühler, C., Schauer, K., Slaghuis, J., Fuchs, T.M., et al., 2006. Identification of *Listeria monocytogenes* genes contributing to intracellular replication by expression profiling and mutant screening. J. Bacteriol. 188, 556–568.

Jydegaard-Axelsen, A.-M., Høiby, P.E., Holmstrøm, K., Russell, N., Knøchel, S., 2004. CO_2- and anaerobiosis-induced changes in physiology and gene expression of different *Listeria monocytogenes* strains. Appl. Environ. Microbiol. 70, 4111–4117.

Karatzas, K.-A.G., Brennan, O., Heavin, S., Morrissey, J., O'Byrne, C.P., 2010. Intracellular accumulation of high levels of gamma-aminobutyrate by *Listeria monocytogenes* 10403S in response to low pH: uncoupling of gamma-aminobutyrate synthesis from efflux in a chemically defined medium. Appl. Environ. Microbiol. 76, 3529–3537.

Karatzas, K.-A.G., Suur, L., O'Byrne, C.P., 2012. Characterization of the intracellular glutamate decarboxylase system: analysis of its function, transcription, and role in the acid resistance of various strains of *Listeria monocytogenes*. Appl. Environ. Microbiol. 78, 3571–3579.

Leclercq, A., Clermont, D., Bizet, C., Grimont, P.A.D., Le Flèche-Matéos, A., Roche, S. M., et al., 2010. *Listeria rocourtiae* sp. nov. Int. J. Syst. Evol. Microbiol. 60, 2210–2214.

Linnan, M.J., Mascola, L., Lou, X.D., Goulet, V., May, S., Salminen, C., et al., 1988. Epidemic listeriosis associated with Mexican-style cheese. New Engl. J. Med. 319 (13), 823–828.

Lou, Y., Yousef, A.E., 1997. Adaptation to sublethal environmental stresses protects *Listeria monocytogenes* against lethal preservation factors. Appl. Environ. Microbiol. 63, 1252–1255.

Madeo, M., O'Riordan, N., Fuchs, T., Utratna, M., Karatzas, K.A.G., O'Byrne, C.P., 2012. Thiamine plays a critical role in the acid tolerance of *Listeria monocytogenes*. FEMS Microbiol. Lett. 326 (2), 137–143.

Maghnouj, a, de Sousa Cabral, T.F., Stalon, V., Vander Wauven, C., 1998. The *arcABDC* gene cluster, encoding the arginine deiminase pathway of *Bacillus licheniformis*, and its activation by the arginine repressor argR. J. Bacteriol. 180, 6468–6475.

Malley, T.J.V., Butts, J., Wiedmann, M., 2015. Seek and destroy process: *Listeria monocytogenes*: process controls in the ready-to-eat meat and poultry industry. J. Food Prot. 78, 436–445.

McLauchlin, J., Rees, C., 2009. Genus I. Listeria Pirie 1940a 383AL. In: De Vos, P., Garrity, G., Jones, D., Krieg, N.R., Ludwig, W., Rainey, F.A., et al. (Eds.), Bergey's Manual of Systematic Bacteriology, vol. 3, second ed. Springer, New York, pp. 244–257.

Milecka, D., Samluk, A., Wasiak, K., Krawczyk-Balska, A., 2015. An essential role of a ferritin-like protein in acid stress tolerance of *Listeria monocytogenes*. Arch. Microbiol. 197, 347–351.

NHS Choices, 2011. Listeriosis—NHS Choices <nhs.uk> (accessed on 05.11.2015).

Rauch, M., Goebel, W., 2005. SigB-dependent in vitro transcription of *prfA* and some newly identified genes of *Listeria monocytogenes* whose expression is affected by PrfA in vivo. J. Bacteriol. 187, 800–804.

Ryan, E.M., Gahan, C.G.M., Hill, C., 2008. A significant role for Sigma B in the detergent stress response of *Listeria monocytogenes*. Lett. Appl. Microbiol. 46 (2), 148–154.

Ryan, S., Begley, M., Gahan, C.G.M., Hill, C., 2009. Molecular characterization of the arginine deiminase system in *Listeria monocytogenes*: regulation and role in acid tolerance. Environ. Microbiol. 11, 432–445.

Sauders, B.D., Wiedmann, M., 2007. Ecology of *Listeria* species and *L.monocytogenes* in the natural environment. In: Ryser, E.T., Marth, E.H. (Eds.), Listeria, *Listeriosis and Food Safety*, third ed. CRC Press, Boca Raton, pp. 21–53.

Schlech, W.F., Lavigne, P.M., Bortolussi, R.A., Allen, A.C., Haldane, E.V., Hightower, A. W., et al., 1983. Epidemic listeriosis—evidence for transmission by food. N. Engl. J. Med. 308, 203–206.

Siegman-Igra, Y., Levin, R., Weinberger, M., Golan, Y., Schwartz, D., Samra, Z., et al., 2002. *Listeria monocytogenes* infection in Israel and review of cases worldwide. Emerg. Infect. Dis. 8, 305–310.

Swaminathan, B., Gerner-Smidt, P., 2007. The epidemiology of human listeriosis. Microbes. Infect. 9, 1236–1243.

Swartz, M.A., Welch, D.F., Narayanan, R.P., Greenfield, R.A., 1991. Catalase-negative *Listeria monocytogenes* causing meningitis in an adult. Clinical and laboratory features. Am. J. Clin. Pathol. 96, 130–133.

van der Veen, S., van Schalkwijk, S., Molenaar, D., de Vos, W.M., Abee, T., Wells-Bennik, M.H.J., 2010. The SOS response of *Listeria monocytogenes* is involved in stress resistance and mutagenesis. Microbiology 156, 374–384.

van Schaik, W., Abee, T., 2005. The role of σ^B in the stress response of Gram-positive bacteria—targets for food preservation and safety. Curr. Opin. Biotechnol. 16, 218–224.

Vázquez-Boland, J. a, Kuhn, M., Berche, P., Chakraborty, T., Domi, G., González-zorn, B., et al., 2001. *Listeria* pathogenesis and molecular virulence determinants *Listeria* pathogenesis and molecular virulence determinants. Clin. Microbiol. Rev. 14, 584–640.

Völker, U., Maul, B., Hecker, M., 1999. Expression of the sigmaB-dependent general stress regulon confers multiple stress resistance in *Bacillus subtilis*. J. Bacteriol. 181, 3942–3948.

Warriner, K., Huber, A., Namvar, A., Fan, W., Dunfield, K., 2009. Recent advances in the microbial safety of fresh fruits and vegetables. Adv. Food. Nutr. Res. 57, 155–208.

Waterman, S.R., Small, P.L., 1998. Acid-sensitive enteric pathogens are protected from killing under extremely acidic conditions of pH 2.5 when they are inoculated onto certain solid food sources. Appl. Environ. Microbiol. 64, 3882–3886.

Wemekamp-Kamphuis, H.H., Karatzas, A.K., Wouters, J.A., Abee, T., 2002. Enhanced levels of cold shock proteins in *Listeria monocytogenes* LO28 upon exposure to low temperature and high hydrostatic pressure. J. Bacteriol. 68, 456–463.

Wemekamp-Kamphuis, H.H., Wouters, J.A., De Leeuw, P.P.L., Hain, T., Chakraborty, T., Abee, T., 2004. Identification of sigma factor σ^B-controlled genes and their impact on acid stress, high hydrostatic pressure, and freeze survival in *Listeria monocytogenes* EGD-e. Appl. Environ. Microbiol. 70, 3457–3466.

Wiedmann, M., Arvik, T.J., Hurley, R.J., Boor, K.J., 1998. General stress transcription factor sigmaB and its role in acid tolerance and virulence of *Listeria monocytogenes*. J. Bacteriol. 180, 3650–3656.

Wong, S.L., Price, C.W., Goldfarb, D.S., Doi, R.H., 1984. The subtilisin E gene of *Bacillus subtilis* is transcribed from a sigma 37 promoter in vivo. Proc. Natl. Acad. Sci. U.S.A. 81, 1184–1188.

Zhang, Q., Feng, Y., Deng, L., Feng, F., Wang, L., Zhou, Q., et al., 2011. SigB plays a major role in *Listeria monocytogenes* tolerance to bile stress. Int. J. Food. Microbiol. 145, 238–243.

CHAPTER

Safety of fresh-squeezed juices

11

Nanci Edid Martínez-Gonzáles[1] and Alejandro Castillo[2]

[1]Departamento de Farmacobiología, Universidad de Guadalajara, Guadalajara, Jalisco, Mexico [2]Texas A&M Center for Food Safety, Department of Animal Science, Texas A&M University, College Station, TX, United States

INTRODUCTION

Fresh-squeezed fruit and vegetable juices are defined as a pure product obtained from a mature fruit or vegetable, which has not been further pasteurized, frozen, or concentrated after extraction. Commercially, fresh juice is mechanically extracted from graded, washed, and sanitized fruit and is commonly distributed for retail once packaged and refrigerated (Parish and Davis, 2001). In developing countries, fresh juice is served to consumers by vendors in local markets, street booths, or restaurants. In these sites, the juice may be maintained at room temperature, cooled on ice, diluted with water, or added to ice. The handling practices during preparation and service will influence the microbial quality of juice. In the last three decades, foodborne disease outbreaks have commonly been linked to fresh orange juice and apple juice or cider contaminated with *Escherichia coli* O157:H7, *Salmonella* spp., *Shigella*, *Cryptosporidium* spp., and hepatitis A virus, both in North America and worldwide (Vojdani et al., 2008). On the other hand, since 2005 there have been reported outbreaks associated with the consumption of juice prepared from exotic fruits such as açai, guava, and sugarcane contaminated with *Trypanosoma cruzi* in Venezuela, Brazil, and Colombia (Alarcón de Noya et al., 2010a). These examples highlight the risks of foodborne illness linked to fresh juices in rural areas of South American countries, and expand the list of pathogens that can cause disease when transmitted by fresh juice. Lack of hygiene, coupled with use of dropped fruits, nontreated manure, inadequate or no washing of the fruit prior to pressing, and the proximity of cattle or deer have been reported as contributing factors in juice-related outbreaks (Mihajlovic et al., 2013; FDA, 2015).

The lack of any antimicrobial treatment makes fresh juices more prone to microbial spoilage, and to act as a vehicle for pathogens. Some studies reveal that pathogenic bacteria can be isolated from fresh juices as a consequence of

P. Kotzekidou (Ed): Food Hygiene and Toxicology in Ready-to-Eat Foods.
DOI: http://dx.doi.org/10.1016/B978-0-12-801916-0.00011-X

inadequate food safety practices during preparation of fresh juices, use of defective fruit for juice extraction, nonskilled or noneducated food handlers, etc. (Castillo et al., 2006; Martínez-Gonzáles et al., 2003).

Until recent decades, it was accepted that acidic fruit juices (pH 3.0–4.0) could not support growth of microbial pathogens. However, although growth of foodborne pathogens is unlikely at low pH, it is well documented that these microorganisms can survive in fruit juices, become adapted to the acidic environment, and cause outbreaks of foodborne illness.

As a result of recurrent outbreaks associated with the consumption of fresh juices processed commercially, some countries have adopted regulations requiring that juice be produced under a Hazard Analysis and Critical Control Point (HACCP) plan (FDA, 2001). Still, consumption of fresh-squeezed juices is common at juice bars. In developing countries, the problem must be addressed from a different perspective due to a general need for applying good hygiene practices during preparation of fresh juice at retail stores, juice bars, and restaurants, as well as in the home.

In this chapter, information will be discussed with regard to foodborne disease outbreaks linked to fresh juice, contamination sources, prevalence, internalization, and survival of pathogens in produce used to obtain juice. Finally, strategies for preventing or reducing contamination with bacterial pathogens will be discussed in an effort to minimize the risks associated with the consumption of fresh juices.

OUTBREAKS ASSOCIATED WITH FRESH-SQUEEZED JUICES

Numerous foodborne outbreaks have been linked to the consumption of fresh vegetable and fruit juices worldwide (Frank et al., 2007; Vojdani et al., 2008). Despite possible food safety risks associated with fresh juices, many advocate their consumption based on potential nutritional advantage, improved taste, and lower price (Abeysinghe et al., 2007; Produce for Better Health Foundation, 2015). Traditionally, these foods were not thought to be hazardous because of their high acidity, but published data reveal that pathogens can survive in low-pH juice for at least 30 days at 4°C (Goverd et al., 1979). A list of juice- and cider-related outbreaks caused by *Salmonella*, *E. coli* O157:H7, and various pathogens is presented in Tables 11.1–11.3, respectively. These pathogens can affect human health, with symptoms ranging from fever and gastrointestinal symptoms to organ failure and even death. Adding all outbreaks and cases listed in Tables 11.1–11.3, it can be concluded that between 1974 and 2014, fresh juices and cider have been implicated in 48 foodborne disease outbreaks involving 5905 cases in various countries. These illnesses were caused by several serovars of *Salmonella*, *Shigella flexneri*, *E. coli* O157:H7, *E. coli* O111, enterotoxigenic *E. coli*, *Cryptosporidium* spp., norovirus genotype II, hepatitis A virus, and in some cases, *T. cruzi*. The involved juice products were

Table 11.1 Reported Outbreaks by *Salmonella* Serovars Associated with the Consumption of Fresh Fruit Juices and Cider

Year	Country	*Salmonella* Serovar	Type of Juice	No. Cases	References
1974	United States	*Salmonella* Typhimurium	Apple cider	296	CDC (1975)
1993	United States	*Salmonella* spp.	Watermelon juice	18	Luedtke and Powell (2000)
1995	United States	*S.* Hartford, Gaminara, and Rubislaw	Orange	62	Cook et al. (1998)
1999	South Australia	*S.* Typhimurium	Orange	405	National Centre for Disease Control (1999)
1999	United States, Canada	*S.* Muenchen	Orange	207	CDC (1999)
1999	United States	*S.* Anatum	Orange	4	Krause et al. (2001)
2000	United States	*S.* Enteritidis	Orange	14	Butler (2000)
2005	United States	*S.* Typhimurium and Saintpaul	Orange	157	Jain et al. (2009)
2008	The Netherlands	*S.* Panama	Orange	15	Noël et al. (2010)

apple cider (22 outbreaks), orange juice (12 outbreaks), açaí (4 outbreaks), guava, sugarcane, and fruit juice unspecified (2 outbreaks each), and watermelon, mango, pumarosa, and passion fruit, açaí, bananas, strawberries, and sugarcane juice (1 outbreak each).

The long history of juice-related outbreaks has revealed the association between *Salmonella* and orange juice (Table 11.1), and *E. coli* O157:H7 and *Cryptosporidium* with apple cider or apple juice (Tables 11.2 and 11.3). Norovirus genotype II has also been found to cause outbreaks related to fresh juice consumption (Visser et al., 2010). In February 2009, 38 cases of gastroenteritis illness occurred among the participants of two Dutch coach trips (A and B) who visited the same hotel in Germany. Norovirus genotype II.4 was detected in stool samples, and further univariate analysis revealed a potential risk for early cases from juice consumption for coach B. This outbreak was probably caused by using large juice containers with a tap for self-service. During self-serving, tap contamination might have occurred from infected customers, transferring the norovirus to other customers. This was the first time that consumption of juice served in containers was demonstrated to be a critical point in the hygiene procedure of hotels (Visser et al., 2010).

Table 11.2 Reported Outbreaks by *Escherichia coli* Types Associated with the Consumption of Fresh Fruit Juices and Cider

Year	Country	Pathogen	Type of Juice	No. Cases	References
1980	Canada	Most likely *E. coli* O157:H7	Apple cider	14	Steele et al. (1982)
1991	United States	*E. coli* O157:H7	Apple cider	23	Besser et al. (1993)
1992	India	Enterotoxigenic *E. coli*	Orange	6	Singh et al. (1995)
1996	United States	*E. coli* O157:H7	Apple cider	14	Hilborn et al. (2000)
1996	United States	*E. coli* O157:H7	Apple cider	6	CDC Foodborne outbreak online database
1996	United States, Canada	*E. coli* O157:H7	Apple cider	70	Cody et al. (1999)
1997	United States	*E. coli* O157:H7	Apple cider	6	Oglesby (2015)
1998	Canada	*E. coli* O157:H7	Apple cider	14	Tamblyn et al. (1999)
1999	United States	*E. coli* O157:H7	Apple cider	16 confirmed, 5 probable	Diallo et al. (2011)
2004	United States	*E. coli* O111 and *C. parvum*	Apple cider	213	Vojdani et al. (2008)
2005	Canada	*E. coli* O157:H7	Apple cider	4	LSDEPC (2005)
2007	United States	*E. coli* O157:H7	Apple cider	9	Mihajlovic et al. (2013)
2008	United States	*E. coli* O157:H7	Apple cider	5	CDC Foodborne outbreak online database
2010	United States	*E. coli* O157:H7	Apple cider	7	CDC Foodborne outbreak online database
2011	United States	*E. coli* O111: NM, *C. parvum*	Apple cider	14	CDC Foodborne outbreak online database
2012	United States	*E. coli* O157: NM (H-)	Apple cider	3	CDC Foodborne outbreak online database

Table 11.3 Reported Outbreaks by Other Microorganisms Associated with the Consumption of Fresh Fruit Juices and Cider

Year	Country	Pathogen	Type of Juice	No. Cases	References
1991	Australia	Norwalk-like virus	Orange reconstituted unpasteurized	>3000	Fleet et al. (2000)
1993	United States	*Cryptosporidium* spp.	Apple cider	160 primary and 53 secondary	Millard et al. (1994)
1995	South Africa	*S. flexneri*	Orange	14	Thurston et al. (1998)
1996	United States	*C. parvum*	Apple cider	20 confirmed, 11 suspected	CDC (1997)
2003	United States	*C. parvum*	Apple cider	144	Blackburn et al. (2006)
2004	Egypt	Hepatitis A virus	Orange	351	Frank et al. (2007)
2005	Brazil	*T. cruzi*	Sugarcane	25	Pereira et al. (2009)
2005	Brazil	*T. cruzi*	Açaí	27	Pereira et al. (2009)
2005	United States	Yeast (suspected)	Fruit juice (unspecified)	21	Vodjani et al. (2008)
2006	Brazil	*T. cruzi*	Sugarcane	94	Pereira et al. (2009)
2006	Brazil	*T. cruzi*	White acaí	17	Pereira et al. (2009)
2007	Brazil	*T. cruzi*	Açaí palm	79	Pereira et al. (2009)
2007	Venezuela	*T. cruzi*	Guava	103	Alarcón de Noya et al. (2010a)
2007	United States	Hepatitis A virus	Açaí, bananas, strawberries, sugarcane	3	CDC Foodborne outbreak online database
2008	Colombia	*T. cruzi*	Orange	7	Hernández et al. (2009)

(*Continued*)

Table 11.3 Reported Outbreaks by Other Microorganisms Associated with the Consumption of Fresh Fruit Juices and Cider *Continued*

Year	Country	Pathogen	Type of Juice	No. Cases	References
2009	The Netherlands	Norovirus genotype 11.4	Fruit juice (unspecified)	38	Visser et al. (2010)
2009	Venezuela	*T. cruzi*	Guava	89	
2010	Venezuela	*T. cruzi*	Passion fruit	22	
2011	United States	*Cryptosporidium* unknown	Apple cider, unpasteurized	4	CDC Foodborne outbreak online database
2013	United States	*Cryptosporidium* unknown; *C. hominis*	Apple cider, unpasteurized	8	CDC Foodborne outbreak online database
2013	United States	*C. parvum*; *Cryptosporidium*	Apple cider, unpasteurized	10	CDC Foodborne outbreak online database
2013	Venezuela	*T. cruzi*	Mango	8	
2014	Venezuela	*T. cruzi*	Pumarosa	3	

Oral infection with *T. cruzi* resulting in Chagas disease has been recognized as an emerging food safety issue due to acquisition of Chagas disease in South America after consuming fresh fruit and berry juices (Rueda et al., 2014). Although not a new issue, these outbreaks have only recently attracted interest due to their frequent association with fruits and berries considered to be good sources of antioxidants, with the corresponding increase in their presence in the international trade scenario. Most of these incidents have occurred in rural communities in Bolivia, Brazil, Colombia, French Guyana, and Venezuela (PAHO, 2009; Hernández et al., 2009; Pereira et al., 2009; Alarcón de Noya et al., 2010b, 2015; Santalla-Vargas et al., 2011) and seem to result from the presence of the triatominae carrying *T. cruzi* on the fruit surface, combined with lack of hygiene during juice extraction. This involves the presence of feces of vectors containing the pathogen, or crushing the vector during extraction and releasing the pathogens into the juice, leading to squeezing the fruits with the vector attached. No outbreaks of orally transmitted Chagas disease have been reported in nonendemic countries.

Various studies of outbreaks and juice-processing facilities indicate that juices may become contaminated with pathogens during extraction, and the source may be the establishment environment or the fruit itself. From the study of juice-related outbreaks, it becomes evident that the most common contamination mechanism seems to be the juice preparation or serving, and the source may be the food handler or the fruit itself. Contamination with pathogens may be less likely after the juice is processed and packaged for sale, but may occur after opening and during serving due to poor hygiene (Visser et al., 2010). Fruit surfaces, equipment, and employees are common sources of pathogens causing outbreaks of illness linked to fresh juice. However, dust, insects (Barro et al., 2006), and other pests such as toads (Parish, 1998) and anything that may come into contact with the fruit, or surfaces that may come into contact with the juice, must also be considered possible sources of pathogens (Martínez-Gonzáles et al., 2003; Parish, 1997).

Many of the outbreaks that have occurred from juice have been attributed to the use of fruit that has been contaminated with animal feces (CDC, 1975; Besser et al., 1993; Parish, 1998). Orchards are often located near livestock or wildlife with potential for microbial contamination. Contamination of the outside of the fruit is more likely to occur if the fruit drops to the ground, hence coming into contact with feces. Drop apples can become contaminated by coming into contact with manure from infected animals. Good agricultural practices require that dropped fruits not be harvested for either direct consumption or juice preparation. In an outbreak of *E. coli* O157:H7 infections in Massachusetts, United States, in 1991, a cider press operator also raised cattle, which grazed in a field adjacent to a mill (Besser et al., 1993). The presence of animals near a cider mill can result in manure inadvertently contacting apples, equipment, or workers' hands (Mihajlovic et al., 2013). An outbreak caused by *Cryptosporidium parvum* including 160 cases at a local fair in Maine in 1993, was associated with the consumption of fresh-pressed cider made from apples picked from an orchard floor contaminated with cattle manure. In one outbreak implicating orange juice, toads in the orange groves were thought to be a source of *Salmonella*, while a general lack of sanitation in the plant was thought to have contributed to the extent of the outbreak (Parish, 1998).

Due to the potential presence of foodborne pathogens on the outside skin or peel of fruits or vegetables, contamination of juice is more likely to occur when the skin or peel of the fruit has intimate contact with the juice during processing (Parish, 1997). However, pathogens can become internalized through natural plant structures or through decayed or damaged sites on the fruit or vegetable. Submerging warm harvested fruit in cold water also increases the potential for pathogens to infiltrate into susceptible produce (FDA, 2001).

Infected food handlers may also be the source of pathogens causing juice-associated disease if best food safety practices are not applied during juice extraction. Two outbreaks linked to fresh orange juice, caused by hepatitis A virus and *S. flexneri* were traced back to inadequate handling by asymptomatic workers carrying the pathogens during manufacturing (Thurston et al., 1998; Frank et al., 2007).

In two documented outbreaks of foodborne illness due to contaminated juice in Australia, one was caused by Norwalk-like virus in 1991, and the other by *Salmonella* Typhimurim phage type 135a in 1999 (National Centre for Disease Control, 1999; Fleet et al., 2000). In both circumstances, the contaminated juice was unpasteurized orange juice. In the 1991 Norwalk-like virus outbreak, the orange juice became contaminated when water was used to reconstitute the juice. The water within the factory became contaminated through faulty piping that allowed the potable water supply to come in contact with a sewerage line. In the 1999 *Salmonella* outbreak, the oranges used to manufacture the juice were contaminated. These oranges were dipped in a tank containing a fungicide dip, which was open to environmental contamination from birds and other pests and was not cleaned. *Salmonella* was cultured from the fungicide dip and wax used to coat the oranges.

Although *Listeria monocytogenes* has not been implicated in any cases of juice-borne illnesses, it has been isolated from oranges (Martínez-Gonzáles et al., 2003); an apple juice and an apple–raspberry blend with a pH of 3.78 and 3.75, respectively. A study carried out by our research group demonstrated that when *L. monocytogenes* is present on the surface of oranges and utensils, the pathogen could be transferred to the orange juice during preparation (Martínez-Gonzáles et al., 2003). This bacterium must be considered in the juice because of its ubiquitous nature, its relatively high resistance to heat, and its ability to survive in adverse conditions.

SOURCES OF PATHOGENS IN FRESH-SQUEEZED JUICES

Considering information provided by documented fresh juice outbreaks, the association of different pathogens with fresh juices can be explained by the fact that these organisms are present on the fruit surfaces, equipment, and in the gastrointestinal contents of animals and employees (Parish, 1998). Pathogens can be spread by different mechanisms to the fruit- and juice-processing site from primary sources, including water, animals, equipment, utensils, and hands of employees during operations related to the growing, harvesting, cleaning, and transportation chain. Also, the opportunities for contamination can occur during service of juice in home, restaurants, and retail establishments.

During the growing and harvesting of fruits, the use of untreated wastewater to irrigate trees, or contaminated water with human and animal wastes, can operate as an important vehicle for pathogens (Beuchat, 1996; Beuchat and Ryu, 1997). Water used in orchards for diluting agrochemicals and washing fruits also represents a possible source of contamination. Similarly, the practice of fertilizing crops with untreated manure has been reported as a risk factor in the contamination of fruits such as apples. *Salmonella*, *E. coli* O157:H7, protozoa, and viruses can enter the fresh fruit through irrigation water and inadequately treated manure (Beuchat and Ryu, 1997). Unpasteurized apple juice has been linked to *E. coli* O157:H7 infections (Besser et al., 1993; Cody et al., 1999). This pathogen can remain viable in bovine feces for up to 70 days, depending on inoculum level and temperature (Beuchat and Ryu, 1997). Fecal contaminants also may enter the fields or juice-processing establishments through insects (Barro et al., 2006), birds, and other wildlife such as reptiles and amphibians (Parish, 1998), or when the entry of animals is allowed in the orchard (Millard et al., 1994). Bare contact of the fruits or vegetables with pickers and handlers during harvest also offers a mechanism by which pathogens in feces can contaminate raw produce that is used for making fresh juices.

Anything that could touch the fruit or food-contact surfaces must also be considered a possible source of contamination during growing, harvesting, packing, processing, or transporting, and further, the contact with juice during processing (Beuchat and Ryu, 1997).

In addition to that, juice-processing establishments may be sources of contamination. Pathogens can be introduced to the juice-processing facility through contaminated fruit, and can persist if proper sanitation standards are not followed. Information on the presence of bacterial pathogens on orange surfaces removed from processing lines at unpasteurized juice processors (Parish et al., 2001), markets, and street booths has been reported (Martínez-Gonzáles et al., 2001a, Castillo et al., 2006). *L. monocytogenes* and *Salmonella* spp. were found in 7% and 5.4% of samples collected from the rind surfaces of oranges used for squeezing juice (Martínez-Gonzáles et al., 2001b). Castillo et al. (2006) reported the presence of *Salmonella* and *Shigella* in fresh oranges and orange juice from markets and street booths in Guadalajara Mexico. Three serovars of *Salmonella* (Agona, Typhimurium, and Anatum) were identified in both types of samples, indicating that oranges may be important sources of these pathogens. Serovars such as Mexico were also isolated from orange surface. These findings indicated that surface-contaminated oranges may be a source of pathogens, which should be considered during the preparation of juice due to the possibility that they may be spread to juice. In a study, Martínez-Gonzáles et al. (2003) demonstrated the potential for cross-contamination of *Salmonella*, *E. coli* O157:H7, and *L. monocytogenes* from contaminated oranges onto all utensils used during orange crushing and into the juice. After inoculation, the average pathogen counts for the orange rind surface were 2.3, 3.6, and 4.4 log CFU/cm^2 for *Salmonella*, *E. coli* O157:H7, and *L. monocytogenes*, respectively. This contamination was spread over all utensils used in orange juice squeezing. Mean pathogen counts for the cutting board, the knife, and the extractor ranged from 0.3 to 2.1 log CFU/cm^2, and the juice contained 1.0, 2.3, and 2.7 log CFU/mL of the above three pathogens, respectively. Producers of unpasteurized juice should realize that any fruit entering the facility may carry pathogens.

Flies are widely recognized as potential reservoirs and vectors of bacteria. Barro et al. (2006) carried out a survey to determine the bacterial carriage of *Musca domestica* and *Lucilia caesar*, and their reservoir in meat/poultry and local fruit juice processing and vending sites in Ouagadougou, Burkina Faso. For both flies, 66.67–100% of feces pools were positive for *Shigella*, *Salmonella*, and streptococci, while 35.41–82.95% of leg and proboscis pools were positive for the same bacteria. The bacterial carriage on organs of houseflies ranged from 10×10^3 to 1.1×10^3 CFU/organ, depending on the bacteria under consideration. *Staphylococcus aureus* was present in low numbers in all organs tested on both flies. The bacterial load on the legs and proboscis indicates that flies may pick up pathogenic bacteria while crawling and feeding. The authors concluded that the bacterial load appears as a function of their living environment. According to this report, flies from meat/poultry-vending sites were more contaminated than those from juice-vending sites.

Other sources of contamination include equipment and utensils that are used to process the juice, such as presses, mills, extractors, pipelines, filling and squeezing machines, and packing materials (Parish et al., 1997). In developing countries, the freshly squeezed fruit juices are frequently sold by street vendors and in booths at public markets. In these establishments, utensils such as knives, cutting boards, and mechanical or electrical juice extractors are employed during squeezing or crushing of fruits used to produce juice (Martínez-Gonzáles et al., 2003). During this process, cloths are used for wiping surfaces, and this practice is of high risk for contamination with pathogens. Castillo et al. (2006), reported the presence of *Salmonella* and *Shigella* in 22.8% and 5% of the cloths used in juice booths and small establishments, respectively. Since juices are commonly served shortly after squeezing at these sites, strict adherence to good manufacturing practices is required to minimize the risk of transmission of pathogens to the consumer.

PREVALENCE OF HUMAN PATHOGENS IN FRESHLY SQUEEZED FRUIT JUICES

Although fresh-squeezed juices have functioned as vehicles of pathogens causing outbreaks in repeated instances, more information is needed about the presence and concentrations of pathogens in fresh juices to generate data for accurate risk assessments. In various reported surveys focused on determining the presence of bacterial pathogens in fresh juices sold at public markets, street booths, restaurants, and cafeterias, *Salmonella* (Castillo et al., 2006; Torres-Vitela et al., 2012), *Shigella* (Castillo et al., 2006), *E. coli* pathotypes (Torres-Vitela et al., 2012), *L. monocytogenes* (Martinez-Gonzáles et al., 2016), *S. aureus*, and *Bacillus cereus* (Ahmed et al., 2009; Rai et al., 2014) have been isolated from fresh juices. The need for more research into the safety of juices becomes evident with the recent interest in exotic fruits and berries and their juices due to their attributed increased nutrient content.

Sado et al. (1998) analyzed 50 samples of fruit juices collected from retail establishments for the presence of *L. monocytogenes*, *E. coli* O157:H7, *Salmonella*, coliforms, and fecal coliforms. Only *L. monocytogenes* was recovered from two samples (0.04%), one of apple juice (pH 3.78) and the other of a mixture of apple and raspberry juices (pH 3.75). A study carried out to evaluate the microbiological quality of orange juice obtained from squeezing machines in foodservice establishments in Spain revealed that 0.5% of the examined lots tested were positive for *Salmonella* spp. and 1% for the presence of *S. aureus*. However, all lots tested were negative for *E. coli* and *L. monocytogenes*.

In a survey of fresh juices conducted by the Government Department of Human Services in Victoria, Australia (2005) collected from retail establishments between March 2004 and May 2004, none of 291 juice samples tested were positive for *Salmonella*. However, *E. coli* as an indicator was detected in seven juice samples, two of which had levels greater than 100 CFU/mL. In this survey, *L. monocytogenes* was detected in one celery juice sample at a level of 25,000 CFU/mL, and was assessed as being potentially hazardous, whereas other *Listeria* spp. were detected in nine juice samples, including two samples of apple juice, four of orange juice, two of celery juice, and one of cantaloupe juice. All juice samples analyzed for coagulase-positive staphylococci contained less than 100 CFU/mL and were assessed as satisfactory. The findings of this survey are comparable to other surveys carried out in the United Kingdom and Ireland. In the UK survey, *Salmonella* spp. or *E. coli* O157:H7 were not detected in any of the unpasteurized fruit and vegetable juice samples examined (Little et al., 2004). *Listeria monocytogenes* was present in 0.7% (2/291) of juices samples in the range of less than 10 to less than 20 CFU/mL. The authors did not describe the type of fruit or vegetable that was used to prepare juice. In the Irish survey of unpasteurized juices carried out in 2007, none of 228 samples tested resulted positive for *Salmonella* spp., *E. coli* O157 or *L. monocytogenes* (FSAI, 2007).

Not all studies have found negative results for the presence of pathogens in juices. Parish et al. (2001) conducted investigations on fruit surface and juice microbiota from individual oranges removed from processing lines at a manufacturer of bulk citrus concentrate, and at a fresh unpasteurized juice processor. Testing included aerobic plate counts (APCs), total coliforms, fecal coliforms, *E. coli*, and *Salmonella*. Fruits were removed from three points in the processing line: pre-grading, post-grading, and culled fruit. Fruit surface APCs were generally in the range of 10^3-10^4 CFU per fruit in both types of processing facilities. On culled fruit surface APC levels were 5–30 times higher than for pre- or post-graded fruit. APCs were in the range of 10^0-10^2 CFU/mL for pre- and post-graded fruit, and 10^5-10^6 CFU/mL for culled fruit. The presence of *Salmonella* was reported in two of 15 pre- and post-graded samples, in four of 15 culled fruit samples and from three of 15 samples of fresh unpasteurized juice squeezed from culled fruit. In general, surfaces and juice of defective culled fruit were much more likely to be contaminated with coliforms, *E. coli*, or *Salmonella* than nondefective fruit. The results suggest the importance of removing defective fruit prior to juice extraction.

Parish (1998) carried out a survey to investigate the presence of *Salmonella* on unpasteurized citrus juices from several processing facilities, including those that produced orange juice involved in an outbreak that occurred during summer of 1995 in the United States. At the processing facility, 70 samples (equipment swabs, fruit surface swabs, juice, and miscellaneous environmental samples) were collected before, during and after processing on two different dates. Bottled juice samples from eight previous extraction dates were also collected. *Salmonella* serovars Hartford, Rubislaw, Saintpaul, and Newport were detected in four sample types of either juice, unwashed fruit surfaces, or amphibians (*Hyla cinerea* and *Bufo terrestris*) captured outside the processing building. *Salmonella* cells in juice were associated with populations levels of fecal coliforms and *E. coli* above the upper most probable number (MPN) limits of detection (>110/mL).

The presence of pathogenic microorganisms in street-sold fruit juice has been reported in developing countries including Bangladesh, Ethiopia, Malaysia, Mexico, Pakistan, and India (Castillo et al., 2006; Tambekar et al., 2009; Diana et al., 2012; Leul and Kibret, 2012). In these countries, a large population of all income and age groups consume freshly squeezed fruit and vegetable juice from street vendors and food service establishments at public markets. Juices are extracted by pressing or squeezing fresh fruits and vegetable such as oranges, pomegranate, pineapple, watermelon, carrot, and papaya, in a diversity of hygiene and storage conditions that may allow entry and activity of microorganisms. Also, studies describe that there are practices in countries for consumption of fresh juices, for example in India they may be diluted at 1:0.1–1.0 ratio with water and added with ice (Lewis et al., 2006; Sharma, 2013); this last practice is common for serving fresh juices in Pakistan. Therefore, the type and levels of microorganisms in fresh-squeezed juice at the moment of consumption could be attributed to the quality of water used for diluting juice or for preparing ice, as well as to prevailing hygiene practices, such as inadequate personnel, washing and decontamination of fruit, and time and temperature of storage.

Results of previous studies conducted in our laboratory indicated that 31% of freshly squeezed orange juice samples obtained from public markets in Guadalajara Mexico were contaminated with *L. monocytogenes* (Martínez-Gonzáles et al., 2011). *Salmonella* has been reported to be present in 2.5% (Martínez-Gonzales et al., 2001a) and 3.9% (Castillo et al., 2006) of freshly squeezed orange juice obtained from public markets, and 14.3% (Castillo et al., 2006) from street vendors in Guadalajara, Mexico. *Salmonella enterica* serotypes Agona, Typhimurium, and Anatum were found in orange juice. In addition, 75 fresh orange samples and 75 wiping cloths were collected from the same establishments from which juices had been collected. This pathogen was isolated from 20% and 23% of these samples, respectively. In this study, *Shigella* spp. was investigated in three types of samples; this bacterium was isolated from 6%, 17%, and 5% samples of orange juice, orange surface, and wiping cloths, respectively. *Shigella sonnei* was found in all three types of samples, *Shigella dysenteriae* was found in juice and orange samples, *Shigella boydii* in orange and wiping cloth samples; and *S. flexneri* on oranges only. The frequency of isolation of *Salmonella* and *Shigella* in samples collected from small juice establishments at public markets was significantly lower ($p < 0.05$) than that found among street vendors.

Torres-Vitela et al. (2012) determined the incidence of coliforms, fecal coliforms, *E. coli*, diarrheagenic *E. coli* pathotypes, and *Salmonella* in 280 samples of fresh carrot juice obtained at restaurants in Pachuca City, Mexico. Of these samples, 100%, 96.8%, 54.3%, 8.9%, and 8.6% contained coliforms, fecal coliforms, *E. coli*, diarrheagenic *E. coli* pathotypes, and *Salmonella*, respectively.

Salmonella, *Shigella*, *S. aureus*, and *Vibrio cholera* have been isolated from freshly squeezed fruit and vegetable juices sold in various locations in India (Lewis et al., 2006; Mahale et al., 2008; Mudgil et al., 2004; Rai et al., 2014; Sharma, 2013; Tambekar et al., 2009; Titarmare et al., 2009). Also, the isolation of *Salmonella*, *S. aureus*, and *B. cereus* from fresh juices obtained from local markets, juice houses, juice vendors, and hawker stalls has been reported in Bangladesh, Ethiopia (Leul and Kribet, 2012), Malaysia (Diana et al., 2012), and Pakistan. The frequency of isolation of these bacteria in fresh juices is variable, as provided in Table 11.4.

A few studies have focused on the enumeration of pathogens in fresh juices. Diana et al. (2012) observed a variable load of *Salmonella* spp., *Salmonella* Typhi, and *Salmonella* Typhimurium in samples of fruit juice (orange, apple, watermelon, starfruit, and carrot) from 0 to 42 MPN/g in Serdang and the Klang Valley, Malaysia.

Factors such as lack of financial resources to invest in safer equipment, lack of training, and unskilled and uneducated food handlers (Castillo et al., 2006;

Table 11.4 Prevalence of Pathogens Bacteria in Freshly Squeezed Fruit and Vegetables Juices Sold in Bangladesh, Ethiopia, India, Malaysia, and Pakistan Cities

Type of Juice	Number of Samples	Pathogens	Positive Samples (%)	Site for Sampling/ Location	References
Mango, papaya, watermelon, pineapple, apple, and bael	124	*S. aureus* *Salmonella* *Bacillus cereus*	7 (6.14) 9 (7.89) 16 (64.91)	Local market/Dhaka City, Bangladesh	Ahmed et al. (2009)
Mango and pineapple	60	*S. aureus* *Salmonella*	57 (95) 7 (20)	Juice houses/Bahir Dar town, Ethiopia	Leul and Kibret (2012)
Orange, apple, watermelon, starfruit/carambola, and carrot	50	*Salmonella* spp. *S.* Typhimurium *S.* Typhi	17 (50) 5 (10) 10 (20)	Juice vendors and hawker stalls/Serdang and Klang, Malaysia	Diana et al. (2012)
Apple, orange, sweet lemon, pineapple, pomegranate, and mix fruit	52	*Salmonella* spp. *S. aureus*	12 (16) 5 (7)	Street vendors/Amravati city, India	Tambekar et al. (2009)
Sugarcane, lime, and carrot	30	*Salmonella* spp. *S. aureus* coagulase positive	0 (0) 7 (23.3)	Street vendors/Mumbai city, India	Mahale et al. (2008)
Pineapple, sweet line, and vegetable juices (carrot, wheat grass, ginger, bitter gourd, cucumber, bottle gourd)	38	*Salmonella* spp.	19 (50)	Street vendors/Nagpur city, India	Titarmare et al. (2009)

Kinnow-mandarin and carrot	150	*S. aureus* coagulase positive	63 (42)	Street vendors/Patiala city, India	Mudgil et al. (2004)
		Salmonella Enteritidis	3 (2)		
Orange, apple, mango, pineapple, and lime	105	*S. aureus*	(23.8)[a]	Local vendors/various cities of Vidarbha/ Wardha, Yavatmal, Amravati, Bhandara, Akola, Chandrapur, and Gondia	Sharma (2013)
		Salmonella	(19.04)		
		Shigella	(17.14)		
Orange, grape, mango, pineapple, and pomegranate	108	*Salmonella*	(38.8)[a]	Street vendors/ Visakhapatnam city, India	Lewis et al. (2006)
		Shigella	(16.6)		
		V. cholerae	(0.0)		
Sweet lime, orange, and carrot	30	*B. cereus*	17 (56.7)	Local markets/ Kurukshetra, India	Rai et al. (2014)
		S. aureus	5 (16.7)		
Sugarcane, apple, and pineapple	36	*Salmonella*	16 (50)	Street outlets/Multan, Pakistan	

[a]*Number of positive samples not specified by the author.*

Leul and Kibret, 2012;) can prevent juice establishments from meeting proper hygiene conditions for preparation of fresh juices, and the lack of hygiene may promote the presence of pathogens in the juice. Since the product is usually consumed within a short period of time after squeezing, pathogens present in contaminated juice may survive and cause illness. Strict hygiene is paramount to prevent juice contamination, while fruit disinfection is recommended to reduce pathogens in raw materials.

INTERNALIZATION OF PATHOGENS IN FRUITS USED TO OBTAIN FRESH JUICE

Any fissure, crack, or bruise in the skin surface of fruits and vegetables may become a port of entry for microorganisms to the fruit, resulting in their allocation in the pulp. These injuries can occur during maturation, harvesting, or processing. The characteristics of some fruits make them less susceptible to infiltration, for example, unlike apple, the citrus fruit is derived from a superior ovary, an ovary completely separate from the calyx (Pao et al., 2001). During development, the citrus fruit does not create a cavity at the blossom end of the fruit at any maturation stage, making the blossom end refractory to microbial infiltration. Nevertheless, *Salmonella* internalization in oranges has been reported when small wounds occur on the pericarp (Eblen et al., 2004).

The internalization is often due to a noncontrolled wash or precooling treatment in cold water. This treatment creates a transient differential in pressure that can result in the diffusion of the wash water and bacteria into the product. Several studies have demonstrated bacterial infiltration within produce tissues that are used to produce fresh juices. Buchanan et al. (1999) studied the susceptibility of four apple varieties (Golden Delicious, McIntosh, Red Delicious, and Braeburn) to waterborne contamination. They found that when apples were submerged in colder water contaminated with 3×10^7 CFU/mL *E. coli* O157:H7, as might occur in processing operations where flume water or dump tanks are not hygienically maintained, the pathogen was occasionally internalized. These results were confirmed with dye studies in which 6% of warm apples immersed in cold dye solution internalized the dye through open channels leading from the blossom end into the core region. This study also evidenced that if bacterial internalization occurs in apples, surface disinfection treatments will not eliminate the microorganism, even by submerging the contaminated apples in 2000 mg/L sodium hypochlorite for 1 min.

Other researchers demonstrated that pathogen internalization in tomatoes (Ibarra-Sánchez et al., 2004; Xia et al., 2012), and mangoes (Penteado et al., 2004) also occurs through the blossom and cavity of the fruit, particularly under conditions that produced a negative pressure differential within the fruit. Penteado et al. (2004) showed that *Salmonella* could be internalized into mangoes as a result of a simulated quarantine treatment using contaminated cooling water, and survive inside the fruit for at least 1 week, increasing the potential for foodborne illness. If pathogens are internalized, the conventional surface decontamination treatments will not reduce or eliminate the potential foodborne illness hazard when the product is employed for juice production, or other minimal processing in industries and the home.

The uptake of a dye solution is used generally to indicate the frequency and extent of pathogen internalization in fruits. However, this method was not a reliable indicator of bacteria penetration in citrus fruit. Pao et al. (2001) showed that a food dye (blue No. 1) penetrated deeper than *E. coli* ATCC 25922 and *Salmonella* Rubislaw. The dye spread laterally and bacteria did not, and bacteria were not observed in every vascular bundle that was infiltrated with the dye. These authors explained the difference in infiltration by the natural sieving effect that limited the lateral and downward spread of bacteria while permitting the penetration of the dye solution. The adhering properties of bacteria cells may also affect their movement in fruit tissues.

SURVIVAL OF PATHOGEN IN FRESH JUICES AND IN PRODUCE USED TO PREPARE JUICES

The results of outbreak investigations due to consumption of fresh juices and laboratory studies suggest that some strains of human pathogens can survive in this type of food for long periods of time, even at low pH. Goverd et al. (1979) showed that several *Salmonella* serotypes were able to survive in apple juice at pH 3.6 for 30 days at 4°C. In this study, the survival of *Salmonella* was greater at 22°C than at 4°C, when the initial population was 2×10^5 CFU/mL. In another study (Parish et al., 1997), *Salmonella* remained present at detectable levels in orange juice for up to 27 days at pH 3.5, 46 days at pH 3.8, 60 days at pH 4.1, and 73 days at pH 4.4, indicating that *Salmonella* present in orange juice may survive for a long enough time to infect and cause illness in susceptible persons.

Various studies indicate that *E. coli* O157:H7 is very resistant to the low pH values of orange juice and apple cider. Fratamico et al. (1997) studied the survival of *E. coli* O157:H7 in orange juice and apple cider. In their study, these authors found that this pathogen survived for 20–25 days at 4°C in both types of juice, although it appeared that *E. coli* O157:H7 was capable of surviving the acidic environment of orange juice for a longer period in comparison to apple cider. Zhao et al. (1993) studied the fate of *E. coli* O157:H7 in apple cider. The pathogen was inoculated at a concentration of 10^5 CFU/mL and the bacterium survived from 10 to 31 days at 8°C and 2 to 3 days at 25°C. Since the expected refrigerated shelf-life of apple cider is typically less than 2–3 weeks, the pathogen is likely to survive during the period at which the juice remains acceptable for consumption. This report also indicated that there was little reduction of the population of the pathogen during the refrigerated storage, and that the addition of 0.1% potassium sorbate, 0.1% sodium benzoate and their combination increased the reduction of *E. coli* O157:H7 in apple cider. The greatest inactivation rate was obtained in the presence of a combination of 0.1% sodium benzoate and 0.1% potassium sorbate.

Others reports indicate that foodborne organisms have a variety of adaptive responses. For example, the populations of *L. monocytogenes* cells that were adapted to acid by exposing to pH 5.7 for 3 h, were able to grow in orange juice at pH 2.6, to almost 11 log CFU/mL in 6 h when incubated at 30°C (Caggia et al., 2009). This highlights the ability of pathogenic bacteria to become adapted to acid and to grow in low-pH foods such as juices. In another study, the heat resistance of *L. monocytogenes* increased significantly after acid adaptation in single-strength apple, orange, and white grape juices adjusted to pH 3.9 (Mazzota, 2001).

Microorganisms possessing unusually high acid resistance, such as *E. coli* O157:H7, are expected to survive in low-pH juices. Ryu and Beuchat (1998) proved the ability of this pathogen to survive the low pH of apple cider and orange juice regardless of the storage temperature (5°C or 25°C) for up to 42 days.

Another feature of pathogenic bacteria is the potential for entering viable but nonculturable (VBNC) state when exposed to environmental conditions such as those in some juices. The VBNC may pose a risk for public health because of the inability to detect these microorganisms during conventional microbiological essays, and their potential for resuscitation after they are in the host. Nicolò et al. (2011) tested the ability of *L. monocytogenes*, *E. coli* O157:H7, *Salmonella* Typhimurium, and *S. flexneri* to go into VBNC stage and resuscitate after exposure to grape juice. The data in this study show that *L. monocytogenes* lost viability and culturability and did not resuscitate within 24 h regardless of inoculum size, whereas *E. coli* O157:H7 was able to resuscitate after 24 h but not after 48 h. *Salmonella* Typhimurium and *S. flexneri*, depending on inoculum size, lost culturability but maintained viability and were able to resuscitate; moreover, *S. flexneri* was still able to form colonies after 48 h at high inoculum size.

Although most bacterial pathogens will not grow in typical fruit juices due to the low pH, they can survive in the juice. The survival time for bacterial pathogens in juice depends on juice pH, storage temperature, and the physiological state of the microorganisms. In general, survival increases at higher pH levels and under chilled storage conditions (El-Safey, 2013). Evidence suggests that pathogens in low-pH systems die more rapidly at temperatures approaching room temperature than under very cold conditions (Parish, 1997).

CONTAMINATION REDUCTION STRATEGIES

Many aspects affect pathogen control in fresh-squeezed juices: agricultural practices, product handling, equipment used, growing location, whether the produce is obtained from below ground (eg, carrots), on the ground (eg, drops), or trees (eg, apples). In addition, factors such as pH, acidulants, method of processing, degree of animal contact, refrigeration, packaging, and the distribution system need to be considered when developing food safety strategies for obtaining fresh juice. In determining the best control mechanisms, it is important to remember that the conditions for microbial survival differ from those for microbial growth. The control has to be focused along the production line starting from the receiving of fruit as the major raw materials. Only fruits produced under Good Agricultural Practices should be used. Sorting must be conducted to remove damaged and spoiled fruits. Adequate storage must be followed, and refrigeration is a necessity. It is important to keep in mind that some pathogens are able to survive, even in low-pH juices (Parish, 1997; Parish et al., 1997). Several guidelines for good hygienic practices in the production of fresh fruit juices are available in *Codex Alimentarius* and specialized agencies of countries.

After fruits or vegetables used for juice production have reached the processing facility or juice booth, juice production methods must minimize contamination of the final product (Parish, 1997). This requires cleaning and sanitation procedures of produce and food-contact surfaces. Active interventions may be used in fruits to obtain a desired reduction in the populations of pathogens of interest. In the United States, the Food and Drug Administration requires that juices be subjected to a control measure that guarantees a 5-log-reduction in the pathogen of interest for the juices. However, compliance with this requirement is difficult to evaluate at small establishments such as juice bars, where fresh-squeezed juice is prepared and served. Chemical substances (200 ppm chlorine, chlorine dioxide, peracetic acid, hydrogen peroxide, lactic acid, etc.), mechanical scrubbing, irradiation, and thermal treatments have been shown to deliver significant reductions to surface-contaminated fruit with pathogens (Kenney and Beuchat, 2002; Venkitanarayanan et al., 2002; Bhagat et al., 2011; Martínez-Gonzáles et al., 2001b). New process technologies are available to accomplish a microbiological reduction in juices, pulsed electric fields, ultraviolet light, minimal thermal processes, and batch and continuous high-pressure processing systems have been shown to inactivate microorganisms in certain juices (Pilavtepe-Çelik, 2013). The applicability of each technology to a specific juice depends of the characteristics of the product and the pathogens of interest, which may be resistant to the process.

Ultimately, the processors of fresh fruit juices must realize that potential health risks from fresh fruit juices cannot be entirely eliminated and can only be minimized to some acceptable level. An HACCP plan assists efforts to reduce risk and improve the safety of fresh fruit juices. On the other hand, the risk assessment approach would be of great help in studies related to pathogen-free fruit juice (Tribst et al., 2009). This approach would be very useful to scientifically support the requirements of 5-log-reduction programs and to better understand the fate of pathogens in each fresh-juice-processing step.

In developing countries where there is contrast in hygiene conditions during the preparation of juices, particularly in street commerce, microbial control measures should be focused on to develop education programs to preparer and consumer, and in monitoring the compliance with Good Manufacturing Practices during the preparation of fresh juices.

CONCLUSIONS

Fresh-squeezed juices are frequently consumed by the population in both developing and industrialized countries. The frequency of outbreaks resulting from consumption of fresh juices is indicative of the importance of fresh-squeezed juices as potential vehicles of pathogens causing foodborne illness. Food safety issues related to fresh juices may be minimized through implementation of Good Manufacturing Practices and, as much as possible, a treatment to reduce pathogenic microorganisms in the juice. More research is needed about the microbiology of fresh juices, including fresh exotic fruit juices, to collect useful data to be used in the development of procedures and programs to reduce microbial risks to the population.

REFERENCES

Abeysinghe, D.C., Li, X., Sun, C.H., Zhang, W., Zhou, C.H., Chen, K., 2007. Bioactive compounds and antioxidant capacities in different edible tissues of citrus fruit of four species. Food Chem. 104, 1338–1344.

Agwa, O.K., Ossai-Chidi, L.N., Ezeani, C.A., 2014. Microbial evaluation of orange fruit juice sold in Port Harcourt, Nigeria. American. J. Food Sci. Nutr. Res. 1, 20–33.

Ahmed, M.S.U., Nasreen, T., Feroza, B., Parveen, S., 2009. Microbiological quality of local market vended freshly squeezed fruit juices in Dhaka City, Bangladesh. Bangladesh J. Sci. Ind. Res. (India) 44, 421–424.

Alarcón de Noya, B., Díaz-Bello, Z., Colmenares, C., Ruiz-Guevara, R., Mauriello, L., Zavala-Jaspe, R., et al., 2010a. Large urban outbreak of orally acquired acute Chagas disease at a school in Caracas, Venezuela. J. Infect. Dis. 201, 1308–1315.

Alarcón de Noya, B., Díaz-Bello, Z., Colmenares, C., Ruiz-Guevara, R., Mauriello, L., Muñoz-Calderón, A., et al., 2010b. Update on oral Chagas disease outbreaks in Venezuela: epidemiological, clinical and diagnostic approaches. Memórias do Instituto Oswaldo Cruz 110, 377–386.

Barro, N., Aly, S., Tidiane, O.C., Sababénédjo, T.A., 2006. Carriage of bacteria by proboscises, legs, and feces of two species of flies in street food vending sites in Ouagadougou, Burkina Faso. J. Food Prot. 69, 2007–2010.

Besser, R.E., Lett, S.M., Weber, J.T., Doyle, M.P., Barret, T.J., Wells, J.G., et al., 1993. An outbreak of diarrhea and hemolytic uremic syndrome from *Escherichia coli* O157: H7 in fresh-pressed apple cider. J. Am. Med. Assoc. 269, 2217–2220.

Beuchat, L.R., 1996. Pathogenic microorganisms associated with fresh produce. J. Food Prot. 59, 204–216.

Beuchat, L.R., Ryu, J.H., 1997. Produce handling and processing practices. Emerg. Infect. Dis. 3, 459–465.

Bhagat, A., Mahmoud, B.S.M., Linton, R.H., 2011. Effect of chlorine dioxide gas on *Salmonella entérica* inoculated on navel orange surfaces and its impact on the quality attributes of treated oranges. Foodborne. Pathog. Dis. 8, 77–85.

Blackburn, B.G., Mazurek, J.M., Hlavsa, M., Park, J., Tillapaw, M., Parrish, M., et al., 2006. Cryptosporidiosis associated with ozonated apple cider. Emerg. Infect. Dis. 12, 684–686.

Buchanan, R.L., Edelson, S.G., Miller, R.L., Sapers, G.M., 1999. Contamination of intact apples after immersion in an aqueous environment containing *Escherichia coli* O157: H7. J. Food Prot. 5, 431–548.

Butler, M.E., 2000. *Salmonella* outbreak leads to juice recall in Western states. Food Chemical News 42, 19.

Caggia, C., Scifò, G.O., Restuccia, C., Randazzo, C.L., 2009. Growth of acid-adapted *Listeria monocytogenes* in orange juice and in minimally processed orange slices. Food Control. 20, 59–66.

Canadian Food Inspection Agency (CFIA). 2012. Code of Practices for the production and distribution of unpasteurized apple and other fruit juice/cider in Canada. Available from: <http://www.inspection.gc.ca/food/processed-products/manuals/code-of-practice/eng/1340636187830/1340637184931> (accessed November, 2015).

Castillo, A., Villarruel-López, A., Navarro-Hidalgo, V., Martínez-González, N.E., Torres-Vitela, M.R., 2006. *Salmonella* and *Shigella* in freshly squeezed orange juice, fresh oranges, and wiping cloths collected from public markets and street booths in Guadalajara, Mexico: incidence and comparison of analytical routes. J. Food Prot. 69, 2595–2599.

CDC, Center for Disease Control and Prevention, 1975. *Salmonella* Typhimurium outbreak traced to a commercial apple cider—New Jersey. Morb. Mortal. Wkly Rep. 24, 87–88.

CDC, undated. Center for Disease Control and Prevention. Foodborne outbreak online database (Food). Available from: <http://wwwn.cdc.gov/foodborneoutbreaks/Default.aspx> (accessed July, 2015).

CDC, 1997. Outbreaks of *Escherichia coli* O157:H7 infection and cryptosporidiosis associated with drinking unpasteurized apple cider—Connecticut and New York, October 1996. Morb. Mortal. Wkly Rep. 124, 4–8.

CDC, 1999. Outbreak of *Salmonella* serotype Muenchen infections associated with unpasteurized orange juice—United States and Canada, June 1999. Morb. Mortal. Wkly Rep. 48, 582–585.

Cody, S.H., Glynn, K., Farrar, J.A., Cairns, K.L., Griffin, P.M., Kobayashi, J., et al., 1999. An outbreak of *Escherichia coli* O157:H7 infection from unpasteurized commercial apple juice. Ann. Intern. Med. 130, 202–209.

Conner, D.E., Kotrola, J.S., 1995. Growth and survival of *Escherichia coli* O157:H7 under acidic conditions. Appl. Environ. Microbiol. 61, 382–385.

Cook, K.A., Dobbs, T.E., Hadly, W.G., Wells, J.G., Barret, T.J., Puhr, N.D., et al., 1998. Outbreak of *Salmonella* serotype Hartford infections associated with unpasteurized orange juice. J. Am. Med. Assoc. 280, 1504–1508.

Diallo, M.O., Bradley, K.K., Crutcher, J.M., Lytle, M., Lee, A., Moolenaar, R.L., 2011. An outbreak of *Escherichia coli* O157:H7 infection linked to unpasteurized apple cider in Oklahoma, 1999. Food Prot. Trends 31, 88–92.

Diana, J.E., Pui, C.F., Son, R., 2012. Enumeration of *Salmonella* spp., *Salmonella* Typhi and *Salmonella* Typhimurium in fruit juices. International Food Research Journal 19, 51–56.

Eblen, B.S., Walderhaug, M.O., Edelson-Mammel, S., Chirtel, S.J., De Jesua, A., Lerker, R.L., et al., 2004. Potential for internalization, growth, and survival of *Salmonella* and *Escherichia coli* O157:H7 in oranges. J. Food Prot. 67, 1578–1584.

Eleftheriadou, M., Quantick, P., Nolan, M., Akkelidou, D., 1997. Factors affecting quality and safety of freshly squeezed orange juice (FSOJ). Dairy Food Environ. Sanitation 18 (1), 14–23.

El-Safey, M.E.-S., 2013. Behaviour of *Salmonella* Heidelberg in fruit juices. Int. J. Nutr. Food Sci. 2, 38–44.

FAO, 2003. Food and Agriculture Organization of the United Nations. Agriculture and Consumer Protection Department. Committee on Agriculture. COAG/2003/5, FAO's Strategy for a Food Chain Approach to Food Safety and Quality: A framework document for the development of future strategic direction. Seventeenth Session. Rome. Available from: <http://www.fao.org/docrep/MEETING/006/Y8350e.HTM> (accessed 24.11.15).

FAO/WHO (2005). Codex General Standard for Fruit Juices and Nectars.

FDA, Food and Drug Administration, 1998. Department of Health and Human Services. Food labeling: warning and notice: labeling of juice products; final rule: 21 CFR Part 101, 63 FR 37030–37056, 8 July 1998. U.S. Food and Drug Administration, Washington, DC.

FDA, 2015. Food and Drug Administration. Analysis and Evaluation of Preventive Control Measures for the Control and Reduction/Elimination of Microbial Hazards on Fresh and Fresh-Cut Produce: Outbreaks Associated with Fresh and Fresh-Cut Produce. Incidence, Growth, and Survival of Pathogens in Fresh and Fresh-Cut Produce. (Chapter 4) (Page last update 08/10/2015). Available from: <http://www.fda.gov/Food/FoodScienceResearch/SafePracticesforFoodProcesses/ucm091265.htm> (accessed 24.11.15).

Fleet, G.H., Heiskanen, P., Reid, I., Buckle, K.A., 2000. Foodborne viral illness—status in Australia. Int. J. Food. Microbiol. 59, 127–136.

Frank, C., Walter, J., Muehlen, M., Jansen, A., van Treeck, U., Hauri, A.M., et al., 2007. Major outbreak of hepatitis A associated with orange juice among tourists, Egypt, 2004. Emerg. Infect. Dis. 13, 156–158.

Fratamico, P.M., Deng, M.Y., Strobaugh, I.P., Palumbo, S.A., 1997. Construction and characterization of *Escherichia coli* O157:H7 strains expressing firefly luciferase and green fluorescent protein and their use in survival studies. J. Food Prot. 60, 1167–1173.

FSAI, Food Safety Authority of Ireland. 2007. Second National Microbiological Survey 2007 (07NS2). Bacteriological Safety of Fruit and/or Vegetable Juices and Smoothies. Available from: <https://www.fsai.ie/uploadedFiles/Monitoring_and_Enforcement/Monitoring/Surveillance/Micro%20Surveillance%2007NS2.pdf> (accessed 09.07.15).

Goverd, K.A., Beech, F.W., Hobbs, R.P., Shannon, R., 1979. The occurrence and survival of coliforms and salmonellas in apple juice and cider. J. Appl. Bacteriol. 46, 521–530.

Hernández, L.M., Ramírez Cano, A.N., Cucunubá, Z., Zambrano, P., 2009. Brote de Chagas agudo en Lebrija, Santander 2008. Revista del Observatorio de Salud Pública de Santander 1, 602–614.

Hilborn, E.D., Mshar, P.A., Fiorentino, T.R., Dembek, Z.F., Barret, T.J., Howard, R.T., et al., 2000. An outbreak of *Escherichia coli* O157:H7 infections an haemolytic uraemic syndrome associated with consumption of unpasteurized apple cider. Epidemiol. Infect. 124, 31–36.

Ibarra-Sánchez, L.S., Alvarado-Casillas, S., Rodríguez-García, M.O., Martínez-Gonzáles, N.E., Castillo, A., 2004. Internalization of bacterial pathogens in tomatoes and their control by selected chemicals. J. Food Prot. 67, 1353–1358.

Jain, S., Bidol, A., Austin, J.L., Berl, E., Elson, F., Williams, M.L., et al., 2009. Multistate outbreak of *Salmonella* Typhimurium and Saintpaul infections associated with unpasteurized orange juice—United States, 2005. Clin. Infect. Dis. 48, 1065–1071.

Kenney, S.J., Beuchat, L.R., 2002. Comparison of aqueous commercial cleaners for effectiveness in removing *Escherichia coli* O157:H7 and *Salmonella* Muenchen from the surface of apples. Int. J. Food. Microbiol. 74, 47–55.

Krause, G., Terzagian, R., Hammond, R., 2001. Outbreak of *Salmonella* serotype Anatum infection associated with unpasteurized orange juice. South. Med. J. 94, 1168–1172.

Leul, A., Kibret, M., 2012. Bacteriological safety of freshly squeezed mango and pineapple juices served in juice houses of Bahir Dar Town, Northwest Ethiopia. Int. J. Sci.: Basic Appl. Res. (IJSBAR). Available from: <http://gssrr.org/index.php?journal=JournalOfBasicAndApplied&page=article&op=view&path%5B%5D=1322> (accessed 10.07.15).

Lewis, J.E., Thompson, P., Rao, B., Kalavati, C., Rajanna, B., 2006. Human bacteria in street vended fruit juices: a case study of Visakhapatnam city, India. Internet J. Food Safety 8, 35–38.

Little, C.L., Mitchell, R.T., Food Standards Agency, Local Authorities Coordinators of Regulatory Services and the Health Protection Agency, 2004. Microbiological quality of pre-cut fruit, sprouted seeds, and unpasteurized fruit and vegetable juices from retail and production premises in the UK, and the application of HACCP. Commun. Dis. Public Health 7, 184–190.

LSDEPC, 2005. Laboratory surveillance data for enteric pathogens in Canada. Annual summary. Available from: <http://publications.gc.ca/collections/collection_2007/phac-aspc/HP57-1-2005E.pdf> (accessed July, 2015).

Luedtke, A.N., and Powell D.A., 2000. Fact sheet: a timeline of fresh juice outbreaks. Available from: <http://www.foodsafetynetwork.ca/food/juice-outbreaks.htm> (accessed 22.4.2003).

Mahale, D.P., Khade, R.G., Vaidya, V.K., 2008. Microbiological analysis of street vended fruit juices from Mumbai city, India. Internet. J. Food Safety 10, 31–34.

Martínez-Gonzáles, N.E., Martínez, G.M.C., Ruiz, R.A.C., Hernández, H.A., Mota de la Garza, L., Castillo, A., 2001a. Frecuencia de *Listeria monocytogenes* y *Salmonella* spp. en la superficie de naranjas. XXXII Congreso Nacional de Microbiología, Guanajuato, Guanajuato, México.

Martínez-Gonzáles, N.E., Ramos, H.M.E., Gómez, I.M.I., Martínez, C.L., Mota de la Garza, L., Castillo, A., 2001b. Frecuencia de *Listeria monocytogenes* y *Salmonella* spp.

en jugo de naranja fresco. XXXII Congreso Nacional de Microbiología, Guanajuato, Guanajuato, México.

Martínez-Gonzáles, N.E., Martínez-Chávez, L., Martínez-Cárdenas, C., Castillo, A., 2011. Comparison of different washing treatments for reducing pathogens on orange surfaces and for preventing the transfer of bacterial pathogens to fresh-squeezed orange juice. J. Food Prot. 74, 1684–1691.

Martínez-Gonzáles, N.E., Hérnandez-Herrera, A., Martínez-Chávez, L., Rodríguez-García, M.O., Torres-Vitela, M.R., Mota de la Garza, L., et al., 2003. Spread of bacterial pathogens during preparation of freshly squeezed orange juice. J. Food Prot. 66, 1490–1494.

Martínez-Gonzáles, N.E., Martínez-Chávez, L., Cabrera-Díaz, E., Martínez-Cárdenas, C., Gutiérrez-González, P., Castillo, A., 2016. Use of a Novel Medium, the Polymyxin Ceftazidime Oxford Medium, for Isolation of *Listeria monocytogenes* from raw or non-pasteurized foods. Food Microbiol. 55, 105–111.

Mazzota, A.S., 2001. Thermal inactivation of stationary-phase and acid-adapted *Escherichia coli* O157:H7, *Salmonella* and *Listeria monocytogenes* in fruit juices. J. Food Prot. 64, 315–320.

Mihajlovic, M., Dixon, B., Couture, H., Farber, J., 2013. Qualitative microbiological risk assessment of unpasteurized fruit juice and cider. Int. Food Risk Anal. J 3, 1–19.

Millard, P.S., Gensheimer, K.F., Addiss, D.G., Sosin, D.M., Beckett, G.A., Houck-Jankoski, A., et al., 1994. An outbreak of cryptosporidiosis from fresh-pressed apple cider. J. Am. Med. Assoc. 272, 1592–1596.

Mudgil, S., Aggarwal, D., Ganguli, A., 2004. Microbiological analysis of street vended fresh squeezed carrot and kinnow-mandarin juices in Patiala city, India. Internet J. Food Safety V.3, 1–3.

National Centre for Disease Control/Communicable Diseases Network Australia New Zealand, Australian Department of Health and Aged Care, 1999. Salmonellosis outbreak, South Australia. Commun. Dis. Intell. 23, 73.

Nicolò, M.S., Gioffrè, A., Carnazza, S., Platania, G., Silvestro, I.D., Guglielmino, S.P., 2011. Viable but nonculturable state of foodborne pathogens in grape fruit: a study of laboratory. Foodborne Pathogen Dis. 8, 11–17.

Noël, H., Hofhuis, A., De Jonge, R., Heuvelin, A.E., De Jong, A., Heck, M.E., et al., 2010. Consumption of fresh fruit juice: how a healthy food practice caused a national outbreak of *Salmonella* Panama gastroenteritis. Foodborne Pathogen Dis. 7, 375–381.

Oglesby A., 2015. Summary of special disease outbreak investigations—1997, Appendix E. Indiana State Department of Health. Available from: <http://www.in.gov/isdh/21186.htm>. (accessed July, 2015).

PAHO (Pan American Health Organization). 2009. Informe final de la 5a Reunión de la Iniciativa Intergubernamental de Vigilancia y Prevención de la Enfermedad de Chagas en la Amazonía (AMCHA), MPPS, OPS/OMS, AECI, 28–30 octubre de 2008, Caracas, Venezuela. Montevideo: OPS; 2009. (OPS/HSD/CD/548-09). Available from <http://www.paho.org/hq/index.php?option = com_docman&task = doc_view&gid = 15601&Itemid => (accessed 27.11.15).

Pao, S., Davis, C.L., 1999. Enhancing microbiological safety of fresh orange juice by fruit immersion in hot water and chemical sanitizers. J. Food Prot. 62, 756–760.

Pao, S., Davis, C.L., Parish, M.E., 2001. Microscopic observation and processing validation of fruit sanitizing treatments for the enhanced microbiological safety of fresh orange juice. J. Food Prot. 64, 310–314.

Parish, M.E., 1997. Public health and nonpasteurized fruit juices. Crit. Rev. Microbiol. 23, 109–119.

Parish, M.E., 1998. Coliforms, *Escherichia coli* and *Salmonella* serovars associated with a citrus-processing facility implicated in a salmonellosis outbreak. J. Food Prot. 61, 280–284.

Parish, M.E., Narciso, J.A., Friedrich, L.M., 1997. Survival of *Salmonellae* in orange juice. J. Food Safety 17, 273–281.

Parish, M.E., Goodrich, E.M., Narciso, J.A., Friedrich, L.M., 2001. Microflorae of orange surfaces and juice from fruit in processing facilities: preliminary results. Proc. Florida State Horticultural Soc. 114, 174–176.

Parish, S., Davis, C.L., 2001. Transfer of natural and artificially inoculated microorganisms from orange fruit to fresh juice during extraction. Lebensmittel-Wissenschaft und Technologie 34, 113–117.

Penteado, A.L., Eblen, B.S., Miller, A.J., 2004. Evidence of *Salmonella* internalization into fresh mangos during simulated postharvest insect disinfection procedures. J. Food Prot. 67, 181–184.

Pereira, K.S., Schmidt, F.L., Guaraldo, A.M.A., Franco, R.M.B., Dias, V.L., Passos, L.A. C., 2009. Chagas' disease as a foodborne illness. J. Food Prot. 72, 441–446.

Pilavtepe-Çelik, M., 2013. High hydrostatic pressure (HHP) inactivation of foodborne pathogens in low-acid juices. Int. J. Food Sci. Technol. 48, 673–677.

Porto-Fett, A.C.S., Call, J.E., Muriana, P.M., Freier, T.A., Luchansky, J.B., 2010. Listeria monocytogenes. In: Juneja, V.K., Sofos, J.N. (Eds.), Pathogens and Toxins in Foods; Challenges and Interventions. ASM Press, Washington, DC.

Produce for Better Health Foundation. State of the Plate, 2015 Study on America's consumption of fruit and vegetables, produce for better health foundation, 2015. Available from: <http://www.pbhfoundation.org/pdfs/about/res/pbh_res/State_of_the_Plate_2015_WEB_Bookmarked.pdf> (accessed 24.11.15).

Rai, A.K., Dhiman, R., Kumar, A.N., Kumar, V., Kaur, M., 2014. Microbes associated with prepared juices of citrus and carrots. International. J. Food. Sci. 2014, 1–7.

Rueda, K., Trujillo, J.E., Carranza, J.C., Vallejo, G.A., 2014. Transmisión oral de *Trypanosoma cruzi*: una nueva situación epidemiológica de la enfermedad de Chagas en Colombia y otros países suramericanos. Biomedica 34, 631–641.

Ryu, J.-H., Beuchat, L.R., 1998. Influence of acid tolerance on survival, growth, and thermal cross-protection of *Escherichia coli* O157:H7 in acidified media and fruit juices. Int. J. Food. Microbiol. 45, 185–193.

Sado, P.N., Jinneman, K.C., Husby, G.J., Sorg, S.M., Omiecinski, C.J., 1998. Identification of Listeria monocytogenes from unpasteurized apple juice using rapid test kits. J. Food Prot. 61, 1199–1202.

Santalla-Vargas, J., Oporto-Carrasco, P., Espinoza, E., Rios, T., Brutus, L., 2011. First reported outbreak of Chagas disease in the Bolivian Amazonean zone: a report of 14 cases of oral transmission of acute *Trypanosoma cruzi* in Guayaramerín, Beni-Bolivia. Biofarbo 19, 52–58.

Sharma, M., Beuchat, L.R., Doyle, M.P., Chen, J., 2001. Survival of *Salmonellae* in pasteurized, refrigerated calcium-fortified orange juice. J. Food Prot. 64, 1299–1304.

Sharma, P.U., 2013. Bacteriological analysis of street vended fruit juices available in Vidarbha. Int. J. Current Microbiol. Appl. Sci. 2, 178–183.

Singh, B.R., Kulshreshtha, S.B., Kapoor, K.N., 1995. An orange juice-borne diarrhoeal outbreak due to enterotoxigenic *Escherichia coli*. J. Food Sci. Technol. 32, 504–506.

Sospedra, I., Rubert, J., Soriano, J.M., Mañes, J., 2011. Incidence of microorganisms from fresh orange juice processed by squeezing machines. Food Control 23, 282–285.

Steele, B.T., Murphy, N., Rance, C.P., 1982. An outbreak of hemolytic uremic syndrome associated with ingestion of fresh apple juice. J. Pediatr. 101, 963–966.

Tambekar, D.H., Jaiswal, V.J., Dhanorkar, D.V., Gulhane, P.B., Dudhane, M.N., 2009. Microbial quality and safety of street vended fruit juices: a case study of Amravati city. Internet J. Food Safety 10, 72–76.

Tamblyn, S., de Grosbois, J., Taylor, D., Stratton, J., 1999. An outbreak of *Escherichia coli* O157:H7 infection associated with unpasteurized non-commercial, custom-pressed apple cider—Ontario, 1998. Canada Commun. Dis. Report 25, 113–117.

Thurston, H., Stuart, J., McDonell, B., Nicholas, S., Cheasty, T., 1998. Fresh orange juice implicated in an outbreak of *Shigella flexneri* among visitors to a South African game reserve. J. Infect. Dis. 36, 350.

Titarmare, A., Dabholkar, P., Godbole, S., 2009. Bacteriological analysis of street vended fresh fruit and vegetable juices in Nagpur city, India. Internet J. Food Safety 11, 1–3.

Torres-Vitela, M.D.R., Gómez-Aldapa, C.A., Cerna-Cortes, J.F., Villarruel-López, A., Rangel-Vargas, E., Castro-Rosas, J., 2012. Presence of indicator bacteria, diarrhoeagenic *Escherichia coli* pathotypes and *Salmonella* in fresh carrot juice from Mexican restaurants. Lett. Appl. Microbiol. 56, 180–185.

Tribst, A.A., Sant'Ana, A., de, S., de Massaguer, P.R., 2009. Review: microbiological quality and safety of fruit juices—past, present and future perspectives. Crit. Rev. Microbiol. 35, 310–339.

U.S. FDA, Food and Drug Administration, 2001. Department of Health and Human Services. Hazard Analysis and Critical Control Point (HACCP); Procedures for the Safe and Sanitary Processing and Importing of Juice Final Rule; final rule, 21 CFR Part 120, 66 FR 6137–6202, 19 January 2001. U.S. Food and Drug Administration, Washington, DC.

Venkitanarayanan, K.S., Lin, C.M., Bailey, H., Doyle, M.P., 2002. Inactivation of *Escherichia coli* O157:H7, *Salmonella* Enteritidis, and *Listeria monocytogenes* on apples, oranges, and tomatoes by lactic acid with hydrogen peroxide. J. Food Prot. 1, 100–105.

Victorian Government Department of Human Services, Food Safety Unit Melbourne, Victoria. 2005. Microbiological survey of freshly squeezed juices from retail businesses across Victoria. Available from: <http://www.health.vic.gov.au/archive/archive2011/foodsafety/archive/downloads/fruit_juice_survey_report_aug05.pdf> (accessed June 23, 2015).

Visser H., Verhoef L., Shop W., Götz H.M. 2010. Outbreak investigation in two groups of coach passengers with gastroenteritis returning from Germany to the Netherlands in February 2009. Euro Surveillance, 15:pii-19615.

Vojdani, J.D., Beuchat, L.R., Tauxe, R.V., 2008. Juice-associated outbreaks of human illness in the United States, 1995 through 2005. J. Food Prot. 71, 356–364.

Xia, X., Luo, Y., Yang, Y., Vinyard, B., Schneider, K., Meng, J., 2012. Effect of tomato variety, temperature differential, and post-stem removal time on internalization of *Salmonella enterica* serovar Thompson in tomatoes. J. Food Prot. 75, 297–303.

Zhao, T., Doyle, M.P., Besser, R.E., 1993. Fate of enterohemorrhagic *Escherichia coli* O157:H7 in apple cider without preservatives. Appl. Environ. Microbiol. 59, 2526–2530.

CHAPTER

12 Safety improvement of fruit juices by novel thermal and nonthermal processing

Vicknesha Santhirasegaram, Zuliana Razali and Chandran Somasundram

Institute of Biological Sciences and Centre for Research in Biotechnology for Agriculture, Faculty of Science, University of Malaya, Kuala Lumpur, Malaysia

INTRODUCTION

Fruit juices have the highest acceptability among beverages, generally due to their natural taste, as well as their nutritional value (Rivera and Cabornida, 2008). Consumption of fruit juice has been linked to the prevention of cardiovascular diseases and cancer, owing to its antioxidant properties. However, the number of outbreaks and cases of illness caused by consumption of contaminated juices, especially unpasteurized juices, has increased over the last decade. Currently, conventional thermal treatment is the preferred technology to inactivate microorganisms and enzymes causing spoilage, thus prolonging the shelf-life of juice. Due to the relatively high temperatures generally needed to inactivate pathogenic microorganisms, thermal processing causes detrimental effects on the nutritional quality of juices (Aguilo-Aguayo et al., 2009; Rawson et al., 2011; Santhirasegaram et al., 2013).

The growing interest for new preservation technology that retains quality and at the same time ensures food safety has promoted the effort for developing innovative nonthermal food processing methods. Some of the nonthermal processing methods extensively studied for juice preservation include high hydrostatic pressure (HHP), pulsed electric field (PEF), ultrasound, ultraviolet-light irradiation, ionizing radiation (IR), dense-phase carbon dioxide, and ozone (Morris et al., 2007).

Nonthermal technologies have different modes of microbial inactivation, therefore being potential options for a hurdle concept. The hurdle technology is a combination of preservation techniques that may have a synergistic effect on microbial destruction, with minimal impact on the quality of the food product (Leistner, 2000). This chapter will provide an overview of issues related to the microbial safety of fruit juices, emerging thermal and nonthermal processing

P. Kotzekidou (Ed): Food Hygiene and Toxicology in Ready-to-Eat Foods.
DOI: http://dx.doi.org/10.1016/B978-0-12-801916-0.00012-1

methods, hurdle technology, and future perspectives of fruit juice processing. This information is necessary to improve the progress of positive implementation of novel processing methods in the fruit juice industry.

FRUIT JUICE DETERIORATION AND SPOILAGE

Freshly squeezed fruit juices are gaining popularity among consumers due to their fresh-like characteristics. However, inappropriate handling and storage conditions may cause fruit juice deterioration and spoilage. Natural fruit juices are susceptible to spoilage, mainly due to their intrinsic properties such as pH, water activity, redox potential, and nutrients (Odumeru, 2012).

Fruit juice deterioration is mostly caused by enzymatic, chemical, and microbial reactions. Enzymes in fruit juices, such as polyphenol oxidase and peroxidase, may react with oxygen, thus contributing to juice browning and off-flavor. In addition, chemical reactions, such as Maillard's reaction, cause nonenzymatic browning of juices. In Maillard reaction, the carbonyl group of reducing sugars reacts with a free amino group of amino acids to form intermediates. Subsequently, these intermediates react further to produce brown-colored pigments, which induce undesirable color changes in juices (Bharate and Bharate, 2012).

The causal agents of microbiological spoilage of fruit juices can be bacteria, as well as yeast and molds. However, yeast and molds are the main spoilage agents due to the low pH of fruit juices. According to the Centers for Disease Control and Prevention (1996), one of the foodborne disease outbreaks has been linked to pathogens, such as *Escherichia coli* O157:H7, where the emphasis was on unpasteurized juices. In addition, bacterial pathogens including *Campylobacter* spp., *Salmonella* spp., and *Listeria monocytogenes* are the top causes of bacterial foodborne infections reported. Usually, foodborne illness resulting from severe infections such as salmonellosis and hemolytic uremic syndrome (HUS) require hospitalization. Infections by *E. coli* O157:H7 cause HUS, which results in kidney failure, abdominal cramps, and eventually death. Meanwhile, salmonellosis caused by *Salmonella* spp. results in diarrhea, nausea, fever, and abdominal pain (Odumeru, 2012).

Certain common molds including *Penicillium* spp., *Aspergillus* spp., *Eurotium* spp., *Alternaria* spp., *Cladosporium* spp., *Paecilomyces* spp., and *Botrytis* spp cause spoilage in fruit products. Besides that, yeasts including *Saccharomyces* spp., *Zygosaccharomyces* spp., *Candida* spp., and *Debaryomyces* spp. are reported to be involved in fruit product spoilage (Bates et al., 2001).

Spoilage of fruit juice and related products as a result of microbial growth may contribute to physical and chemical changes in food products. These alterations include unacceptable flavor and odor, changes in color and turbidity, gas production, and formation of slime. Usually, growth of microorganisms to high numbers is necessary before spoilage becomes noticeable. Hence, it is important to control the growth of spoilage organisms in order to inhibit microbial spoilage (Odumeru, 2012).

FRUIT JUICE PROCESSING

The increase in outbreaks and cases of illness related to consumption of unpasteurized juices have urged the development of a more effective food safety control program, known as the hazard analysis critical control point (HACCP) program. HACCP is a systematic approach to identify, assess, and control microbiological, chemical, and physical hazards of public health concern (Odumeru, 2012). Currently, there are a number of fruit juice preservation technologies for controlling microbial growth and survival. These preservation methods must be evaluated to avoid significant organoleptic changes in food products. Hence, the main objective of fruit juice processing is to prevent microbiological spoilage while assuring safety and maintaining quality characteristics. Fruit juice processing technologies can be divided into two groups, namely thermal and nonthermal processing (Bates et al., 2001).

THERMAL PROCESSING

Conventional thermal treatment is the preferred technology to inactivate microorganisms and enzymes causing spoilage, thus prolonging the shelf-life of juice. Traditional thermal processing depends on the generation of heat outside the product to be heated and its transfer into the product via convection and conduction mechanisms (Pereira and Vicente, 2010). Pasteurization is an example of thermal treatment that is commonly practiced in the food industry. There are two types of pasteurization methods, which include low-temperature long time (LTLT) and high-temperature short time (HTST). LTLT, also known as batch pasteurization, was initially established for the preservation of milk, whereby it is conducted at 63°C for 30 min. The most common pasteurization method for fruit juice is HTST, known as flash pasteurization (David, 1996).

According to Nagy et al. (1993), HTST treatment for fruit juices range from 90°C to 95°C for 15–60 s to assure at least a 5 log reduction in microbial count. This method is faster and more efficient than LTST. However, the short holding time of HTST may require special equipment, thus it is less cost-effective than LTST (David, 1996). The time and temperature variables for pasteurization of juice depend on the type of juice, initial microbial count, pH, water activity, and thermal inactivation kinetics of microorganisms present in juice. Hence, pasteurization conditions should be selected appropriately to avoid overprocessing. However, underprocessing may not completely inactivate microorganism growth, thus resulting in juice spoilage (Rawson et al., 2011).

Table 12.1 Effects of Thermal Processing on the Reduction of Microbial Load in Fruit Juices

Fruit Juices	Conditions	Target Microbial Inactivation	References
Orange and carrot juice blend	98°C, 21 s	TPC[a], yeasts, molds	Rivas et al. (2006)
Tomato juice	80°C, 1 min	Aerobic bacteria	Dede et al. (2007)
Apple juice	72°C and 94°C, 26 s	TPC[a]	Noci et al. (2008)
Orange juice	72°C, 20 s	TPC[a], *E. coli*, lactic acid bacteria, yeasts, molds	Timmermans et al. (2011)

[a]*TPC, total plate count.*

High temperatures cause adverse effects on microbes by damaging organic molecules (nucleic acids and proteins) required for the proper functioning of cells, thus causing cytolytic effects. In addition, heat disrupts the integrity of the cellular membrane, eventually inactivating microbial growth (Rawson et al., 2011). This is consistent with the study conducted by Noci et al. (2008) on heat-treated apple juice (at 72°C and 94°C for 26 s), where microbial count was reduced to below the detection limit (<1 log CFU/mL). Correspondingly, successful reduction of microbial load and prolonged shelf-life were observed in thermally pasteurized orange and carrot juice blend (98°C for 21 s), as reported by Rivas et al. (2006). In addition, thermal treatment regardless of mild or high-heat pasteurization (at 90°C for 30 and 60 s) completely inactivated microbial growth in the Chokanan mango juice samples (Santhirasegaram et al., 2013). Table 12.1 summarizes the effect of thermal processing on the reduction of microbial load in fruit juices.

Due to the relatively high temperatures generally needed to inactivate food-poisoning and spoilage microorganisms, conventional pasteurization can adversely affect the quality of food products, by reducing their nutritional value or altering sensory attributes, such as color and flavor (Rawson et al., 2011). Some studies on thermally treated fruit juices, such as orange (Cortes et al., 2008) and strawberry (Aguilo-Aguayo et al., 2009), reported significant loss of quality and degradation of bioactive compounds such as ascorbic acid. In addition, Rattanathanalerk et al. (2005) reported significant color degradation in thermal-treated pineapple juice (at 85°C and 95°C for 60 s). Similarly, Pala and Toklucu (2013) observed significant loss of phenolic compounds in orange juice subjected to HTST treatment (at 90°C for 2 min).

NONTHERMAL PROCESSING

The growing interest in fresh-like products has promoted the effort for developing innovative nonthermal food preservation methods. Nonthermal processing techniques have been explored for their efficacy to extend shelf-life and enhance the safety of fresh juice while preserving organoleptic and nutritional qualities. In addition, these preservation methods are considered to be more energy-efficient and provide better retention of quality when compared to conventional thermal processing. Some of the nonthermal processing methods extensively studied for juice preservation include HHP, PEF, ultrasound, ultraviolet-light irradiation, IR, dense-phase carbon dioxide, and ozone (Morris et al., 2007). These treatments are classified as "nonthermal" as they eliminate the use of high temperature to achieve microbial inactivation, thus avoiding the adverse effects of heat on physical, nutritional, and sensory properties of foods. However, some of these treatments, such as high pressure and PEF, may involve heat due to generation of internal energy (Pereira and Vicente, 2010).

High hydrostatic pressure

HHP, also known as ultrahigh pressure, is a cold pasteurization technique that entails the transmission of pressure ranging from 100 to 1000 MPa to food. However, pressures between 300 and 700 MPa are commonly applied for food processing, from a millisecond pulse to over 20 min, in a batch or semicontinuous system. Food products (as solid, liquid, or packages) are loaded into steel vessels filled with hydraulic fluid (normally water). Then, pressurization takes place through direct or indirect compression. For every 100 MPa of pressure applied, there is an increase in temperature (3°C) of the product (Morris et al., 2007; Lopez-Gomez et al., 2009). Since the pressure is transmitted uniformly throughout the product and reaches isostatic equilibrium, it is important to highlight that this process is not influenced by the size and shape of product, unlike thermal processing.

HHP inactivates microbial cells by breaking noncovalent bonds and damaging cell membrane. In addition, this process causes protein denaturation, thus interrupting cellular functions, and eventually resulting in cell death (Morris et al., 2007; Lopez-Gomez et al., 2009). Effective microbial inactivation was reported in HHP-processed tomato juice (at 250 MPa, 35°C for 15 min) and pomegranate juice (at 400–600 MPa, 25–50°C, for 5 and 10 min) (Dede et al., 2007; Ferrari et al., 2010). Table 12.2 summarizes the effect of HHP processing on the reduction of microbial load in fruit juices.

Besides that, there is minimal alteration in nutritional and sensory quality of HHP-processed food due to its limited effects on the covalent bonds (Ferrari et al., 2010). Indeed, better retention of ascorbic acid and antioxidant capacity was reported in HHP-treated juices when compared to conventional thermal processing, such as was observed in Valencia and Navel orange (Bull et al., 2004) and tomato juice (Dede et al., 2007).

Pulsed electric field

PEF treatment applies short pulses (1–100 μs) with high voltage (10–50 kV/cm) to liquid products in a continuous system. A simple PEF system consists of a high-voltage power supply, a pulse generator, a treatment chamber, and a switch to

Table 12.2 Effects of HHP Processing on the Reduction of Microbial Load in Fruit Juices

Fruit Juices	Conditions	Target Microbial Inactivation	References
Valencia and Navel orange juice	600 MPa, 20°C, 60 s	Aerobic bacteria, yeasts, molds, *Salmonella*	Bull et al. (2004)
Tomato juice	250 MPa, 35°C, 15 min	Aerobic bacteria	Dede et al. (2007)
Pomegranate juice	400–600 MPa, 25–50°C, 5 and 10 min	TPC[a]	Ferrari et al. (2010)
Orange juice	600 MPa, 17°C, 1 min	TPC[a], *E. coli*, lactic acid bacteria, yeasts, molds	Timmermans et al. (2011)

[a]*TPC, total plate count.*

discharge energy to electrodes. In addition, there is a cooling system to balance moderate temperature rise during treatment. Some of the limitations of PEF are that it cannot be used on products that have variable electrical conductivity as well as products that contain or could form air bubbles. Besides that, PEF has limited effects on microbial spores (Morris et al., 2007; Lopez-Gomez et al., 2009). The effectiveness of PEF processing is dependent on variables such as pulse width, electric field strength, flow rate, treatment temperature, and time of exposure.

PEF creates pores in the cell membrane of microbial cells. This electroporation process causes leakage of cellular contents, eventually resulting in microbial cell disruption. Indeed, a longer pulse width and higher-intensity pulse fields were more effective in inactivating microbial growth (Morris et al., 2007; Rawson et al., 2011). The reduction in microbial loads greater than 5 log was observed in PEF-treated orange juice (40 kV/cm, 56°C, 100 μs) (McNamee et al., 2010). Correspondingly, effective reduction of microbial count was observed in thermally pasteurized orange and carrot juice blend (25 kV/cm, 280 and 330 μs), as reported by Rivas et al. (2006). Table 12.3 summarizes the effect of PEF processing on the reduction of microbial load in fruit juices.

Some studies reported that juices treated with lower-intensity pulse fields and shorter pulse width exhibited higher retention of ascorbic acid, such as was observed in orange (Elez-Martinez and Martin-Belloso, 2007). In addition, enhancement of antioxidant capacity was observed in PEF-processed juices due to increased extraction yield of secondary metabolites and generation of free radicals (Rawson et al., 2011).

Ultrasound

Power ultrasound (10–1000 W/cm^2) is used to alter food properties, either physically or chemically, such as by disrupting cells and inactivating enzymes.

Table 12.3 Effects of PEF Processing on the Reduction of Microbial Load in Fruit Juices

Fruit Juices	Conditions	Target Microbial Inactivation	References
Orange and carrot juice blend	25 kV/cm, 280 μs, 68°C and 25 kV/cm, 330 μs, 70°C	TPC[a], yeasts, molds	Rivas et al. (2006)
Orange juice	40 kV/cm, 56°C, 100 μs	*Pichia fermentans*, *Listeria innocua*, and *E. coli* k12	McNamee et al. (2010)
Orange juice	23 kV/cm, 58°C, 2 μs	TPC[a], *E. coli*, lactic acid bacteria, yeasts, molds	Timmermans et al. (2011)

[a]*TPC, total plate count.*

Ultrasonic processing equipment includes ultrasonic bath and probe systems (Carcel et al., 2012). The mechanism of action for ultrasonic processing, commonly known as sonication, is explained in three different approaches, which are cavitation, localized heating, and formation of free radicals. When high-power ultrasound at low frequencies (20–100 kHz) propagates in liquid, cavitation (formation and collapse of bubbles) occurs. As a result, there is elevation of localized pressure (up to 500 MPa) and temperature (up to 5000°C). These "tiny hotspots" provide the energy to alter the properties of food product either physically or chemically. Accordingly, these cavitation bubbles induce microstreaming and shear stress, resulting in the disintegration of the microbial cells. Besides that, cavitation causes intracellular micromechanical shock that disrupts the functional components of cell, thus inactivating enzymes (O'Donnell et al., 2010; Abid et al., 2013). Another mode of action of microbial inactivation is the formation of free radicals due to sonolysis of water. Therefore, hydrogen peroxide, OH^-, and H^+ ions which have bactericidal effects are produced (O'Donnell et al., 2010).

The cavitation phenomenon is affected by the frequency and amplitude of ultrasound waves. At lower frequencies (20 kHz), the bubbles produced are larger in size and when they collapse higher energies are produced. Additionally, higher-intensity cavitation can be produced by using higher-amplitude ultrasound.

Sonication is a potential technology to achieve the US FDA condition of a 5 log reduction in foodborne pathogens in fruit juices. Previous studies have reported that sonication is effective in reducing the microbial count of foodborne pathogens, such as in orange (500 kHz, 240 W, for 15 min), apple (25 kHz for 60 and 90 min), and carrot juice (20 kHz, 750 W, for 2 min) (Valero et al., 2007; Abid et al., 2013; Jabbar et al., 2014). This could be explained by the collapse of cavitation-induced microbubbles resulting in a localized decontamination effect. Table 12.4 summarizes the effect of ultrasonic processing on the reduction of microbial load in fruit juices.

Table 12.4 Effects of Ultrasonic Processing on the Reduction of Microbial Load in Fruit Juices

Fruit Juices	Conditions	Target Microbial Inactivation	References
Orange juice	500 kHz, 240 W, 15 min	TPC[a], yeasts, molds	Valero et al. (2007)
Apple juice	25 kHz, 20°C, 60 and 90 min	TPC[a], yeasts, molds	Abid et al. (2013)
Chokanan mango juice	40 kHz, 130 W, 25°C, 15, 30 and 60 min	Aerobic bacteria, coliforms, yeasts, molds	Santhirasegaram et al. (2013).
Carrot juice	20 kHz, 750 W, 15°C, 2 min	TPC[a], yeasts, molds	Jabbar et al. (2014)

[a]*TPC, total plate count.*

Several studies using ultrasonic processing on fruit juice reported minimal effect on the degradation of quality parameters, and improved functionalities such as in kasturi lime (Bhat et al., 2011a), apple (Abid et al., 2013), and carrot juice (Jabbar et al., 2014). Besides that, Rawson et al. (2011) reported that sonication provides better retention of bioactive compounds. In addition, Tiwari et al. (2009) reported that ultrasonic processing (25 kHz for 2 min) improves the cloud value and stability of orange juice during storage. However, significant color degradation was observed in orange juice subjected to sonication (Tiwari et al., 2009).

Ultraviolet-C light

Ultraviolet-C (UV-C) light is a part of the electromagnetic spectrum with wavelengths between 200 and 280 nm, exhibiting germicidal properties as it inactivates bacterial and viral microorganisms. Although UV-C radiation technology is considered to be an effective method for food preservation, consumers' misconceptions about this process have delayed many of its potential applications in the food industry. Actually, UV-C radiation is a physical treatment which does not result in chemical residues. Hence, consumption of UV-C-treated food products is not harmful to humans.

One of the limitations of the application of UV in juice processing is associated with the high absorbance coefficients of juice. According to Koutchma et al. (2004), UV-C penetration largely depends on the presence of dissolved organic solutes (suspended solids) and colored compounds which act as a barrier, thus exhibiting UV-C attenuation effects. Hence, an appropriate UV reactor should be designed to reduce the interference of UV absorbance and improve microbial inactivation efficiency. The reactor design should include a narrow laminar flow or conditions with high turbulence, where juices are mixed, resulting in all parts being exposed to the UV light source (Koutchma et al., 2004).

The UV-C light, with a peak emission at the wavelength of 254 nm, exhibits a germicidal effect on bacteria, yeasts, molds, and viruses. This could be explained

Table 12.5 Effects of UV-C Processing on the Reduction of Microbial Load in Fruit Juices

Fruit Juices	Conditions	Target Microbial Inactivation	References
Orange juices	123 mJ/cm^2 dose	TPC[a], yeasts, molds	Tran and Farid (2004)
Apple juice	20°C, 30 min	*Staphylococcus aureus*	Walkling-Ribeiro et al. (2008)
Starfruit juice	2.2 J/m^2 dose, 25°C	TPC[a], yeasts, molds	Bhat et al. (2011b)
Orange juice	36.1 kJ/L dose	TPC[a], yeasts, molds, *E. coli* O157:H7	Pala and Toklucu (2013).
Chokanan mango juice	3.525 J/m^2, 25°C	Aerobic bacteria, coliforms, yeasts, molds	Santhirasegaram et al. (2015a)

[a]*TPC, total plate count.*

by the characteristic of microbial DNA to absorb UV-C light photons, thus generating crosslinks between neighboring cytosine and thymine (pyrimidine) bases in the same DNA strand (Tran and Farid, 2004). Therefore, these pyrimidine dimers prevent DNA transcription and translation. Subsequently, this condition prevents microorganisms from reproducing, and eventually results in cell death. However, yeast and mold are more resistant to UV-C. This could be due to the difference in thickness of the cell wall and size of microorganisms, thus influencing the passage of UV-C light. In addition, lesser pyrimidine bases on the DNA strand of yeast and mold contribute to less probability of crosslink formation, and thus higher resistance to UV-C (Miller et al., 1999).

The US FDA criterion of a 5 log reduction of chosen pathogen in fruit juices can be achieved by UV-C radiation. Previous studies have reported that UV-C treatment causes a significant reduction in microbial load, such as in apple (Walkling-Ribeiro et al., 2008), starfruit (Bhat et al., 2011b), and orange juices (Tran and Farid, 2004; Pala and Toklucu, 2013). Table 12.5 summarizes the effect of UV-C processing on the reduction of microbial load in fruit juices.

Several studies using short-wave UV-C light treatment on fruit juices reported better retention of nutritional and quality attributes, such as in starfruit (Bhat et al., 2011b) and orange juice (Pala and Toklucu, 2013). Besides that, Bhat et al. (2011b) reported that UV-C processing (for 30 and 60 min) induces a significant increase in polyphenol and flavonoid content of starfruit juice. However, there is an increasing trend in browning degree and color changes corresponding to increased UV-C treatment time, as previously reported by Bhat et al. (2011b).

Ionizing radiation

IR involves the exposure of food products to IR sources including high-energy electrons, X-rays (machine-generated), and gamma rays (from radioisotopic source cobalt-60 or cesium-137). The World Health Organization has declared

Table 12.6 Effects of IR Processing on the Reduction of Microbial Load in Fruit Juices

Fruit Juices	Conditions	Target Microbial Inactivation	References
Pomegranate juice	2 kGy (gamma rays)	TPC[a], yeasts, molds	Alighourchi et al. (2008)
Pomegranate juice	3 kGy (gamma rays)	*E. coli*, *Saccharomyces cerevisiae*	Alighourchi et al. (2014)

[a]*TPC, total plate count.*

that IR exposure of any food commodity, up to 10 kGy, does not produce any toxicological hazards or nutritional or microbiological problems in food (Lopez-Gomez et al., 2009; Sorour et al., 2014). The IR plant consists of a conveyer belt, an irradiation chamber containing the radiation source, with separate loading and unloading areas.

IR damages microbial DNA, therefore terminating reproduction and cellular functions, and eventually resulting in microbial inhibition. Besides that, IR indirectly promotes the formation of radiolytic products such as free radicals (H^+ and OH^-), which inactivate spoilage microorganisms in juices (Sorour et al., 2014). IR reduces spoilage and pathogenic microorganisms, such as *Salmonella*, *E. coli* O157:H7, *Listeria monocytogenes*, and *Campylobacter jejuni*, by DNA fragmentation (Morris et al., 2007). Alighourchi et al. (2008) reported a significant reduction in microbial growth in pomegranate juice exposed to a 2-kGy dose of gamma rays. Table 12.6 summarizes the effect of IR on the reduction of microbial load of fruit juices. Low doses of IR cause minimal modification of nutritional and organoleptic (flavor and color) qualities of food. However, doses of IR above 10 kGy induce a reduction in total phenolics of irradiated product (Morris et al., 2007).

Ozone

Ozone is a triatomic allotrope of oxygen and is characterized by its high bactericidal and virucidal properties. The FDA approved ozone as a direct additive to food due to its wide antimicrobial spectrum for food preservation. Ozone can be generated by reaction of free oxygen radicals (by ultraviolet radiation or electric discharge method) with diatomic oxygen. Ozone processing can be conducted on solid foods by either washing with ozonated water or by gaseous treatment (Rawson et al., 2011; Sorour et al., 2014).

Ozone inactivates microbial cells by oxidation, thus making it an environmentally friendly antimicrobial agent. In addition, ozone causes disruption of cell membrane, thus leading to microbial cell lysis. The strong biocidal action of ozone in fruit juice processing was reported in apple cider and orange juice (Steenstrup and Floros, 2004; Patil et al., 2009). However, ozone treatment causes

Table 12.7 Effects of Ozone Processing on the Reduction of Microbial Load in Fruit Juices

Fruit Juices	Conditions	Target Microbial Inactivation	References
Apple cider	860 ppm	*E. coli* O157:H7	Steenstrup and Floros (2004)
Orange juice	75–78 μg/mL	*E. coli* ATCC 25922, *E. coli* NCTC 12900	Patil et al. (2009)

a significant difference in color of apple cider (Choi and Nielsen, 2005). Table 12.7 summarizes the effect of ozone processing on the reduction of microbial load in fruit juices.

COMBINATION OF PROCESSING METHODS OR HURDLE CONCEPT

The hurdle concept is based on the combination of different preservation methods or factors ("hurdles") to produce minimally processed food with fresh-like characteristics, safe, nutritious, stable, and economical. These hurdles can be classified into three groups, namely physical, physicochemical, and microbially derived hurdles (Leistner and Gorris, 1995). Generally, by placing a combination of sublethal stress (hurdles) on a microbial cell, the organism makes an effort to overcome the hostile environment, thus leading to metabolic exhaustion and eventually death (Leistner and Gorris 1995; Leistner, 2000).

According to the hurdle technology, a combination of two or more nonthermal processing techniques at lower individual intensities may have an additive or, even, a synergistic effect on microbial destruction, with minimal impact on the quality of the food product (Leistner, 2000). Nonthermal technologies have different modes of microbial inactivation, therefore being potential options for a hurdle concept. Some studies have demonstrated that fruit juices were successfully preserved by combining nonthermal technologies (Noci et al., 2008, Walkling-Ribeiro et al., 2008; Santhirasegaram et al., 2015b). Combination of nonthermal processing methods in a hurdle sequence can speed up the rate of sterilization of food products. Table 12.8 summarizes the effect of hurdle technology on the reduction of microbial load in fruit juices.

Previous studies have reported promising retention of sensory quality of fruit juices by combined nonthermal processing methods (Caminiti et al., 2011; Caminiti et al., 2012). In addition, Caminiti et al. (2011) and Santhirasegaram et al. (2015b) reported that apple and cranberry juice blend and Chokanan mango juice subjected to combined nonthermal processing methods exhibited no accumulation of brown pigments.

Table 12.8 Effects of Hurdle Technology on the Reduction of Microbial Load in Fruit Juices

Fruit Juices	Combination of Processing Methods	Target Microbial Inactivation	References
Apple juice	PEF (40 kV/cm, 57°C, 1 μs) and UV-C (30 min)	TPC[a]	Noci et al. (2008)
Apple juice	PEF (34 kV/cm, 63 μs, 44°C) and UV-C (20°C, 30 min)	*S. aureus*	Walkling-Ribeiro et al. (2008)
Chokanan mango juice	UV-C (3.525 J/m^2, 25°C, 30 min) and US (40 kHz, 130 W, 25°C, 30 min)	Aerobic bacteria, coliforms, yeasts, molds	Santhirasegaram et al. (2015b)

[a]*TPC, total plate count.*

CONCLUSION AND FUTURE PERSPECTIVES

The value for fruit juices has been increasing in the global market due to their health benefits. Conventional thermal pasteurization is the preferred technology used to achieve microbial inactivation, and extend the shelf-life of juices. Lately however, consumers' demand for a new preservation technology that retains freshness and at the same time ensures food safety, has resulted in growing interest in nonthermal processing methods. Nonetheless, only HHP processing has been used on a pilot scale. Therefore, there is need for other nonthermal processing techniques to be tested on a pilot scale, so that these methods can be developed as an alternative to thermal pasteurization. With the information obtained from this chapter, further research work is needed to develop models to optimize some of the critical processing variables during fruit juice processing. Besides that, additional research studies are needed to evaluate the effects of processing methods on the microbial stability of nonthermal-treated juice (as a standalone and combined treatment) during storage. In future, the combination of nonthermal processing methods as a hurdle concept would be a new trend in the preservation of fruit juices that improves the microbiological quality and safety with minimal impact on the quality of the food product.

REFERENCES

Abid, M., Jabbar, S., Wu, T., Hashim, M.M., Hu, B., Lei, S., et al., 2013. Effect of ultrasound on different quality parameters of apple juice. Ultrason. Sonochem. 20, 1182–1187.

Aguilo-Aguayo, I., Oms-Oliu, G., Soliva-Fortuny, R., Martin-Belloso, O., 2009. Changes in quality attributes throughout storage of strawberry juice processed by high-intensity pulsed electric fields or heat treatments. LWT—Food Sci. Technol. 42, 813–818.

Alighourchi, H., Barzegar, M., Abbasi, S., 2008. Effect of gamma irradiation on the stability of anthocyanins and shelf-life of various pomegranate juices. Food. Chem. 110, 1036–1040.

Alighourchi, H., Barzegar, M., Sahari, M.A., Abbasi, S., 2014. The effects of sonication and gamma irradiation on the inactivation of *Escherichia coli* and *Saccharomyces cerevisiae* in pomegranate juice. Iran J. Microbiol. 6, 51–58.

Bates, R.P., Morris, J.R., Crandall, P.G., 2001. Principles and practices of small- and medium-scale fruit juice processing. Food and Agriculture Organization (FAO). FAO, Rome, Agricultural Services Bulletin Issue 146.

Bharate, S.S., Bharate, S.B., 2012. Non-enzymatic browning in citrus juice: chemical markers, their detection and ways to improve product quality. J. LWT—Food Sci. Technol. 51, 2271–2288.

Bhat, R., Ameran, S.B., Voon, H.C., Karim, A.A., Tze, L.M., 2011b. Quality attributes of starfruit (*Averrhoa carambola* L.) juice treated with ultraviolet radiation. Food. Chem. 127, 641–644.

Bhat, R., Kamaruddin, N.S., Min-Tze, L., Karim, A.A., 2011a. Sonication improves kasturi lime (*Citrus microcarpa*) juice quality. Ultrason. Sonochem. 18, 1295–1300.

Bull, M., Zerdin, M., Howe, E., Goicoechea, D., Paramanandhan, P., Stockman, R., 2004. The effect of high pressure processing on the microbial, physical and chemical properties of Valencia and Navel orange juice. Innovative LWT—Food Sci. Technol. 5, 135–149.

Caminiti, I.M., Noci, F., Morgan, D.J., Cronin, D.A., Lyng, J.G., 2012. The effect of pulsed electric fields, ultraviolet light or high intensity light pulses in combination with manothermosonication on selected physico-chemical and sensory attributes of an orange and carrot juice blend. Food Bioprod. Process. 90, 442–448.

Caminiti, I.M., Noci, F., Munoz, A., Whyte, P., Morgan, D.J., Cronin, D.A., et al., 2011. Impact of selected combinations of non-thermal processing technologies on the quality of an apple and cranberry juice blend. Food. Chem. 124, 1387–1392.

Carcel, J.A., Garcia-Perez, J.V., Benedito, J., Mulet, A., 2012. Food process innovation through new technologies: use of ultrasound. J. Food Eng. 110, 200–207.

Centers for Disease Control and Prevention (CDC), 1996. Outbreak of Escherichia coli O157:H7 infections associated with drinking unpasteurised commercial apple juice: British Columbia, California, Colorado, and Washington. Morb. Mortal. Wkly Rep. 45, 975.

Choi, L.H., Nielsen, S.S., 2005. The effects of thermal and non-thermal processing methods on apple cider quality and consumer acceptability. J. Food. Qual. 28, 13–29.

Cortes, C., Esteve, M., Frigola, A., 2008. Color of orange juice treated by high intensity pulsed electric fields during refrigerated storage and comparison with pasteurized juice. Food. Control. 19, 151–158.

David, J.R.D., 1996. Principles of thermal processing and optimization. In: David, J.R.D., Graves, R.H., Carlson, V.R. (Eds.), Aseptic processing and packaging of food: a food industry perspective. CRC Press LLC, Boca Raton, FL, p. 14.

Dede, S., Alpas, H., Bayindirli, A., 2007. High hydrostatic pressure treatment and storage of carrot and tomato juices: antioxidant activity and microbial safety. J. Sci. Food. Agric. 87, 773–782.

Elez-Martinez, P., Martin-Belloso, O., 2007. Effects of high intensity pulsed electric field processing conditions on vitamin C and antioxidant capacity of orange juice and gazpacho, a cold vegetable soup. Food. Chem. 102, 201–209.

Ferrari, G., Maresca, P., Ciccarone, R., 2010. The application of high hydrostatic pressure for the stabilization of functional foods: pomegranate juice. J. Food Eng. 100, 245–253.

Jabbar, S., Abid, M., Hu, B., Wu, T., Hashim, M.M., Lei, S., et al., 2014. Quality of carrot juice as influenced by blanching and sonication treatments. LWT—Food Sci. Technol. 55 (1), 16–21.

Koutchma, T., Keller, S., Chirtel, S., Parisi, B., 2004. Ultraviolet disinfection of juice products in laminar and turbulent flow reactors. Innovative Food Sci. Emerg. Technol. 5, 179–189.

Leistner, L., 2000. Basic aspects of food preservation by hurdle technology. Int. J. Food. Microbiol. 55, 181–186.

Leistner, L., Gorris, L.G.M., 1995. Food preservation by hurdle technology. Trends Food Sci. Technol. 6, 41–46.

Lopez-Gomez, A., Fernandez, P.S., Palop, A., Periago, P.M., Martinez-Lopez, A., Marin-Iniesta, F., et al., 2009. Food safety engineering: an emergent perspective. Food Eng. Rev. 1, 84–104.

McNamee, C., Noci, F., Cronin, D.A., Lyng, J.G., Morgan, D.J., Scannell, A.G.M., 2010. PEF based hurdle strategy to control *Pichia fermentans*, *Listeria innocua* and *Escherichia coli* k12 in orange juice. Int. J. Food. Microbiol. 138 (1–2), 13–18.

Miller, R., Jeffrey, W., Mitchell, D., Elasri, M., 1999. Bacteria responses to ultraviolet light. Am. Soc. Microbiol. 65, 535–541.

Morris, C., Brody, A.L., Wicker, L., 2007. Non-thermal food processing/preservation technologies: a review with packaging implications. Packaging Technol. Sci. 20, 275–286.

Nagy, S., Chen, C.S., Shaw, P.E., 1993. Fruit juice processing technology. Agscience, Auburndale, FL.

Noci, F., Riener, J., Walkling-Ribeiro, M., Cronin, D.A., Morgan, D.J., Lyng, J.G., 2008. Ultraviolet irradiation and pulsed electric fields (PEF) in a hurdle strategy for the preservation of fresh apple juice. J. Food Eng. 85, 141–146.

O'Donnell, C.P., Tiwari, B.K., Bourke, P., Cullen, P.J., 2010. Effect of ultrasonic processing on food enzymes of industrial importance. Trends Food Sci. Technol. 21, 358–367.

Odumeru, J.A., 2012. Microbial safety of food and food products. In: Simpson, B.K., Nollet, L.M.L., Toldra, F., Benjakul, S., Paliyath, G., Hui, Y.H. (Eds.), Food Biochemistry and Food Processing, second edition John Wiley & Sons, UK, pp. 787–797.

Pala, C.U., Toklucu, A.K., 2013. Microbial, physicochemical and sensory properties of UV-C processed orange juice and its microbial stability during refrigerated storage. LWT—Food Sci. Technol. 50, 426–431.

Patil, S., Bourke, P., Frias, J.M., Tiwari, B.K., Cullen, P.J., 2009. Inactivation of *Escherichia coli* in orange juice using ozone. Innovative Food Sci. Emerg. Technol. 10 (4), 551–557.

Pereira, R.N., Vicente, A.A., 2010. Environmental impact of novel thermal and non-thermal technologies in food processing. Food Res. Int. 43, 1936–1943.

Rattanathanalerk, M., Chiewchan, N., Srichumpoung, W., 2005. Effect of thermal processing on the quality loss of pineapple juice. J. Food Eng. 66, 259–265.

Rawson, A., Patras, A., Tiwari, B.K., Noci, F., Koutchma, T., Brunton, N., 2011. Effect of thermal and non-thermal processing technologies on the bioactive content of exotic fruits and their products: review of recent advances. Food Res. Int. 44, 1875–1887.

Rivas, A., Rodrigo, D., Martinez, A., Barbosa-Canovas, G.V., Rodrigo, M., 2006. Effect of PEF and heat pasteurization on the physical–chemical characteristics of blended orange and carrot juice. LWT—Food Sci. Technol. 39, 1163–1170.

Rivera, A.P., Cabornida, I.M.B., 2008. Development of ready-to-drink green mango juice. Univ. Southern Mindanao R&D J. 16, 71–77.

Santhirasegaram, V., Zuliana, R., Chandran, S., 2013. Effects of thermal treatment and sonication on quality attributes of Chokanan mango (*Mangifera indica* L.) juice. Ultrason. Sonochem. 20, 1276–1282.

Santhirasegaram, V., Zuliana, R., Chandran, S., 2015b. Effects of sonication and ultraviolet-C treatment as a hurdle concept on quality attributes of Chokanan mango (*Mangifera indica* L.) juice. Food Sci. Technol. Int. 21 (3), 232–241.

Santhirasegaram, V., Zuliana, R., Dominic, S.G., Chandran, S., 2015a. Comparison of UV-C treatment and thermal pasteurization on quality of Chokanan mango (*Mangifera indica* L.) juice. Food Bioprod. Process. 94, 313–321.

Sorour, H., Tanaka, F., Uchino, T., 2014. Impact of non-thermal processing on the microbial and bioactive content of foods. Global J. Biol. Agric. Health Sci. 3 (1), 153–167.

Steenstrup, D.L., Floros, J.D., 2004. Inactivation of *E. coli* O157:H7 in apple cider by ozone at various temperatures and concentrations. J. Food Process. Preserv. 28, 103–116.

Timmermans, R.A.H., Mastwijk, H.C., Knol, J.J., Quataert, M.C.J., Vervoort, L., Van der Plancken, I., et al., 2011. Comparing equivalent thermal, high pressure and pulsed electric field processes for mild pasteurization of orange juice. Part I: Impact on overall quality attributes. Innovative Food Sci. Emerg. Technol. 12 (3), 235–243.

Tiwari, B.K., O'Donnell, C.P., Muthukumarappan, K., Cullen, P.J., 2009. Ascorbic acid degradation kinetics of sonicated orange juice during storage and comparison with thermally pasteurised juice. LWT—Food Sci. Technol. 42, 700–704.

Tran, M.T.T., Farid, M., 2004. Ultraviolet treatment of orange juice. Innovative Food Sci. Emerg. Technol. 5, 495–502.

Valero, M., Recrosio, N., Saura, D., Munoz, N., Martic, N., Lizama, V., 2007. Effects of ultrasonic treatments in orange juice processing. J. Food Eng. 80, 509–516.

Walkling-Ribeiro, M., Noci, F., Cronin, D.A., Riener, J., Lyng, J.G., Morgan, D.J., 2008. Reduction of *Staphylococcus aureus* and quality changes in apple juice processed by ultraviolet irradiation, pre-heating and pulsed electric fields. J. Food Eng. 89, 267–273.

CHAPTER

Safety of ready-to-eat seafood

13

Juan José Rodríguez Herrera

Seafood Microbiology and Technology Group, Marine Research Institute, Agencia Estatal Consejo Superior de Investigaciones Científicas, Vigo, Spain

INTRODUCTION

Seafood has traditionally been an important source of proteins in low-income countries, whereas it is increasingly appreciated as an important element in the diverse and nutritious diets of high-income countries. However, the consumption of seafood is not risk-free, and a wide range of chemical and microbial hazards can be occasionally present (Mozaffarian and Rimm, 2006; Iwamoto et al., 2010; EFSA and ECDC, 2015). Thus, seafood is one of the most important vehicles of foodborne outbreaks in the European Union, accounting for over 15% of strong-evidence outbreaks in 2013 (EFSA and ECDC, 2015). Similarly, it is one of the food categories most commonly linked to foodborne diseases in the United States (Pires et al., 2009; Jami et al., 2014), and it even involves the highest risk for public health when ingestion exposure is considered (CSPI, 2014). This situation is largely accounted for by the increased relevance of ready-to-eat seafood in industrialized countries (Jaroni et al., 2010). Noncompliance of safety regulations by different ready-to-eat food categories was found to be highest in seafood (EFSA and ECDC, 2015).

Following Codex Alimentarius Commission (2009), ready-to-eat food is considered to be "any food which is normally eaten in its raw state or any food handled, processed, mixed, cooked, or otherwise prepared into a form which is normally eaten without further [bactericidal] steps." Ready-to-eat food has thus been a relevant part of human diet since ancient times, with examples such as bread, cheese, yoghurt, ice cream, butter or, more recently, canned food, so that relevant hazards and associated risks are commonly known and control measures can be generally implemented. However, a series of social, economic, and cultural changes taking place in the last few decades, as well as globalization, have driven important changes in patterns of food consumption and production (Warde et al., 2007; Broglia and Kapel, 2011). Modern consumers prefer fresh and more natural products, additive-free, which are nutritious and healthy, as well as fast and easy (convenience) to prepare. Paradoxically, it is not unusual to believe that less processed or natural foods are healthier, even though some products actually

P. Kotzekidou (Ed): Food Hygiene and Toxicology in Ready-to-Eat Foods.
DOI: http://dx.doi.org/10.1016/B978-0-12-801916-0.00013-3

represent a significant risk to public health (Oliveira et al., 2011). To satisfy this demand, the food industry has provided an increasingly huge selection of ready-to-cook, ready-to-serve and ready-to-eat food products.

Some ready-to-eat seafood has been consumed since ancient times too, but they were generally confined to particular geographical areas (ie, sashimi and sushi-like products in Japan, ceviche in Perú, or gravlax in Nordic countries). Nowadays, however, ready-to-eat seafood comprises a wide array of products too. These include smoked fish, marinated and pickled products, precooked shellfish, surimis, fermented products, seafood salads, fish roes, and many other processed products, but also live molluscs (majorly bivalves) and raw fish products, as well as canned seafood. The latter is frequently excluded from the category of ready-to-eat seafood, but it will be taken under consideration in this chapter.

That heating (or any other bactericidal treatment) is not applied prior to consumption of ready-to-eat food precludes the most important barrier against the entry of live pathogens into the human body. As a result, the consumption of ready-to-eat seafood can cause a number of diseases, such as gastroenteritis (the most usual form), hepatitis, or invasive infections, and also long-term sequelae and even death. Relevant hazards which are specific of ready-to-eat seafood, products acting as vehicles of such hazards, sources of contamination, and control measures will be dealt with within the framework "from farm-to-fork" (or "from sea-to-fork"). Accordingly, it will be shown how the safety of any seafood depends on its journey through all stages of the food supply chain.

ENVIRONMENTAL HAZARDS

Following the principle "from farm to fork," the safety of any seafood is firstly dependent on the environmental conditions that the fish or shellfish has lived in until the moment that it is caught or slaughtered. However, there is a major difference between seafood and most other food products in this respect. That is, though aquaculture production has increased rather steadily during the last two or three decades, about half of seafood production for human consumption is still based on the use of wild fish and shellfish as raw materials (FAO, 2014). The lifecycle of these animals is not under control and it is well known that controlling microbial and chemical hazards in fishing grounds so that associated risks are negligible can be very difficult, when not impracticable.

The marine environment contains remarkable levels of microorganisms, which makes fish and shellfish commonly bear a relatively important microbial load. Generally, this load is harmless for humans, but small numbers of pathogenic bacteria and viruses can be occasionally present, particularly in heavily populated coastal areas. Fortunately, fresh seafood is usually subjected to a thermal treatment prior to consumption which destroys microbial pathogens. As a consequence, the consumption of thermally treated seafood caught in noncontaminated or minimally contaminated waters entails a very low initial microbial risk. However, the situation can be significantly different in some ready-to-eat seafood, which become potential vehicles of pathogens present in the marine environment. Thus, the consumption of live bivalve molluscs and the presence of helminth parasites in the flesh of some fish are of great concern at present.

HAZARDS IN LIVE BIVALVE MOLLUSCS

The consumption of live bivalve molluscan shellfish inhabiting coastal areas can be particularly hazardous (Shumway and Rodrick, 2009; Oliveira et al., 2011). As filter feeders, they can concentrate pathogenic bacteria, such as *Vibrio parahaemolyticus* and *Vibrio vulnificus*, as well as a number of algal biotoxins (saxitoxin, okadaic acid, domoic acid, and brevetoxins, among others) that can be naturally present in the marine environment and reach levels threatening food safety (up to 1000-fold higher than in surrounding waters). The situation becomes more hazardous in waters bordering highly populated and industrialized zones which receive sewage from land-based sources, since enteric bacteria such as *Escherichia coli* and *Salmonella*, and viruses (mainly norovirus) are added to the pathogen burden that can be concentrated.

Heat treatments can destroy pathogenic microorganisms but, unlike fish and crustaceans, oysters, clams, and some other bivalves are frequently consumed raw (or undercooked) as a whole (with viscera). The merging of bioconcentration by filter-feeding and raw consumption is responsible for a large proportion of seafood-borne diseases in industrialized countries.

Being aware of these hazardous conditions, molluscan shellfish and production areas are subject to official surveillance programs in which parameters such as marine biotoxins and *E. coli* (the latter as an indicator of fecal contamination), among others, are regularly monitored. Official surveillance programs are also used to classify and monitor production areas as being of class A, B, or C in EU (or approved, conditionally approved, restricted, conditionally restricted and prohibited in the United States). The classification is used to define the requirements to place live bivalve molluscs in the market, for instance, the adoption of preventive control measures such as treatment in purification centers, relaying or precautionary closures of production areas (Regulation (EC) No. 854/2004).

However, official surveillance programs occasionally fail as a result of illegal harvesting and sale (ie, poaching), failure of preventive measures of control, or no time–temperature control after harvesting. This is clearly shown in Table 13.1, in which the highest proportion of the alerts for microbial hazards affecting live bivalves was ascribed to high counts of *E. coli* and, to a lower extent, *Salmonella* as reported by the Rapid Alert System for Food and Feed (RASFF) from 2007 to mid-2015. Both *E. coli* and *Salmonella* are set as microbiological criteria for live bivalves in EU. In addition, the density of *E. coli* does

Table 13.1 Alerts for Microbial Hazards in Bivalve Molluscs Reported by the Rapid Alert System for Food and Feed (RASFF) Concerning Ready-to-Eat Seafood from 2007 to Mid-2015 in the European Union

Etiologic Agent	Seafood	Number of Alerts	Proportion
Escherichia coli	Live/fresh bivalves	89	54.27
	Frozen raw bivalves	1	0.61
	Frozen prepared bivalves	1	0.61
	Frozen cooked bivalves	1	0.61
Norovirus	Live/fresh oysters	37	22.56
	Other live/fresh bivalves	3	1.83
	Frozen cooked bivalves	3	1.83
	Frozen raw bivalves	1	0.61
Salmonella	Frozen cooked bivalves	13	7.93
	Live/fresh bivalves	9	5.49
Hepatitis A	Frozen raw bivalves	1	0.61
Clostridium botulinum	Frozen raw bivalves	1	0.61

Notes: Proportion shows the percentage of cases in respect of the total number of alerts by pathogenic bacteria and virus due to consumption of bivalve molluscs. All products suited for consumption in raw state or undercooked were included.

not match the levels of other pathogenic agents potentially present in shellfish, such as pathogenic vibrios, norovirus, and hepatitis A virus (Oliveira et al., 2011). And depuration and even relaying, which are commonly used control measures to reduce the number of *E. coli* in live bivalves down to acceptable levels, have been found not to be effective for the removal of vibrios and norovirus (SCVPH, 2001; Romalde et al., 2002; EFSA, 2012).

As shown in Table 13.1, norovirus has been a major cause for alerts due to microbial hazards in bivalves (mostly raw oysters) reported by RASFF of the European Union from 2007 to mid-2015. In fact, norovirus is currently the most common causative agent of viral gastroenteritis in humans, and shellfish is the most important vehicle of foodborne outbreaks (Bitler et al., 2013; EFSA and ECDC, 2015). To prevent the situation, EFSA (2012) has recommended that an acceptable limit for norovirus in oysters should be established through the support of a quantitative reverse transcriptase polymerase chain reaction method that has been recently standardized (CEN ISO/TS 15216-1:2013 and CEN ISO/TS 15216-2:2013). This approach could also be used to improve risk management in production areas.

A high prevalence of *V. parahaemolyticus* (and to a lower extent of *V. vulnificus*) is also found in Japan, South East Asia, and the United States, where it is the major cause of gastroenteritis, basically associated with the consumption of raw or improperly cooked shellfish (Letchumanan et al., 2014). Safety levels have been therefore included in some national regulations. Several analytical schemes for enumeration are presented by Kaysner and DePaola (2004). PCR-based methods reduced significantly the time for identification (from 5–10 days by culture methods to 2 days). However, the lack of reliable and consistent analytical methods for detection and enumeration, particularly of virulent strains of *V. parahaemolyticus*, is a constraint still to be resolved (Wang et al., 2015).

HELMINTH PARASITES IN FISH FLESH

Helminth parasites living in commercially important species of fish and shellfish are another environmental hazard potentially found in seafood not subject to any heat treatment (or other anti-worm treatment) from sea to fork (Chai et al., 2005; Dorny et al., 2009). These parasites can thus be present in raw or undercooked fishery products, but also in cold-smoked, marinated, salted and other ready-to-eat seafoods, which are processed under conditions that do not kill them (EFSA, 2010; FDA, 2011).

Among such parasites, some nematodes of the *Anisakidae* family (mainly belonging to *Anisakis* and also *Pseudoterranova* genera) have acquired an increasing relevance for public health worldwide (Audicana and Kennedy, 2008; Baird et al., 2014). Several factors have raised this concern. Firstly, *Anisakis* is the most prevalent macroparasite in seafood from major stocks around the world (Llarena-Reino et al., 2015). Secondly, although the great majority of foodborne parasitic zoonoses occur in Asia, the prevalence of anisakiasis has increased worldwide markedly over the last 30 years (and it is likely underestimated) (EFSA, 2010; Baird et al., 2014). This increase has been mainly due to an increasing preference for raw or lightly cooked food by consumers, improved transportation and distribution systems resulting from globalization of food supply, and a more widespread use of better diagnostic techniques (Macpherson, 2005; Broglia and Kapel, 2011). Thirdly, *A. simplex* can cause sensitization to parasite antigens, which can lead to highly aggressive allergic reactions upon subsequent exposures, even if parasites are destroyed (Audicana and Kennedy, 2008; Baird et al., 2014). And lastly, the presence of viable parasites can have an adverse effect on the commercial value of seafood (Pozio, 2013; Llarena-Reino et al., 2015).

European regulations state that fishery products obviously contaminated with parasites cannot be placed on the market for human consumption (Regulation (EC) No. 853/2004). However, it is not clear yet how to manage the risk (D'amico et al., 2014; Llarena-Reino et al., 2015). A lack of standardization raises questions such as to how visual inspection has to be done, what obvious contamination means, or which visible parasites are. Moreover, visual inspection and candling are unreliable and not suitable as critical control points in food production (Levsen et al., 2005). Although several methods have been developed for the detection of anisakids, none of them is suitable for nonintrusive mass screening of industrially produced fish.

The adoption of measures to prevent the presence of viable forms of *Anisakis* in seafood has thus become a priority. However, this is unfeasible in wild live fish, since no fishing area can be considered free from *A. simplex* at present (EFSA, 2010). Consequently, control measures have to be applied after capture. For instance, removal of viscera and belly flaps as soon as possible is highly recommended to prevent the migration of parasites and subsequent infestation of muscle (Pozio, 2013), but this neither completely eliminates parasites nor minimizes the number to an acceptable level (FDA, 2011). Also, it is mandatory that all potentially parasitized ready-to-eat seafood is subject to complete freezing (as raw or finished product) at a temperature of −20°C or lower or to another process that kills viable parasites (Regulation (EC) No. 853/2004; FDA, 2011). This strategy prevents infestation of consumers from industrially processed ready-to-eat seafood, but allergens from dead parasites, which are highly resistant to freezing and heating, are still of concern, particularly for sensitized persons. Audicana and Kennedy (2008) raised many important questions on allergic sensitization, most of which are still to be answered.

POSTHARVEST HAZARDS

PROCESSING: NEW SCENARIOS AND NEW RISKS

The presence of human pathogenic microorganisms in live fish and shellfish is a matter of concern when they are to be used as raw materials for the production of ready-to-eat seafood. However, this is not highly alarming in most cases, since microbial pathogens, if present, are generally in low numbers prior to harvesting (fishing). The safety of ready-to-eat seafood products is conditioned by the microbiological quality of live fish and shellfish, but majorly by handling, processing, packaging, transportation, and storage.

Processing modifies some of the characteristics of raw materials, such as pH, water activity, or microbial load, or environmental properties such as temperature or redox potential (eg, vacuum-packaging). New scenarios with a longer shelf-life are thus created in the form of processed food. Basically, processing can follow two different and complementary strategies:

- To create new conditions that must be continuously active. This strategy is applied to most ready-to-eat seafood. The effectiveness is based on slowing down bacterial growth (and nonmicrobial reactions) or, in the case of frozen products, stopping it.
- To apply highly intense conditions allowing part of (or all) the microbial load (and enzymes) to be destroyed in a short time. This strategy is applied to some ready-to-eat seafood, such as those subject to canning, cooking, or hot smoking.

Processing modifies the microbial ecosystem of raw materials. Microbial behavior is affected by temperature, but also by other factors (eg, pH, a_w, redox potential). Also, different microorganisms exhibit different responses to processing conditions. Consequently, different bacterial communities (biocenosis) develop in different seafood products. If a process based on the former strategy is applied, those microorganisms that adapt to new conditions will outgrow, whereas if highly intense treatments are applied, only those that resist them will be able to survive. Should microbial safety of ready-to-eat seafood therefore be a matter of concern?

Ready-to-eat seafood is subjected to bacteriostatic (eg, refrigeration) or bactericidal (eg, heating) processes. Furthermore, in some products, a targeted inhibition strategy is applied against some particular microorganisms, for instance, modified atmosphere packaging is used against Gram-negative spoilage bacteria. As a result, spoilage organisms are inhibited or partly eliminated, which reduces microbial competition and extends shelflife. As a result, the growth of pathogenic bacteria or toxin formation may become possible in some products.

To prevent the presence and outgrowth of pathogenic microorganisms, food operators must have a hazard analysis and critical control points (HACCP)-based program in place with the aim of establishing control measures that allow all relevant product-specific hazards to be removed, prevented or reduced below acceptable limits. Critical limits for cumulative time of exposure to unrefrigerated conditions, heat processing (time and temperature of treatment), cooling rate after heat processing, salt content, pH, among many others, are set at critical control points. However, 21st century consumers prefer fresh and more natural products, which are additive-free, nutritious, and healthy, and they also spend less time in food preparation (Warde et al., 2007; Broglia and Kapel, 2011). Furthermore, minimal processing based on the use of mild processing conditions (ie, sublethal conditions) has emerged in order to retain original sensory and nutritional properties, and this makes control measures difficult to implement. In fact, multiple factors must be applied following the principles of hurdle technology, so that a cumulative effect that prevents bacterial growth and minimizes organoleptic changes is achieved (Leistner and Gould, 2005; Rosnes et al., 2011). Typical examples of minimal processing are cold-smoking and low-intensity thermal treatments (eg, *sous-vide* technology). However, the chance of bacterial adaptation increases under sublethal stress conditions, and adaptation to a particular stress may allow bacteria to become adapted to other stresses (cross-protection). Capozzi et al. (2009) underlined the need for research on how to use cumulative sublethal hurdles without inducing stress response in bacterial pathogens.

POSTHARVEST CONTAMINATION

Generally, pathogenic bacteria and viruses are either absent or present in very small numbers in live fish and shellfish. However, given the ubiquitous nature of microorganisms, seafood can suffer microbial contamination after harvesting, at any stage of the food supply chain. The production process, in particular, provides multiple opportunities for cross-contamination and recontamination. More importantly, postharvest microbial contamination not only increases bacterial load but it can introduce pathogens rarely found in the marine environment and the microbiota of live fish and shellfish.

Table 13.2 Alerts for Pathogenic Microorganisms in Ready-to-Eat Seafood (Prepared From Fish and Crustanceans) Reported by the Rapid Alert System for Food and Feed (RASFF) from 2007 to Mid-2015 in the European Union

Etiologic Agent	Seafood	Number of Alerts	Proportion
Listeria monocytogenes	Smoked fish	80	67.23
	Fish roe	4	3.36
	Cooked crustaceans	2	1.68
	Fish salad	2	1.68
	Pickled fish	1	0.84
	Fish pate	1	0.84
	Semipreserved anchovies	1	0.84
	Frozen crustaceans	1	0.84
	Crustaceans in brine	1	0.84
Salmonella	Frozen cooked crustaceans	2	1.68
	Cooked crustaceans	1	0.84
	Frozen cooked seafood cocktail	1	0.84
	Smoked fish	1	0.84
Vibrio parahaemolyticus	Cooked crustaceans	1	0.84
Clostridium botulinum (type E) (suspicion)	Smoked fish	1	0.84

Note: Proportion shows the percentage of cases in respect of the total number of alerts by pathogenic microorganisms due to consumption of ready-to-eat seafood produced from fish and crustaceans.

Contamination of ready-to-eat seafood with pathogens is particularly dangerous when it takes place after processing, particularly if the process has reduced the bacterial load significantly or has imposed inhibitory conditions on microbial growth. The former situation would be in the case of surimi-based products and precooked shellfish, whereas the latter would be that of enzymatically matured products and cold-smoked fish. Microbial competition is much reduced and the chance for some pathogens to outgrow increases notably. Of particular significance is canned seafood left at room temperature after being opened, since contaminating microorganisms do not meet competitors.

Though it can be occasionally present in the marine environment, *Listeria monocytogenes* is the paradigm of postharvest microbial contamination of ready-to-eat seafood (Luber et al., 2011; Jami et al., 2014). In fact, it has been the major etiologic agent of alerts due to microbial hazards in ready-to-eat seafood reported by the RASFF from 2007 to mid-2015 in the European Union (Table 13.2). This is a psychrotrophic bacterium that grows well at low temperatures, particularly in moist or wet environments, which makes seafood-processing plants an excellent niche for its presence. The consumption of food contaminated with *L. monocytogenes* can cause listeriosis in humans. This is a disease with a high mortality rate, ranging from 20% to 30% in high-risk groups (pregnant women, newborns, the elderly, and immune-suppressed individuals), which makes it a serious threat for public health. Also, it has been identified as the cause of several foodborne disease outbreaks (Jaroni et al., 2010).

Microbiological criteria for *L. monocytogenes* in ready-to-eat food have been thus set in regulations of most countries. However, there is no consensus on the "acceptable levels" (Jami et al., 2014). European regulations set a limit of 100 colony-forming units (CFU)/g in ready-to-eat products during shelf-life and zero tolerance in those supporting its growth by the time they leave the control of the food processor (Commission Regulation (EC) No. 2073/2005), whereas no presence during shelf-life is mandatory in some other countries, including the United States (FDA, 2011). The latter, however, is not always possible for some ready-to-eat seafood (Luber et al., 2011; Tocmo et al., 2014).

Postharvest microbial contamination results from a lack of hygiene, particularly in food-processing plants (Luber et al., 2011; Tocmo et al., 2014). To prevent contamination, food operators must implement a number of prerequisite programs aimed at preventing hazards potentially present in the working environment, such as cleaning and sanitation of equipment and facilities, and personnel hygiene and training. However, bacteria can form biofilms in not properly cleaned industrial settings. Biofilms are heterogeneous and architecturally complex bacterial communities which are attached to and grow on substrates embedded into an extracellular polymeric matrix that bacteria themselves produce. Biofilms protect bacteria from the application of antimicrobials. The high resistance of biofilms to biocides used for disinfection (Brooks and Flint, 2008; Bridier et al., 2015) makes the removal of bacteria difficult and, as a consequence, the risk of contamination increases significantly.

In the food industry, biofilms can form on the surfaces of equipment and facilities (Srey et al., 2013). Hard- or impossible-to-clean sites are excellent harborage sites for the persistence of bacteria (Carpentier and Cerf, 2011). Drains and floors, which are particularly difficult to clean, have been found to contain the highest levels of *L. monocytogenes* (Gudbjornsdottir et al., 2004; Jami et al., 2014). Humans are an important reservoir of bacteria such as *E. coli* or *Staphylococcus aureus*. Therefore, appropriate design and maintenance of food-processing equipment and premises, detailed working instructions for employees, and frequent and proper cleaning and disinfection followed by monitoring of food-contact surfaces to validate the effectiveness have been recommended to prevent places where biofilms can be developed and disinfectants are diluted down to sublethal levels (Carpentier and Cerf, 2011; Luber et al., 2011; Jami et al., 2014).

DISTRIBUTION AND STORAGE: THE LIFE-TIME

Generally, pathogenic microorganisms are only in small numbers in live fish and shellfish. Postharvest contamination also involves small numbers of pathogenic bacteria initially (Gombas et al., 2003; Jami et al., 2014). Therefore, most pathogenic bacteria need to grow to reach a number large enough to cause disease. Fortunately, most foodborne pathogenic bacteria grow poorly at low temperatures. For instance, *S. aureus*, *E. coli*, and *V. parahaemolyticus* cannot grow under 7°C. Time–temperature control is essential to delay microbial growth and subsequent spoilage and ensure food safety during distribution and storage. The control must be started at the moment of fishing (harvesting) and kept continuously until consumption, except for shelf-stable ready-to-eat seafood (canned, highly salted, and highly dried products).

However, some psychrotrophic microorganisms, such as *L. monocytogenes*, can grow at refrigeration temperatures and reach levels threatening food safety. For instance, for *L. monocytogenes* the infectious dose causing illness in 10% and 90% of the population, respectively, is estimated to range from 10^7 to 10^9 cells for normal individuals, and from 10^5 to 10^7 for high-risk people (Farber et al., 1996). The risk associated with food supporting the growth of *L. monocytogenes* is thus reported to be 100–1000-fold higher than with those that do not support it (WHO/FAO, 2004). A number of control measures can be implemented as a part of HACCP programs in order to prevent the growth of *L. monocytogenes* in ready-to-eat seafood, such as pH $\leq$ 4.4, water activity $\leq$0.92 (ie, salt content $>$8%), or the presence of inhibitory substances (FDA, 2011). Also, it is selected as the target pathogen for thermal processes of packaged ready-to-eat products where outgrowth of spore-forming pathogens is not likely, because it is regarded as the most heat-tolerant nonspore-forming foodborne bacterial pathogen. However, *L. monocytogenes* is still of special concern in cold-smoked fish (EFSA, 2013; Tocmo et al., 2014; Løvdal, 2015). Unlike hot smoking, which is carried out at a temperature between 60°C and 80°C, cold-smoking is performed at about 30°C, which allows *L. monocytogenes* to survive in final products. Changes in water activity and redox potential do not prevent the growth, nor inhibitory compounds from smoke. Ensuring that *L. monocytogenes* contamination in cold-smoked fish meets regulatory limits is a considerable challenge for the seafood industry (EFSA and ECDC, 2015).

Any seafood spends most of its lifetime in distribution and storage. Therefore the chance of growth of pathogenic bacteria or toxin formation depends mostly on the conditions that seafood is subjected to during these stages. Unfortunately, proper maintenance of refrigeration temperatures (<3°C) cannot be ensured throughout distribution and storage. In fact, time–temperature abuses are not unusual. This makes the growth of psychrotrophic bacteria faster and may also allow mesophilic bacteria, such as *E. coli* or *S. aureus*, outgrow and compromise food safety. Of particular concern are live bivalves held with no time–temperature control after harvesting, since pathogenic vibrios and enteric bacteria can multiply and reach high levels, which increases still further the risk of food safety incidents associated with the consumption of raw or not properly cooked bivalves (Su and Liu, 2010; FDA, 2011).

Time–temperature abuses are also considered to be the major cause of scombrotoxin poisoning (Hungerford, 2010). Scombrotoxin poisoning is one of the most relevant hazards associated with fishery products (EFSA, 2015). This is an allergy-like intoxication caused by the consumption of seafood with a relatively high content of histamine, which is formed by decarboxylation of free histidine by bacterial enzymes prior to sensory spoilage becoming evident. Histamine was therefore set as a microbiological criterion for fishery products from fish species with a high content of free histidine. However, fish species belonging to the families *Scombridae* (tunas and mackerels), *Clupeidae* (eg, sardines, herrings), and *Engraulidae* (eg, anchovies) continue being frequently involved in scombrotoxin poisoning (FAO and WHO, 2012; EFSA and ECDC, 2015).

Histamine can be formed during inappropriate distribution and storage of ready-to-eat seafood (and even during processing) if bacteria or enzymes have not been inactivated or recontamination takes place. Of particular attention is the case of seafood that has undergone enzyme maturation treatment in brine (eg, anchovies). However, this hazard is not specific of ready-to-eat seafood. In fact, histamine is most commonly formed during distribution and storage of fresh fish (Lehane and Olley, 2000; Prester, 2011). The relevance of histamine comes particularly from its high resistance to heat treatment, even to canning. In fact, canned products have been the ready-to-eat seafood most involved in alerts for histamine reported by the RASFF from 2007 to mid-2015 (Table 13.3).

Mesophilic bacteria such as *Clostridium perfringens*, *Morganella morganii*, *Hafnia alvei*, and *Raoultella planticola* are major biogenic amine producers (FAO and WHO, 2012). Rapid chilling and continuous maintenance of the cold chain is thus considered the most convenient strategy to prevent histamine formation (FDA, 2011). However, psychrotrophic bacteria such as *Morganella psychrotolerans* and *Photobacterium phosphoreum* have been found to be strong histamine formers at refrigeration temperatures (Emborg et al., 2006; Kanki et al., 2004), which can become a problem in products with a long shelf-life.

CONCLUSIONS

Trends in food consumption have raised increasing interest in ready-to-eat food. Seafood is highly popular and provides excellent nutrition, and regular consumption is recommended by health authorities. However, a large part of production uses wild fish and shellfish as raw materials, which makes the control of some hazards very difficult or impracticable in some ready-to-eat seafood.

Table 13.3 Alerts for Histamine in Ready-to-Eat Seafood Reported by the Rapid Alert System for Food and Feed (RASFF) from 2007 to Mid-2015 in the European Union

Seafood	Number	Proportion
Canned seafood	24	35.8
Semipreserved anchovies	9	13.4
Fish paté	2	2.99
Smoked fish	2	2.99
Marinated fish	1	1.49

Note: Proportion shows the percentage of cases in respect of the total number of alerts for histamine in seafood.

This risk of eating ready-to-eat seafood is clear in live bivalves. Official surveillance programs of control occasionally fail due to a number of reasons. Currently, there is a need for complementary indicators of microbial contamination in bivalves to be eaten raw or lightly cooked, as well as for reliable and consistent analytical methods for detection and enumeration, in order to set adequate microbiological criteria in regulations. Also, some preventive measures of control still need to be implemented, for instance, using viral contamination as a tool for risk management of shellfish harvesting areas, optimizing conditions of depuration allowing a more sensitive and reliable viral and bacterial reduction to be achieved, and mainly preventing the discharge of untreated or undertreated domestic and industrial sewage into coastal waters. Similarly, the presence of *Anisakis* in fish muscle is not always under proper control in spite of regulatory measures. In fact, it is not clear yet how to manage the risk, with two issues making it more difficult. On one side, no fishing area is considered free from *A. simplex*. On the other side, no method for detection of anisakids is suitable for nonintrusive mass screening of industrially produced fish.

The safety of ready-to-eat seafood is however mostly determined by conditions of processing, distribution, and storage, as well as by postharvest contamination, all of which can make outgrowth of pathogenic bacteria or toxin formation possible. Two scenarios focus the interest of academia, industry, and governments at present:

1. *L. monocytogenes* as the paradigm of postharvest microbial contamination in minimally processed ready-to-eat seafood, affecting mainly cold-smoked fish.
2. Scombrotoxin poisoning generally resulting from improper time–temperature control after harvesting.

With the aim of improving the prevention and control of *L. monocytogenes* in ready-to-eat food, over 30 recommendations covering general factors, policy aspects, industry approaches, education, training, and communications were generated by an international group of experts (Luber et al., 2011). Similarly, several recommendations and research needs were recently listed to prevent and monitor histamine formation (FAO and WHO, 2012). It seems clear that plenty of resources, time, and good intentions, as well as an increased awareness-raising by stakeholders, are needed for implementation of control strategies. On the contrary, it can be easily foreseen that both hazards will still remain as persistent and global problems.

ACKNOWLEDGMENTS

The author wishes to thank Mrs Eva Eiriz for her help in the collection of data and making of tables.

REFERENCES

Audicana, M.T., Kennedy, M.W., 2008. *Anisakis simplex*: from obscure infectious worm to inducer of immune hypersensitivity. Clin. Microbiol. Rev. 21, 360–379.

Baird, F.J., Gasser, R.B., Jabbar, A., Lopata, A.L., 2014. Foodborne anisakiasis and allergy. Mol. Cellular Probes 28, 167–174.

Bitler, E.J., Matthews, J.E., Dickey, B.W., Eisenberg, J.N.S., Leon, J.S., 2013. Norovirus outbreaks: a systematic review of commonly implicated transmission routes and vehicles. Epidemiol. Infect. 14, 1563–1571.

Bridier, A., Sanchez-Vizuete, P., Guilbaud, M., Piard, J.C., Naïtali, M., Briandet, R., 2015. Biofilm-associated persistence of food-borne pathogens. Food Microbiol. 45, 167–178.

Broglia, A., Kapel, C., 2011. Changing dietary habits in a changing world: emerging drivers for the transmission of foodborne parasitic zoonoses. Vet. Parasitol. 182, 2–13.

Brooks, J.D., Flint, S.H., 2008. Biofilms in the food industry: problems and potential solutions. Int. J. Food Sci. Technol. 43, 2163–2176.

Capozzi, V., Fiocco, D., Amodio, M.L., Gallone, A., Spano, G., 2009. Bacterial stressors in minimally processed food. Int. J. Mol. Sci. 10, 3076–3105.

Carpentier, B., Cerf, O., 2011. Review—persistence of *Listeria monocytogenes* in food industry equipment and premises. Int. J. Food Microbiol. 145, 1–8.

CEN ISO/TS 15216-1:2013 Microbiology of food and animal feed. Horizontal method for determination of hepatitis A virus and norovirus in food using real-time RT-PCR. Part 1: Method for quantification.

CEN ISO/TS 15216-2:2013. Microbiology of food and animal feed. Horizontal method for determination of hepatitis A virus and norovirus in food using real-time RT-PCR. Part 2: Method for qualitative detection.

Center for Science in the Public Interest (CSPI), 2014. Outbreak alert! 2014. A review of foodborne illness in America from 2002-2011. Written by C. Smith DeWaal and M. Glassman.

Chai, J.Y., Murrell, K.D., Lymbery, A.J., 2005. Fish-borne parasitic zoonoses: status and issues. Int. J. Parasitol. 35, 1233–1254.

Codex Alimentarius Commission (CAC), 2009. Guidelines on the application of general principles of food hygiene to the control of *Listeria monocytogenes* in ready-to-eat foods.

Commission Regulation (EC) No. 2073/2005 of 15 November 2005 on microbiological criteria for foodstuffs.

Dalgaard, P., Emborg, J., Kjølby, A., Sørensen, N.D., Ballin, N.Z., 2008. Histamine and biogenic amines: formation and importance in seafood. In: Børresen, T. (Ed.), Improving Seafood Products for the Consumer. Woodhead Publishing Ltd, Cambridge, pp. 292–324.

D'amico, P., Malandra, R., Costanzo, F., Castigliego, L., Guidi, A., Gianfaldoni, D., et al., 2014. Evolution of the *Anisakis* risk management in the European and Italian context. Food Res. Int. 64, 348–362.

Dorny, P., Praet, N., Deckers, N., Gabriel, S., 2009. Emerging food-borne parasites. Vet. Parasitol. 163, 196–206.

EFSA (European Food Safety Authority), 2013. Analysis of the baseline survey on the prevalence of *Listeria monocytogenes* in certain ready-to-eat (RTE) foods in the EU, 2010-2011. Part A: *Listeria monocytogenes* prevalence estimates. EFSA J. 11, 3241.

EFSA (European Food Safety Authority), 2015. Scientific and technical assistance on the evaluation of the temperature to be applied to pre-packed fishery products at retail level. EFSA J. 13, 4162.

EFSA (European Food Safety Authority) and ECDC (European Centre for Disease Prevention and Control), 2013. The European Union summary report on trends and sources of zoonoses, zoonotic agents and food-borne outbreaks in 2011. EFSA J. 11, 3129.

EFSA (European Food Safety Authority) and ECDC (European Centre for Disease Prevention and Control), 2015. The European Union summary report on trends and sources of zoonoses, zoonotic agents and food-borne outbreaks in 2013. EFSA J. 13, 3991.

EFSA Panel on Biological Hazards (BIOHAZ), 2010. Scientific opinion on risk assessment of parasites in fishery products. EFSA J. 8, 1543.

EFSA Panel on Biological Hazards (BIOHAZ), 2012. Scientific opinion on norovirus (NoV) in oysters: methods, limits and control options. EFSA J. 10, 2500.

Emborg, J., Dalgaard, P., Ahrens, P., 2006. *Morganella psychrotolerans* sp. nov., a histamine producing bacterium isolated from various seafoods. Int. J. Syst. Evolut. Microbiol. 56, 2473–2479.

FAO Fisheries and Aquaculture Department, 2014. Fisheries and aquaculture topics. The state of world fisheries and aquaculture (SOFIA). Text by Pulvenis J.F., Rome.

FAO (Food and Agriculture Organization of the United Nations) and WHO (World Health Organization), 2012. Joint FAO/WHO expert meeting on the public health risks of histamine and other biogenic amines from fish and fishery products: meeting report [126 pp.].

Farber, J.M., Ross, W.H., Harwig, J., 1996. Health risk assessment of *Listeria monocytogenes* in Canada. Int. J. Food Microbiol. 30, 145–156.

FDA (US Food and Drug Administration), 2011. Fish and Fisheries Products Hazards and Controls Guidance, fourth ed FDA, Silver Spring, MD.

Gombas, D.E., Chen, Y.H., Clavero, R.S., Scott, V.N., 2003. Survey of *Listeria monocytogenes* in ready-to-eat foods. J. Food Protect. 66, 559–569.

Gudbjornsdottir, B., Suihko, M.L., Gustavsson, P., Thorkelsson, G., Salo, S., Sjoberg, A. M., et al., 2004. The incidence of *Listeria monocytogenes* in meat, poultry and seafood plants in the Nordic countries. Food Microbiol. 21, 217–225.

Hungerford, J.M., 2010. Scombroid poisoning: a review. Toxicon 56, 231–243.

Iwamoto, M., Ayers, T., Mahon, B.E., Swerdlow, D.L., 2010. Epidemiology of seafood-associated infections in the United States. Clin. Microbiol. Rev. 23, 399–411.

Jami, M., Ghanbari, M., Zunabovic, M., Domig, K.J., Kneifel, W., 2014. *Listeria monocytogenes* in aquatic food products—a review. Comprehen. Rev. Food Sci. Food Safety 13, 798–813.

Jaroni, D., Ravishankar, S., Juneja, V., 2010. Microbiology of ready-to-eat foods. In: Hwang, A., Huang, L. (Eds.), Ready-to-Eat Foods. Microbial Concerns and Control Measures. CRC Press, Boca Ratón, FL, pp. 1–59.

Kanki, M., Yoda, T., Ishibashi, M., Tsukamoto, T., 2004. *Photobacterium phosphoreum* caused a histamine fish poisoning incident. Int. J. Food Microbiol. 92, 79–87.

Kaysner, C.A., DePaola, A., 2004. *Vibrio*. Bacteriological Analytical Manual, Chapter 9. US Food and Drug Administration, Arlington, VA.

Lehane, L., Olley, J., 2000. Histamine fish poisoning revisited. Int. J. Food Microbiol. 58, 1–37.

Leistner, L., Gould, G.W., 2005. Update on hurdle technology approaches to food preservation. In: third ed Davidson, P.M., Sofos, J.N., Branen, A.L. (Eds.), Antimicrobials in Food, vol. 143. CRC Press, Boca Ratón, FL, pp. 621–631.

Letchumanan, V., Chan, K.G., Lee, L.H., 2014. *Vibrio parahaemolyticus*: a review on the pathogenesis, prevalence, and advance molecular identification techniques. Front. Microbiol. 5, 705.

Levsen, A., Lunestad, B.T., Berland, B., 2005. Low detection efficiency of candling as a commonly recommended inspection method for nematode larvae in the flesh of pelagic fish. J. Food Protect. 68, 828–832.

Llarena-Reino, M., Abollo, E., Regueira, M., Rodriguez, H., Pascual, S., 2015. Horizon scanning for management of emerging parasitic infections in fishery products. Food Control 49, 49–58.

Løvdal, T., 2015. The microbiology of cold smoked salmon. Food Control 54, 360–373.

Luber, P., Crerar, S., Dufour, C., Farber, J., Datta, A., Todd, E., 2011. Controlling *Listeria monocytogenes* in ready-to-eat foods: working towards global scientific consensus and harmonization—recommendations for improved prevention and control. Food Control 22, 1535–1549.

Macpherson, C.N.L., 2005. Human behaviour and the epidemiology of parasitic zoonoses. Int. J. Parasitol. 35, 1319–1331.

Mozaffarian, D., Rimm, E.B., 2006. Fish intake, contaminants, and human health—evaluating the risks and the benefits. J. Am. Med. Assoc. 296, 1885–1899.

Oliveira, J., Cunha, A., Castilho, F., Romalde, J.L., Pereira, M.J., 2011. Microbial contamination and purification of bivalve shellfish: crucial aspects in monitoring and future perspectives—a mini-review. Food Control 22, 805–816.

Pires, S.M., Evers, E.G., van Pelt, W., Ayers, T., Scallan, E., Angulo, F.J., et al., 2009. Attributing the human disease burden of foodborne infections to specific sources. Foodborne Pathogen Disease 6, 417–424.

Pozio, E., 2013. Integrating animal health surveillance and food safety: the example of *Anisakis*. Revue Scientifique et Technique-Office International des Epizooties 32, 487–496.

Prester, L., 2011. Biogenic amines in fish, fish products and shellfish: a review. Food Addit. Contam. Part A Chem. Anal. Control Expo. Risk Assess. 28, 1547–1560.

Regulation (EC) No. 853/2004 of the European Parliament and of the Council of 29 April 2004 laying down specific hygiene rules for on the hygiene of foodstuffs.

Regulation (EC) No. 854/2004 of the European Parliament and of the Council of 29 April 2004 laying down specific rules for the organisation of official controls on products of animal origin intended for human consumption.

Romalde, J.L., Area, E., Sánchez, G., Ribao, C., Torrado, I., Abad, X., et al., 2002. Prevalence of enterovirus and hepatitis A virus in bivalve molluscs from Galicia (NW Spain): inadequacy of the EU standards of microbiological quality. Int. J. Food Microbiol. 74, 119–130.

Rosnes, J.T., Skara, T., Skipnes, D., 2011. Recent advances in minimal heat processing of fish: effects on microbiological activity and safety. Food Bioprocess Technol. 4, 833–848.

SCVPH, 2001. Opinion of the Scientific Committee on Veterinary Measures relating to public health on *Vibrio vulnificus* and *Vibrio parahaemolyticus* in raw and undercooked seafood.

Shumway, S.E., Rodrick, G.E., 2009. Shellfish Safety and Quality. Woodhead Publishing Limited, Cambridge.

Srey, S., Jahid, I.K., Ha, S.D., 2013. Biofilm formation in food industries: a food safety concern. Food Control 31, 572–585.

Su, Y.C., Liu, C., 2010. Seafood and restructured seafood. In: Hwang, A., Huang, L. (Eds.), Ready-to-Eat Foods. Microbial Concerns and Control Measures. CRC Press, Boca Raton, FL, pp. 125–161.

Tocmo, R., Krizman, K., Khoo, W.J., Phua, L.K., Kim, M., Yu, H.G., 2014. *Listeria monocytogenes* in vacuum-packed smoked fish products: occurrence, routes of contamination, and potential intervention measures. Comprehen. Rev. Food Sci. Food Safety 13, 172–189.

Wang, R., Zhong, Y., Gu, X., Yuan, J., Saeed, A., Wang, S., 2015. The pathogenesis, detection, and prevention of *Vibrio parahaemolyticus*. Front. Microbiol. 6, 144.

Warde, A., Cheng, S., Olsen, W., Southerton, D., 2007. Changes in the practice of eating—a comparative analysis of time-use. Acta Sociol. 50, 363–385.

WHO/FAO (World Health Organization/Food and Agriculture Organization of the United Nations). 2004. Risk assessment of *Listeria monocytogenes* in ready-to-eat foods. Technical report. Microbiological risk assessment series, No. 5.

CHAPTER

14 Staphylococcal enterotoxins in processed dairy products

Wanchun Jin and Keiko Yamada

Department of Bacteriology, Nagoya University Graduate School of Medicine, Showa-ku, Nagoya, Aichi, Japan

STAPHYLOCOCCAL FOOD POISONING

Staphylococcus aureus has been identified as one of the most prevalent mastitic pathogens and a major causal bacterium in human and animal diseases. Staphylococcal food poisoning (SFP) is one of the most common foodborne diseases and is caused by the ingestion of food contaminated by one or more staphylococcal enterotoxins (SEs) produced by *S. aureus* strains (Argudin et al., 2010). SFP symptoms—including nausea, vomiting, abdominal cramps and pain, myalgia, diarrhea, giddiness, and headache—typically have a rapid onset, occurring 2–8 h following ingestion of the toxin-contaminated food (Argudin et al., 2010).

Although the prevalence of SFP has decreased in the last few decades, it remains a global public health problem. Data from the European Food Safety Authority indicate that food poisoning caused by enterotoxins is one of the leading causes of foodborne outbreaks in the EU (Kraushaar and Fetsch, 2014). According to the Centers for Disease Control and Prevention (CDC), there have been approximately 20 SFP outbreaks per year for last 5 years in the United States (http//wwwn.cdc.gov/foodborneoutbreaks/). In Brazil, *S. aureus* was the second most common cause of foodborne diseases between 1999 and 2011 (Lima et al., 2013). In China, *S. aureus* was the third most frequently detected pathogen in foodborne outbreaks after *Salmonella* and *Vibrio parahaemolyticus* (Li et al., 2013). In Korea, *S. aureus* has been the fourth most commonly encountered pathogen after *Escherichia coli*, *Salmonella*, and *Vibrio* spp. in the past 10 years (2002–12), accounting for 7.38% of the total number of cases (Hyeon et al., 2013). In Japan, a minimum of 26 (in 2014) and a maximum of 92 (in 2001) SFP outbreaks were reported (with the major source being boxed lunches or rice balls) according to the latest statistics from the Ministry of Health, Labour and Welfare of Japan, although the incidence has been gradually decreasing (http://www.mhlw.go.jp/stf/seisakunitsuite/bunya/kenkou_iryou/shokuhin/syokuchu/04.html).

P. Kotzekidou (Ed): Food Hygiene and Toxicology in Ready-to-Eat Foods.
DOI: http://dx.doi.org/10.1016/B978-0-12-801916-0.00014-5

Approximately 95% of SFP outbreaks are caused by classical SE types (SEA–SEE), with unidentified SEs accounting for the remaining 5% (Tang et al., 2011).

CHARACTERIZATION OF SEs

In total, 23 types of enterotoxins have been reported and can be classified as either SEs with confirmed emetic activities or staphylococcal enterotoxin-like proteins (SE*l*s) with unconfirmed emetic activities and effects on food poisoning. Emetic activity has been characterized in SEA, SEB, SEC (C1–C3), SED, SEE, SEG, SEH, SEI, SER, SES, and SET (Argudin et al., 2010). Recently, the emetic potential of other SE*l*s was assessed using a monkey-feeding assay, and the emetic activity of SEK, SEL, SEM, SEN, SEO, SEP, and SEQ was confirmed (Omoe et al., 2013). However, the roles of SE*l*s such as SE*l*J, SE*l*U, SE*l*V, SE*l*W, and SE*l*X in food poisoning remain unclear (Hu and Nakane, 2014), given that their emetic activity is nonexistent or unconfirmed. SE*l*W are characterized based on window analysis, and their functions are largely unknown (Okumura et al., 2012). One enterotoxin-like toxin, originally classified as SEF, was found to lack emetic activity but possess strong superantigenic activity, and was renamed toxic shock syndrome toxin (TSST)-1 (Hu and Nakane, 2014). The characteristics of 23 enterotoxins are summarized in Table 14.1.

SE GENETICS AND GROUPING

Most SE and SE*l* genes are encoded by mobile genetic elements such as prophages, staphylococcal pathogenicity (genomic) islands (SaPIs), transposons, staphylococcal cassette chromosomes, or plasmids (Table 14.1), suggesting that they could have spread among *S. aureus* isolates; *S. aureus* has been found to simultaneously harbor *se* and *sel* genes (Hu and Nakane, 2014).

SEs and SE*l*s have been classified based on amino acid sequence comparisons into four or five groups, depending on the inclusion of SEH within group 1 (Argudin et al., 2010). All 23 types of mature or full-length enterotoxins and TSST-1 were grouped using ClustalX and NJplot (Fig. 14.1), with SEH distinct from the SEA group and SE*l*X and SE*l*W belonging to the SET group. SE and SE*l* families exhibit remarkable sequence variations; for instance, SEA and SEE show 81.7% identity in amino acid sequence, whereas SEM and SE*l*V show 86.7% and SE*l*X and SE*l*W show 96.6% amino acid sequence identity. The most distantly related are SED and SE*l*X, which have only 10.5% amino acid sequence identity. SE*l*X shares 13.4% identity with SET in the same group and 16.5% identity with TSST-1.

Table 14.1 Characteristics of Staphylococcal Enterotoxins

Toxin	Genomic Element(s)	Length (a.a.)	Molecular Weight (Da)	Superantigenic Activity	Emetic Activity	References
SEA	Prophage (φSa3mu, φSa3mw)	233	27.1	+	+	Schelin et al. (2011)
SEB	SaPI3, chromosome, plasmid	239	28.4	+	+	Schelin et al. (2011)
SEC1	SaPI, plasmid	239	27.6	+	+	Schelin et al. (2011)
SEC2	SaPI, plasmid	239	27.6	+	+	Schelin et al. (2011)
SEC3	SaPI, plasmid	239	27.6	+	+	Schelin et al. (2011)
SED	Plasmid (pIB485)	207	24.1	+	+	Bayles and Iandolo (1989)
SEE	Prophage (φSa)	230	26.5	+	+	Schelin et al. (2011)
SEG	*egc*, chromosome	233	27.1	+	+	Schelin et al. (2011)
SEH	Transposon	217	25.2	+	+	Ren et al. (1994), Su and Wong (1995)
SEI	*egc*, chromosome	218	24.9	+	+	Schelin et al. (2011)
SE/J	Plasmid (pIB485, pF5)	245	28.6	+	nk	Schelin et al. (2011)
SEK	SaPI1, SaPI3	219	25.6	+	+	Schelin et al. (2011)
SEL	$SaPI_{bov}$	215	24.6	+	+	Omoe et al. (2013), Schelin et al. (2011)
SEM	*egc*, chromosome	217	24.9	+	+	Omoe et al. (2013), Schelin et al. (2011)
SEN	*egc*, chromosome	227	26.1	+	+	Omoe et al. (2013), Schelin et al. (2011)
SEO	*egc*, chromosome	232	26.8	+	+	Omoe et al. (2013), Schelin et al. (2011)

(*Continued*)

Table 14.1 Characteristics of Staphylococcal Enterotoxins *Continued*

Toxin	Genomic Element(s)	Length (a.a.)	Molecular Weight (Da)	Superantigenic Activity	Emetic Activity	References
SEP	Prophage (φSa3n)	260[a]	27.0 (29.7[b])	+	+	Omoe et al. (2013), Schelin et al. (2011)
SEQ	SaPI3	242[a]	25.2 (28.2[b])	+	+	Omoe et al. (2013), Schelin et al. (2011)
SER	Plasmid (pIB485, pF5)	259[a]	27.1 (30.1[b])	+	+	Omoe et al. (2013), Schelin et al. (2011)
SES	Plasmid (pIB485, pF5)	257[a]	26.3 (29.8[b])	+	+	Schelin et al. (2011)
SET	Plasmid (pIB485, pF5)	216	22.7 (25.7[b])	+	+	Schelin et al. (2011)
SE/U	*egc*, chromosome	261[a]	27.1 (30.6[b])	+	nk	Schelin et al. (2011)
SE/U2	*egc*, chromosome	256[a]	30.0[b]	+	nk	Thomas et al. (2006)
SE/V	*egc*, chromosome	239[a]	27.6[b]	+	nk	Schelin et al. (2011)
SE/W	Chromosome	203[a]	23.2[b]	nk	nk	Okumura et al. (2012)
SE/X	Chromosome	168 (203[a])	19.3 (23.3[b])	+	nk	Hu and Nakane (2014), Wilson et al. (2011)
TSST-1	Chromosome	194	22.1	+	−	Kuroda et al. (2001)

Note: egc, enterotoxin gene cluster; SaPI, staphylococcal pathogenicity island; nk, not known.
[a]*a.a. length of estimated precursor.*
[b]*http://web.expasy.org/protparam/.*

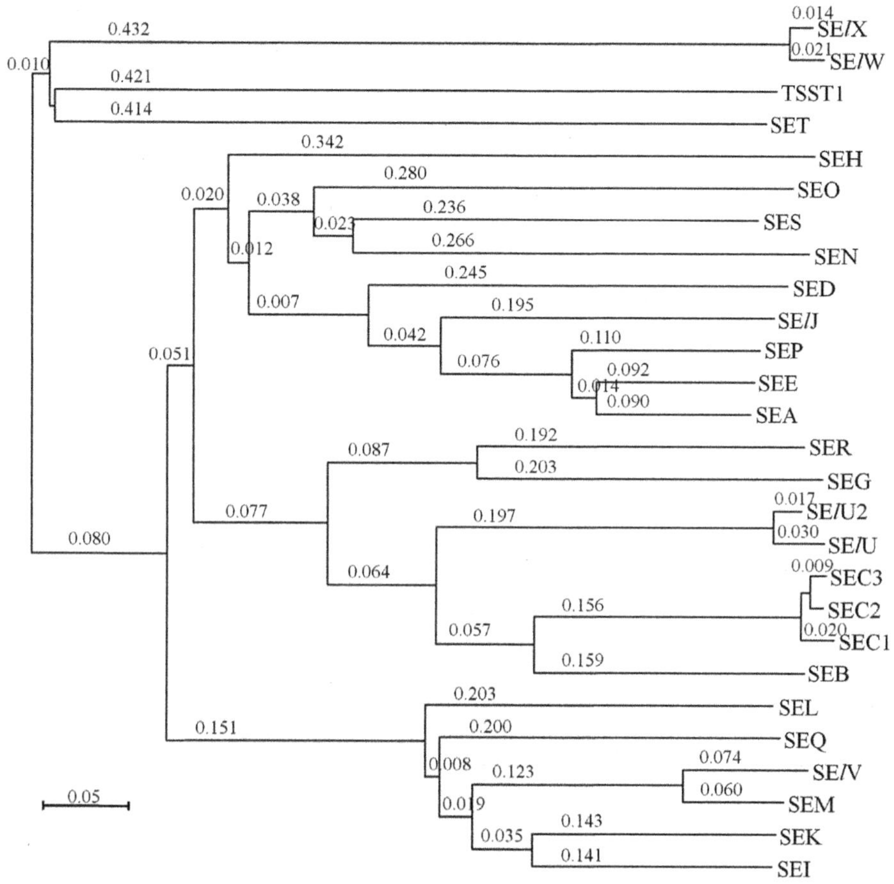

FIGURE 14.1

Alignment of SE and SE*l* amino acid sequence, including SEA (GenBank accession no. M18970), SEB (M11118), SEC1 (X05815), SEC2 (P34071), SEC3 (POAOL5), SED (M28521), SEE (M21319), SEG (AF064773), SEH (U11702), SEI (AF064774), SE*l*J (AF053140), SEK (AF410775), SEL (AF217235), SEM (AF285760), SEN (AF285760), SEO (AF285760), SEP (NP_375057), SEQ (AF410775), SER (AB075606), SES (AB330135), SET (AB330135), SE*l*U (AY205307), SE*l*U2 (EF030428), SE*l*V (EF030427), SE*l*W (CP000046), SE*l*X (HQ850970), and TSST-1 (BA000018).

SUPERANTIGENIC ACTIVITY AND STABILITY OF SEs AND SE*l*s

SEs and SE*l*s are relatively small proteins (molecular weight: 22–30 kDa) belonging to the superantigen family that activates nonspecific T cells via a specific receptor, thereby inducing an upregulation in the expression of cytokines (Balaban and Rasooly, 2000), which stimulate the secretomotor neuron reflex in the enteric nervous system, leading to vomiting (Hu and Nakane, 2014). SEs are relatively stable under changing environmental conditions, such as heat treatment (eg, 20 min at 120°C), freezing, and variations in pH (Argudin et al., 2010). Therefore, SEs cannot be inactivated by heat produced during cooking even though bacterial cells themselves are destroyed, such that food remains contaminated and can still cause food poisoning. Moreover, SEs are resistant to inactivation by proteases such as trypsin and pepsin, and are stable at low pH, which allows them to retain their activity during transport through the digestive tract (Ortega et al., 2010).

BACTERIAL CARRIER

PREVALENCE OF *S. AUREUS* IN HUMAN

Approximately 20–60% of humans are permanent or intermittent carriers of *S. aureus*, which harbors SE genes in one- to two-thirds of cases (Hennekinne et al., 2012). Food handlers carrying enterotoxin-producing *S. aureus* on their skin or noses are likely to contaminate food through direct contact. In a study in Egypt, 38% of individuals working in food-processing plants were positive for *S. aureus*, and SEA–SED was produced in 14% of isolates (Moustafa et al., 2014). Among workers at a milk and dairy product-processing plant, 45% were positive for *S. aureus*, with 68%, 36%, 46%, and 18% producing SEA, SEB, SEC, and SED, respectively (Moustafa et al., 2014). In a study in 2009 in Japan, the prevalence of superantigenic toxin genes was 98.1%, with *sel*, *sec*, *sen*, *seg*, *sem*, and *seo* genes showing rates of 76.4–94.2% and *sea* detected at a rate of 9.6% in meticillin-resistant *S. aureus* (MRSA) isolates in hospitals nationwide (Ohkura et al., 2009).

PREVALENCE OF *S. AUREUS* IN ANIMALS

Domestic animals carrying enterotoxin-producing *S. aureus* or affected by subclinical mastitis are another potential source of contamination. *S. aureus* is among the most prevalent pathogens causing mastitis, and there have been studies investigating SE genes in *S. aureus* strains from subclinical bovine mastitis cases in various countries. One report showed that 93.6% of *S. aureus* isolates from the milk of cows with mastitis were positive for one or more enterotoxin genes, including *sen* (84.6%), *sem* (71.8%), *sei* (60.3%), and *sed* (52.6%) (Srinivasan et al., 2006). Another study found that 25.5% of *S. aureus* subclinical mastitis isolates were enterotoxigenic, and of these, 23.6% and 1.9% were positive for SEA and SEB, respectively (Boynukara et al., 2008). It was also demonstrated that 46% of *S. aureus* isolates from subclinical mastitis were positive for one or more enterotoxin genes; the most frequently found enterotoxin genes were *seg* (16.2%) and *sei* (16.2%), followed by *sec* (15.4%), *sed* (10.8%), and *sej* (10.8%)

(Gunaydin et al.,2011). A total of 65% of *S. aureus* isolates collected from the milk of cows with subclinical mastitis at two major dairy production regions in China contained at least one toxin gene, with *sea* (36.0%), *sei* (31.8%), *seg* (31.4%), and *sem* (26.9%) most frequently observed (Wang et al., 2009). *S. aureus* isolates from bovine mastitis cases in Beijing and the surrounding area harbored the SE genes *she* (36.2%), *sei* (12.1%), *seg* (11.2%), *ser* (4.3%), *sec* (3.4%), *sea* (2.6%), and *sed* (1.7%) (Liu et al., 2014). These results indicate that the prevalence of virulence factors varies depending on geographical location (Pexara et al., 2013).

PREVALENCE OF *S. AUREUS* OR SE IN MILK AND DAIRY PRODUCTS

S. aureus is frequently isolated from bulk milk or dairy products. Several studies carried out in Norway have reported a high prevalence of *S. aureus* in bovine and caprine bulk milk: according to one report, 58% of *S. aureus* isolates from 81 different cows in 79 Norwegian dairy herds in 10 counties produced one or more enterotoxins, predominantly a combination of SEC and TSST-1 (Tollersrud et al., 2000). Another study found that 65% and 85% of *S. aureus* isolates from Norwegian bovine and caprine bulk milk, respectively, produced SE (Jorgensen et al., 2005b). *S. aureus* was shown to be highly prevalent in Norwegian bulk milk; *S. aureus* was detected in 75% of bovine and 96.2% of caprine bulk milk samples and in 37.8% of raw milk products (Jorgensen et al., 2005b). Moreover, enterotoxin production was detected in 22.1% and 57.3% of *S. aureus* isolates, while SE genes were detected in 52.5% and 55.8% of isolates from bovine and caprine bulk milk, respectively. In northeastern Switzerland, 23 *S. aureus* strains obtained from milk samples of 34 dairy cows with mastitis from 34 different farms were identified as SEC-producing (Stephan et al., 2001). Typical Italian semihard and hard cheeses made with raw milk were contaminated by *S. aureus* at a frequency of 78%; 80% of these isolates harbored at least one SE gene, with *sed* being the most highly represented at 63%, followed by *ser*, *sed*, *seg*, and *sem* (Poli et al., 2007). Additionally, 12.6% of strains from cows tested in Austria in 2009 were positive for enterotoxins, a figure that is similar to that obtained in France in 1997 (15.9%), and in all of these tested strains, SEC was the most abundant SE (Gonano et al., 2009; Stephan et al., 2001). Of SE-producing *S. aureus* isolates tested in Brazil, 13.6% were from mastitic cow's milk, 41.7% were from bulk tank milk, and 72.9% were from minas frescal cheese; *seg* and *sei* genes were most frequently detected alone or in combination with other toxin genes, and were both present in 60.8% of cases (Argudin et al., 2010). In another report, the *seh* gene was detected in 29.8% of *S. aureus* isolates from six dairy farms in northeastern Brazil (Silveira-Filho et al., 2014). A prevalence of 70% was reported for *S. aureus* isolates from bulk tank milk in Hungary (Peles et al., 2007). In Korea, 5.6% of milk samples from individual cow udders were positive for *S. aureus* (Moon et al., 2007), which represented the majority (12.2%) of bacteria isolated from mastitic bovine milk samples (Nam et al., 2010). Enterotoxin genes were detected in 68.3% of *S. aureus* isolates collected from bovine mastitis cases in France in 2007 and 2008, with 54% having a combination of two to four genes (Haenni et al., 2011). In bovine, ovine, and caprine milk and milk filtrates from 78 dairy production facilities supplying the farmhouse cheese sector in Ireland, 18.6% were positive for *S. aureus* (Murphy et al., 2010). Samples positive for *S. aureus* comprised 51.4% of bulk milk samples collected from 18 farms that practice traditional Austrian raw milk cheese production; moreover, the number of *S.-aureus*-positive bulk milk samples increased to 94.3% when a culture-independent approach was applied (Walcher et al., 2014).

According to the National Monitoring Network of China, 21.94% of *S. aureus* isolates were detected in dairy products in China in 2008 (Xiang et al., 2015). In 2013 in Shanghai, 26.3% of milk samples were positive for *S. aureus* (Li et al., 2013), whereas 11.3% of powdered infant milk formula samples collected in the Shaanxi Province in China were positive, with the predominant enterotoxin genes being *seg* and *sec*, followed by *sea*, *seb*, *sed*, and *see* (Wang et al., 2012).

A possible reason for this variability in detection rates is differences in methods used to detect SE genes—such as PCR, enzyme-linked immunosorbent assay (ELISA), and the SE-reversed passive latex agglutination (SET-RPLA) assay—as well as regional differences.

MRSA IN COWS AND DAIRY PRODUCTS

There have been a few outbreaks associated with enterotoxin-producing MRSA, although there are no data for their association with dairy products. MRSA has been detected in milk from cows or small ruminants as well as various dairy products in several countries. In Italy, MRSA comprised 0.4% of strains isolated from bovine milk and cheese (Normanno et al., 2007), while it constituted 1.1% of *S. aureus* isolates from bovine milk samples originating from 260 dairy farms located in 21 prefectures investigated between 1998 and 2005 in Japan (Hata et al., 2010). MRSA represented 6.2% of *S. aureus* isolates from bovine mastitis milk collected between 2003 and 2009 in Korea (Nam et al., 2010) and in France, it comprised 0.7% of *S. aureus* isolates from mastitic cow milk sampled between 2007 and 2008 (Haenni et al., 2011). The prevalence of MRSA in cow bulk tank milk samples was 4.4% in Germany between 2009 and 2010 (Kreausukon et al., 2012) and 4% in Minnesota in 2009 (Haran et al., 2012). In China, MRSA was not detected in raw milk samples collected from three Chinese Holstein herds (Wang et al., 2012); however, a high prevalence (47.6%) was observed among dairy cows with clinical mastitis in four provinces in China in 2014 (Pu et al., 2014). The prevalence of MRSA in traditional cheeses produced in the Azerbaijan region of northwest Iran during 2012 was about 16%; 21% of isolates were *mecA*-positive; and 60.8% were positive for SEs (Shanehbandi et al., 2014).

Significant differences in MRSA prevalence in milk or dairy products have been reported in different regions (Pexara et al., 2013). Because some *S. aureus* isolates from SFP cases may not be routinely tested for antibiotic susceptibility or presence of the *mecA* or *mecC* gene in livestock (Paterson et al., 2014), outbreaks may not be recognized as being caused by MRSA, thereby increasing the risk of foodborne MRSA transmission (Doyle et al., 2012).

CASES OF SFP FROM DAIRY PRODUCTS AND PREVALENCE OF SE IN HUMAN

The above-cited studies demonstrate the variable occurrence of *S. aureus* in humans and animals as well as in dairy foods such as cheese and raw milk, all of which are linked to the risk of food poisoning. Traditional dairy products contain a variety of microorganisms that can cause food poisoning (Shanehbandi et al., 2014). The incidence of SFP caused by dairy products varies by country, as this depends on differences in dietary habits (Hait et al., 2014).

Reports from 15 European countries indicate that milk and dairy products represented 19% of foods implicated in staphylococcal outbreaks (Balaban and Rasooly, 2000; Bianchi et al., 2014). Approximately 2.1% of outbreaks reported in 1998–2010 in the United States were traced to dairy products (Doyle et al., 2012) and in Canada, 14% of SFP outbreaks in 2007–10 were attributed to dairy products (Crago et al., 2012). In contrast, the Ministry of Health, Labour, and Welfare of Japan reported only one SFP outbreak associated with dairy products in last 10 years (http://www.mhlw.go.jp/stf/seisakunitsuite/bunya/kenkou_iryou/shokuhin/syokuchu/04.html). Major SFP outbreaks caused by classical SE types in dairy products were reported in nine countries from 1980 to 2014 (Table 14.2).

Five types of SE—namely SEA, SEB, SEC (C1, C2, and C3), SED, and SEE—are most frequently detected in SFP outbreaks. SEA is the most commonly reported, followed by SED, SEH, SEE, and SEC. In the United Kingdom between 1969 and 1990, it was estimated that 79% of food poisoning cases involving SEs were due to SEA alone (56.9%) or in combination with SED (15.4%), SEB (3.4%), SEC (2.5%), or both SEB and SED (1.1%) (Wieneke et al., 1993). SEA has been linked to 77.8% of all SFP outbreaks in the United States, followed by SED (37.5) and SEB (10%) (Argudin et al., 2010; Kadariya et al., 2014). SEA was also the most common SE in SFP in Japan (Shimizu et al., 2000) and was responsible for 69.7% and 90% of cases in France (Kérouanton et al., 2007) and Korea (Cha et al., 2006), respectively.

In 1985, an SFP caused by SEA (94–184 ng/mL) in contaminated chocolate milk in the United States affected over 1000 individuals (Evenson et al., 1988). The gravest SFP incident in Japan (mainly in Osaka) was in 2000 and was caused by SEA and SHE. Toxin concentrations were 0.38 ng/mL and 3.7 ng/g SEA in low-fat milk and skim milk powder, respectively, and 2.8–18.8 ng/g SEH in skim milk powder (Asao et al., 2003; Ikeda et al., 2005); in this case, the original source of contamination was suspected as *S. aureus* in fresh raw milk (http//www.mhlw.go.jp/topics/0012/tp1220-2.html). The separation of cream from fresh raw

Table 14.2 Major Foodborne Outbreaks Caused by SE in Dairy Products

SFP Year	Location	SE	Dairy Products	Number of Cases	References
1985	Kentucky (United States)	SEA (94–184 ng/mL)	2% Chocolate milk	>1000	Evenson et al. (1988)
2000	Japan	SEA (0.38 ng/mL), SEH (2.8–18.8 ng/mL)	Low-fat milk, skim milk	14,780	Asao et al. (2003), Ikeda et al. (2005), LWJ[a]
2003	Norway	SEH (55.5 ng/g)	Mashed potato prepared with raw milk	8	Jorgensen et al. (2005a)
2004	United States	nk	Ice cream	132	CDC[b]
2007	Austria	SEA, SED	Pasteurized milk	166	Schmid et al. (2009)
2007	Paraguay	SEC, SED	Ultrapasteurized milk	400	Weiler et al. (2011)
2009	France	SEE (0.36–1.14 ng/g)	Soft raw milk cheese	23	Ostyn et al. (2010)
2013	Germany	SEA	Ice cream	13	Fetsch et al. (2014)
2014	Switzerland	SEA (>6 ng/mL), SED (>200 ng/mL)	Raw milk cheese	14	Johler et al. (2015)

Note: nk, not known.

[a] *http//www.mhlw.go.jp/topics/0012/tp1220-2.html.*

[b] *http//www.n.cde.gov/foodborneoutbreaks/.*

milk is the first step in the production of the skim milk powder, followed by heating, centrifugal separation, and cooling. An electrical failure during this step in the factory in the northern island caused the temperature of 650 L of milk to remain at 20–50°C for about 4 h, which was thought to result in *S. aureus* growth. On the other hand, approximately 800 L of line milk was not sufficiently cooled, and was maintained for more than 9 h in the milk concentration process at ambient temperatures, although it was unclear whether this resulted in contamination. SEA and SEH may be produced during the cream separation and/or the line milk concentration processes. This contaminated skim milk powder did not conform to Ministerial Ordinance on Milk and Milk Products Concerning Compositional Standards in Japan; however, it was reused to produce another lot of skim milk by adding supplemental raw milk, which was in turn used to produce low-fat milk, yogurt, and so on. This triggered the largest foodborne outbreak ever reported in Japan, affecting 14,780 people and causing major concerns among food manufacturers, food hygiene officers, and researchers (Yamane, 2006). In the aftermath of this incident, the system of Hazard Analysis and Critical Control Points (HACCP) for dairy products was strengthened in Japan.

In October/November 2009, the first SFP caused by SEE in soft cheese with six cases in six different metropolitan areas was reported in France (Ostyn et al., 2010). Recently, a large outbreak involving 400 cases was reported in Paraguay that was traced to ultrapasteurized milk; *S. aureus* isolated from patients and milk samples carried *sec* and *sed* genes (Weiler et al., 2011).

Only two food-poisoning outbreaks involving SEH were reported in 2000 and 2005, which were associated with the consumption of low-fat milk and mashed potato prepared with raw milk (Table 14.2) (Ikeda et al., 2005; Jorgensen et al., 2005a). On the other hand, *seh* was recently found to be the most prevalent SE gene (36.2%) in *S. aureus* isolates from mastitis cases in Beijing, China and surrounding areas (Liu et al., 2014).

Raw milk and dairy products are high-risk substrates for *S. aureus* growth (Morandi et al., 2007). Contaminated dairy products processed and stored under conditions favorable to *S. aureus* growth and SE production can pose a significant risk for consumers.

METHODS FOR SE DETECTION IN DAIRY PRODUCTS

As described above, SE-producing *S. aureus* often spread in dairy farms. To prevent SFP, it is important to determine whether dairy products are contaminated with SEs from *S. aureus* and to evaluate the degree of contamination.

Various methods have been developed to date for detecting SEs, including molecular-biology-based and immunological assays (Hennekinne et al., 2010). Some immunological kits are commercially available; for example, SET-RPLA (Denka Seiken, Tokyo, Japan), RIDASCREEN SET kit (R-Biopharm, Darmstadt,

Germany), VIDAS SET2, and Transia Plate SET kits. The detection limits of these kits range from 0.2 to 2 ng/mL and testing times range from 120 min to 20 h. The outbreak in Japan was caused by small amounts of SEA in low-fat milk (0.38 ng/mL) that were undetectable using the kits, especially since SEA can only be detected after milk delipidation and concentration (Asao et al., 2003; Ikeda et al., 2005). Several recent studies have investigated more rapid and simple, yet highly sensitive and specific methods for detecting SEs. One study reported an ELISA system using monoclonal antibodies (Sasaki et al., 2005) with sensitivity of 0.78–25 ng/mL for SEA, SEB, and SEC1, while another detected SEB in serum samples and culture supernatants by ELISA using a monoclonal antibody (Cook et al., 2007). An SEA detection limit of 0.25 ng/mL was reported using IgY as a capture antibody and anti-SEA IgG in ELISA (Mizutani et al., 2012); these assays could be particularly useful for testing serum or milk samples, and require 2–8 h to generate results. In addition, immunochromatographic approaches using a lateral flow device (LFD) have also been used to detect SEB concentrations below 5 and 0.25 ng/mL in milk and milk products within 20 min (Boyle et al., 2010). An LFD with polyclonal and monoclonal anti-SEA IgY has been applied for the detection in milk at concentrations as low as 0.2 ng/mL, and an immuno-pillar chip with specific anti-SE IgY provided rapid, specific, and highly sensitive detection of SEA–SEE in milk with sensitivity of 0.01–0.1 ng/mL (Jin et al., 2013). Multiplex immuno-pillar devices have a wide range of detection (15.6 pg/mL–100 ng/mL) for SEA, SEB, and SED (Kasama et al., 2015). These methods are therefore useful for screening SEs in milk and dairy products and are promising technologies in the area of food safety.

DAIRY PRODUCT SAFETY

Dairy products contaminated with *S. aureus* underlie many SFP outbreaks. In some regions, dairy products were the primary source of *S.-aureus*-related illness (30.3%), followed by grain and beans (12.9%), meat (10.5), and other sources (7.0%) (Lima et al., 2013). Interestingly, the main location of food preparation associated with foodborne outbreaks in Brazil was private homes (43.7–61.5%), followed by food services (13.7–25.2%) (Lima et al., 2013).

SEs produced by *S. aureus* are an important aspect of dairy product safety and public health, and, as such, more research is required to establish reasonable surveillance of carrier status in cows and employee food-handing practices, and to develop methods that reduce the risk of food poisoning and trace sources of contamination, especially for isolates obtained from bovine and caprine mastitis cases.

A science-based HACCP is recognized as the most effective means of controlling foodborne diseases. In many developed countries, relatively large-scale companies dealing with dairy products are regulated under standard prerequisite programs and the HACCP system. This is due to the fact that microbial contamination of raw milk is unavoidable, as previously noted. A common CCP in dairy processing is pasteurization; the appropriate heating and cooling of the raw milk can eliminate most pathogens. However, various incidents have provided a great deal of information about the importance of education in terms of the risks associated with those who are engaged in dairy food production and the continuous remediation of the system.

The improvement of dairy product safety depends on the following factors: (1) preventing and more efficiently controlling human and animal diseases (such as bovine mastitis); (2) detecting the presence of *S. aureus* contaminants in fresh milk through strengthened HACCP procedures; (3) controlling storage and processing temperatures to prevent *S. aureus* growth and toxin production; (4) educating and training handlers in dairy product factories on the HACCP; (5) developing better detection methods for toxins; and (6) identifying the minimum amount of SE that causes SFP and clarifying the underlying mechanisms.

Understanding of the toxicology of enterotoxins has preventative benefits that differ from quality control. Novel methods for the safe production of dairy products are desired that could be introduced even in small-scale productions or plants. Improving the safety of milk and dairy products and decreasing the risk of SFP requires on-site effort and further studies.

REFERENCES

Argudin, M.A., Mendoza, M.C., Rodicio, M.R., 2010. Food poisoning and *Staphylococcus aureus* enterotoxins. Toxins (Basel) 2, 1751–1773.

Asao, T., Kumeda, Y., Kawai, T., Shibata, T., Oda, H., Haruki, K., et al., 2003. An extensive outbreak of staphylococcal food poisoning due to low-fat milk in Japan: estimation of enterotoxin A in the incriminated milk and powdered skim milk. Epidemiol. Infect. 130, 33–40.

Balaban, N., Rasooly, A., 2000. Staphylococcal enterotoxins. Int. J. Food Microbiol. 61, 1–10.

Bayles, K.W., Iandolo, J.J., 1989. Genetic and molecular analyses of the gene encoding staphylococcal enterotoxin D. J. Bacteriol. 171, 4799–4806.

Betley, M.J., Mekalanos, J.J., 1988. Nucleotide sequence of the type A staphylococcal enterotoxin gene. J. Bacteriol. 170, 34–41.

Bianchi, D.M., Gallina, S., Bellio, A., Chiesa, F., Civera, T., Decastelli, L., 2014. Enterotoxin gene profiles of *Staphylococcus aureus* isolated from milk and dairy products in Italy. Lett. Appl. Microbiol. 58, 190–196.

Boyle, T., Njoroge, J.M., Jones Jr.R.L., Principato, M., 2010. Detection of staphylococcal enterotoxin B in milk and milk products using immunodiagnostic lateral flow devices. J. AOAC. Int. 93, 569–575.

Boynukara, B., Gulhan, T., Alisarli, M., Gurturk, K., Solmaz, H., 2008. Classical enterotoxigenic characteristics of *Staphylococcus aureus* strains isolated from bovine subclinical mastitis in Van, Turkey. Int. J. Food Microbiol. 125, 209–211.

Cha, J.O., Lee, J.K., Jung, Y.H., Yoo, J.I., Park, Y.K., Kim, B.S., et al., 2006. Molecular analysis of *Staphylococcus aureus* isolates associated with staphylococcal food poisoning in South Korea. J. Appl. Microbiol. 101, 864–871.

Chen, T.R., Chiou, C.S., Tsen, H.Y., 2004. Use of novel PCR primers specific to the genes of staphylococcal enterotoxin G, H, I for the survey of *Staphylococcus aureus* strains isolated from food-poisoning cases and food samples in Taiwan. Int. J. Food Microbiol. 92, 189–197.

Cook, E., Wang, X., Robiou, N., Fries, B.C., 2007. Measurement of staphylococcal enterotoxin B in serum and culture supernatant with a capture enzyme-linked immunosorbent assay. Clin. Vaccine Immunol. 14, 1094–1101.

Crago, B., Ferrato, C., Drews, S.J., Svenson, L.W., Tyrrell, G., Louie, M., 2012. Prevalence of *Staphylococcus aureus* and methicillin-resistant *S. aureus* (MRSA) in food samples associated with foodborne illness in Alberta, Canada from 2007 to 2010. Food Microbiol. 32, 202–205.

Doyle, M.E., Hartmann, F.A., Lee Wong, A.C., 2012. Methicillin-resistant staphylococci: implications for our food supply? Anim. Health. Res. Rev. 13, 157–180.

Evenson, M.L., Hinds, M.W., Bernstein, R.S., Bergdoll, M.S., 1988. Estimation of human dose of staphylococcal enterotoxin A from a large outbreak of staphylococcal food poisoning involving chocolate milk. Int. J. Food Microbiol. 7, 311–316.

Fetsch, A., Contzen, M., Hartelt, K., Kleiser, A., Maassen, S., Rau, J., et al., 2014. *Staphylococcus aureus* food-poisoning outbreak associated with the consumption of ice-cream. Int. J. Food Microbiol. 187, 1–6.

Gonano, M., Hein, I., Zangerl, P., Rammelmayr, A., Wagner, M., 2009. Phenotypic and molecular characterization of *Staphylococcus aureus* strains of veterinary, dairy and human origin. Epidemiol. Infect. 137, 688–699.

Gunaydin, B., Aslantas, O., Demir, C., 2011. Detection of superantigenic toxin genes in *Staphylococcus aureus* strains from subclinical bovine mastitis. Trop. Anim. Health. Prod. 43, 1633–1637.

Haenni, M., Galofaro, L., Ponsin, C., Bes, M., Laurent, F., Madec, J.Y., 2011. Staphylococcal bovine mastitis in France: enterotoxins, resistance and the human Geraldine methicillin-resistant *Staphylococcus aureus* clone. J. Antimicrob. Chemother. 66, 216–218.

Hait, J., Tallent, S., Melka, D., Keys, C., Bennett, R., 2014. Prevalence of enterotoxins and toxin gene profiles of *Staphylococcus aureus* isolates recovered from a bakery involved in a second staphylococcal food poisoning occurrence. J. Appl. Microbiol. 117, 866–875.

Haran, K.P., Godden, S.M., Boxrud, D., Jawahir, S., Bender, J.B., Sreevatsan, S., 2012. Prevalence and characterization of *Staphylococcus aureus*, including methicillin-resistant *Staphylococcus aureus*, isolated from bulk tank milk from Minnesota dairy farms. J. Clin. Microbiol. 50, 688–695.

Hata, E., Katsuda, K., Kobayashi, H., Uchida, I., Tanaka, K., Eguchi, M., 2010. Genetic variation among *Staphylococcus aureus* strains from bovine milk and their relevance to methicillin-resistant isolates from humans. J. Clin. Microbiol. 48, 2130–2139.

Hennekinne, J.A., De Buyser, M.L., Dragacci, S., 2012. *Staphylococcus aureus* and its food poisoning toxins: characterization and outbreak investigation. FEMS. Microbiol. Rev. 36, 815–836.

Hennekinne, J.A., Ostyn, A., Guillier, F., Herbin, S., Prufer, A.L., Dragacci, S., 2010. How should staphylococcal food poisoning outbreaks be characterized? Toxins (Basel) 2, 2106–2116.

Hu, D.L., Nakane, A., 2014. Mechanisms of staphylococcal enterotoxin-induced emesis. Eur. J. Pharmacol. 722, 95–107.

Hyeon, J.Y., Chung, G.T., Bing, S.H., Kwon, K.S., Lee, H.H., Kim, S.J., et al., 2013. A foodborne outbreak of *Staphylococcus aureus* associated with fried chicken in Republic of Korea. J. Microbiol. Biotechnol. 23, 85–87.

Ikeda, T., Tamate, N., Yamaguchi, K., Makino, S., 2005. Mass outbreak of food poisoning disease caused by small amounts of staphylococcal enterotoxins A and H. Appl. Environ. Microbiol. 71, 2793–2795.

Jin, W., Yamada, K., Ikami, M., Kaji, N., Tokeshi, M., Atsumi, Y., et al., 2013. Application of IgY to sandwich enzyme-linked immunosorbent assays, lateral flow devices, and immunopillar chips for detecting staphylococcal enterotoxins in milk and dairy products. J. Microbiol. Methods 92, 323–331.

Johler, S., Weder, D., Bridy, C., Huguenin, M.C., Robert, L., Hummerjohann, J., et al., 2015. Outbreak of staphylococcal food poisoning among children and staff at a Swiss boarding school due to soft cheese made from raw milk. J. Dairy Sci. 98, 2944–2948.

Jorgensen, H.J., Mathisen, T., Lovseth, A., Omoe, K., Qvale, K.S., Loncarevic, S., 2005a. An outbreak of staphylococcal food poisoning caused by enterotoxin H in mashed potato made with raw milk. FEMS. Microbiol. Lett. 252, 267–272.

Jorgensen, H.J., Mork, T., Hogasen, H.R., Rorvik, L.M., 2005b. Enterotoxigenic *Staphylococcus aureus* in bulk milk in Norway. J. Appl. Microbiol. 99, 158–166.

Kadariya, J., Smith, T.C., Thapaliya, D., 2014. *Staphylococcus aureus* and staphylococcal food-borne disease: an ongoing challenge in public health. BioMed Res. Int. 2014, 827965.

Kasama, T., Ikami, M., Jin, W., Yamada, K., Kaji, N., Atsumi, Y., et al., 2015. Rapid, highly sensitive, and simultaneous detection of staphylococcal enterotoxins in milk by using immuno-pillar devices. Analyt. Methods in press.

Kérouanton, A., Hennekinne, J.A., Letertre, C., Petit, L., Chesneau, O., Brisabois, A., et al., 2007. Characterization of Staphylococcus aureus strains associated with food poisoning outbreaks in France. Int. J. Food Microbiol 115, 369–375.

Kraushaar, B., Fetsch, A., 2014. First description of PVL-positive methicillin-resistant *Staphylococcus aureus* (MRSA) in wild boar meat. Int. J. Food Microbiol. 186, 68–73.

Kreausukon, K., Fetsch, A., Kraushaar, B., Alt, K., Muller, K., Kromker, V., et al., 2012. Prevalence, antimicrobial resistance, and molecular characterization of methicillin-resistant *Staphylococcus aureus* from bulk tank milk of dairy herds. J. Dairy Sci. 95, 4382–4388.

Kuroda, M., Ohta, T., Uchiyama, I., Baba, T., Yuzawa, H., Kobayashi, I., et al., 2001. Whole genome sequencing of meticillin-resistant *Staphylococcus aureus*. Lancet 357, 1225–1240.

Li, Z., Shi, C., Song, M., Shi, X., 2013. Distribution status of food-borne *Staphylococcus aureus in* Shanghai. Chin. J. Food Sci. 34, 268–271.

Lima, G.C., Loiko, M.R., Casarin, L.S., Tondo, E.C., 2013. Assessing the epidemiological data of *Staphylococcus aureus* food poisoning occurred in the State of Rio Grande do Sul, Southern Brazil. Braz. J. Microbiol. 44, 759–763.

Liu, Y., Chen, W., Ali, T., Alkasir, R., Yin, J., Liu, G., et al., 2014. Staphylococcal enterotoxin H induced apoptosis of bovine mammary epithelial cells in vitro. Toxins (Basel) 6, 3552–3567.

Mizutani, N., Sugita-Konishi, Y., Omoe, K., Shinagawa, K., Kawakami, H., Kanno, S., et al., 2012. Advantages of immunoglobulin Y for the detection of Staphylococcal enterotoxin A in a double-antibody sandwich enzyme-linked immunosorbent assay. Int. J. Food Sci. Technol. 47, 155–159.

Moon, J.S., Lee, A.R., Kang, H.M., Lee, E.S., Kim, M.N., Paik, Y.H., et al., 2007. Phenotypic and genetic antibiogram of methicillin-resistant staphylococci isolated from bovine mastitis in Korea. J. Dairy Sci. 90, 1176–1185.

Morandi, S., Brasca, M., Lodi, R., Cremonesi, P., Castiglioni, B., 2007. Detection of classical enterotoxins and identification of enterotoxin genes in *Staphylococcus aureus* from milk and dairy products. Vet. Microbiol. 124, 66–72.

El-Shenawy, M., Tawfeek, M., El-Hosseiny, L., El-Shenawy, M., Farag, A., Baghdadi, H., et al., 2014. Cross sectional study of skin carriage and enterotoxigenicity of *Staphylococcus aureus* among food handlers. Open J. Med. Microbiol. 4, 16.

Murphy, B.P., O'Mahony, E., Buckley, J.F., O'Brien, S., Fanning, S., 2010. Characterization of *Staphylococcus aureus* isolated from dairy animals in Ireland. Zoonoses. Public Health. 57, 249–257.

Nam, H.M., Kim, J.M., Lim, S.K., Jang, K.C., Jung, S.C., 2010. Infectious aetiologies of mastitis on Korean dairy farms during 2008. Res. Vet. Sci. 88, 372–374.

Normanno, G., Corrente, M., La Salandra, G., Dambrosio, A., Quaglia, N.C., Parisi, A., et al., 2007. Methicillin-resistant *Staphylococcus aureus* (MRSA) in foods of animal origin product in Italy. Int. J. Food Microbiol. 117, 219–222.

Ohkura, T., Yamada, K., Okamoto, A., Baba, H., Ike, Y., Arakawa, Y., et al., 2009. Nationwide epidemiological study revealed the dissemination of meticillin-resistant *Staphylococcus aureus* carrying a specific set of virulence-associated genes in Japanese hospitals. J. Immunol. Methods 58, 1329–1336.

Okumura, K., Shimomura, Y., Murayama, S.Y., Yagi, J., Ubukata, K., Kirikae, T., et al., 2012. Evolutionary paths of streptococcal and staphylococcal superantigens. BMC Genom. 13, 404.

Omoe, K., Hu, D.L., Ono, H.K., Shimizu, S., Takahashi-Omoe, H., Nakane, A., et al., 2013. Emetic potentials of newly identified staphylococcal enterotoxin-like toxins. Infect. Immun. 81, 3627–3631.

Omoe, K., Imanishi, K., Hu, D.L., Kato, H., Takahashi-Omoe, H., Nakane, A., et al., 2004. Biological properties of staphylococcal enterotoxin-like toxin type R. Infect. Immun. 72, 3664–3667.

Omoe, K., Imanishi, K., Hu, D.L., Kato, H., Fugane, Y., Abe, Y., et al., 2005. Characterization of novel staphylococcal enterotoxin-like toxin type P. Infect. Immun. 73, 5540–5546.

Ortega, E., Abriouel, H., Lucas, R., Galvez, A., 2010. Multiple roles of *Staphylococcus aureus* enterotoxins: pathogenicity, superantigenic activity, and correlation to antibiotic resistance. Toxins (Basel) 2, 2117–2131.

Ostyn, A., De Buyser, M.L., Guillier, F., Groult, J., Felix, B., Salah, S., et al., 2010. First evidence of a food poisoning outbreak due to staphylococcal enterotoxin type E, France, 2009. Eurosurveillance 15, 10–13.

Paterson, G.K., Harrison, E.M., Holmes, M.A., 2014. The emergence of *mecC* methicillin-resistant *Staphylococcus aureus*. Trends Microbiol. 22, 42–47.

Peles, F., Wagner, M., Varga, L., Hein, I., Rieck, P., Gutser, K., et al., 2007. Characterization of *Staphylococcus aureus* strains isolated from bovine milk in Hungary. Int. J. Food Microbiol. 118, 186–193.

Pexara, A., Solomakos, N., Govaris, A., 2013. Prevalence of methicillin-resistant *Staphylococcus aureus* in milk and dairy products. J. Hellenic Vet. Med. Soc. 64, 17–34.

Poli, A., Guglielmini, E., Sembeni, S., Spiazzi, M., Dellaglio, F., Rossi, F., et al., 2007. Detection of *Staphylococcus aureus* and enterotoxin genotype diversity in Monte Veronese, a protected designation of Origin Italian cheese. Lett. Appl. Microbiol. 45, 529–534.

Pu, W., Su, Y., Li, J., Li, C., Yang, Z., Deng, H., et al., 2014. High incidence of oxacillin-susceptible mecA-positive *Staphylococcus aureus* (OS-MRSA) associated with bovine mastitis in China. PLoS One 9, e88134.

Ren, K., Bannan, J.D., Pancholi, V., Cheung, A.L., Robbins, J.C., Fischetti, V.A., et al., 1994. Characterization and biological properties of a new staphylococcal exotoxin. J. Exp. Med. 180, 1675–1683.

Rosec, J.P., Guiraud, J.P., Dalet, C., Richard, N., 1997. Enterotoxin production by staphylococci isolated from foods in France. Int. J. Food Microbiol. 35, 213–221.

Sasaki, T., Terano, Y., Shibata, T., Kawamoto, H., Kuzuguchi, T., Kohyama, E., et al., 2005. Establishment of highly specific and quantitative immunoassay systems for staphylococcal enterotoxin A, B, and C using newly-developed monoclonal antibodies. Microbiol. Immunol. 49, 589–597.

Schelin, J., Wallin-Carlquist, N., Cohn, M.T., Lindqvist, R., Barker, G.C., Radstrom, P., 2011. The formation of *Staphylococcus aureus* enterotoxin in food environments and advances in risk assessment. Virulence 2, 580–592.

Schmid, D., Fretz, R., Winter, P., Mann, M., Hoger, G., Stoger, A., et al., 2009. Outbreak of staphylococcal food intoxication after consumption of pasteurized milk products, June 2007, Austria. Wien. Klin. Wochenschr. 121, 125–131.

Shanehbandi, D., Baradaran, B., Sadigh-Eteghad, S., Zarredar, H., 2014. Occurrence of methicillin resistant and enterotoxigenic *Staphylococcus aureus* in traditional cheeses in the north west of Iran. ISRN Microbiol. 2014, 129580.

Shimizu, A., Fujita, M., Igarashi, H., Takagi, M., Nagase, N., Sasaki, A., et al., 2000. Characterization of *Staphylococcus aureus* coagulase type VII isolates from staphylococcal food poisoning outbreaks (1980–1995) in Tokyo, Japan, by pulsed-field gel electrophoresis. J. Clin. Microbiol. 38, 3746–3749.

Silveira-Filho, V.M., Luz, I.S., Campos, A.P., Silva, W.M., Barros, M.P., Medeiros, E.S., et al., 2014. Antibiotic resistance and molecular analysis of *Staphylococcus aureus* isolated from cow's milk and dairy products in northeast Brazil. J. Food Prot. 77, 583–591.

Srinivasan, V., Sawant, A.A., Gillespie, B.E., Headrick, S.J., Ceasaris, L., Oliver, S.P., 2006. Prevalence of enterotoxin and toxic shock syndrome toxin genes in *Staphylococcus aureus* isolated from milk of cows with mastitis. Foodborne Pathog. Dis. 3, 274–283.

Stephan, R., Annemuller, C., Hassan, A.A., Lammler, C., 2001. Characterization of enterotoxigenic *Staphylococcus aureus* strains isolated from bovine mastitis in north-east Switzerland. Vet. Microbiol. 78, 373–382.

Su, Y.C., Wong, A.C., 1995. Identification and purification of a new staphylococcal enterotoxin, H. Appl. Environ. Microbiol. 61, 1438–1443.

Tang, Y.W., Himmelfarb, E., Wills, M., Stratton, C.W., 2010. Characterization of three *Staphylococcus aureus* isolates from a 17-year-old female who died of tampon-related toxic shock syndrome. J. Clin. Microbiol. 48, 1974–1977.

Tang, J., Tang, C., Chen, J., Du, Y., Yang, X.N., Wang, C., et al., 2011. Phenotypic characterization and prevalence of enterotoxin genes in *Staphylococcus aureus* isolates from outbreaks of illness in Chengdu City. Foodborne Pathog. Dis. 8, 1317–1320.

Thomas, D.Y., Jarraud, S., Lemercier, B., Cozon, G., Echasserieau, K., Etienne, J., et al., 2006. Staphylococcal enterotoxin-like toxins U2 and V, two new staphylococcal superantigens arising from recombination within the enterotoxin gene cluster. Infect. Immun. 74, 4724–4734.

Tollersrud, T., Kenny, K., Caugant, D.A., Lund, A., 2000. Characterisation of isolates of *Staphylococcus aureus* from acute, chronic and subclinical mastitis in cows in Norway. Acta Pathol. Microbiol. Immunol. Scand. 108, 565–572.

Walcher, G., Gonano, M., Kummel, J., Barker, G.C., Lebl, K., Bereuter, O., et al., 2014. *Staphylococcus aureus* reservoirs during traditional Austrian raw milk cheese production. J. Dairy Res. 81, 462–470.

Wang, S.C., Wu, C.M., Xia, S.C., Qi, Y.H., Xia, L.N., Shen, J.Z., 2009. Distribution of superantigenic toxin genes in *Staphylococcus aureus* isolates from milk samples of bovine subclinical mastitis cases in two major dairy production regions of China. Vet. Microbiol. 137, 276–281.

Wang, X., Meng, J., Zhang, J., Zhou, T., Zhang, Y., Yang, B., et al., 2012. Characterization of *Staphylococcus aureus* isolated from powdered infant formula milk and infant rice cereal in China. Int. J. Food Microbiol. 153, 142–147.

Weiler, N., Leotta, G.A., Zarate, M.N., Manfredi, E., Alvarez, M.E., Rivas, M., 2011. Foodborne outbreak associated with consumption of ultrapasteurized milk in the Republic of Paraguay. Rev. Argent. Microbiol. 43, 33–36.

Wieneke, A.A., Roberts, D., Gilbert, R.J., 1993. Staphylococcal food poisoning in the United Kingdom, 1969–90. Epidemiol. Infect. 110, 519–531.

Wilson, G.J., Seo, K.S., Cartwright, R.A., Connelley, T., Chuang-Smith, O.N., Merriman, J.A., et al., 2011. A novel core genome-encoded superantigen contributes to lethality of community-associated MRSA necrotizing pneumonia. PLoS. Pathog. 7, e1002271.

Xiang, H., Zhou, L., Liao, C., Chen, G., 2015. Progress of *Staphylococcus aureus* and food poisoning. Chin. J. Food Hygiene 27, 196–199.

Yamane, I., 2006. Epidemics of emerging animal diseases and food-borne infection problems over the last 5 years in Japan. Ann. NY Acad. Sci. 1081, 30–38.

CHAPTER

Sporeforming bacterial pathogens in ready-to-eat dairy products 15

Carlo Spanu

Department of Veterinary Medicine, University of Sassari, Sassari, Italy

INTRODUCTION

Sporeformers are low G + C Gram-positive bacteria belonging to the phylum *Firmicutes*, characterized by the ability to form dormant and highly resistant cells referred to as spores (endospores). Due to the greater resistance to harsh conditions as compared to their vegetative counterparts, endospores represent a survival strategy adopted by bacteria in a hostile environment. Sporulation occurs in response to adverse conditions such as starvation, osmotic pressure, extreme temperature variation, drought, ultraviolet radiation, and chemical agents (Nicholson et al., 2000). Spores can maintain a metabolic inactive state, even for centuries. When favorable conditions for bacterial growth are reestablished, the spores germinate breaking the dormant state and restoring the bacterial vegetative growth (Russel, 1990).

Sporeformers can be detected at different stages along the dairy chain, from dairy farm environments to dairy-product-processing plants. Due to the large distribution in the natural environment and the spore's resistance, sporeforming bacteria can easily contaminate foods, thus representing a concern for food quality or food safety.

Sporeformers include bacteria characterized by a wide variety of phenotypic and genotypic features, which allow them to grow under different conditions. Of particular interest for the dairy sector are the psychrotrophic thermophilic sporeformers (PTS) characterized by multiplication at refrigeration temperature and heat resistance of spores. Among PTS, members of the aerobic genus *Bacillus* and of the anerobic genus *Clostridium* are constantly detected in milk and dairy products (Doyle et al., 2015). The most important foodborne sporeformers associated with ready-to-eat (RTE) dairy products are *Bacillus cereus* and *Clostridium botulinum*, respectively, belonging to the aerobic and anaerobic groups. An overview of the pathways of contamination and of the occurrence of sporeforming pathogens in dairy products and the strategies for their control will be discussed.

P. Kotzekidou (Ed): Food Hygiene and Toxicology in Ready-to-Eat Foods.
DOI: http://dx.doi.org/10.1016/B978-0-12-801916-0.00015-7

PATHWAYS OF CONTAMINATION OF DAIRY PRODUCTS

The main source of sporeforming bacteria contamination in dairy products is raw milk, although contamination may also originate from the processing environment (Griffiths and Phillips, 1990). At the farm level, spores can be isolated from different environmental matrices including soil, water, sediments, silage, bedding materials, or the feces of infected animals (Lindström et al., 2010; Driehuis, 2013). The primary source of spores in the farm environment is the soil, from which they can be transferred to milk through a number of direct and indirect routes (Fig. 15.1). A direct way of contamination is from the soiled udder of lactating animals during milking. The following routes have been proposed for indirect contamination (Heyndrickx, 2011; Driehuis, 2013):

- crops contaminated with soil are used for silage production and animal feeding;
- spores ingested with silage survive the passage through the gastrointestinal tract of dairy animals and are eliminated in the environment with the feces;
- spores shed with feces contaminate the bedding materials where animals are housed;
- from the bedding, spores can attach to the udder and teats and contaminate the milk during the milking process.

Poor udder hygiene of lactating animals expose to spore contamination, which cannot be completely removed from the skin with the application of any teat-cleaning routine (Magnusson et al., 2006). Insufficiently cleaned milking equipment represents another possible route of sporeformers contamination. In fact, some are able to produce biofilms that can attach to stainless steel and

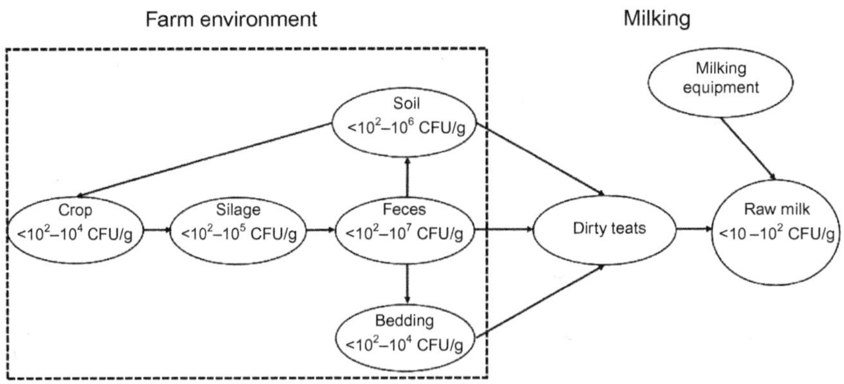

FIGURE 15.1

Contamination route of sporeforming bacteria at the farm level.

Adapted from Driehuis (2013).

release spores into milk during machine milking (Simões et al., 2010). The level of spores in soil at farm level varies largely depending on the management system, ranging from 2 up to 6 log spores per gram of soil (Vissers et al., 2007; Julien et al., 2008). The number of spores in silages and feed depends mainly on the level of soil contamination of the crops and ranges from 1 to >5 log per gram, with an average concentration of 2 log per gram (teGiffel et al., 2002; Driehuis, 2013). The concentration of spores in feces is about three times higher than the concentration ingested with feeds (Vissers et al., 2007). Sporeforming bacteria found in silage belong to the genera *Clostridium* and *Bacillus*. Despite pathogenic *C. botulinum* having limited acid tolerance and not growing in well-fermented silage, the presence of *C. botulinum* spores cannot be totally excluded. The greater the contamination of the silage the greater is the level of spores in raw milk. Vreman et al. (2001) estimated that a silage containing 10^7 *Bacillus* spores resulted in 100 spores per mL of raw milk. The contamination of milk is almost impossible to avoid but the number of spores observed in raw milk is generally low, $<1-10^2$ CFU/mL (Vissers et al., 2007). Seasonal variation has been reported with counts as high as 10^4 CFU/mL (teGiffel et al., 2002; Coorevits et al., 2008).

Residual milk remaining in storage and piping system of milking equipment can favor the formation of biofilm, which protect bacteria from temperature and sanitation, making them more resistant to cleaning agents and other antimicrobial substances (Peng et al., 2001). Spores can be successively detached from the biofilm and released into milk (Wijman et al., 2007).

Once in raw milk sporeformers can represent a potential concern for spoilage and safety of dairy products. While vegetative cells are inactivated by heating, spores are able to resist heat treatments commonly used in industrial processes (Scheldeman et al., 2006). The application of mild pasteurization treatment, such as high-temperature short-time, may activate dormant spore germination and, if favorable conditions exist in the food matrix, lead to the subsequent growth of the vegetative form (Griffiths and Phillips, 1990). Even severe heat treatment such as ultra-high temperature and commercial sterilization do not guarantee the inactivation of all spores. Furthermore, the competitive background microbiota inactivated by the heat treatment does not contrast with the germination and multiplication of the survived spores, which could result in toxin production if appropriate control measures are not taken (Hauschild and Dodds, 1993). The growth of the surviving spores depends upon a number of food-intrinsic and environmental factors such as pH, a_w, chemical composition, surrounding microbiota, presence of preservatives in the formulation, gaseous atmosphere in the packaging, and temperature during storage and distribution. In the dairy industry, the presence of PTS (ie, sporeforming bacteria that can withstand pasteurization and grow at refrigeration temperature) represents a major concern (Doyle et al., 2015). Spores that survived the treatment can germinate and grow at temperatures even below 5°C, thus representing a serious risk for food safety and a limiting factor for shelf-life extension (Huck et al., 2007). Due to possible temperature abuse during storage, not only psychrotrophic but also sporeformers with minimal growth temperature above 10°C may represent a potential risk (Carlin et al., 2000). Gram-positive sporeformers are the main contributors to the spoilage of milk products manufactured under conditions of good hygiene and for which a long storage period is expected. Besides spoilage microorganisms, the possible presence of pathogenic *B. cereus* and proteolytic and nonproteolytic *C. botulinum* should also be taken into account in dairy products obtained from raw (unpasteurized) and heat-treated milk. The number of spores present in processed foods is generally low (EFSA, 2005), making the growth in the food matrix necessary to cause foodborne illness. Therefore, the presence of sporeformers in dairy products is a concern not just for refrigerated RTE foods with a long shelf-life, but also for those with a shorter shelf-life (ie, milk-based desserts) if exposed to temperature abuse.

ANAEROBIC SPOREFORMING BACTERIA

Anaerobic sporeformers of interest in the dairy sector are almost exclusively obligate or strictly anaerobic nonsulfate reducing bacteria, comprised in the genus *Clostridium*. Within the genus *Clostridium* a group of species, referred to as *Clostridium sensu stricto* (Wiegel et al., 2006), has been responsible for food spoilage and foodborne illness associated with the dairy industry. Spoilage of dairy products typically is caused by *C. butyricum* and *C. tyrobutyricum*, which are microorganisms producing gas and/or putrid odors. Gas production is responsible for a defect during ripening of semihard cheeses such as Gouda, Emmental, and Provolone, referred to as "late-blowing." However, they have not been associated with human illness. Among anaerobes responsible for foodborne illness are *C. perfringens* and *C. botulinum*. *C. perfringens* food poisoning is generally linked with the consumption of meat, poultry, gravies, and dried or precooked foods. *C. perfringens* outbreaks are rarely reported in the literature to be associated with dairy products (Galbraith et al., 1982). *C. botulinum* is an important foodborne pathogen causing a severe, sometimes fatal, neuroparalytic disease referred to as botulism. Foodborne botulism is an intoxication since it is the consequence of the ingestion of preformed toxin formed when the microorganism grows in the food. The neurotoxin of *C. botulinum* (BoNT) is among the most toxic substances known in nature, making this microorganism one of the major hazards for public health and a control priority in many food industries. Based on the serological specificity, seven types of toxin (A–G) are produced, but only four types (A, B, E, and F) have been associated with human foodborne botulism (Rusnak and Smith, 2014). Although *C. botulinum* could be divided in four metabolically and genetically distinct groups, designated I–IV, cases of human botulism have been associated with Group I and Group II strains (Peck, 1997; Peck et al., 2011). Group I (proteolytic) *C. botulinum* includes mesophilic strains with a minimal temperature growth of 10–12°C and spores characterized by high heat resistance. The so-called botulinum cook (121°C for 3 min) is required for their inactivation. Group II (nonproteolytic) *C. botulinum* can grow down to 2.5–3.5°C and their spores are less heat-resistant. They are inactivated by heating at 90°C for 10 min (Peck et al., 2011).

C. BOTULINUM IN RTE DAIRY PRODUCTS

A wide variety of foods have been associated with botulism outbreaks, including fruit, vegetables, meat and poultry, fish, honey, and dairy products (EFSA, 2005). Even though the presence of *C. botulinum* spores in foods and the environment is frequent, the incidence of botulism is very low when compared with other foodborne pathogens (Hauschild and Dodds, 1993). In healthy individuals, the ingestion of botulinum spores per se with foods rarely represents a concern, since foodborne botulism occurs following the ingestion of BoNT that is preformed in foods. In order to have toxin production, spores of *C. botulinum* have to germinate and grow in food. Due to the extreme toxicity of BoNT, severe cases of illness have been reported as a consequence of the mere ingestion of contaminated food after dipping one finger into it. Despite the widespread distribution of *C. botulinum* in the environment, only a small percentage of outbreaks that occurred in over a century (<1%) involved dairy products (Lindström et al., 2010). Outbreaks of botulism have been associated with the following dairy products: cottage cheese, curd cheese (Meyer and Eddie, 1965), spread cheese (de Làgarde, 1974), brie cheese (Sebald et al., 1974), soft cheese (Billon et al., 1980), hazelnut yogurt (O'Mahony et al., 1990), mascarpone (Aureli et al., 1996), and a locally made cheese preserved in oil (Pourshafie et al., 1998). Botulism outbreaks have been associated with both home-preserved and commercially prepared foods (EFSA, 2005). "Homemade" botulism generally occurs when correct procedures are not followed during preparation of foods, while in industrial products it is usually associated with failure in thermal processing or package integrity (EFSA, 2005). In 1951, cheese spreads were associated with a mortal case of botulism (Meyer and Eddie, 1965). The authors reported that the toxin was recovered from the jar of the cheese that was involved in the case. Despite the isolation of *C. botulinum* spores from cheese jars of the same batch, no preformed toxins were found, suggesting as a possible explanation the incorrect storage of the product. The epidemiological investigation following the Brie cheese outbreaks that occurred in France and Switzerland in 1973, involving *C. botulinum* type B, showed that the origin was the consumption of cheese contaminated from straw on which the cheese was allowed to ripen, which was in turn contaminated from animal feces (Sebald et al., 1974; Billon et al., 1980). The 1989 outbreak that occurred in England and Wales (Critchley et al., 1989), involved a yogurt product, which was flavored with a hazelnut conserve. It was confirmed that the

Table 15.1 Growth Limit for *Clostridium botulinum* and *Bacillus cereus* Strains

Strains	Temperature (°C)			pH	a_w
	Minimum	Optimum	Maximum		
C. botulinum Group I (proteolytic)	10–12	35–40	40–42	<4.6	<0.93
C. botulinum Group II (nonproteolytic)	3–4	28–30	34–35	<5.0	<0.97
B. cereus	4–5[a]/10–12[b]	30–37	37–42	<4.5	<0.92

[a]Psychrotrophic strains.
[b]Mesophilic strains.

source of the contamination was the hazelnut puree added to the yogurt after it was submitted to a heat treatment inadequate for the destruction of *C. botulinum* spores (O'Mahony et al., 1990). Industrial-produced mascarpone cheese that was used as an ingredient in the preparation of tiramisu dessert, was responsible for a serious outbreak that occurred in 1996 in Italy (Aureli et al., 1996). The epidemiological investigation showed that the presence of preformed botulinum toxin type A in the cheese was the consequence of a break in the cold-chain at retail level, which caused the germination of *C. botulinum* spores and the production of the toxin. Despite *C. botulinum* being able to occur in infant formula and other dried milk products (Barash et al., 2010), no confirmed outbreaks have been reported as a consequence of the consumption of milk powders (Brett et al., 2005; Johnson et al., 2005).

The reported botulism intoxications involving dairy products were in most of the cases consequences of spore contamination originating from ingredients added in the product formulation, materials in contact with the food or the inadequate destruction of spores during processing (eg, failure in thermal processing). However, the mere contamination of spores does not necessarily cause botulism, because conditions supporting the growth and toxin production by *C. botulinum* (eg, low acidity, high water activity, or temperature abuse) are also needed (Table 15.1).

The preformed BoNT can be effectively inactivated by heating at 80°C for 10 min. This makes RTE foods or foods in which dairy products are used without heating (such as in the tiramisu case) risky foods for human health.

AEROBIC SPOREFORMING BACTERIA

The prevalent aerobic sporeformers associated with dairy products are included in the genus *Paenibacillus* and the genus *Bacillus*. *Paenibacillus* spp. are generally responsible for spoilage, while *B. cereus*, within the *Bacillus* group, is the most important foodborne pathogen found in milk and other dairy products

(Ranieri et al., 2009). The *Bacillus* group, besides *B. cereus sensu stricto*, comprises *B. thuringiensis* (an insect pathogen used as a biopesticide), *B. anthracis* (the causative agent of anthrax), the rhizoid *B. mycoides*, *B. pseudomycoides,* and *B. weihenstephanensis* (Montville and Matthews 2005) and the thermotolerant *B. cytotoxicus* (Guinebretière et al., 2013). With the exception of *B. antracis*, these species in most of the cases are not distinguished with routine laboratory methods, and are referred to as *B. cereus sensu lato*. Some strains of *B. thuringiensis* can produce enterotoxin and have been implicated in gastroenteritis outbreaks (Jackson et al., 1995). *B. cereus* is a Gram-positive rod-shaped, motile bacterium widely distributed in the environment and well adapted to live in the intestinal tract of mammals. In the dairy industry, *B. cereus* can lead to spoilage of food products due to the production of proteinase, lipase, and phospholipases that cause defects such as off-flavors, sweet curdling, and bitty cream (Heyndrickx, 2011). Because of its pathogenic potential, *B. cereus* is a serious risk for human health, causing two clinical forms of food poisoning: the emetic and the diarrheal syndromes (Granum and Lund, 1997).

The emetic form is an intoxication following the ingestion of a low-molecular-weight cyclic toxin (cerulide) that is preformed in the food. Of relevance for human health are *B. cereus* strains capable of producing cerulide, referred to as "emetic *B. cereus.*" The cerulide is produced in the late exponential and the early stationary phases of growth of *B. cereus*. The preformed cerulide is not inactivated by heating of the foods or by gastric acidity; therefore, the intoxication can also occur after the death of *B. cereus* cells (Agata et al., 1995). The onset of the symptoms, characterized by nausea, vomiting, and malaise, is 0.5–5 h after the ingestion and generally last for 6–24 h. The diarrheal form is an infection following the ingestion of spores or vegetative cells that produce enterotoxins during growth in the small intestine (Granum and Lund, 1997). Different enterotoxin complexes are involved in diarrheal poisoning, among these are: hemolysin (Hbl), nonhemolytic enterotoxin (Nhe), and cytotoxic K (CytK) (Granum and Lund, 1997). Abdominal pain, watery diarrhea, and nausea occur after 8–16 h of incubation with a duration of 12–24 h. The emetic and diarrheal forms share similar features: (1) the symptoms of illness are generally mild in healthy people and are not considered life-threatening; (2) they are the result of the consumption of cooked foods in which surviving spores can germinate and grow as a consequence of slow cooling and the absence of competitive microbiota (Granum and Lund, 1997).

Within the *B. cereus* group, strains may vary largely in their growth and survival characteristics. Psychrotrophic strains can grow below 8–10°C, which are temperatures that are just above refrigeration (4–5°C), with the optimal temperature growth of these strains between 30°C and 37°C (Wijnands et al., 2006). The mesophilic strain's optimal growth temperature is 37°C but they can grow between 10°C and 42°C. However, some strains can grow at temperatures as high as 55°C (Kramer and Gilbert, 1989). The level of contamination of food matrices with *B. cereus* is usually $\leq 10^2$ CFU/g (Messelhäusser et al., 2014). It is difficult to have a reliable picture of the incidence of foodborne *B. cereus* outbreaks for a series of reasons: in many countries it is not a notifiable disease; cases are underreported due to the short duration of symptoms and self-limiting evolution of the illness; the intoxication with the emetic toxin cerulide could also be misdiagnosed with *S. aureus* (EFSA, 2005). In 2013, the European Food Safety Authority (EFSA) reported an overall reporting rate of *Bacillus* outbreaks in the EU of 0.1 cases per 100,000 inhabitants (EFSA, 2015).

B. CEREUS IN RTE DAIRY PRODUCTS

B. cereus is frequently isolated from raw milk and pasteurized milk and several processed dairy products (Messelhäusser et al., 2010). In raw milk the prevalence varies from 9% to 30% while in pasteurized milk it is from 2% to 71.4% (Singh et al., 2015). The occurrence of *B. cereus* in different types of cheese was 14%, 52.9%, and 17% in cheddar, paneer, and pasta filata cheese, respectively (Singh et al., 2015). In cream samples the presence of *B. cereus* was observed from 29.41% to 48% of the samples, while spreading butter accounted for 65.1% (Singh et al., 2015), ice creams 35–52% (Wong et al., 1998; Messelhäusser et al., 2010), and milk powder and dried milk products from 29% to 54% (Wong et al., 1998). The reported prevalence of *B. cereus* in yogurt and other fermented milk products ranges from 2% to 17% (Wong et al., 1998). This can be attributed to the low pH (~4.2–4.5) and to antimicrobial agents (ie, bacteriocins) produced by lactic acid bacteria which make the fermented milk an unsuitable environment (Wong et al., 1998). Despite being frequently recovered from foods, *B. cereus* outbreaks account for only a small percentage of the total foodborne disease outbreaks (EFSA, 2015). Outbreaks associated with dairy products have been linked with milkshakes at a fast-food restaurant in Ontario in 1988 and milk at a school in Quebec in 1989. In both episodes, the cause of the illness was the consumption of the product after it was temperature-abused. Messelhäusser et al. (2014) reported an outbreak of emetic *B. cereus* that occurred in 2007 at a school kitchen in Bavaria (Germany) associated with the consumption of a hard cheese. Pasteurized milk was responsible for emetic intoxication of 280 persons in the Netherlands (Van Netten et al., 1990). No confirmed *B. cereus* outbreaks where dairy foods were implicated occurred in the United States between 1998 and 2008 (Bennett et al., 2013). This low incidence is explained by the fact that a series of conditions must be in place for an outbreak to occur:

- presence of residual spores in the food, as under normal circumstances are generally in low numbers ($<10^3$/g but mostly $<10^2$/g, which is below the estimated infectious dose);
- activation of spores by heat treatment, such as pasteurization or cooking, triggers spore germination;

- absence of competitive microbiota as *B. cereus* is a poor competitor, so other spoilage microorganisms can overgrow and spoil dairy products before *B. cereus* becomes a risk;
- *B. cereus* multiplication as ideal growth conditions should be met in order to reach the infective dose of 10^5-10^8 cells or spores per gram of food (Granum and Lund, 1997).

The ability of *B. cereus* to cause illness is related to the strain and how the food is handled before the ingestion. The most important contaminants of dairy products are mesophilic and psychrotrophic *B. cereus* but emetic toxin production is usually associated with the mesophilic ones. It is necessary for the growth of the microorganism to produce the cerulide, a condition that requires storage at temperature >10°C. As a consequence, even mild temperature abuse may render the product unsafe, making the unbroken cold-chain the essential strategy to ensure the safety of dairy products. In properly refrigerated dairy products, a concern may be *B. cereus* psychrotropic strains that are able to grow at 4–5°C. Considering that only few strains can grow between 4°C and 10°C and even so, at refrigeration temperature it is necessary for a longer period for the microorganism to reach a level that may represent a hazard to human health, this represents a concern especially for products with an extended shelf-life. Furthermore, psychrotrophic strains are considered not able to produce emetic toxins (Carlin et al., 2006). While psychrotropic strains can potentially produce enterotoxins in refrigerated dairy products or in reconstituted milk-based infant formulae their growth rate and enterotoxin production is low. At 37°C during the passage in the small intestine (c. 4 h), the growth of such psychotropic strains is so low that they do not reach the infective dose necessary to cause disease (Wijnands et al., 2006). Therefore, they represent a threat to consumer safety as a consequence of improper and prolonged cold storage.

CONTROL OF SPOREFORMING BACTERIA IN DAIRY INDUSTRY

Due to the ubiquitous presence of sporeforming bacteria in the environment, it is almost impossible to avoid the contamination of dairy products. Several factors contribute to make the control of these contaminants a very difficult task in the dairy industry:

- spores are very resistant to extreme environmental stresses, allowing them to enter the dairy chain with raw milk and other dry ingredients;
- the hydrophobic properties of spores influence the adhesion properties to inert surfaces such as pipelines of dairy industry;
- spores are resistant to chlorine and other chemicals used in the sanitation of food-processing facilities, making it almost impossible to eradicate them from industrial dairy plants;

- heat treatment applied in industrial dairy products, such as thermization and pasteurization, is able to inactivate vegetative cells but fails in effectively killing heat-resistant endospores;
- spores activated by heat treatment germinate and grow in the product where other competing microbiota is eliminated by pasteurization (Griffiths and Phillips, 1990; Heyndrickx, 2011; Scheldeman et al., 2006).

Representative sporeforming microorganisms, which can negatively affect both the quality and safety of dairy products, include the *B. cereus* group and *Clostridium* genera. The control of sporeformers in the dairy industry needs the application of a systematic and comprehensive approach that includes all the steps from farm to processing plant. The strategy could be oriented on one side in the prevention of introduction of sporeformers in the dairy chain and on the other side in removing, killing, or inhibiting the growth of spores in milk and dairy products.

CONTROL AT FARM LEVEL

As mentioned above, there are a number of factors at farm level that contribute to the presence of spores in milk, such as increased soil contamination on pastures, poor quality of the silage, contamination of bedding materials with feces. At the farm, the initial content of spores in raw milk can be reduced through the implementation of good farm management practices. The reduction of spores in the soil where crops are grown can minimize the level of spores prior to ensiling. High numbers of spores in the soil also depend on the use of cattle manure as fertilizer, which could contain high level of spores. The concentration of spores in animal feces in turn depends on the quality of the silage used to feed the animals. The control of the fermentation process during silage preparation using lactic acid bacteria or chemical additives (ie, propionic acid and benzoic acid) will rapidly achieve a low pH, preventing outgrowth of spores (TeGiffel et al., 2002).

The effective cleaning of teats through the consistent implementation of milking routines, such as the use of disinfectant agents to dip the teat in the pre- and postmilking operation, associated with the use of single towels to dry teats before cluster attachment, can reduce the spore load in raw milk (Visser et al., 2007). However, even the best cleaning procedure cannot completely avoid the presence of spores from teats (Magnusson et al., 2006). At the milking parlor, appropriate cleaning and disinfection of milking equipment prevent the formation of biofilms on stainless steel surfaces of storage and piping systems (Wijman et al., 2007). The improvement of cleaning and sanitizing procedures of milking equipment should be oriented either in their frequency and in the use of methods aimed at effectively removing biofilms.

CONTROL AT PLANT LEVEL

In order to reduce the load of spores entering the plant, raw milk can be submitted to cross-flow microfiltration (CF-MF) or to bactofugation to physically remove spores. With the application of these methods, a reduction of more than 90% of spores present in milk can be achieved. Most of the residual spores are adhered to the fat globules of the cream, which can be pasteurized prior to mixing with the retentate from microfiltration and the bactofugate from bactofugation. Another possible strategy to control spores in dairy products is the use of high-pressure processing technologies. Van Opstal et al. (2004) demonstrated that the germination and subsequent inactivation of *B. cereus* spores in milk can be effectively reduced by the application of mild pressure and heath treatments, contributing to the safety of minimally processed foods.

The hydrophobic properties of endospores allow the attachment to the surfaces of equipment (ie, pipelines) used in the dairy industry, making more difficult their elimination from industrial plants (Peng et al., 2001). Sporeforming bacteria can contribute to biofilm formation, in some circumstances interacting with other bacteria species. The genus *Bacillus* can represent c. 40% of the constitutive microbiota of biofilms present in commercial dairy plants (Wijman et al., 2007). The ability of sporeformers to produce biofilm varies from strain to strain and environmental conditions such as medium and temperature. Wijman et al. (2007) reported that the formation of biofilm was greater in the air–liquid interface, indicating that all the conditions where piping systems are partially filled during operation, or some residual fluid is present after the production cycle, are predisposing factors for biofilm formation. The implementation of an effective cleaning and sanitization procedure is essential to remove biofilms from equipment. The reduction of spore viability and spore adhesion can be achieved using temperatures above 60°C and concentration of NaOH over 0.5% during clean-in-place procedures (Faille et al., 2010). A possible origin of sporeformers in dairy products is recontamination of the finished product from the processing environment. However, this should be considered as a minor source as compared with raw milk.

In dairy industrial processing, the use of nitrate (E251 and E252) and lysozyme is permitted in the milk during manufacture to control germination and outgrowth of sporeforming clostridia responsible for the late blowing defect in cheeses. Other antimicrobial compounds including bacteriocin nisin, benzoate, and sorbates are used to inhibit the growth of *B. cereus* (Jenson and Moir, 2003).

FOOD HANDLING PRIOR TO CONSUMPTION

Prevention and control of food poisoning are generally based on proper reheating and rapid cooling of foods. However, RTE foods by definition are consumed without heating before their consumption. The increasing demand for minimally processed foods that preserve their organoleptic attributes and that require little preparation time increased the market of refrigerated processed foods of extended durability (REPFEDs). These are foods typically processed with light heat treatment (usually 65–95°C), prepared with low levels of preservatives, stored at refrigeration temperature, and with a shelf-life from a few days up to 3 months (Peck, 1997). As a consequence of gentle heat treatment and refrigeration temperature REPFEDs enable the activation of spores and the growth of psychrotolerant sporeforming bacteria which, associated with the prolonged shelf-life, provides sufficient time for toxin production (Peck, 1997). When contamination is due to *C. botulinum* Group I strains, which need temperature greater than 10°C to grow, they do not represent a problem for foods stored below this temperature. When psychotropic *B. cereus* or *C. botulinum* Group II strains are involved, foods should be rapidly chilled below 4°C to avoid possible multiplication and toxin production (Table 15.1).

CONCLUSIONS

Soilborne sporeforming bacteria are responsible for food spoilage and foodborne disease. Among sporulate pathogens found in RTE dairy products the most dangerous is *C. botulinum*, while the most frequent is *B. cereus*. The presence of residual spores in the dairy product is almost impossible to avoid. Since the level of contamination in the finished product is generally below the infective dose, the growth of the microorganism in the food is needed to cause illness. Their control can rely on the reduction of spore load in raw milk and on the final product but almost exclusively in preventing growth of residual spores. Heat treatment used in the dairy industry is not only ineffective in killing all spores but activates their germination. If rapid cooling of the product followed by storage at refrigeration temperature is not adopted, the vegetative cells can grow to levels potentially dangerous for human health. Minimally processed foods or chilled foods with extended durability are at risk. However, an accurate risk categorization of RTE dairy products should consider several combined factors influencing the ability of the food matrices to support the growth of sporeforming pathogens; in particular, food pH, a_w, content in NaCl, preservatives, food packaging (ie, vacuum or modified atmosphere packed food), storage time, and temperature and presence of competitive microbiota. Despite the frequent recovery of sporeformers, RTE dairy products have seldom been associated with foodborne *C. botulinum* or *B. cereus* outbreaks. Outbreaks are generally the consequence of failure in industrial processes or of abuse temperatures during refrigerated storage.

REFERENCES

Agata, N., Ohta, M., Mori, M., Isobe, M., 1995. A novel dodecadepsipeptide, cereulide, is an emetic toxin of *Bacillus cereus*. FEMS Microbiol. Lett. 129, 17–19.

Aureli, P., Franciosa, G., Pourshaban, M., 1996. Foodborne botulism in Italy. Lancet 348, 1594.

Barash, J.R., Hsia, J.K., Arnon, S.S., 2010. Presence of soil-dwelling clostridia in commercial powdered infant formulas. J. Pediatr. 156, 402–408.

Bennett, S.D., Walsh, K.A., Gould, L.H., 2013. Foodborne disease outbreaks caused by *Bacillus cereus, Clostridium perfringens*, and *Staphylococcus aureus*—United States, 1998–2008. Clin. Infect. Diseases 57, 425–433.

Billon, J., Guerin, J., Sebald, M., 1980. Etude de la toxinogenese de *Clostridium botulinum* type B aucours de la maturation de fromages a pate molle. Lait 60, 392.

Brett, M., McLauchlin, J., Harris, A., O'Brien, S., Black, N., Forsyth, R., et al., 2005. A case of infant botulism with a possible link to infant formula milk powder: evidence for the presence of more than one strain of *Clostridium botulinum* in clinical specimens and food. J. Med. Microbiol. 54, 769–776.

Carlin, F., Fricker, M., Pielaat, A., Heisterkamp, S., Shaheen, R., Salonen, M.S., et al., 2006. Emetic toxin-producing strains of *Bacillus cereus* show distinct characteristics within the *Bacillus cereus* group. Int. J. Food Microbiol. 109, 132–138.

Carlin, F., Girardin, H., Peck, M.W., Stringer, S.C., Barker, G.C., Martinez, A., et al., 2000. Research on factors allowing a risk assessment of spore-forming pathogenic bacteria in cooked chilled foods containing vegetables: a FAIR collaborative project. Int. J. Food Microbiol. 60, 117–135.

Coorevits, A., De Jonghe, V., Vandroemme, J., Reekmans, R., Heyrman, J., Messens, W., et al., 2008. Comparative analysis of the diversity of aerobic spore-forming bacteria in raw milk from organic and conventional dairy farms. Syst. Appl. Microbiol. 31, 126–140.

Critchley, E., Hayes, P., Isaacs, P., 1989. Outbreak of botulism in north west England and Wales, June, 1989. Lancet 334 (8667), 849–853.

De Làgarde, E.A., 1974. Boletin Informativo del Centre Paramericano de Zoonosis, Vol. 1. CentroParamericano de Zoonosis, Buenos Aires.

Doyle, C.J., Gleeson, D., Jordan, K., Beresford, T.P., Ross, R.P., Fitzgerald, G.F., et al., 2015. Anaerobic sporeformers and their significance with respect to milk and dairy products. Int. J. Food Microbiol. 197, 77–87.

Driehuis, F., 2013. Silage and the safety and quality of dairy foods: a review. Agric. Food Sci. 22, 16–34.

EFSA, ECDC (European Food Safety Authority and European Centre for Disease Prevention and Control), 2015. The European Union Summary Report on trends and sources of zoonoses, zoonotic agents and food-borne outbreaks in 2013. EFSA J. 13 (1), 3991, 162 pp.

EFSA, European Food Safety Authority, 2005. Opinion of the scientific panel on biological hazards of *Bacillus cereus* and other Bacillus spp. in foodstuff. EFSA J. 175, 1–48.

Faille, C., Sylla, Y., Le Gentil, C., Bénézech, T., Slomianny, C., Lequette, Y., 2010. Viability and surface properties of spores subjected to a cleaning-in-place procedure: consequences on their ability to contaminate surfaces of equipment. Food Microbiol. 27, 769–776.

Galbraith, N.S., Forbes, P., Clifford, C., 1982. Communicable disease associated with milk and dairy products in England and Wales 1951–80. Br. Med. J. 284, 1761–1765.

Granum, P.E., Lund, T., 1997. *Bacillus cereus* and its food poisoning toxins. FEMS Microbiol. Lett. 157, 223–228.

Griffiths, M.W., Phillips, J.D., 1990. Incidence, source and some properties of psychrotrophic *Bacillus* found in raw and pasteurised milk. Int. J. Dairy Technol. 43, 62–66.

Guinebretière, M.H., Auger, S., Galleron, N., Contzen, M., De Sarrau, B., De Buyser, M. L., et al., 2013. *Bacillus* cytotoxicus sp nov. is a new thermotolerant species of the *Bacillus cereus* group occasionally associated with food poisoning. Int. J. Syst. Evol. Microbiol. 63, 31–40.

Hauschild, A.H.W., Dodds, K.L., 1993. *Clostridium botulinum*: Ecology and Control in Foods. Marcel Dekker, New York.

Heyndrickx, M., 2011. The importance of endospore-forming bacteria originating from soil for contamination of industrial food processing. Appl. Environ. Soil Sci.2011, http://dx.doi.org/10.1155/2011/561975.

Huck, J.R., Hammond, B.H., Murphy, S.C., Woodcock, N.H., Boor, K.J., 2007. Tracking spore-forming bacterial contaminants in fluid milk-processing systems. J. Dairy Sci. 90, 4872–4883.

Jackson, S.G., Goodbrand, R.B., Ahmed, R., Kasatiya, S., 1995. *Bacillus cereus* and *Bacillus thuringiensis* isolated in a gastroenteritis outbreak investigation. Lett. Appl. Microbiol. 21, 103–105.

Jenson, I., Moir, C.J., 2003. *Bacillus cereus* and other *Bacillus* species. In: Hocking, A.D. (Ed.), Foodborne Microorganisms of Public Health Significance, sixth ed. Australian Institute of Food Science and Technology (NSW Branch), Sydney, pp. 445–478.

Johnson, E.A., Tepp, W.H., Bradshaw, M., Gilbert, R.J., Cook, P.E., McIntosh, E.D.G., 2005. Characterization of *Clostridium botulinum* strains associated with an infant botulism case in the United Kingdom. J. Clin. Microbiol. 43 (6), 2602–2607.

Julien, M.C., Dion, P., Lafreniere, C., Antoun, H., Drouin, P., 2008. Sources of clostridia in raw milk on farms. Appl. Environ. Microbiol. 74, 6348–6357.

Kramer, J.M., Gilbert, R.J., 1989. *Bacillus cereus* and other *Bacillus* species. Foodborne Bacterial Pathogens 19, 21–70.

Lindström, M., Myllykoski, J., Sivelä, S., Korkeala, H., 2010. *Clostridium botulinum* in cattle and dairy products. Crit. Rev. Food Sci. Nutr. 50, 281–304.

Magnusson, M., Christiansson, A., Svensson, B., Kolstrup, C., 2006. Effect of different premilking manual teat-cleaning methods on bacterial spores in milk. J. Dairy Sci. 89, 3866–3875.

Messelhäusser, U., Kämpf, P., Fricker, M., Ehling-Schulz, M., Zucker, R., Wagner, B., et al., 2010. Prevalence of emetic *Bacillus cereus* indifferent ice creams in Bavaria. J. Food Prot. 73, 395–399.

Messelhäusser, U., Frenzel, E., Blöchinger, C., Zucker, R., Kämpf, P., Ehling-Schulz, M., 2014. Emetic *Bacillus cereus* are more volatile than thought: recent foodborne outbreaks and prevalence studies in bavaria (2007–2013). BioMed Res. Int. 2014, 465603.

Meyer, K.F., Eddie, B., 1965. Sixty Five Years of Human Botulism in the United States and Canada: Epidemiology and Tabulations of Reported Cases 1899 Through 1964. George Williams Hooper Foundation, San Francisco, CA.

Montville, T.J., Matthews, K.R., 2005. Food Microbiology: An Introduction. American Society for Microbiology, Washington, DC.

Nicholson, W.L., Munakata, N., Horneck, G., Melosh, H.J., Setlow, P., 2000. Resistance of *Bacillus* endospores to extreme terrestrial and extraterrestrial environments. Microbiol. Mol. Biol. Rev. 64, 548–572.

O'Mahony, M.O., Mitchell, E., Gilbert, R.J., Hutchinson, D.N., Begg, N.T., Rodhouse, J.C., et al., 1990. An outbreak of foodborne botulism associated with contaminated hazelnut yogurt. Epidemiol. Infect. 104, 389–395.

Peck, M.W., 1997. *Clostridium botulinum* and the safety of refrigerated processed foods of extended durability. Trends Food Sci. Technol. 8, 186–192.

Peck, M.W., Stringer, S.C., Carter, A.T., 2011. *Clostridium botulinum* in the post-genomic era. Food Microbiol. 28, 183–191.

Peng, J.S., Tsai, W.C., Chou, C.C., 2001. Surface characteristics of *Bacillus cereus* and its adhesion to stainless steel. Int. J. Food Microbiol. 65, 105–111.

Pourshafie, M.R., Saifie, M., Shafiee, Vahdani, P., Aslani, M., Salemian, J., 1998. An outbreak of food-borne botulism associated with contaminated locally made cheese in Iran. Scand. J. Infect. Dis. 30, 92–94.

Ranieri, M.L., Huck, J.R., Sonnen, M., Barbano, D.M., Boor, K.J., 2009. High temperature, short time pasteurization temperatures inversely affect bacterial numbers during refrigerated storage of pasteurized fluid milk. J. Dairy Sci. 92, 4823–4832.

Rusnak, J.M., Smith, L.A., 2014. Botulinum neurotoxins from *C. botulinum*. In: Dongyou, L. (Ed.), Manual of Security Sensitive Microbes and Toxins. CRC Press, Taylor and Francis Group, Boca Raton, FL.

Russel, A.D., 1990. Bacterial spores and chemical sporicidal agents. Clin. Microbiol. Rev. 3, 99–119.

Scheldeman, P., Herman, L., Foster, S., Heyndrickx, M., 2006. *Bacillus sporothermodurans* and other highly heat-resistant spore formers in milk. J. Appl. Microbiol. 101, 542–555.

Sebald, M., Jouglard, J., Gilles, G., 1974. Botulism ehumaine de type B apres ingestion de fromage. Ann. Microbiol. (Inst. Pasteur) 125A, 349–357.

Simões, M., Simoes, L.C., Vieira, M.J., 2010. A review of current and emergent biofilm control strategies. LWT-Food Sci. Technol. 43, 573–583.

Singh, V.K., Shukla, S., Chaturvedi, A., 2015. Study the incidence of *Bacillus cereus* isolates from dairy foods. Pharma Innovat. J. 3 (12), 41–43.

TeGiffel, M.C., Wagendorp, A., Herrewegh, A., Driehuis, F., 2002. Bacterial spores in silage and raw milk. Antonie. Van. Leeuwenhoek. 81, 625–630.

Van Netten, P., Van de Moosdijk, A., Van Hoensel, P., Mossel, D.A.A., Perales, I., 1990. Psychrotrophic strains of *Bacillus cereus* producing enterotoxin. J. Appl. Bacteriol. 69, 73–79.

Van Opstal, I., Bagamboula, C.F., Vanmuysen, S.C., Wuytack, E.Y., Michiels, C.W., 2004. Inactivation of *Bacillus cereus* spores in milk by mild pressure and heat treatments. Int. J. Food Microbiol. 92, 227–234.

Vissers, M.M.M., TeGiffel, M.C., Driehuis, F., De Jong, P., Lankveld, J.M.G., 2007. Predictive modeling of *Bacillus cereus* spores in farm tank milk during grazing and housing periods. J. Dairy Sci. 90, 281–292.

Vreman, K., Spoelstra, S.F., Oude Elferink, S.J.W.H., 2001. Aerobic spores occur in vast quantities in silages from laboratory and farm silos. Das Wirtschafteigene Futter 44, 1–8.

Wiegel, J., Tanner, R., Rainey, F.A., 2006. An introduction to the family Clostridiaceae. Prokaryotes 4, 654–678.

Wijman, J.G., de Leeuw, P.P., Moezelaar, R., Zwietering, M.H., Abee, T., 2007. Air-liquid interface biofilms of *Bacillus cereus*: formation, sporulation, and dispersion. Appl. Environ. Microbiol. 73, 1481–1488.

Wijnands, L.M., Dufrenne, J.B., Zwietering, M.H., van Leusden, F.M., 2006. Spores from mesophilic *Bacillus cereus* strains germinate better and grow faster in simulated gastrointestinal conditions than spores from psychrotrophic strains. Int. J. Food Microbiol. 112, 120–128.

Wong, H.C., Chang, M.H., Fan, J.Y., 1998. Incidence and characterization of *Bacillus cereus* isolates contaminating dairy products. Appl. Environ. Microbiol. 54, 699–702.

CHAPTER

16 Sporeforming bacteria associated with bread production: spoilage and toxigenic potential

Paola Lavermicocca[1], Francesca Valerio[1], Palmira De Bellis[1], Angelo Sisto[1] and Ivan Leguérinel[2]

[1]*Institute of Sciences of Food Production, National Research Council, Bari, Italy*

[2]*Université de Brest, Laboratoire Universitaire de Biodiversité et Ecologie Microbienne, Quimper, France*

INTRODUCTION

Bread is a cereal-based product consumed in most countries as a basic food providing nutrients and calories. Even if bread staling remains the main reason for product losses, microbiological spoilage affects bread quality causing heavy economic damage (Saranraj and Geetha, 2012). The dominant spoilage microbiota is constituted by molds, but sporeforming bacteria also represent a source of industrial concern because of their ability to form endospores when environmental conditions are not favorable for growth. The extreme resistance of endospores to various treatments encountered in food industries explains the persistence of sporeforming bacteria in industrial plants and their natural prevalence in a wide range of foodstuffs. In particular, dehydrated powders, spices, or other ingredients may constitute major contamination sources for various industries (Postollec et al., 2012). While no visible trace of spore contamination is noticed in dehydrated ingredients, even the potential presence of one spore may lead to food spoilage during storage of formulated products when environmental conditions allow germination, outgrowth, and vegetative growth.

Sporeforming bacteria belonging to the *Bacillus* genus may represent a risk for the occurrence of "bread rope spoilage" when they are present in raw materials such as flours, brewer's yeast, or bread improvers as well as on processing equipment (Thompson et al., 1993; Pepe et al., 2003; Valerio et al., 2012). Sporeformers contaminating flours originate from cereals: they are generally concentrated in the outer layers of grains and then associated with the bran-rich fractions after the milling process. As a consequence, bacterial counts in the resulting

P. Kotzekidou (Ed): Food Hygiene and Toxicology in Ready-to-Eat Foods.
DOI: http://dx.doi.org/10.1016/B978-0-12-801916-0.00016-9

flour should be lower than in the corresponding wheat, even though some conditioning treatments (ie, increasing of moisture to prevent kernel fracturing) could raise the bacterial load. Berghofer et al. (2003) found *Bacillus* spp. (vegetative cells and spores) as the most frequently detected microorganisms in all milling process samples. Even in the presence of low sporeformer load in flour (about 2.0 log spores/g), the baking process and the subsequent storage conditions of bread favor the germination of heat-resistant spores and their increase up to about 7.0 log CFU/g in bread crumb within 2 days, causing the worsening of the bread quality (Rosenkvist and Hansen, 1995; Viedma et al., 2011). Rope spoilage is first detected by an odor similar to that of pineapple. Later, the crumb becomes discolored, soft and sticky to the touch, which makes the bread inedible. The deterioration also includes slime formation as a result of the combined effect of proteolytic and amylolytic enzymes produced by contaminant bacterial strains (Viljoen and von Holy, 1997; Sorokulova et al., 2003). The full extent of losses of bread caused by ropy spoilage is difficult to quantify, because the condition is often misidentified as sour or rotten spoilage caused by failed dough leavening or an insufficient bake. However, it must be taken into consideration that the type and rate of spoilage are complex and depend on the strain, spore heat resistance, spoilage ability, and a number of intrinsic and extrinsic factors related to the food and to its storage environment (Baril et al., 2012a,b; Valerio et al., 2012; Pepe et al., 2003). In fact, even if during the bread cooking process the temperature reaches 100°C for a few minutes in the center of the crumb, heat-resistant spores can survive the baking process (Chhanwal and Anandharamakrishnan, 2014; Purlis and Salvadori, 2009). Under favorable conditions (water activity (a_w) ≥ 0.95, pH >5, temperature above 25°C), spores may germinate and vegetative cells cause rope spoilage which can develop very rapidly under warm and humid conditions, becoming a remarkable problem in geographical regions characterized by warm climates such as Mediterranean countries, Africa, and Australia (Rosenkvist and Hansen, 1995).

It should also be highlighted that microbial contamination levels of bread higher than 5 log CFU/g are associated with the risk of foodborne illness when the causative agent is a toxigenic species (Kramer and Gilbert, 1989). In fact, bakery products were implicated in foodborne disease outbreaks involving *Bacillus* species (Smith et al., 2004). *Bacillus cereus* is well-known as a cause of food poisoning and much more is now understood about its toxins and their involvement in infections and intoxications. Foodborne illnesses may be caused by a number of *Bacillus* species other than *B. cereus* which are also known to produce toxic metabolites. Therefore, the knowledge of species occurring in bread and ingredients could help to evaluate the risk of spoilage and of contamination of the final product by toxic compounds.

The strategy to reduce damage caused by spoilage must take into account the prevention of the contamination risk, the destruction of contaminants and the control of microbial growth in food. Bacterial behavior can be envisaged by predictive modeling and modulated by modifying different factors such as food formulation and environmental conditions. In fact, predictive microbiology, that has recently emerged as a comprehensive and systematic approach for addressing the risk of microbial food spoilage, enables to assess microbial behavior for both microbial destruction during a given inactivation process and microbial growth throughout shelf-life, for relevant conditions encountered in food industries for several model foods (Membré et al., 2005). Therefore, the use of mathematical modeling to predict the contamination risk should be considered among the strategies to reduce damage caused by bacterial contamination. These strategies also include the evaluation of the microbiological quality of raw materials and the use of chemical preservatives, acidifiers, or biopreservatives.

Food preservation is regulated by a European Directive (1995/2/CE) which approves for bread-making the use of ascorbic acid, propionic acid, and ethanol as preservatives, lactic and acetic acid as acidifying additives, and considers allowable 0.2% wt/wt of sorbate and 0.3% wt/wt of propionate. Nevertheless, the possible toxic effect of propionic acid led to prohibition of the use of calcium propionate in some European countries and/or to a limited use of organic acids as preservatives in food (Pattison et al., 2004; Dengate and Ruben, 2002). Therefore, in the last decade researches have been directed to develop biopreservatives for reducing the use of chemical preservatives while improving the physicochemical and textural quality of bread.

SPOREFORMING BACTERIAL SPECIES ASSOCIATED WITH ROPE SPOILAGE OF BREAD

In recent years some studies were aimed at the identification of sporeforming bacteria usually contaminating bakery products and their ingredients responsible for rope spoilage. The bacterial species more frequently isolated from ingredients and/or spoiled bread and considered responsible for rope spoilage were mainly *Bacillus subtilis* and *Bacillus licheniformis*, even if other species such as *Bacillus pumilus*, *Bacillus amyloliquefaciens, Bacillus megaterium*, and *B. cereus* were also more seldomly identified (Bailey and von Holy, 1993; Collins et al., 1991; Rosenkvist and Hansen, 1995; Thompson et al., 1998). However, it should be considered that the identifications were based on phenotypic techniques which in some cases were recognized by the authors themselves as providing ambiguous results (Collins et al., 1991). Subsequently, the frequent isolation of *B. subtilis* and *B. licheniformis* from flour and spoiled bread was confirmed in other studies in which molecular techniques were used (Pepe et al., 2003; Sorokulova et al., 2003). However, also in those studies, the unambiguous identification of some *Bacillus* species was not definitely assessed. A more recent study (Valerio et al., 2012) was focused on the identification of sporeforming bacteria contaminating different bread ingredients and breads in southern Italy with the aim of ascertaining bacterial species responsible for rope spoilage. In this study, a high number of sporeforming bacterial isolates was first characterized by repetitive element sequence-based polymerase chain reaction (rep-PCR) and then identified on the basis of an almost complete sequence of the 16S rRNA gene. The cases of unclear identification were solved by sequencing also the *gyrA* or the *gyrB* genes or using differential phenotypic tests. Bacterial strains belonging to *B. cereus* were also assigned to phylogenetic groups by partial sequencing of the *panC* gene according to a proposal of Guinebretière et al. (2010). The ability of the different species to cause rope spoilage was ascertained by inoculating bread slices with heat-treated bacterial cultures. Based on this test, bacterial species were considered more or less potentially responsible for rope spoilage depending on the number and the percentage of bacterial isolates able to cause rope spoilage after the heat treatment (Fig. 16.1).

Even if a bacterial species can be considered potentially responsible for rope spoilage, single strains may show a different spoiling behavior. In fact, the results of the test highlighted that, although all the isolates were able to spoil bread, only a variable percentage of bacterial isolates of each species maintained this capability after the heat treatment. Actually, the spoilage ability is strain-related, depending not only on the strain heat resistance and growth parameters but also on strain enzymatic activity. As a consequence, high cell concentrations in bread could not be always associated with the occurrence of rope spoilage, because of the lack of the required enzymatic activity in the contaminating *Bacillus* strain (Valerio et al., 2012). Moreover, the study also indicated for the first time that *B. amyloliquefaciens* was the species mainly related to the occurrence of rope spoilage. In this regard, it should be considered that the relevance of *B. amyloliquefaciens* was

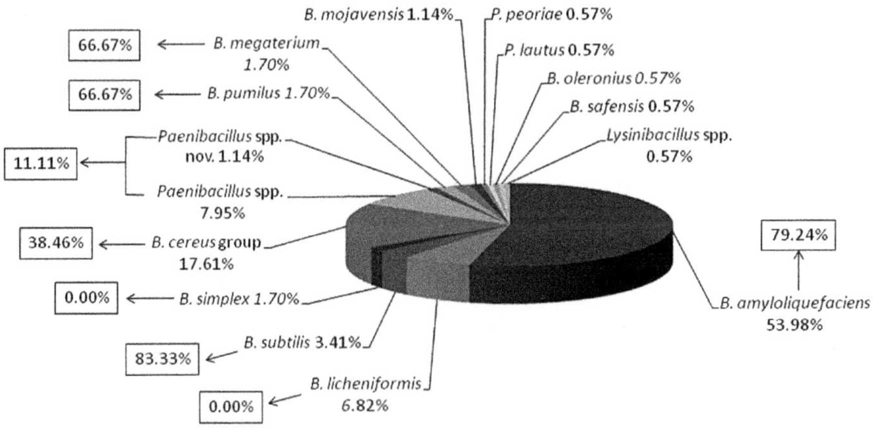

FIGURE 16.1

Occurrence in percentage of sporeforming bacteria isolated from bread and ingredients. The numbers in the box indicate the percentage of isolates able to cause rope spoilage after the heat treatment (100°C) of the main bacterial species (Valerio et al., 2012).

acknowledged for the first time possibly because of its misidentification as *B. subtilis* in previous works. Actually, *B. subtilis* and *B. pumilus* were able to cause rope in bread after the heat treatment but they were isolated in a low number as well as *B. megaterium* and *Bacillus mojavensis*. The involvement of *B. licheniformis* in rope spoilage of bread remains questionable; in fact, its alterative ability was previously reported (Sorokulova et al., 2003) but all the strains tested by Valerio et al. (2012) lost their ability to spoil bread after the heat treatment. These contrasting results could be possibly explained considering the strain variability within a bacterial species and that different strains were tested in those works. In contrast, a potential relevant role in rope spoilage should be recognized to the *B. cereus* group because of the significant number of isolates obtained from bread and its ingredients, even if a lower percentage of them (38.5%), in comparison to *B. amyloliquefaciens*, maintained the ability to cause rope after the heat treatment. The recurrent occurrence of *B. cereus* in bread and its ingredients should also be taken into consideration because of the toxigenic potential of strains belonging to this group.

TOXIGENIC SPOREFORMING BACTERIA

Bacillus cereus was reported as the fourth cause of foodborne outbreaks in France and 50% of the outbreaks between 2006 and 2010 were associated with the consumption of starchy foods (Cadel Six et al., 2012). Concerning bakery products, *B. cereus*, *B. subtilis*, and *B. licheniformis* were the bacterial species more frequently associated with foodborne illness also due to the consumption of ropy bread (Smith et al., 2004). As raw materials used in the bread-making process can be an important source of sporeforming bacteria, the presence of these bacteria in bread and ingredients was evaluated in a number of studies (Pepe et al., 2003, Valerio et al., 2012). The results indicated the presence of a high diversity of bacterial strains, mainly belonging to species of the genus *Bacillus*, including those of the *B. cereus* group. *Bacillus cereus* can induce two distinct types of foodborne diseases: a diarrheal and an emetic syndrome. The diarrheal illness is due to the production of heat-labile enterotoxins, that is the hemolytic enterotoxin hemolysin BL (HBL), the non-hemolytic enterotoxin (NHE), and the cytotoxin K (CytK), by vegetative cells in the small intestine (Senesi and Ghelardi, 2010; Logan, 2012). Moreover, *B. cereus* strains can produce other cytotoxins, hemolysins, collagenases, and phospholipases (Ceuppens et al., 2011). The emetic syndrome is due to a cyclic dodecadepsipeptide, known as cereulide, resistant to high temperatures and proteolysis, and stable at pH 2–11; therefore it is not affected by food processing or by the passage through the gastrointestinal tract (Logan, 2012). The production of cereulide by *B. cereus* was quantified in various foods; high levels were detectable in farinaceous foods and higher rates were found in rice and pasta than in bread and cakes (Agata et al., 2002). Jääskeläinen et al. (2003) studied the toxin production in bakery products and demonstrated that cereulide was produced at concentrations of *B. cereus* $\geq 10^6$ log CFU/g of product, $a_w > 0.953$ and $pH > 5.6$. In a recent study the toxigenic potential and heat survival of *B. cereus* strains isolated from bread and its ingredients were investigated to evaluate whether they could represent a risk to consumer health (De Bellis et al., 2015). The potential toxigenicity of the strains was evaluated by multiple approaches including a screening test of cytotoxic activity on HT-29 cells using bacterial culture filtrates after growing bacterial cells in brain heart infusion medium (BHI) and in the bread-based medium bread extract broth (BEB), immunoassays, and PCR detection of toxin genes. *Bacillus cereus* strains showed a high cytotoxic activity, although it was lower after the growth of the strains in BEB medium; the presence of genes involved in the production of NHE, HBL, or CytK toxins was also highlighted. The production of NHE and HBL toxins was also confirmed by specific immunoassays. Importantly, none of the *B. cereus* strains tested by De Bellis et al. (2015) contained the gene for the production of cereulide; nevertheless, they were assigned to phylogenetic groups III and IV, both associated with a high food poisoning risk in accordance with Guinebretière et al. (2010). However, when spores were subjected to heat treatment reproducing the bread cooking process, strains of group III showed a higher thermal resistance than spores of group IV, indicating the actual risk of the persistence of strains of group III after bread cooking.

Bacillus species outside the *B. cereus* group were also, but more rarely, identified as agents of foodborne diseases characterized by diarrhea and/or vomiting (Logan, 2012). In this regard, the production of toxins was demonstrated for *B. licheniformis, B. subtilis*, *B. mojavensis*, and *B. pumilus*, species implicated in food poisoning. *Bacillus licheniformis*, producing lichenysin A, was associated with one case of fatal illness (Madslien et al., 2013). *Bacillus subtilis* was shown to produce amylosin, a heat-stable nonprotein toxin, but also heat-labile enterotoxins (Ouoba et al., 2008). Amylosin can also be produced by *B. amyloliquefaciens* and *B. mojavensis* strains. *Bacillus pumilus* produces pumilacidin, a heat-stable cyclic lipopeptide; moreover, a heat-stable toxin, considered similar to cereulide was produced by strains of *Bacillus firmus*, *B. megaterium*, and *Bacillus simplex* (Logan, 2012). De Bellis et al. (2015) also detected cytotoxic activity in culture filtrates of *B. amyloliquefaciens* (seven strains), *Paenibacillus* spp. (three), *B. mojavensis* (one), *B. simplex* (one), and *B. pumilus* (one) isolated from bread and its ingredients. Interestingly, the *B. amyloliquefaciens* strains were shown to almost completely survive the thermal bread cooking process, although they induced low cytotoxic activity. Single strains belonging to *B. mojavensis* and *B. simplex* showed higher cytotoxic activity after growth in BEB than in BHI and spores survived the thermal treatment. Actually, sporeforming bacterial strains which can represent a risk for consumer health are those jointly able to produce toxic substances and to survive the bread cooking conditions. In this regard, it is noteworthy that the risk of food poisoning should be considered even higher for those toxigenic strains able to contaminate bread without causing a visible alteration, as spoiled bread is often too unpleasant to be consumed.

SPORE HEAT RESISTANCE OF *BACILLUS* SPECIES

During baking the temperature at the dough core reaches 100°C and then remains constant until the end of baking (Chhanwal and Anandharamakrishnan, 2014; Purlis and Salvadori, 2009). Like any heat treatment, bread baking thermally inactivates microorganisms. At 100°C vegetative bacteria and molds are completely inactivated. However bacterial spores, in particular species responsible for rope spoilage, resist the heat treatment and develop during shelf-life.

The reduction rate of the bacterial spores depends on temperature and heating time. The heat resistance is quantified by the classical decimal reduction time D at 100°C (D_{100}) as reference temperature as well as temperature sensitivity quantified by the z parameter (temperature increase needed to obtain a 10-fold reduction of D_{100}). Heat resistance values of species associated with ropy bread are reported in Table 16.1.

Table 16.1 Range Values of Decimal Reduction Time at 100°C (D_{100}) and Temperature Sensitivity (z) of Sporeforming Bacterial Species Associated with Rope Spoilage

Species	D_{100} (min)	z (°C)	References
B. amyloliquefaciens	7.21÷507.2	6.23÷14.2	Ahn and Balasubramaniam (2007), Berendsen et al. (2015), Witthuhn et al. (2011)
B. subtilis	0.2÷5123	5.8÷15	André et al. (2013), Berendsen et al. (2015), Kort et al. (2005), Leuschner et al. (1998), Nakayama et al. (1996), Palop et al. (1996), Sala et al. (1995)
B. licheniformis	0.9÷56	6.86÷16.8	André et al. (2013), Janstová and Lukasova (2001), Leguérinel et al. (2007), Leuschner et al. (1998), Montville and Sapers (1981), Nakayama et al. (1996), Palop et al. (1996), Raso et al. (1995), Rodríguez-Lozano et al. (2010)
B. cereus	0.2÷11.4	5.7÷18.1	Janstová and Lukasova (2001), Kaur (1986), González et al. (1999), Luu-Thi et al. (2014)
B. pumilus	0.2÷10.0	9.7÷11.2	Janstová and Lukasova (2001), Gibriel and Abd-el Al (1973), Leuschner et al. (1998)
Paenibacillus spp. (including *P. macerans* and *P. polymyxa*)	0.4÷195.1	6.2÷49.2	André et al. (2013), Casadei et al. (2000), Huo et al. (2012)
B. megaterium	0.2÷2.3	8.3÷13.3	Gibriel and Abd-el Al (1973), Janstová and Lukasova (2001), Nakayama et al. (1996), Rodríguez-Lozano et al. (2010)

Regarding *B. amyloliquefaciens*, data are relevant to strains mainly isolated from pizza and ropy bread or dairy products. The D_{100} values of strains isolated from ropy bread varied from 23.3 to 44.6 min (Ahn and Balasubramaniam, 2007); while values ranged from 15 to 507 min for strains isolated from dairy products (Berendsen et al., 2015; Witthuhn et al., 2011).

Regarding *B. subtilis*, a considerable variability of spore heat resistance is reported in the literature for different strains. A large difference in the heat resistance values was reported in the case of strains isolated from ropy bread (D_{100} 14 min) (Leuschner et al., 1998) or flour (D_{100} 300 min) (Berendsen et al., 2015).

For *B. licheniformis* spores, D_{100} values obtained from literature for 13 strains isolated from different origins, showed a median value of 2.8 min associated with a z median value of 8.61°C. The higher reported D_{100} value (56 min) corresponded to spores of a *B. licheniformis* strain isolated from brown soda bread dough with heat resistance parameters determined in this food matrix (Leuschner et al., 1998).

Spore heat resistance values (D_{100}) of 50 *B. cereus* strains from different origins (food, dairy environment, soil, or outbreak) were determined by Janstová and Lukasova (2001) and Luu-Thi et al. (2014). These D_{100} values ranged from 0.71 to 1.48 min for 6 strains isolated from milk and the farm environment (Janstová and Lukasova, 2001), while for 39 strains isolated from other sources (Luu-Thi et al., 2014) values ranged from 0.04 to 9.44 min with a mean value of 1.43. Kaur (1986) determined D_{100} values higher than 10 min for two strains isolated from wheat flour. Luu-Thi et al. (2014) observed a relationship between heat resistance as well as growth cardinal values and the *B. cereus* groups as defined by Guinebretière et al. (2008).

Spores of *B. pumilus* isolated from milk presented a low heat resistance with D_{100} of 0.8 min (Janstová and Lukasova, 2001), while a strain isolated from brown soda bread dough (Leuschner et al., 1998) showed a higher D_{100} value (10 min). Although spores of *Paenibacillus* spp. can be responsible for spoilage of bread, no heat resistance value of strains isolated from bread or flour, was published. However, spores of *Paenibacillus* spp. strains, including *Paenibacillus polymyxa* and *Paenibacillus macerans*, mainly isolated from canned food, presented a large variability in the heat resistance values (D_{100}). *Bacillus megaterium* spores isolated from fruit juice or milk (six strains) presented relatively low heat resistance since the higher D_{100} value registered was 2.3 min (Janstová and Lukasova, 2001).

All these data highlight a large variation in the heat resistance of sporeforming bacteria associated with the bread rope, both at species and strain levels. Therefore, the bread cooking process can differently affect spore survival even if the knowledge of heat resistance parameters can be an aid in determining the number of survived spores. Temperature kinetics in the bread core during baking present a temperature increase to 100°C in 15 min and then the temperature becomes stable at 100°C regardless of the oven temperature (Chhanwal and Anandharamakrishnan, 2014; Purlis and Salvadori, 2009). For a baking time of 30 min the heat treatment is constant during 15 min at 100°C and may be used to determine decimal reductions of various spore populations, despite their origin, knowing their D_{100} values. The value of decimal reduction ratio "n" is determined by Eq. (16.1) where t_T is the treatment time at constant temperature (100°C) and D_T is the decimal reduction time at temperature 100°C.

$$n = \frac{t_T}{D_T} \tag{16.1}$$

For bacterial spores with D_{100} values ranging from 1 to 3 min (*B. cereus, B. megaterium*), the decimal reduction ratio varies from 15 to 5 yielding spore inactivation during bread baking, even if among *B. cereus*, some strains isolated from flour (Kaur, 1986) showed D_{100} higher than 10 yielding a decimal reduction ratio of about 1.5. For *B. amyloliquefaciens* spores with D_{100} values ranging from 7.21 to 44.6 min, the calculated decimal reduction number is lower than 1, yielding no thermal inactivation of spores during bread baking. Similar conclusions can be drawn for spores of *B. licheniformis* (D_{100} 0.9–56 min) and spores of *B. subtilis* strains showing the highest heat resistance. These data support the findings obtained by Valerio et al. (2012), which indicated *B. amyloliquefaciens* as a species frequently associated with the rope spoilage.

PREDICTIVE MICROBIOLOGY TO EVALUATE THE CONTAMINATION RISK IN BREAD

Bacterial spore populations are not always completely inactivated during bread baking. Certain species or strains can resist the baking thermal treatments and once activated by the heat stress, can germinate and multiply. Nevertheless, depending on the environmental bread conditions the bacterial germination and development can be reduced or inhibited. Growth of sporeforming bacteria in bread can be evaluated by challenge test or using predictive microbiology to take into account the environmental factor values in the prediction of bacterial behavior. Predictive microbiology is based on mathematical equations developed to simulate and predict the growth or inactivation of microorganisms considering the values of environmental factors such as temperature, pH, a_w, organic acids, atmosphere, or antimicrobial compounds. These mathematical models can be used to estimate food shelf-life or as an aid to optimize the food formulation.

The predictive microbiology approach is performed in two steps. In the first step primary models describe the bacterial growth kinetics which are characterized and quantified by the lag time (λ), growth rate (μ_{max}), and maximum population reached. The main models used for describing the growth kinetics are the Gompertz model, the Baranyi et al. (1993) model, or the Rosso et al. (1995) models (Eq. (16.2))

$$f(t,\Theta_1)=\begin{cases}\ln x_0 & ,t\le \text{lag}\\ \ln x_{\max}-\ln\left(1+\left(\frac{x_{\max}}{x_0}-1\right)\cdot\exp(-\mu_{\max}\cdot(t-\text{lag}))\right) & ,t>\text{lag}\end{cases} \tag{16.2}$$

where x_0 is the initial population; x_{max}, maximal population; lag, lag time (h); μ_{max}, maximal growth rate (h^{-1}). The second step or secondary modeling describes the effect of environmental factors on the primary model parameter values. The growth rate μ_{max} is the main parameter modeled for microbial growth. The approach proposed by Zwietering et al. (1992), called the gamma concept (Eqs. (16.3)–(16.4)), enables the use of the gamma function (γ) to account for the multiplicative effect of each environmental factor affecting bacterial growth. These inhibitory functions or gamma function are robust, simple and parsimonious models. For temperature, pH, and water activity factors, the parameters of the cardinal models Rosso et al. (1995) (Eq. (16.5)) present a biological significance and correspond for studied factors to the minimum, optimum and maximum growth condition values.

$$\gamma=\gamma(T)\cdot\gamma(\text{pH})\cdot\gamma(a_w) \tag{16.3}$$

$$\mu_{\max}=\gamma\cdot\mu_{\text{opt}} \tag{16.4}$$

$$\gamma(X)=\frac{(X-X_{\max})(X-X_{\min})^n}{(X_{\text{opt}}-X_{\min})^{n-1}\left((X_{\text{opt}}-X_{\min})(X-X_{\text{opt}})-(X_{\text{opt}}-X_{\max})(X_{\text{opt}}+X_{\min}-nX)\right)} \tag{16.5}$$

For a given strain, the cardinal values are determined from growth rates obtained in synthetic medium. Factor growth limit values (X_{min}, X_{opt}, and X_{max}) could also be obtained from bibliographic data.

The μ_{opt} parameter corresponds to the growth rate obtained in the food in the optimal environmental conditions. It depends on both the microorganism and the food matrix in which it develops. It is determined by a growth kinetic carried out in the food or challenge test. Knowing the food environmental characteristics it is possible to calculate μ_{opt} value using the corresponding gamma values. The cardinal model parameter values quantifying the growth limit values could be included as a parameter value in models to perform simulations. Numerous softwares are already available to facilitate the use of such recognized mathematical models for various applications. The modular approaches described here are compiled in decision-making tools applied to food industrial issues such as Sym'Previus (www.symprevius.org) or Food Spoilage and Safety Predictor (FSSP) (http://sssp.dtuaqua.dk).

These predictive microbiology tools can be applied to quantify growth and heat destruction in order to optimize food formulation and process and control bread spoilage by sporeforming bacteria during the shelf-life. Thus, it is essential to further increment the database with physiological characteristic temperature, pH, and a_w cardinal values of sporeforming species yielding ropy bread and further determine the associated μ_{opt} in bread by using challenge testing.

The growth cardinal parameters of the species associated with rope spoilage (*B. subtilis, B. amyloliquefaciens, B. cereus, B. licheniformis, B. megaterium*) even if isolated from different origins, are reported in Table 16.2.

In particular, for *B. amyloliquefaciens*, Valerio et al. (2015) determined the growth cardinal values for three strains isolated from ropy bread, flour, and grain, while artificial spore inoculations were performed in dough for various pan bread recipes in order to calculate the μ_{opt} (1.61 h^{-1}) in bread and to compare experimental counts with in silico growth simulations.

Regarding the bacterial strains of the species *B. cereus*, Carlin et al. (2013) presented the cardinal values for all the groups defined by Guinebretière et al. (2008). A wide variability of growth temperature limit values was observed between groups while the minimum growth pH values were very close, ranging from 4.59 (group III) to 4.96 (group IV) with a pH_{opt} ranging from 6.39 to 6.73 (Table 16.2). *Bacillus licheniformis* is a mesophilic bacterium, however, it can grow at high temperatures (T_{min} 11°C, T_{opt} 49°C and T_{max} 57°C) with a minimum pH value at 4.6 and an optimal pH at 8.1 (Trunet et al., 2015), a minimum water activity value at 0.915 (Baril et al., 2012a) and a μ_{opt} 3.12 h^{-1} as determined in laboratory media. Certain strains of *B. megaterium* are psychrotrophic (T_{min} 2°C, T_{opt} 30°C, and T_{max} 40°C) (Suutari and Laakso, 1992) while other strains are thermophilic (T_{min} 38°C, T_{opt} 64°C, and T_{max} 73°C) (Ståhl and Olsson, 1977). No bibliographic data for pH and water activity cardinal values are available for this species.

Knowing cardinal values for target species and using the models (Eqs. (16.1)–(16.4)) it is easy to calculate and predict the growth of this spoilage flora in bread where water activities range from 0.93 to 0.95 and the pH values vary from 5.87 for classical bread to 4.92 for bio-acidified bread (Rosenkvist and Hansen, 1995; Valerio et al., 2015). As an example, pH and water activity values of bio-acidified bread, being lower than the pH and a_w boundary growth conditions for previously presented bacterial species, can hamper the development of survived spores and the potential occurrence of the rope spoilage (Valerio et al., 2015).

In conclusion, the use of mathematical models to predict bacterial ability to growth is a useful tool to optimize the bread formulation or process by impacting on the physicochemical characteristics of bread to limit the development of sporeforming bacteria in bread and increase its shelf-life with, for instance, the addition of lactic acid bacteria-based bioingredients to acidify the food environment.

Table 16.2 Range Values of Growth Cardinal Parameters of Sporeforming Bacterial Species Associated with Rope Spoilage

Species	μ_{opt} (h^{-1})	T_{min} (°C)	T_{opt} (°C)	T_{max} (°C)	pH_{min}	pH_{opt}	$a_{w\,min}$	a_{wopt}	References
B. amyloliquefaciens	1.95÷2.13	10.49÷11.79	42÷43.77	52.3÷55.84	4.28÷4.57	6.87÷6.88	0.917÷0.955	0.992÷0.994	Valerio et al. (2015)
B. subtilis	3÷3.4	3.5÷12	37÷47.3	50÷55.7	4.9÷5	6.68÷7	0.91÷0.93	–	Suutari and Laakso (1992), Hoffman and Bremer (2013), Pandey et al. (2013), Pant et al. (2015), Nichols et al. (1995), Holtmann and Bremer (2004), Scott (1957), Gauvry et al., unpublished results
B. licheniformis	3.12	8.6÷11.3	49.1÷50.9	56÷57.87	4.63÷5.3	6.9÷8.17	0.915	1	Baril (2011), Trunet et al. (2015)
B. cereus (groups II÷VII)	1.83÷3.89	1.4÷19.1	31÷43	41÷55	4.59÷4.96	6.39÷6.73	0.941÷0.973	–	Carlin et al. (2013), Guinebretière et al. (2008)
B. megaterium	–	0÷38	30÷64	40÷73	–	–	–	–	Suutari and Laakso (1992), Ståhl and Olsson (1977)

MICROBIAL BIOTECHNOLOGY TO REDUCE ROPE SPOILAGE

Bacillus spore numbers can be controlled by ensuring raw material quality, good sanitation, and cooling of production and storage environments (Viljoen and von Holy, 1997). Spore germination and growth in bread can be inhibited by chemical preservatives such as propionic and acetic acids, although the current trend is to reduce the levels of these substances (Pattison et al., 2004; Marin et al., 2002). Acetic acid adversely affects the organoleptic quality of baked products, while propionic acid was reported to cause irritability, restlessness, inattention, and sleep disturbance in some children (Dengate and Ruben, 2002). Alternative biopreservative systems to prevent bread spoilage are therefore required. The most important microorganisms used in the biotechnology of bakery products to provide a natural preservation, are lactic acid bacteria (LAB) of sourdough and their antimicrobial metabolites. These bacteria operate directly in the microbiological quality of bread by producing bioactive molecules (such as organic acids, fatty acids, hydrogen peroxide, diacetyl, cyclic dipeptides, and bacteriocins), but they also operate indirectly, reducing the pH value of the dough and increasing the total titratable acidity (TTA), thus inhibiting spoilage microorganisms and pathogens (Ryan et al., 2008; Crowley et al., 2013). Among organic acids produced by LAB, a key role in the antimicrobial activity of these bacteria, was demonstrated for phenyllactic (PLA) and *p*-hydroxy-phenyllactic (OH-PLA) acids (Lavermicocca et al., 2000; Cortés-Zavaleta et al., 2014; Gerez et al., 2013). Numerous studies of the microbiota in sourdough and flours used for bakery product manufacturing individuated different LAB strains with peculiar biotechnological properties which allow improvement of the overall quality of products (Valerio et al., 2009; De Vuyst and Neysens, 2005). Well studied is the application of sourdough that, because of the fermentation activity played by LAB species, including *Lactobacillus sanfranciscensis*, *Lactobacillus brevis*, and *Lactobacillus plantarum*, exerts a control activity on molds and bacterial spoilers (Di Cagno et al., 2014; Katina et al., 2002; Denkova et al., 2014; Pepe et al., 2003). Actually, LAB lower the pH of bread dough, inhibit the growth of contaminants, and improve the flavor of the baked product.

The desirable effects of LAB can also be exploited in the bread-making processes that use *Saccharomyces cerevisiae* as the leavening agent (Valerio et al., 2008; Di Biase et al., 2014). Bread contamination by *Bacillus* spores can be efficiently controlled by using LAB-derived acidifiers as biopreservatives which allow to obtain pH conditions (<5) that hamper spore germination and toxin production by toxigenic strains in yeast-leavened bread. It was demonstrated that the use of a fermentation product of a sourdough *L. plantarum* strain, containing PLA and OH-PLA in addition to lactic and acetic acids, resulted effective as calcium propionate 0.3% (w/v) preventing the rope spoilage in yeast-leavened bread (Valerio et al., 2008). Moreover, this study suggested the potential involvement of some unknown metabolites in the inhibitory activity of the fermentation product. Subsequently, a *L. brevis*-based bioingredient, acting as a sourdough, improved shelf-life, enhanced physicochemical properties and crumb hardness of a wheat bread containing bran by modulating the intrinsic product's characteristics and thus counteracting the negative effects of bran addition (Valerio et al., 2014). The study also indicated that bread formulation can modulate bacterial behavior during storage and consequently bread spoilage. Actually, the growth kinetics of a *B. amyloliquefaciens* strain producing rope in yeast-leavened bread were determined under five different intrinsic (pH and a_w) and environmental (storage temperature) conditions (Valerio et al., 2015). The results demonstrated that the *B. amyloliquefaciens* growth was efficiently reduced by applying the *L. brevis*-based acidic bioingredient which allowed bread to reach a pH of 4.9, a value below the pH_{opt} of the strain.

CONCLUDING REMARKS

This chapter highlights that an effective risk evaluation requires multiple approaches in which different factors should be considered with specific regard to the investigated food. In the case of rope spoilage of bread, multiple factors contribute to its occurrence, such as, on the one hand, the physiological and genetic traits of several species of sporeformers contaminating raw materials and, on the other hand, methodologies to counteract the alterative phenomenon and tools for risk evaluation. The current knowledge of species associated with bread rope spoilage clearly indicates that single strains could represent a risk for consumer health if specific characteristics of the strains and of the food matrix as well as conditions encountered during process and storage coexist. Finally, in order to prevent rope occurrence, available decision-making tools which are largely recognized to predict the behavior of pathogenic strains should also be used with spoilage strains to optimize food composition and processes, also applying microbial biotechnologies that can preserve bread quality and extend its shelf-life.

REFERENCES

Agata, N., Ohta, M., Yokoyama, K., 2002. Production of *Bacillus cereus* emetic toxin (cereulide) in various foods. Int. J. Food Microbiol. 73, 23–27.

Ahn, J., Balasubramaniam, B.V.M., 2007. Inactivation kinetics and injury recovery of *Bacillus amyloliquefaciens* spores in low-acid foods during pressure-assisted thermal processing. Food Sci. Biotechnol. 23, 1851–1857.

André, S., Zuber, F., Remize, F., 2013. Thermophilic spore-forming bacteria isolated from spoiled canned food and their heat resistance. Results of a French ten-year survey. Int. J. Food Microbiol. 165, 134–143.

Bailey, C.P., von Holy, A., 1993. *Bacillus* spore contamination associated with commercial bread manufacture. Food Microbiol. 10, 287–294.

Baranyi, J., McClure, P.J., Sutherland, J.P., Roberts, T.A., 1993. Modeling bacterial growth responses. J. Ind. Microbiol. 12, 190–194.

Baril, E., 2011. Quantification de l'influence de l'environnement sur la formation et la thermo-résistance des spores bactériennes. Thesis Université de Bretagne Occidentale, France.

Baril, E., Coroller, L., Couvert, O., Leguerinel, I., Postollec, F., Boulais, C., et al., 2012a. Modeling heat resistance of *Bacillus weihenstephanensis* and *Bacillus licheniformis* spores as a function of sporulation temperature and pH. Food Microbiol. 30, 29–36.

Baril, E., Coroller, L., Couvert, O., El Jabri, M., Leguerinel, I., Postollec, F., et al., 2012b. Sporulation boundaries and spore formation kinetics of *Bacillus* spp. as a function of temperature, pH and aw. Food Microbiol. 32, 79–86.

Berendsen, E.M., Zwietering, M.H., Kuipers, O.P., Wells-Bennik, M.H., 2015. Two distinct groups within the *Bacillus subtilis* group display significantly different spore heat resistance properties. Food Microbiol. 45, 18–25.

Berghofer, L.K., Hocking, A.D., Miskelly, D., Jansson, E., 2003. Microbiology of wheat and flour milling in Australia. Int. J. Food Microbiol. 85, 137–149.

Cadel Six, S., De Buyser, M.L., Vignaud, M.L., Dao, T.T., Messio, S., Pairaud, S., et al., 2012. Toxi-infections alimentaires collectives à *Bacillus cereus*: bilan de la caractérisation des souches de 2006 à 2010. Bulletin épidémiologique, santé animale et alimentation/Spécial Risques alimentaires microbiologiques 50, 57–61.

Carlin, F., Albagnac, C., Rida, A., Guinebretière, M.H., Couvert, O., Nguyen-The, C., 2013. Variation of cardinal growth parameters and growth limits according to phylogenetic affiliation in the *Bacillus cereus* Group. Consequences for risk assessment. Food Microbiol. 33, 69–76.

Casadei, M.A., Ingram, R., Skinner, R.J., Gaze, J.,E., 2000. Heat resistance of *Paenibacillus polymyxa* in relation to pH and acidulants. J. Appl. Microbiol. 89, 801–806.

Ceuppens, S., Rajkovic, A., Heyndrickx, M., Tsilia, V., Van De Wiele, T., Boon, N., et al., 2011. Regulation of toxin production by *Bacillus cereus* and its food safety implications. Crit. Rev. Microbiol. 37, 188–213.

Chhanwal, N., Anandharamakrishnan, C., 2014. Temperature- and moisture-based modeling for prediction of starch gelatinization and crumb softness during bread-baking Process. J. Texture Stud. 45, 462–476.

Collins, N.E., Kirschner, L.M., von Holy, A., 1991. Characterization of *Bacillus* isolated from ropey bread, bakery equipment and raw material. S. Afr. J. Sci. 87, 62–66.

Cortés-Zavaleta, O., López-Malo, A., Hernández-Mendosa, A., García, H.S., 2014. Antifungal activity of lactobacilli and its relationship with 3-phenyllactic acid production. Int. J. Food Microbiol. 173, 30–35.

Crowley, S., Mahonya, J., van Sinderen, D., 2013. Current perspectives on antifungal lactic acid bacteria as natural bio-preservatives. Trends Food Sci. Technol. 33, 93–109.

De Bellis, P., Minervini, F., Di Biase, M., Valerio, F., Lavermicocca, P., Sisto, A., 2015. Toxigenic potential and heat survival of spore-forming bacteria isolated from bread and ingredients. Int. J. Food Microbiol. 197, 30–39.

De Vuyst, L., Neysens, P., 2005. The sourdough microflora: biodiversity and metabolic interactions. Trends Food Sci. Technol. 16, 43–56.

Dengate, S., Ruben, A., 2002. Controlled trial of cumulative behavioural effects of a common bread preservative. J. Paediatr. Child Health 38, 373–376.

Denkova, R., Ilieva, S., Denkova, Z., Georgieva, L., Yordanova, M., 2014. Production of wheat bread without preservatives using sourdough starters. Biotechnol. Biotechnol. Equip. 28, 889–898.

Di Biase, M., Lavermicocca, P., Lonigro, S.L., Valerio, F., 2014. *Lactobacillus brevis*-based bioingredient inhibits *Aspergillus niger* growth on pan bread. Ital. J. Agron. 9 (614), 146–151.

Di Cagno, R., Pontonio, E., Buchin, S., De Angelis, M., Lattanzi, A., Valerio, F., et al., 2014. Diversity of the lactic acid bacteria and yeast microbiota switching from firm to liquid sourdough fermentation. Appl. Environ. Microbiol. 80, 3161–3172.

Gerez, C.L., Torres, M.J., Font de Valdez, G., Rollán, G., 2013. Control of spoilage fungi by lactic acid bacteria. Biol. Control 64, 231–237.

Gibriel, A.Y., Abd-el Al, A.T., 1973. Measurement of heat resistance parameters for spores isolated from canned products. J. Appl. Bacteriol. 36, 321–327.

González, I., López, M., Martıenez, S., Bernardo, A., González, J., 1999. Thermal inactivation of *Bacillus cereus* spores formed at different temperatures. Int. J. Food Microbiol. 51, 81–84.

Guinebretière, M.H., Thompson, F.L., Sorokin, A., Normand, P., Dawyndt, P., Ehling-Schulz, M., et al., 2008. Ecological diversification in the *Bacillus cereus* Group. Environ. Microbiol. 10, 851–865.

Guinebretière, M.H., Velge, P., Couvert, O., Carlin, F., Debuyser, M.L., Nguyen-The, C., 2010. Ability of *Bacillus cereus* group strains to cause food poisoning varies according to phylogenetic affiliation (Groups I to VII) rather than species affiliation. J. Clin. Microbiol. 48, 3388–3391.

Hoffman, T., Bremer, E., 2011. Protection of *Bacillus subtilis* against cold stress via compatible-solute acquisition. J. Bacteriol. 193, 1153–1562.

Holtmann, G., Bremer, E., 2004. Thermoprotection of *Bacillus subtilis* by exogenously provided glycine betaine and structurally related compatible solutes: involvement of Opu transporters. J. Bacteriol. 186, 1683–1693.

Huo, Z., Zhang, N., Raza, W., Huang, X., Yong, X., Liu, Y., et al., 2012. Comparison of the spores of *Paenibacillus polymyxa* prepared at different temperatures. Biotechnol. Lett. 34, 925–933.

Jääskeläinen, E.L., Häggblom, M.M., Andersson, M.A., Vanne, L., Salkinoja-Salonen, M.S., 2003. Potential of *Bacillus cereus* for producing an emetic toxin, cereulide, in bakery products: quantitative analysis by chemical and biological methods. J. Food Prot. 66, 1047–1054.

Janstová, B., Lukasova, J., 2001. Heat Resistance of *Bacillus* spp. spores isolated from cow's milk and farm environment. Acta Veterinaria Brno 70, 179–184.

Katina, K., Sauri, M., Alakomi, H.-L., Mattila-Sandholm, T., 2002. Potential of lactic acid bacteria to inhibit rope spoilage in wheat sourdough bread. Lebensmittel Wissenschaft und Technologie 35, 38–45.

Kaur, P., 1986. Survival and growth of *Bacillus cereus* in bread. J. Appl. Bacteriol. 60, 513–516.

Kort, R., O'Brien, A.C., van Stokkum, I.H., Oomes, S.J., Crielaard, W., Hellingwerf, K.J., et al., 2005. Assessment of heat resistance of bacterial spores from food product isolates by fluorescence monitoring of dipicolinic acid release. Appl. Environ. Microbiol. 71, 3556–3564.

Kramer, J.M., Gilbert, R.J., 1989. *Bacillus cereus* and other *Bacillus* species. In: Doyle, M. P. (Ed.), Foodborne Bacterial Pathogens. Marcel Dekker, New York, NY, pp. 22–70.

Lavermicocca, P., Valerio, F., Evidente, A., Lazzaroni, S., Corsetti, A., Gobbetti, M., 2000. Purification and characterization of novel antifungal compounds from the sourdough *Lactobacillus plantarum* strain 21B. Appl. Environ. Microbiol. 66, 4084–4090.

Leguérinel, I., Couvert, O., Mafart, P., 2007. Modelling the influence of the sporulation temperature upon the bacterial spore heat resistance, application to heating process calculation. Int. J. Food Microbiol. 114, 100–104.

Leuschner, R.G.K., O'Callaghan, M.J.A., Arendt, E.K., 1998. Bacilli spoilage in part-baked and rebaked brown soda bread. J. Food Sci. 63, 915–918.

Logan, N.A., 2012. *Bacillus* and relatives in foodborne illness. J. Appl. Microbiol. 112, 417–429.

Luu-Thi, H., Khadka, D.B., Michiels, C.W., 2014. Thermal inactivation parameters of spores from different phylogenetic groups of *Bacillus cereus*. Int. J. Food Microbiol. 189, 183–188.

Madslien, E., Rønning, H., Lindbäck, T., Hassel, B., Andersson, M., Granum, P.E., 2013. Lichenysin is produced by most *Bacillus licheniformis*. J. Appl. Microbiol. 115, 1068–1080.

Marin, S., Guynot, M.E., Neira, P., Bernado, M., Sanchis, V., Ramos, A.J., 2002. Risk assessment of the use of sub-optimal levels of weak-acid preservatives in the control of *mould growth on bakery* products. Int. J. Food Microbiol. 79, 203–211.

Membré, J.M., Leporq, B., Vialette, M., Mettler, E., Perrier, L., Thuault, D., et al., 2005. Temperature effect on bacterial growth rate: quantitative microbiology approach including cardinal values and variability estimates to perform growth simulations on/in food. Int. J. Food Microbiol. 100, 179–186.

Montville, T.J., Sapers, G.M., 1981. Thermal resistance of spores from pH elevating strains of *Bacillus licheniformis*. J. Food Sci. 46, 1710–1712.

Nakayama, A., Yano, Y., Kobayashi, S., Ishikawa, M., Sakai, K., 1996. Comparison of pressure resistances of spores of six bacillus strains with their heat resistances. Appl. Environ. Microbiol. 62, 3897–3900.

Nichols, D.S., Nichols, P.D., McMeekin, T.A., 1995. Ecology and physiology of psychrophilic bacteria from Antarctic saline lakes and ice-sea. Sci. Prog. 78, 311–348.

Ouoba, L.I., Thorsen, L., Varnam, A.H., 2008. Enterotoxins and emetic toxins production by *Bacillus cereus* and other species of *Bacillus* isolated from Soumbala and Bikalga, African alkaline fermented food condiments. Int. J. Food Microbiol. 124, 224–230.

Palop, A., Raso, J., Pagán, R., Condón, S., Sala, F.J., 1996. Influence of pH on heat resistance of *Bacillus licheniformis* in buffer and homogenized foods. Int. J. Food Microbiol. 29, 1–10.

Pandey, R., Ter Beek, A., Vischer, N.O.E., Smelt, J.P.P.M., Brul, S., Manders, E.M.M., 2013. Live cell imaging of germination and outgrowth of individual *Bacillus subtilis* spores; the effect of heat stress quantitatively analyzed with SporeTracker. PLoS ONE 10.1371/journal.pone.0058972.

Pant, G., Prakash, A., Pavani, J.V.P., Bera, S., Deviram, G.V.N.S., Kumar, A., et al., 2015. Production, optimization and partial purification of protease from *Bacillus subtilis*. J. Taibah Univ. Sci. 9, 50–55.

Pattison, T.-L., Lindsay, D., von Holy, A., 2004. Natural antimicrobials as potential replacements for calcium propionate in bread. S. Afr. J. Sci. 100, 342–348.

Pepe, O., Blaiotta, G., Moschetti, G., Greco, T., Villani, F., 2003. Rope-producing strains of *Bacillus* spp. from wheat bread and strategy for their control by lactic acid bacteria. Appl. Environ. Microbiol. 69, 2321–2329.

Postollec, F., Mathot, A.-G., Bernard, M., Divanac'h, M.-L., Pavan, S., Sohier, D., 2012. Tracking spore-forming bacteria in food: from natural biodiversity to selection by processes. Int. J. Food Microbiol. 158, 1–8.

Purlis, E., Salvadori, V.O., 2009. Bread baking as a moving boundary problem. Part 2: model validation and numerical simulation. J. Food Eng. 91, 434–442.

Raso, J., Palop, A., Bayarte, M., Condón, S., Sala, F.J., 1995. Influence of sporulation temperature on the heat resistance of a strain of *Bacillus licheniformis* (Spanish type culture collection 4523). Food Microbiol. 12, 357–361.

Rodríguez-Lozano, A., Campagnoli, M., Jewel, K., Monadjemi, F., Gaze, J.E., 2010. *Bacillus* spp. thermal resistance and validation in soups. In: Mendez-Vilas, A. (Ed.), Current Research, Technology and Education Topics in Applied Microbiology and Microbial Biotechnology. Microbiology Book Series Number 2. Formatex Research Center, Badajoz, pp. 537–544.

Rosenkvist, H., Hansen, Å., 1995. Contamination profiles and characterization of *Bacillus* species in wheat bread and raw materials for bread production. Int. J. Food Microbiol. 26, 353–363.

Rosso, L., Lobry, J.R., Bajard, S., Flandrois, J.P., 1995. Convenient model to describe the combined effects of temperature and pH on microbial growth. Appl. Environ. Microbiol. 61, 610–616.

Ryan, L.A., Dal Bello, F., Arendt, E.K., 2008. The use of sourdough fermented by antifungal LAB to reduce the amount of calcium propionate in bread. Int. J. Food Microbiol. 125, 274–278.

Sala, F.J., Ibarz, P., Palop, A., Raso, J., Condón, S., 1995. Sporulation temperature and heat resistance of *Bacillus subtilis* at different pH's. J. Food Prot. 58, 239–243.

Saranraj, P., Geetha, M., 2012. Microbial spoilage of bakery products and its control by preservatives. Int. J. Pharmacol. Biol. Arch. 3, 38–48.

Scott, W.J., 1957. Water relations of food spoilage microorganisms. Adv. Food Res. 7, 83–127.

Senesi, S., Ghelardi, E., 2010. Production, secretion and biological activity of *Bacillus cereus* enterotoxins. Toxins 2, 1690–1703.

Smith, J.P., Daifas, D.P., El-Khoury, W., Koukoutsis, J., El-Khoury, A., 2004. Shelf life and safety concerns of bakery products—a review. Crit. Rev. Food Sci. Nutr. 44, 19–55.

Sorokulova, I.B., Reva, O.N., Smirnov, V.V., Pinchuk, I.V., Lapa, S.V., Urdaci, M.C., 2003. Genetic diversity and involvement in bread spoilage of *Bacillus* strains isolated from flour and ropy bread. Lett. Appl. Microbiol. 37, 169–173.

Ståhl, S., Olsson, O., 1977. Temperature range variants of *Bacillus megaterium*. Arch. Microbiol. 113, 221–229.

Suutari, M., Laakso, S., 1992. Unsaturated and branched chain-fatty acids in temperature adaptation of *Bacillus subtilis* and *Bacillus megaterium*. Biochim. Biophys. Acta 1126, 119–124.

Thompson, J.M., Dodd, C.E.R., Waites, W.M., 1993. Spoilage of bread by *Bacillus*. Int. Biodeterior. Biodegradation 32, 55–66.

Thompson, J.M., Waites, W.M., Food, C.E.R., 1998. Detection of rope spoilage in bread caused by *Bacillus* species. J. Appl. Microbiol. 85, 481–486.

Trunet, C., Mtimet, N., Mathot, A.G., Postollec, F., Leguerinel, I., Sohier, D., et al., 2015. Modeling the recovery of heat-treated *Bacillus licheniformis* Ad978 and *Bacillus weihenstephanensis* KBAB4 spores at suboptimal temperature and pH using growth limits. Appl. Environ. Microbiol. 81, 562–568.

Valerio, F., De Bellis, P., Lonigro, S.L., Visconti, A., Lavermicocca, P., 2008. Use of *Lactobacillus plantarum* fermentation products in bread-making to prevent *Bacillus subtilis* ropy spoilage. Int. J. Food Microbiol. 122, 328–332.

Valerio, F., Favilla, M., De Bellis, P., Sisto, A., De Candia, S., Lavermicocca, P., 2009. Antifungal activity of strains of lactic acid bacteria isolated from a semolina eco system against *Penicillium roqueforti*, *Aspergillus niger* and *Endomyces fibuliger* contaminating bakery products. Syst. Appl. Microbiol. 32, 438–448.

Valerio, F., De Bellis, P., Di Biase, M., Lonigro, S.L., Giussani, B., Visconti, A., et al., 2012. Diversity of spore-forming bacteria and identification of *Bacillus amyloliquefaciens* as a species frequently associated with the ropy spoilage of bread. Int. J. Food Microbiol. 156, 278–285.

Valerio, F., Di Biase, M., Caputo, L., Creanza, T.M., Ancona, N., Visconti, A., et al., 2014. Effect of *Lactobacillus brevis*—based bioingredient and bran on microbiological, physico-chemical and textural quality of yeast leavened bread during storage. Innovative Food Sci. Emerg. Technol. 25, 2–8.

Valerio, F., Di Biase, M., Huchet, V., Desriac, N., Lonigro, S.L., Lavermicocca, P., et al., 2015. Comparison of three *Bacillus amyloliquefaciens* strains growth behaviour and evaluation of the spoilage risk during bread shelf-life. Food Microbiol. 45, 2–9.

Viedma, P.M., Abriouel, H., Omar, N.B., López, R.L., Gálvez, A., 2011. Inhibition of spoilage and toxigenic *Bacillus* species in dough from wheat flour by the cyclic peptide enterocin AS-48. Food Control 22, 756–761.

Viljoen, C.R., von Holy, A., 1997. Microbial populations associated with commercial bread production. J. Basic Microbiol. 37, 439–444.

Witthuhn, M., Lücking, G., Atamer, Z., Ehling-Schulz, M., Hinrichs, J., 2011. Thermal resistance of aerobic spore formers isolated from food products. Int. J. Dairy Technol. 64, 486–493.

Zwietering, M.H., Wijtzes, T., de Wit, J.C., van't Riet, K., 1992. A decision support system for prediction of the microbial spoilage in foods. J. Food Prot. 55, 973–979.

CHAPTER

Molds and mycotoxins in nuts

17

Sonia Marín and Antonio J. Ramos
Food Technology Department, XaRTA-UTPV, Agrotecnio Center, University of Lleida, Lleida, Spain

INTRODUCTION

Tree nuts are dry fruits with one seed (rarely two), in which the ovary wall becomes hard at maturity. The most popular edible tree nuts are almonds (*Prunus dulcis*), walnuts (*Juglans regia*), pistachios (*Pistacia vera*), and cashews (*Anacardium occidentale*). Other common edible nuts are hazelnuts (*Corylus avellana*), chestnuts (*Castanea sativa*), pecans (*Carya illinoinensis*), macadamias (*Macadamia integrifolia*), pine nuts (*Pinus pinea*), and Brazil nuts (*Bertholletia excelsa*). In a broad, not botanical sense, peanuts (*Arachis hypogaea*) are also designated as nuts, although botanically are groundnuts or legumes. For the purpose of this text, the term "nuts" includes all common tree nuts plus peanuts.

The International Nut and Dried Fruit Council Foundation has informed that, in season 2014/15, tree nut production achieved 3.7 millions metric tons, 8.5% more than the previous season and 56% higher than in 2004/05, which confirms the upward trend of consumption of this snack over the last 10 years. Almond was the tree nut that experienced the greatest growth, as its production was doubled in only 10 years. Walnuts and pistachios were the second and third most produced tree nuts in 2014 (INC, 2015).

Nuts contain high amounts of proteins and fat, mostly unsaturated fatty acids, and a variety of other nutrients as dietary fiber, vitamins, minerals, and many other bioactive constituents such as phytosterols and phenolic compounds. Nowadays, human intake of this kind of food is increasingly supported by its proven health benefits, since it has been reported that frequent nut consumption can reduce the incidence of coronary heart disease, blood cholesterol, gallstones, and diabetes (Ros, 2010).

Nuts are suitable substrates for fungal growth if conditions are appropriate. Sometimes, molds which colonize nuts can be mycotoxigenic. Mycotoxins are secondary metabolites produced by fungi that may be toxic to vertebrates upon ingestion, inhalation, or skin contact. The diseases they cause, known as

P. Kotzekidou (Ed): Food Hygiene and Toxicology in Ready-to-Eat Foods.
DOI: http://dx.doi.org/10.1016/B978-0-12-801916-0.00017-0

mycotoxicosis, do not need to involve the toxin-producing fungus. Thus, they are abiotic hazards but with a biotic origin (Marin et al., 2013).

According to the 2014 annual report of the Rapid Alert System for Food and Feed (RASFF, 2015), mycotoxins were the main hazard in border rejection notifications in the European Union (EU). In the report, aflatoxins (AFs) were the primary mycotoxins associated to the notifications, and nuts and nut products were two of the most affected categories.

FUNGAL CONTAMINATION OF NUTS

Due to the extremely high fat and protein, and low water content of nuts, these products are quite resistant to spoilage by microorganisms. However, molds can grow in these products if the conditions during nut production and trade are inappropriate.

Usually, fungal contamination of nuts occurs at three different stages (Abdulla, 2013). In the case of tree nuts, the first stage of contamination may occur at the field, prior to harvest, while nuts are still on the tree. At this stage, especially when the nuts have ripened and the hulls have opened, they are often infected by airborne and insect-borne fungal spores. If nuts fall from the tree and are collected from the ground, the probability of fungal contamination increases (Baquião et al., 2012). In the same sense, peanuts, that produce seed-bearing pods below the soil surface, are in direct contact with soil microorganisms and the seeds are frequently invaded by soil fungi before harvest (Youssef et al., 2008). On the other hand, several environmental and agronomic factors or practices at field may predispose nuts to colonization by fungi (Mutegi et al., 2012). For example, in the case of peanuts, contamination has been found to be higher when peanuts are grown under rainfed conditions compared to those grown under irrigation (Reddy et al., 2003). Drought stress and elevated soil temperatures also promote the growth of mycotoxin-producing fungi on peanuts, while differential cultivar susceptibility has also been documented (Bankole et al., 2006).

Fungal contamination at field shows that, frequently, *Aspergillus* species are the most common fungi. Bayman et al. (2002) found that on almond, pistachio, and walnut the most common genera found were *Aspergillus*, *Rhizopus*, and *Penicillium*, followed by *Alternaria*, *Cladosporium*, *Cunninghamella*, *Fusarium*, *Trichoderma*, and *Verticillium*, with zero to nine fungal species isolated from each nut, and an overall average of 1.9 fungi species/nut. The most common *Aspergillus* species were *Aspergillus niger*, *Aspergillus flavus*, *Aspergillus nidulans*, *Aspergillus tamarii*, and *Aspergillus ochraceus*. These results agree with fungal contamination found in field-collected Portuguese almond samples, where *Aspergillus* and *Penicillium*, together with *Cladosporium*, were shown to be important superficial contaminants of these nuts, with *Aspergillus* sections *Nigri* and *Flavi* being the most frequent (Rodrigues et al., 2013).

The second stage of fungal contamination of nuts may occur at postharvest, when nuts are dehulled, washed, and sorted. Water used for washing can be a source of contamination, and if the nuts are allowed to remain wet, they will be highly susceptible to mold growth and mycotoxin production.

The third place of contamination by molds is during storage, especially when nuts are stored under adverse conditions of temperature and moisture. Most research on fungal contamination of nuts has been conducted sampling nuts at the postharvest stage, mainly at the retail market level. Again, *Aspergillus* is one of the genus more frequently found, followed by *Penicillium*, *Rhizopus* and, to a lesser extent, *Fusarium*.

Many of the fungal genera found both at pre- and postharvest level include species that can produce mycotoxins. Among them, *A. flavus* and *Aspergillus parasiticus*, two widely distributed saprophyte fungi, and opportunistic pathogens of plants, are especially frequent, with many isolates that are able to produce AFs, as well as a neurotoxin named cyclopiazonic acid. Among known mycotoxins, AFs pose the greatest threat to human and animal health. In a similar way, in nuts, fungal contamination by other *Aspergillus* mycotoxigenic species, like *A. ochraceus* or *Aspergillus carbonarius*, that are able to produce a nephrotoxin named ochratoxin A (OTA) is often found.

MYCOTOXINS IN NUTS

Mycotoxins are a wide group of fungal secondary metabolites that exert multiple toxic effects on humans and animals. Some mycotoxins can cause autoimmune illnesses, have allergenic properties, and some of them are teratogenic, carcinogenic, mutagenic, nephrotoxic, or estrogenic (CAST, 2003). The major mycotoxin-producing fungal genera are *Aspergillus*, *Penicillium*, *Fusarium*, *Alternaria*, and *Claviceps*. Although hundreds of mycotoxins have been described, the most relevant groups of mycotoxins found in food and feeds are AFs, OTA, fumonisins (FBs), zearalenone (ZEA), and the wide group of the trichothecenes. Besides, patulin, the emerging *Fusarium*-related mycotoxins (fusaproliferin, moniliformin, beauvericin, and enniatins), the ergot alkaloids produced by *Claviceps*, and the *Alternaria* mycotoxins (mainly altenuene, alternariol, alternariol methyl ether, altertoxin, and tenuazonic acid) are other mycotoxins of relative concern (Marin et al., 2013).

AFs are difuranocoumarins mainly produced by *A. flavus* and *A. parasiticus*, primarily found in hot, humid climates, often those in the areas of production of many kinds of nuts. Aflatoxin B_1 (AFB1) is considered the most toxic of the AFs, and the most carcinogenic compound naturally produced. AFs have been shown to be extremely potent carcinogens in all animal species investigated, and in several organs, the liver being the primary target. Furthermore, AFs are genotoxic compounds.

OTA is a phenylalanyl derivative of a substituted isocoumarin. The main producing species belong to the *Aspergillus* sections *Circumdati*, *Nigri*, and *Flavi*, and to the species *Penicillium verrucosum* and *Penicillium nordicum*. Dietary direct sources of OTA for humans are cereals and derived products, nuts, pulses, dried fruits, coffee, cocoa, spices, wines, and beer. OTA is nephrotoxic, hepatotoxic, teratogenic, immunosuppressive, and carcinogenic in animals, and has been related to Balkan endemic nephropathy in humans and with urinary tract tumors.

FBs are a group of mycotoxins with a strong structural similarity to sphinganine, the backbone precursor of sphingolipids. These mycotoxins are mainly produced by *Fusarium* species from the *Liseola* section. *Fusarium verticillioides* and *Fusarium proliferatum* are the main producing species. FBs are the most important mycotoxins found in maize, but they have been found also in nuts. This group of mycotoxins, especially fumonisin B_1 (FB1), strongly inhibits the enzyme ceramide synthase resulting in an increase in sphinganine and other sphingoid bases, which are highly cytotoxic compounds.

Trichothecenes are a group of structurally related compounds with a common tetracyclic sesquiterpenoid 12,13-epoxytrichot-hec-9-ene ring system. The approximately 170 identified trichothecenes have been divided into four types (A–D) according to variations in the functional hydroxyl and acetoxy side groups. Type A is represented by HT-2 and T-2 toxins, and type B is frequently represented by deoxynivalenol (DON). The types A and B are mainly produced by several *Fusarium* species, but also by some species of *Trichoderma*. These fungi are soil fungi and are important plant pathogens that grow on the crop in the field. These mycotoxins are frequent in grains and derivatives, although they can be found also in other substrates such as nuts. T-2 and HT-2 toxins are inhibitors of protein synthesis and mitochondrial function, and show immunosuppressive and cytotoxic effects, whereas DON disrupts normal cell function by inhibiting protein synthesis. Trichothecenes cause a wide range of toxic effects in animals and humans including feed refusal, vomiting, diarrhea, hemorrhage, anemia, and immunosuppression.

ZEA is a mycotoxin produced by several *Fusarium* species, mainly *Fusarium graminearum*. Generally, *Fusarium* species grow and invade crops in moist cool field conditions during blooming, but growth and toxin production may also occur at postharvest under poor storage conditions. ZEA, and some of its related metabolites, competitively bind to estrogen receptors, and therefore, the toxicity of this toxin is associated with reproductive problems.

Finally, emerging *Fusarium* mycotoxins include other toxic secondary metabolites such as enniatins (ENN), beauvericin (BEA), fusaproliferin, and moniliformin. There are limited available data on these metabolites, not only because of their late recognition, but particularly due to the late understanding of their role as mycotoxins (Marin et al., 2013). This lack of information hinders the establishment of regulations and the development of a comprehensible monitoring plan capable of assessing risk and defining the limits of acceptability of these compounds (Escrivá et al., 2015).

Regarding carcinogenicity of these compounds, toxicological studies led the International Agency for Research on Cancer (IARC) to consider the classification of mycotoxins in different groups. AFs belong to group 1, as there is sufficient evidence in humans for their carcinogenicity. OTA and FB1 were classified as group 2B (possibly carcinogenic to humans), whereas DON, T-2, and ZEA were placed in group 3 (not classifiable according to its carcinogenicity to humans) (IARC, 1993, 2002).

PEANUTS

Peanuts are one of the most frequently AF-contaminated commodities. Most developing countries that produce peanuts lie in the tropics, where the temperature and relative humidity favor mold growth and subsequent AF contamination. Infection by *Aspergillus* spp. and AF production in groundnut occurs at preharvest, during harvest, postharvest, in storage, as well as during processing and transport along the value chain.

In recent years a high number of publications about AF contamination of peanut or peanut derivatives (such as peanut butter) have been published, mainly focused in contamination at the postharvest level, and especially from developing countries. While AF peanut contamination in developed countries is quite well controlled through strict surveillance programs, this is not the case in developing countries. In Table 17.1, a non-exhaustive summary of published results obtained in the last 5 years in some developing countries is presented. As can be seen, AF contamination of peanuts and derivatives is still quite common, with levels far exceeding those established in the EU legislation for peanuts or peanut derivatives intended for direct human consumption (see section "Legislation"), which evidences the serious public health problem that AF contamination represents.

Besides AFs, other mycotoxins have also been detected in peanuts. Thus, mycotoxins like FB1, ZEA and DON (Njobeh et al., 2010), OTA (Ediage et al., 2011), BEA (Warth et al., 2012), and ENN have also been found.

TREE NUTS

Mycotoxin contamination in tree nuts has been frequently described, pistachios being the main source of human dietary AFs from tree nuts worldwide. The most important mycotoxins found in pistachio are AFs and, secondarily, OTA. Several studies have shown that occurrence of AFB1 in pistachio occasionally occurs at high levels. For instance, a 52% incidence of AF-contaminated pistachios was found in Tunisia, with an average contamination level of 21.8 μg/kg (Ghali et al., 2009). Such high contamination constitutes a threat to public health. In Iran, the largest pistachio-producing country, in a study conducted with pistachios exported to the EU during the period 2009–11, amongst 8203 subsamples analyzed, AFB1 was detected in 1921 cases (23.4%) with a mean of 2.18 μg/kg (Dini et al., 2013). However, in studies where samples were taken in retail markets in Iran, levels

Table 17.1 Aflatoxin (AF) Contamination in Peanuts and Derivatives in Developing Countries (2010–15)

Foodstuff	Country	Sampling Year	No. Samples	% Positives	Range AF Contamination (μg/kg)	References
Peanuts	Haiti	2012–13	21	14%	AFs: <2–787	Schwartzbord and Brown (2015)
Peanut butter	Haiti	2012–13	32	97%	AFs: <2–2720	Schwartzbord and Brown (2015)
Peanuts	Mali	2009–10	2160	n.g.	AFB1: 14.3–310.9	Waliyar et al. (2015)
Peanut meal	Cameroon	2012	4	100%	AFs: 39–950	Kana et al. (2013)
Fresh and processed peanuts	Malawi	2008–09	1397	46% (2008) 23% (2009)	AFB1: <l.d.-3240	Monyo et al. (2012)
Peanuts and peanut products	Brazil	2009–10	101	14%	AFB1: 24–87.5	Hoeltz et al. (2012)
Peanut butter	Haiti and Kenya	2009–10	–	100%	AFs: 7.9–799.8	Filbert and Brown (2012)
Peanuts	China	2009–10	1040	25%	AFB1: <l.d.-720	Ding et al. (2012)
Peanuts	D.R. Congo	2009	60	43%	AFB1: <1–937	Kamika and Takoy (2011)
Traditionally processed peanut butter	Sudan	2005–06	36	100%	AFB1: 17–170	Elshafie et al. (2011)

n.g., data not given; l.d., limit of detection.

found were much higher, as incidences of 55% of positive samples and means of 22 μg/kg (and maximum levels of 52 μg/kg) were found in salted roasted pistachios (Ostadrahimi et al., 2014).

AFs, and to a lesser extent OTA, are also the main mycotoxins found in other tree nuts. However, other mycotoxins have been occasionally found, such as alternariol, alternariol monomethyl ether, citreoviridin, citrinin, cyclopiazonic acid, mycophenolic acid, patulin, sterigmatocystin, roquefortine C, and ZEA. Several emerging *Fusarium* mycotoxins (BEA, ENN), have also been recently described in some of these products (Tolosa et al., 2013).

PREVENTION OF FIELD MYCOTOXIN CONTAMINATION

Prevention of infection by *Aspergillus* species at preharvest stage is an important step in reducing AF contamination in nuts. Most research in preharvest prevention of AFs has been carried out in peanut crop, where the strategies may be very different from those for true nuts. Several strategies including good agricultural practices, the use of biocontrol agents, plant breeding for resistance, and prediction of AF risk for early harvesting have been employed to prevent and/or reduce AF contamination in peanuts at preharvest stage. However, preharvest mycotoxin contamination is still highly dependent on climatic conditions.

Drought and high soil temperatures of 29–31°C in the pod zone during the final 3–6 weeks of the growing season have been shown to be determinant. Developing and maturing peanuts are not susceptible to colonization by *A. flavus* and *A. parasiticus* until kernel water activity (a_w) begins to decrease in response to late-season drought conditions with increased soil temperature (Dorner et al., 1989). Maintaining high kernel a_w until the time of harvest maintains the natural defense mechanism (phytoalexin production) of peanuts against growth by aflatoxigenic fungi, even if fungal invasion occurs. There are negative linear relations between infection and the average fraction of extractable soil water (FESW) between flowering and harvest, and between AF concentration and FESW in the last 25 days of pod-filling (Craufurd et al., 2006). Thus preharvest AF contamination of peanuts essentially can be eliminated with proper and adequate irrigation.

Moreover, among agricultural practices, timely planting, maintaining optimal plant densities, providing adequate plant nutrition and weed control, controlling plant pathogens and insect pests, and following proper harvesting practices should be employed to reduce the incidence of AF contamination in the field. In addition, preventing a high aflatoxigenic inoculum in the field is important, thus again weed and insect control, crop rotation, and management of crop residues also play an important role in controlling *A. flavus* infection (Bhatnagar et al., 2015).

Most research efforts have been devoted to biological control strategies. Most of the success in biological control of AF contamination in crops has been achieved through application of competitive non-toxigenic strains of *A. flavus* and/or *A. parasiticus*. The effectiveness of biocontrol using atoxigenic *A. flavus* strains was based on the logic that these are predominantly asexual, genetically stable, and aggressive as competitors coupled with their inability to recombine with native toxigenic strains (Abbas et al., 2011; Ehrlich and Cotty, 2004). Significant and consistent reductions in AF contamination (70–90%) have been observed in peanut field experiments using such non-toxigenic *Aspergillus* strains (Dorner, 2008). A commercial biopesticide product (called Afla-Guard) has been developed based on the *A. flavus* strain NRRL21882. Application of Afla-Guard changed the composition of *A. flavus* soil populations from an average of 71.1% toxigenic strains in untreated fields to only 4.0% in treated soils. AF averaged 78.9 ng/g in untreated peanuts compared with 11.7 ng/g in treated peanuts, an 85.2% reduction. The results indicate that field application of the non-toxigenic strains had a carryover effect and reduced AF contamination that occurred in storage (Dorner and Lamb, 2006). Different solid-substrates, such as wheat and rice, can be used to produce the biocontrol formulation. When the fermented grains are applied to the field, the non-toxigenic strains resume growth and produce numerous conidia on the surface of the grains. There is a strong relationship between the effectiveness of biocontrol agents and inoculum rate and retreatment with biocontrol agents in subsequent years.

Other microorganisms have been tested with some success for antiaflatoxigenic activity in peanuts, such as rhizobacteria and *Trichoderma* spp. (Navya et al., 2015) and *Burkholderia* sp. (Vijayasamundeeswari et al., 2010).

Breeding for resistance is also a key point. Crop resistance to *A. flavus* invasion and AF contamination is not only genotype-specific, but may also be attributed to several biochemical, environmental and physical factors, with drought playing an important role in host–pathogen interactions. The resistance to pod infection is attributed to pod-shell structure, which acts as a physical barrier, as well as to the production of fungistatic phenolic compounds. Under drought stress, peanut seeds lose the capacity to produce phytoalexins. Drought resistance traits including specific leaf area, relative water content, chlorophyll density, and drought stress ratings are useful as selection tools for improving resistance (Girdthai et al., 2010).

Availability of transgenic varieties with antifungal traits is extremely valuable as a breeding tool. The growing concerns regarding environmental safety and groundwater quality demand less dependence on agrochemicals. Disease-resistant transgenic crops would not only control mycotoxin-producing organisms such as *A. flavus*, *A. parasiticus*, and *Fusarium* spp., but also other microbial (fungal, bacterial, and viral) diseases that cause significant economic losses in crop production. Identification of potent antifungal proteins and peptides from different sources will be valuable in transferring them through genetic engineering to susceptible crops to combat fungal infection resulting in preharvest AF contamination (Rajasekaran et al., 2009).

Finally, AF risk may be forecasted for early harvesting. Preharvest AF contamination of peanuts can be prevented if peanuts are harvested before AF is actually produced. Both in the United States and Australia, prediction models have been built based on geographical and meteorological data, including daily rainfall and soil and ambient temperatures, from which an AF risk index can be predicted (Chauhan et al., 2010; Doster et al., 2014).

For pistachio, early harvest has been suggested (Panahi and Khezri, 2011). Application of the AF36 atoxigenic *A. flavus* in the soil of pistachio orchards for several years reduced AF contamination in treated orchards from 20% to 45%, depending on the year (Doster et al., 2014).

POSTHARVEST STRATEGIES

Nuts, and in particular groundnuts, are highly sensitive to contamination by toxigenic fungi in the field, before and after harvest, and during drying and storing. As described before, certain AF contamination in the field is almost unavoidable, thus it is critical to minimize postharvest AF accumulation, where conditions may be more easily controlled. In addition, good practices concerning agriculture, hygiene, storage, and manufacturing should be taken into account. Prevention of the infestation of nuts with aflatoxigenic fungi should be a priority, rather than the elimination of AFs once produced.

HARVEST AND DRYING

For tree nuts, an accurate harvesting method, together with quick transfer to units by suitable vehicles would be preferred. Sun drying is to be avoided. In the drying units dehulling plus wet sorting before drying usually takes place. Regarding peanuts, after peanut plants are pulled from the ground, slow drying in the sun in windrows should be avoided. After nuts are threshed from the plants while they are still wet, precleaning and artificial drying should be applied. Low-temperature dryers are preferred because low temperature is essential to prevent damage of texture and the onset of rancidity. Any damage in the shell or kernel in the field, during transportation, processing, and handling or storage enhances the prospect for fungal invasion in nuts.

IN-SHELL STORAGE

Despite the low level of a_w usually maintained throughout the storage period (0.40–0.55), fungal populations may remain constant. Moreover, if temperature and moisture gradients occur in the store by the biological activity of insects and rodents, localized pockets of higher moisture may develop, opening the way for mold germination, growth and, consequently, AF production (Hocking, 2003). During storage, AF production is affected by several factors, among which temperature and a_w are especially relevant. AF synthesis by *A. flavus* occurs between

15°C and 37°C, being most pronounced in the range 20–30°C, whereas a_w below 0.83 prevents AF formation. It is then necessary to maintain a safe storage moisture until peanuts are processed. The presence of insect vectors of aflatoxigenic fungi presents a potential risk for AF production during the peanut storage period. Integrated management of fungi and insect vectors is then required (Nesci et al., 2011). Thus the main preventive measure relies on humidity and temperature control during storage, with aeration, or rapid processing as corrective actions in case of detected heating.

The storage time is to be limited, depending on storage conditions. For example, analysis of stored peanut hulls through 1 year showed that 6.7% of the samples were contaminated with AFB1 (mean levels = 15–23.9 μg/kg) and aflatoxin B_2 (AFB2) (mean levels = 3.3–5.6 μg/kg); in kernels, 33.3% of the samples were contaminated with AFB1 (mean levels = 7.0–116 μg/kg) and 28.3% were contaminated with AFB2 (mean levels = 3.3–45.5 μg/kg), while increasing *A. flavus* populations were reported, an increase in AF presence was not reported, probably due to the limitations of the sampling plan (Nakai et al., 2008). Predictive models may be used to estimate the safe storage time depending on the moisture/temperature conditions (Marín et al., 2012).

The same rules apply to long-distance transport and shipment when final processing occurs in importing countries.

Antiaflatoxigenic postharvest treatments have been also assayed. For example, the effect of a food-grade antioxidant formulation containing butylated hydroxyanisole, butylated hydroxytoluene, and propylparaben in stored peanuts inhibited the development of natural peanut mycoflora and, particularly, *Aspergillus* section *Flavi* populations (Passone et al., 2009). Other antiaflatoxigenic treatments have been experimentally tested with some success, including essential oils from plants, ferulic acid, cinnamic acid, salycillic acid, citric acid, and *Saccharomyces* extracts. Peanuts have also been treated with non-toxigenic strains both in the field and prior to storage for 3–5 months under high temperature and relative humidity conditions designed to promote AF contamination. Although the postharvest treatment worked properly (78 vs 48 μg/kg), the results demonstrated that the carryover effect of field treatments through storage was much more important than the inhibitory effect of a postharvest treatment (Dorner and Cole, 2002). Finally, ozonation has also been studied as a postharvest treatment for safe storage.

PROCESSING: LOT SEGREGATION, SORTING, AND ROASTING

The main critical points in processing plants consist of the reception of raw nuts, storage, sorting, and roasting. Suggested preventive measures include the determination of AF levels, moisture, and a_w at the reception and storage of nuts, accurate sorting and control of temperature and roasting time, for a guarantee of food safety.

Lot segregation by AF level

At intake to shelling plants (or previously, at the storage facilities), the incoming lots are sampled by a random sampling system, then assayed for AF and other defects and segregated into grades depending on AF content, or rejected. In the EU (Regulation 165/2010) there are maximum limits for the total AF and AFB1 levels present in nuts before sorting, thus if those limits are exceeded the lot cannot be processed and must be rejected (see section "Legislation"). It is a common difficulty to obtain a representative sample from a truckload or a ship container of nuts for testing. Representative sampling is important because of nonuniform distribution of AF in a batch, and decision on rejection relies on this. The EU sampling plan for official control of mycotoxins (Regulation 178/2010) may be used as a guideline for food safety management in processing companies, however, despite its complexity, it may still fail in correct lot segregation (García-Cela et al., 2013).

Sorting

It is recognized that sorting and physical segregation significantly reduce the AF content of consignments of nuts. Mycotoxins are mainly linked to moldy, damaged by insects, small, deformed and discolored nuts.

In the case of peanut pods, a first screening separates certain high-AF-risk components (loose shelled kernels, small, immature pods). After shelling, density segregation allows for separation of lighter, moldy kernels and foreign material. Kernel sizing is not intended for AF management, however, the smaller ones are byproducts used in oil production which contain higher AF concentrations than larger categories. Finally, electronic color sorting allows for separation of discolored kernels, as discoloration is primarily due to mold growth, with a reduction of 70% of the existing AFs. Blanching (removing the seed coat helped by a thermal treatment) before electronic color sorting enhances its performance (Dorner, 2008).

Regarding pistachio nuts, sorting by weight and size also takes place, with efficient separation of contaminated nuts. Arjmand et al. (2006) found 286, 210, 140, and 1.04 μg/kg in very small screened nuts, stained nuts, rehulled pistachio nuts, and final pistachio, respectively. Results indicated that it is necessary not to mix rehulled pistachio nuts with pistachios that hulled easily on first stage, and separating stained and very small nuts to decrease AF in final pistachio. The early-split nuts are the most contaminated nuts, growth-split nuts are less contaminated, and pistachios with sound hulls are almost clean. Brown-stained pistachios usually contain AFs. As the brown kernels are a subdivision of fluorescent fraction (Hadavi, 2005), discrepancies have been found on the use of UV light to detect contaminated nuts, as fluorescent pistachios do not always contain AFs (Steiner, 1992). However, a strong relationship between Bright Greenish-Yellow Fluorescence (BGYF) and AF content at harvest exists (Hadavi, 2005). The average AF content of the yellowish-brown stained nuts may be 120.3–632.4 μg/kg, while dark-grayish-stained nuts may contain 7.3–292.8 μg/kg, respectively. Thus the amount of AF could be significantly reduced by 94.6–98.8%, if all the stained nuts were removed from the lot during sorting (Shakerardekani et al., 2012). Industrial sorting caused a decrease in contamination of two to four times in processed pistachios compared to non-processed pistachios (Schatzki, 1995) or a 40–80% reduction of AFs (Park, 2002).

Similarly, in Brazil nuts all individually analyzed kernels containing AFs showed yellow fluorescence at the cut surface, although not all fluorescent kernels were contaminated with AFs. In this case, brown spots in kernels were not a criterion to recognize contaminated kernels (Steiner, 1992). When kernels and shells were sorted in five fractions: good kernels, rotten kernels, good shells with kernel residue, good shells without kernel residue, and rotten shells, and analyzed for AFs, the Brazil nut shell was found to be contaminated with AF; rotten nuts contributed only 4.2% of the sample mass (kg), but contributed 76.6% of the total AF mass (μg) in the in-shell test sample. High correlations were found between the AF concentration in in-shell Brazil nuts samples and the AF concentration in all defective fractions (Vargas et al., 2011).

As a decontamination process, experimental ultraviolet irradiation of ground and tree nuts at 265 nm for 15, 30, and 45 min showed a proportional decrease in AFs level with increase in exposure time. After 45 min exposure to UVC, AFB1 showed maximum reduction of 96.5% in almond and pistachio (Jubeen et al., 2012).

Roasting

AFs have high decomposition temperatures, ranging from 237°C to 306°C and AFB1 is quite stable to dry heating (Betina, 1989; Rustom, 1997). Although these temperatures are higher than those actually used by the nut industry, it is usually accepted that the heat treatment decreases the concentration of AFs to some extent. However, conflicting results have been published about the effect of the heat treatments on peanuts and pistachios. In general, the extent of the destruction achieved was very dependent on the initial level of contamination, heating temperature, and time. The effects of heat in naturally contaminated peanuts by oven roasting at 150°C for 30 min caused a 30–45% reduction of AFB1, while in artificially contaminated peanuts treated under the same conditions, the inactivation was 48–61% (Pluyer et al., 1987). Degradation of AFs in peanuts roasted at 150°C for 30 min increased with the addition of ionic salts in a range from 38%, 41.5%, and 47.6% in unsalted peanuts, and salted with 20 and 50 μg/kg, respectively (Ozkarsli, 2003). In pistachio, the results regarding degradation of AFs due to roasting are also contradictory. Yazdanpanah et al. (2005) studied the effect of roasting for 30, 60, and 90 min at different temperatures (90°C, 120°C, and 150°C). The milder treatment (90°C for 30 min) resulted in the slightest effect while the most extreme treatment resulted in the degradation of over 95% of AFB1 but the pistachio showed a burned appearance. The roasting process at 150°C for 30 min showed significant reduction of AFB1 and AFB2 without any noticeable change in taste of sample. Percentages of AF reduction of 87.62%, 81.05%, and 86.74% for AFB1, AFB2, and total AFs, respectively, were obtained in industrially roasted pistachio nuts (García-Cela et al., 2013). On the other hand, Ariño et al. (2009) studied the effect of roasting on AFs: four commercial batches of raw pistachios in-shell from Iran were salted (1% salt content) and roasted at 120°C for 20 min in a roasting industry in Spain. This study did not obtain significant differences in relation to AF reduction after roasting. However the level of contamination of the starting material was low, ranging from 0.12 to 0.18 μg/kg. Analyzing the existing results on the effect of time and temperature in connection with the degradation of AFs, it can be observed that high temperatures (200–400°C) produce higher mycotoxin reduction. Moreover, lower temperatures need longer exposition time than higher temperatures to obtain the same reduction percentages. Thus the percentage of reduction depends on temperature, time, as well as the initial mycotoxin contamination.

After roasting aflatoxigenic fungi are inactivated, thus no further variation in AF content is expected once packaged.

LEGISLATION

Many countries have adopted regulations to limit mycotoxin exposure. Their presence is not only related to the effect they might have on consumer health, but may also have an impact on world trade. For nuts, in the EU there is only mycotoxin regulation related to AFs (AFB1 or the sum of AFB1 + AFB2 + AFG1 + AFG2). At present, Regulation 165/2010, that amends Regulation 1881/2006, sets the maximum levels for AFs in foodstuffs. Table 17.2 shows the maximum allowable levels for AFB1 and total AFs in nuts and groundnuts. The maximum levels in Table 17.2 refer to the edible part of groundnuts and nuts. Regarding sampling, Commission Regulation (EC) No. 401/2006 laying down the methods of sampling and analysis for the official control of the levels of mycotoxins in foodstuff, and Commission Regulation (EU) No 178/2010, amending Regulation 401/2006 as regards groundnuts (peanuts) and tree nuts, are applicable. Recently, the Commission Regulation (EU) No 519/2014, has amended Regulation 401/2006 regarding the methods of sampling of large lots, spices, and food supplements, performance criteria for T-2, HT-2 toxin, and citrinin, and screening methods of analysis (including tree nuts and peanuts as commodity categories for the validation of screening methods).

On the other hand, other countries have regulated the maximum allowable level of AFs in nuts. Thus, in the United States, the maximum total AF level allowed is 20 μg/kg (the same level as other foods), and the same level has been set in Mexico and in the countries belonging to MERCOSUR. In Canada the maximum admissible level is 15 μg/kg, in Brazil limits ranged from 10 (nuts, except Brazil nuts) to 10–20 μg/kg (Brazil nuts), and in Chile the level is 5 μg/kg (Ramos et al., 2011).

Table 17.2 EU Maximum Allowable Levels for Aflatoxin B_1 (AFB1) and Total Aflatoxins (AFs—Sum of AFB1, AFB2, AFG1, and AFG2) in Nuts and Groundnuts

Foodstuffs	AFB1 (μg/kg)	Total AFs (μg/kg)
Groundnuts (peanuts) to be subjected to sorting, or other physical treatment, before human consumption or use as an ingredient in foodstuffs, with the exception of: • groundnuts (peanuts) for crushing for refined vegetable oil production	8	15
Almonds and pistachios to be subjected to sorting, or other physical treatment, before human consumption or use as an ingredient in foodstuffs	12	15
Hazelnuts and Brazil nuts, to be subjected to sorting, or other physical treatment, before human consumption or use as an ingredient in foodstuffs	8	15
Tree nuts, other than almonds, pistachios, hazelnuts, and Brazil nuts, to be subjected to sorting, or other physical treatment, before human consumption or use as an ingredient in foodstuffs	5	10
Groundnuts (peanuts) and processed products thereof, intended for direct human consumption or use as an ingredient in foodstuffs, with the exception of: • crude vegetable oils destined for refining • refined vegetable oils	2	4
Almonds and pistachios, intended for direct human consumption or use as an ingredient in foodstuffs	8	10
Hazelnuts and Brazil nuts, intended for direct human consumption or use as an ingredient in foodstuffs	5	10
Tree nuts, other than almonds, pistachios, hazelnuts, and Brazil nuts, and processed products thereof, intended for direct human consumption or use as an ingredient in foodstuffs	2	4

In some circumstances the EU has set special conditions for foodstuffs imported from third countries due to contamination risk of these products by AFs. It is very common that this circumstance occurs in the import of pistachios (from Iran and Turkey), peanuts (China, Egypt, India, and Ghana), hazelnuts (Turkey), and Brazil nuts (Brazil) (Commission Implementing Regulation 884/2014).

CONCLUDING REMARKS

Mycotoxins, mainly AFs, are a public health concern, with peanuts and pistachio nuts as the main source of AF intake through the diet. Strict risk management decisions, such as the issued regulations limiting AFs concentrations, are required to minimize the associated risk. In developed countries, food safety management, including preharvest good agricultural practices, safe storage, contaminated lots segregation, and sorting allow for a reasonable exposure to AFs through nut ingestion, however, the situation in developing countries is of high concern. Research is ongoing for further pre- and postharvest control strategies.

REFERENCES

Abbas, H.K., Weaver, M.A., Horn, B.W., Carbone, I., Monacell, J.T., Shier, W.T., 2011. Selection of *Aspergillus flavus* isolates for biological control of aflatoxins in corn. Toxin Rev. 30, 59–70.

Abdulla, N.Q.F., 2013. Evaluation of fungal flora and mycotoxin in some important nut products in Erbil local markets. Res. J. Environ. Earth Sci. 5, 330–336.

Ariño, A., Herrera, M., Estopañan, G., Rota, M.C., Carramiñana, J.J., Juan, T., et al., 2009. Aflatoxins in bulk and pre-packed pistachios sold in Spain and effect of roasting. Food Control 20, 811–814.

Arjmand, M., Alipour, H., Mirdamadiha, F., 2006. Comparison between aflatoxin contaminations in different stages of pistachio processing. Acta Hortic. 726, 643–646.

Bankole, S., Schollenberger, M., Drochner, W., 2006. Mycotoxins in food systems in Sub Saharan Africa: a review. Mycotoxin Res. 22, 163–169.

Baquião, A.C., Zorzete, P., Reis, T.A., Assunção, E., Vergueiro, S., Correa, B., 2012. Mycoflora and mycotoxins in field samples of Brazil nuts. Food Control 28, 224–229.

Bayman, P., Baker, J.L., Mahoney, N.E., 2002. *Aspergillus* on tree nuts: incidence and associations. Mycopathologia 155, 161–169.

Betina, V., 1989. Mycotoxins: chemical, biological and environmental aspects. Bioactive Molecules 9, 114–150.

Bhatnagar-Mathur, P., Sunkara, S., Bhatnagar-Panwar, M., Waliyar, F., Sharma, K.K., 2015. Biotechnological advances for combating *Aspergillus flavus* and aflatoxin contamination in crops. Plant Sci. 234, 119–132.

CAST, 2003. Mycotoxins: Risks in Plant, Animal and Human Systems. Task Force Report No 139. Council for Agricultural Science and Technology, Ames, IA.

Chauhan, Y.S., Wright, G.C., Rachaputi, R.C.N., Holzworth, D., Broome, A., Krosch, S., et al., 2010. Application of a model to assess aflatoxin risk in peanuts. J. Agric. Sci. 148, 341–351.

Craufurd, P.Q., Prasad, P.V.V., Waliyar, F., Taheri, A., 2006. Drought, pod yield, preharvest *Aspergillus* infection and aflatoxin contamination on peanut in Niger. Field Crops Res. 98, 20–29.

Ding, X., Li, P., Bai, Y., Zhou, H., 2012. Aflatoxin B1 in post-harvest peanuts and dietary risk in China. Food Control 23, 143–148.

Dini, A., Khazaeli, P., Roohbakhsh, A., Madadlou, A., Pourenamdari, M., Setoodeh, L., et al., 2013. Aflatoxin contamination level in Iran's pistachio nut during years 2009–2011. Food Control 30, 540–544.

Dorner, J.W., 2008. Management and prevention of mycotoxins in peanuts. Food Addit. Contam. 25, 203–208.

Dorner, J.W., Cole, R.J., 2002. Effect of application of nontoxigenic strains of *Aspergillus flavus* and *A. parasiticus* on subsequent aflatoxin contamination of peanuts in storage. J. Stored. Prod. Res. 38, 329–339.

Dorner, J.W., Lamb, M.C., 2006. Development and commercial use of afla-Guard, an aflatoxin biocontrol agent. Mycotoxin Res. 22, 33–38.

Dorner, J.W., Cole, R.J., Sanders, T.H., Blankenship, P.D., 1989. Interrelationship of kernel water activity, soil temperature, maturity, and phytoalexin production in pre-harvest aflatoxin contamination of drought-stressed peanuts. Mycopathologia 105, 117–128.

Doster, M.A., Cotty, P.J., Michailides, T.J., 2014. Evaluation of the atoxigenic *Aspergillus flavus* strain AF36 in pistachio orchards. Plant Dis. 98, 948–956.

Ediage, E.M., Di Mavungu, J.D., Monbaliu, S., Van Peteghem, C., De Saeger, S., 2011. A validated multianalyte LC-MS/MS method for quantification of 25 mycotoxins in cassava flour, peanut cake and maize samples. J. Agric. Food. Chem. 59, 5173–5180.

Ehrlich, K.K., Cotty, P.J., 2004. An isolate of *Aspergillus flavus* used to reduce aflatoxin contamination in cottonseed has a defective polyketide synthase gene. Appl. Microbiol. Biotechnol. 65, 473–478.

Elshafie, S.Z.B., ElMubarak, A., El-Nagerabi, S.A.F., Elshafie, A.E., 2011. Aflatoxin B1 contamination of traditionally processed peanuts butter for human consumption in Sudan. Mycopathologia 171, 435–439.

Escrivá, L., Font, G., Manyes, L., 2015. In vivo toxicity studies of fusarium mycotoxins in the last decade: a review. Food Chem. Toxicol. 78, 185–206.

García-Cela, E., Ramos, A.J., Sanchis, V., Marin, S., 2012. Emerging risk management metrics in food safety: FSO, PO. How do they apply to the mycotoxin hazard? Food Control 25, 797–808.

García-Cela, E., Ramos, A.J., Sanchis, V., Marin, S., 2013. Risk management towards food safety objective achievement regarding to mycotoxins in pistachio: the sampling and measurement uncertainty issue. Food Control 31, 392–402.

Ghali, R., Belouaer, I., Hdiri, S., Ghorbel, H., Maaroufi, K., Hedili, A., 2009. Simultaneous HPLC determination of aflatoxins B1, B2, G1 and G2 in Tunisian sorghum and pistachios. J. Food Composit. Anal. 22, 751–755.

Girdthai, T., Jogloy, S., Vorasoot, N., Akkasaeng, C., Wongkaew, S., Holbrook, C., et al., 2010. Associations between physiological traits for drought tolerance and aflatoxin contamination in peanut genotypes under terminal drought. Plant Breed. 129, 693–699.

Hadavi, E., 2005. Several physical properties of aflatoxin-contaminated pistachio nuts: application of BGY fluorescence for separation of aflatoxin-contaminated nuts. Food Addit. Contam. 22, 1144–1153.

Hocking, A.D., 2003. Microbiological Facts and Fictions in Grain Storage. Stored Grain in Australia. CSIRO Stored Grain Research Laboratory, Canberra, pp. 55–58.

Hoeltz, M., Einloft, T.C., Oldoni, V.P., Dottori, H.A., Noll, I.B., 2012. The occurrence of aflatoxin B1 contamination in peanuts and peanut products marketed in Southern Brazil. Braz. Arch. Biol. Technol. 55, 313–317.

IARC (International Agency for Research on Cancer), 1993. IARC Monographs on the evaluation of carcinogenic risks to humans, some naturally occurring substances, food items and constituents, heterocyclic aromatic amines and mycotoxins, vol. 56. International Agency for Research on Cancer, Lyon.

IARC (International Agency for Research on Cancer), 2002. IARC Monographs on the evaluation of carcinogenic risks to humans, vol. 82. International Agency for Research on Cancer, Lyon.

INC (International Nut and Dried Fruit Council Foundation), 2015. Nuts and dried fruits global statistical review 2014–2015. Available at, <https://www.nutfruit.org/global-statistical-review-2014-2015_101779.pdf> (accessed 13.07.15).

Jubeen, F., Bhatti, I.A., Khan, M.Z., Zahoor-Ul-Hassan, Shahid, M., 2012. Effect of UVC irradiation on aflatoxins in ground nut (*Arachis hypogea*) and tree nuts (*Juglans regia*, *Prunus duclus* and *Pistacia vera*). J. Chem. Soc. Pak. 34, 1366–1374.

Kamika, I., Takoy, L.L., 2011. Natural occurrence of aflatoxin B1 in peanut collected from Kinshasa, Democratic Republic of Congo. Food Control 22, 1760–1764.

Kana, J., Gnonlonfin, B.G.J., Harvey, J., Wainaina, J., Wanjuki, I., Skilton, R.A., et al., 2013. Assessment of aflatoxin contamination of maize, peanut meal and poultry feed mixtures from different agroecological zones in Cameroon. Toxins 5, 884–894.

Marín, S., Ramos, A.J., Sanchis, V., 2012. Modelling *Aspergillus flavus* growth and aflatoxins production in pistachio nuts. Food Microbiol. 32, 378–388.

Marin, S., Ramos, A.J., Cano-Sancho, G., Sanchis, V., 2013. Mycotoxins: occurrence, toxicology, and exposure assessment. Food Chem. Toxicol. 60, 218–237.

Monyo, E.S., Njoroge, S.M.C., Coe, R., 2012. Occurrence and distribution of aflatoxin contamination in groundnuts (*Arachis hypogaea* L) and population density of aflatoxigenic Aspergilli in Malawi. Crop Prot. 42, 149–155.

Mutegi, C.K., Ngugi, H.K., Hendriks, S.L., Jones, R.B., 2012. Factors associated with the incidence of *Aspergillus* section *Flavi* and aflatoxin contamination of peanuts in the Busia and Homa bay districts of western Kenya. Plant Pathol. 61, 1143–1153.

Nakai, V.K., de Oliveira Rocha, L., Gonçalez, E., Fonseca, H., Ortega, E.M.M., Corrêa, B., 2008. Distribution of fungi and aflatoxins in a stored peanut variety. Food Chem. 106, 285–290.

Navya, H.M., Naveen, J., Hariprasad, P., Niranjana, S.R., 2015. Beneficial rhizospheric microorganisms mediated plant growth promotion and suppression of aflatoxigenic fungal and aflatoxin contamination in groundnut seeds. Ann. Appl. Biol. 167, 225–235.

Nesci, A., Montemarani, A., Etcheverry, M., 2011. Assessment of mycoflora and infestation of insects, vector of *Aspergillus* section *Flavi*, in stored peanut from Argentina. Mycotoxin Res. 27, 5–12.

Njobeh, P.B., Dutton, M.F., Koch, S.H., Chuturgoon, A.A., Stoev, S.D., Mosonik, J.S., 2010. Simultaneous occurrence of mycotoxins in human food commodities from Cameroon. Mycotoxin Res. 26, 47–57.

Ostadrahimi, A., Ashrafnejad, F., Kazemi, A., Sargheini, N., Mahdavi, R., Farshchian, M., et al., 2014. Aflatoxin in raw and salt-roasted nuts (pistachios, peanuts and walnuts) sold in markets of Tabriz, Iran. Jundishapur J. Microbiol. 7, e8674.

Ozkarsli, M., 2003. Effect of traditional-roasting and microwave-roasting on aflatoxin B1 in peanuts. MSc Thesis, Cukurova University Institute of Natural and Applied Sciences, Adana.

Panahi, B., Khezri, M., 2011. Effect of harvesting time on nut quality of pistachio (*Pistacia vera* L.) cultivars. Sci. Hortic. (Amsterdam) 129, 730–734.

Park, D.L., 2002. Effect of processing on aflatoxin. Adv. Exp. Med. Biol. 504, 173–179.

Passone, M.A., Ruffino, M., Ponzio, V., Resnik, S., Etcheverry, M.G., 2009. Post-harvest control of peanut *Aspergillus* section *Flavi* populations by a formulation of food-grade antioxidants. Int. J. Food Microbiol. 131, 211–217.

Pluyer, H.R., Ahmed, E.M., Wei, C.I., 1987. Destruction of aflatoxins on peanuts by oven- and microwave-roasting. J. Food Prot. 50, 504–508.

Rajasekaran, K., De Lucca, A.J., Cary, J.W., 2009. Aflatoxin control through transgenic approaches. Toxin Rev. 28, 89–101.

Ramos, A.J., Da Rocha Rosa, C.A., Cavaglieri, L.R., Guedes, C.A., 2011. Legislación e impacto económico de las micotoxinas. In: "Micotoxinas y micotoxicosis", A.M.V. Ediciones (Eds.), Madrid, pp. 427–462.

Rapid Alert System for Food and Feed (RASFF), 2015. Annual reports 2014. Available at, <http://ec.europa.eu/food/safety/rasff/docs/rasff_annual_report_2014.pdf> (accessed 12.07.15).

Reddy, T.Y., Sulochanamma, B.N., Subramanyam, A., Balaguravaiah, D., 2003. Influence of weather, dry spells and management practices on aflatoxin contamination in groundnut. Indian Phytopathol. 56, 262–265.

Rodrigues, P., Venâncio, A., Lima, M., 2013. Incidence and diversity of the fungal genera *Aspergillus* and *Penicillium* in Portuguese almonds and chestnuts. Eur. J. Plant Pathol. 137, 197–209.

Ros, E., 2010. Health benefits of nut consumption. Nutrients 2, 652–682.

Rustom, I.Y.S., 1997. Aflatoxin in food and feed: occurrence, legislation and inactivation by physical methods. Food Chem. 59, 57–67.

Schatzki, T.F., 1995. Distribution of aflatoxin in pistachios. 1. Lot distributions. J. Agric. Food Chem. 43, 1561–1565.

Schwartzbord, J.R., Brown, D.L., 2015. Aflatoxin contamination in Haitian peanut products and maize and the safety of oil processed from contaminated peanuts. Food Control 56, 114–118.

Shakerardekani, A., Karim, R., Mirdamadiha, F., 2012. The effect of sorting on aflatoxin reduction of pistachio nuts. J. Food Agric. Environ. 10, 459–461.

Steiner, W.E., 1992. Aflatoxins and fluorescence in Brazil nuts and pistachio nuts. J. Agric. Food Chem. 40, 2453–2457.

Tolosa, J., Font, G., Mañes, J., Ferrer, E., 2013. Nuts and dried fruits: natural occurrence of emerging *Fusarium* mycotoxins. Food Control 33, 215–220.

Vargas, E.A., dos Santos, E.A., Whitaker, T.B., Slate, A.B., 2011. Determination of aflatoxin risk components for in-shell Brazil nuts. Food Addit. Contam. Part A Chem. Anal. Control Expo. Risk Assess. 28, 1242–1260.

Vijayasamundeeswari, A., Vijayanandraj, S., Paranidharan, V., Samiyappan, R., Velazhahan, R., 2010. Integrated management of aflatoxin B1 contamination of groundnut (*Arachis hypogaea* L.) with *Burkholderia* sp. and zimmu (*Allium sativum* L. × *Allium cepa* L.) intercropping. J. Plant Interact. 5, 59–68.

Waliyar, F., Umeh, V.C., Traore, A., Osiru, M., Ntare, B.R., Diarra, B., et al., 2015. Prevalence and distribution of aflatoxin contamination in groundnut (*Arachis hypogaea* L.) in Mali, West Africa. Crop Prot. 70, 1–7.

Warth, B., Parich, A., Atehnkeng, J., Bandyopadhyay, R., Schuhmacher, R., Sulyok, M., et al., 2012. Quantitation of mycotoxins in food and feed from Burkina Faso and Mozambique using a modern LC-MS/MS multitoxin method. J. Agric. Food Chem. 60, 9352–9363.

Yazdanpanah, H., Mohammadi, T., Abouhossain, G., Cheraghali, A.M., 2005. Effect of roasting on degradation of aflatoxins in contaminated pistachio nuts. Food Chem. Toxicol. 43, 1135–1139.

Youssef, M.S., El-Maghraby, O.M.O., Ibrahim, Y.M., 2008. Mycobiota and mycotoxins of Egyptian peanut (*Arachis hypogaea* L.) seeds. Int. J. Botany 4, 349–360.

CHAPTER

Potential industrial applications of decontamination technologies for fresh produce

18

Ramandeep Garg[1], Daniela Abela[1], Brijesh Tiwari[2] and Vasilis Valdramidis[1]

[1]Department of Food Studies and Environmental Health, Faculty of Health Sciences, University of Malta, Msida, Malta [2]Department of Food Biosciences, Teagasc Food Research Centre, Dublin, Ireland

INTRODUCTION

The need for large quantities of fresh, often even exotic, produce is satisfied by the harvest and production in many different countries with very diverse practices of harvest, water irrigation, cleaning and decontamination, with a consequent variable quality and safety of the products.

For several decades, chlorination (50–200 mg/L chlorine) was a widely used decontamination technology by the fresh produce industry to reduce the microbial load on fresh fruits and vegetables, however because of the formation of highly carcinogenic byproducts such as chloramines, haloketones, and trihalomethanes (Akbaş and Ölmez, 2007; Ölmez and Kretzschmar, 2009), resistance of some pathogens (Alvaro et al., 2009), chlorine and chlorinated compounds are banned in some European countries (Rico et al., 2007; Issa-Zacharia et al., 2010). Therefore, emphasis is given on more innovative and effective technologies in reducing the microbial populations, as well as maintaining the quality of fresh produce, while reducing any negative impact to the environment (Ölmez and Akbaş, 2009; Singh et al., 2002). In recent years, studies evaluating the application of novel physical and chemical decontamination technologies such as ultrasound (US), pulsed light, irradiation, photosensitization, gas plasma, high-pressure processing, organic acids, ozone, and electrolyzed water have been widely studied for the fresh produce industry. This chapter reviews the application of novel decontamination technologies for the fresh produce industry. The focus is on applications that can be considered for industrial use or have the potential to be scaled-up.

P. Kotzekidou (Ed): Food Hygiene and Toxicology in Ready-to-Eat Foods.
DOI: http://dx.doi.org/10.1016/B978-0-12-801916-0.00018-2

DECONTAMINATION TECHNOLOGIES FOR FRESH VEGETABLES AND FRUITS

CHEMICAL INTERVENTION TECHNOLOGIES

Gaseous and aqueous chlorine dioxide

Huang et al. (2006) reported that the efficacy of ClO_2 is not decreased by organic matter, its efficacy is not highly affected by pH, and it does not generate toxic byproducts by reacting with organic matter, unlike chlorine. Aqueous chemical sanitizers do not reach all sections of the surface; pathogens which are found in crevices, etc., are not effectively removed. Also, treatment with aqueous solutions may leave residual moisture on fruit which may promote growth of molds (Huang et al., 2006; Sy et al., 2005; Wu and Kim, 2007). Gaseous ClO_2 may be an alternative to aqueous chemical sanitizers including aqueous ClO_2 for fruit disinfection, since, for example, it penetrates deeper than aqueous ClO_2 (Wu and Kim, 2007) and it has 2.5 times stronger oxidation capacity compared to chlorine. ClO_2 was investigated at concentrations of 4.1, 6.2, and 8.0 mg/L of air within treatment times of 30, 60, and 120 min at 23 (± 1)°C and 75–90% relative humidity. When treated with 8.0 mg/L ClO_2 *Salmonella* on blueberries, strawberries, and raspberries was reduced by 2.4–3.7 log CFU/g, 3.8–4.4 log CFU/g and 1.5 log CFU/g, respectively (Sy et al., 2005). These results indicate that gaseous ClO_2 is a potential disinfectant for small fruits. However, further research is required in order to determine the effectiveness of applying gaseous ClO_2 as a disinfectant on an industrial scale. Another limitation of using gaseous ClO_2 industrially is related to its handling, which is more complex than aqueous ClO_2. For example, in the case of the latter, there is no need for a chamber to produce it on site. Additionally, aqueous ClO_2 may be easily applied to an existing washing process without the need to alter any steps. Thus, this method provides an efficient and relatively cheap decontamination method for industries. When certain concentrations are used, the visual appearance of fruit is not affected, for example, when blueberries were treated with 15 ppm ClO_2 for 2 h, maximum reduction (4.88 log CFU/g) of *Listeria monocytogenes* was achieved without affecting its visual appearance (Wu and Kim, 2007). One of the drawbacks of aqueous ClO_2 is that its oxidizing power decreases when it comes into contact with organic matter.

Chung et al. (2011) compared the efficacy of ClO_2 with sodium hypochlorite. Both chlorine dioxide and sodium hypochlorite were more effective than washing with tap water, whereby the latter failed to meet the required standards. The initial total bacterial count of fresh-cut Fuji apples was 6.23 log CFU/g. When treated with 100 ppm chlorine dioxide or 200 ppm sodium hypochlorite the total bacteria counts decreased to less than 5 log CFU/g. ClO_2 inhibited enzyme activity and hence prevented apple browning. Sodium hypochlorite bleached fresh-cut apples. Hence, in general, it can be concluded that chlorine dioxide is better than sodium hypochlorite solution (Chung et al., 2011).

Ozone

Ozonated water, a very strong antimicrobial agent, has been commonly used at low concentrations such as 1–5 ppm for 1–5 min for decontaminating food materials (Olmez and Kretzschmar, 2009), but it may affect the product quality. Previous reports have shown that if it is used at levels higher than 5 ppm it can cause surface damage and browning of certain food materials (Kim et al., 2006; Koseki and Isobe, 2006). Therefore, ozone levels should be as low as possible if used in the food industry for the decontamination of fresh produce. Ozone has higher antimicrobial efficacy than chlorine (Khadre et al., 2001; Kim et al., 1999), leaving no residue (gained GRAS by United States) (Kim et al., 2003b; Singer, 1994; FDA, 1997; Smilanick et al., 1999; White, 1992) enabling its recycling. It has also been reported to result in microbial reductions comparable with that of 100 ppm of chlorine, therefore ozone could be widely used for fresh produce decontamination. Other advantages of using ozone treatment include that excess ozone decomposes to oxygen and therefore does not leave any harmful byproducts on the food. Also, relatively low concentrations are required for decontamination purposes (Khadre et al., 2001). Nevertheless, it is not fully adapted by the fresh food industry due to certain limitations including its instability, higher corrosiveness, and initial capital costs.

Ozonated water concentrations are generally applied in the range of 0.03–20.0 ppm and 20,000 ppm in the gaseous form (Manganaris et al., 2006; Martin-Diana et al., 2005; Rico et al., 2007; Saftner et al., 2003). Ozonated water was generally used to reduce the microbial load and extend the shelf-life of fresh produce such as lettuce, asparagus, broccoli, cherry tomatoes, spinach, cucumbers, green beans, onions, celery, rocket leaves, carrots, watercress, beans, and bell peppers (Akbas and Olmez, 2007; Alegria et al., 2009; Alexandre et al., 2011, 2013; Beltrán et al., 2005a; Chauhan et al., 2011; Ketteringham et al., 2006; Kim et al., 2006; Long et al., 2011; Martínez-Sánchez et al., 2006; Ölmez and Akbas, 2009; Qiang et al., 2005; Rodgers et al., 2004; Selma et al., 2007). Although it was used to inactivate a large number of the bacterial population, yeasts and molds, most of the studies focused on the resistance of *Escherichia coli* to determine its effectiveness (Table 18.1). As presented in Tables 18.1 and 18.2, the reduction of total bacterial counts varied between 1.42 and 2.44 log CFU/g with aqueous ozone, and 0.28–1.04 log CFU/g if ozone was used in gaseous form. When the effectiveness of aqueous ozone was compared with gaseous ozone very inconsistent results were obtained probably due to the different kind of food materials used in each study (Miller et al., 2013). Combinations of ozone with chlorine and organic acids effectively reduced the microbial flora on shredded lettuce by 1.4 log CFU/g when 7.5 mg/L ozone was combined with 150 mg/L chlorine (Garcia et al., 2003) while 2.31 and 1.84 log reductions were achieved for *E. coli* and *L. monocytogenes* on iceberg lettuce at a concentration of 3 ppm of ozone with combined use of citric acid (1%) for 1 min (Yuk et al., 2006). In package, ozonation effectively reduced the microbial population by 2–3 log CFU/g for

Table 18.1 Decontamination of Fresh Produce by Aqueous Ozone

Food Product	Scale of Operation	Operational Conditions	Microorganisms Affected	Microbial Reductions (Log Cycles)	References
Tomatoes	7 L tank	0.05–0.1 ppm and 0.1–0.2 ppm ozone	*S. enterica, E. coli*	1.9 CFU/g (coliform count) and more than 2 CFU/g (*E. coli*)	Long et al. (2011)
Asparagus	10 L reactor	8 mg/L for 30 min	Total bacteria counts	0.85	Qiang et al. (2005)
Carrots	Medium scale	10 mg/L	AMB, coliforms, yeasts/molds	Reduced SPC by 1–2	Chauhan et al. (2011)
	120 L reservoir	1 ppm/120 L for 5 min at 5°C	Microbial load reductions	0.4 for AMB and 0.6–0.7 for yeasts and molds	Alegria et al. (2009)
Cucumber	10 L reactor	8 mg/L for 30 min	Total bacteria counts	2.44	Qiang et al. (2005)
Green beans	10 L reactor	8 mg/L for 30 min	Total bacteria counts	0.81	Qiang et al. (2005)
Lettuce	–	0–10 mg/L for 3 min	*E. coli* O157:H7, *Salmonella enterica* serovar Typhimurium, *Listeria monocytogenes* and *Staphylococcus aureus*	3–4 log reduction for total cell counts and coliform counts after 3 mg/L ozonated water	Kim et al. (2006)
	–	Ozonated water (2 ppm), chlorinated water (100 ppm), organic acid (0.25 g/100 g citric acid plus 0.50 g/100 g ascorbic acid), and cold water (control) (in 5 L for 2 min) at 10°C	*Listeria monocytogenes*, AMB, Enterobactericeae, psychrotrophic bacteria	Ozone treatment: AMB, psychrotrophic counts and Enterobacteriaceae decreased by about 1.5, 1.1, and 1.5 log CFU/g	Olmez and Akbas (2009)
	–	4 mg/L	Mesophilic bacteria, psychrotrophic bacteria	1.7 and 1.5 log CFU/g	Akbas and Olmez (2007)
	100 L tank	(1) 1 ppm for 3 min; (2) 2 ppm for 5 min; and (3) 5 ppm for 5 min (mg/L)	*Shigella sonnei*	Reduced by 0.7, 0.9, and 1.8 log units when exposed to condition (1), (2) and (3), respectively	Selma et al. (2007)

	–	1, 3, 6, and 10 ppm for 3, 5, and 10 min (effective 10 mg/L)	Mesophilic bacteria, psychrotrophic bacteria, yeasts/molds	0.79, 1.17, and 0.99 log reductions on mesophiles, psychotrophes and yeasts/molds, respectively	Wei et al. (2007)
	–	3, 5, and 10 ppm for 5 min	AMB	1.4 log CFU/g	Koseki and Isobe (2006)
	100 L dissolution tank	10–20 mg/L ozone activated by ultraviolet C	Mesophilic bacteria, psychrotrophic bacteria, coliforms	1.6 and 3.2 log reduction (ozone at 20 ppm and ozone 10 ppm activated by UV-C)	Beltran et al. (2005a)
Onions (green)	4 L tank	(1) 0.05–0.1 ppm and (2) 0.1–0.2 ppm	AMB, *S. enterica*, *E. coli*	1.31–1.71 (for aerobic mesophilic count)	Long et al. (2011)
Peppers (red bell)	Pilot plant 30 L tank	0.3 mg/L	*L. innocua*	2.27 ± 0.25 (at −7°C), 2.11 ± 0.32 (at −30°C)	Alexandre et al. (2013)
Peppers (green)	160 L	0.30–3.95 mg ozone/L water for 20 s, 1, 5, 15 or 30 min	AMB	0.72	Ketteringham et al. (2006)
Potato (peeled strips)	100 L dissolution tank	(1) water containing ozone (20 ppm) and (2) water containing ozone, followed by dipping in Tsunami solution (300 ppm) up to 5 min	AMB, total psychrotrophics, coliforms, lactic acid bacteria (LAB), yeasts/molds, total anaerobic bacteria	3.3, 3.0, and 1.2 log reductions for LAB, coliforms and anaerobic bacteria after ozone plus tsunami	Beltrán et al. (2005b)
Rocket leaves	100 L dissolution tank	10 mg/L	Mesophilic and psychrophilic bacteria, coliforms, LAB, yeasts/molds	0.5–1	Martínez-Sánchez et al. (2006)
Spinach	10 L reactor	8 mg/L for 30 min	Total bacteria counts	0.42	Qiang et al. (2005)
Watercress	Pilot plant (30 L)	0.3 and 2.0 ppm for 1, 2, and 3 min	Total coliforms	1.7 ± 0.4 log-cycles reduction during 3 min	Alexandre et al. (2011)
Peppers (red bell)		0.3 ppm for 1, 2, and 3 min	*L. innocua*	2.8 ± 0.5 log-cycles reduction during 3 min	Alexandre et al. (2011)
Vegetables	1–4 L tank	1.4 mg/L for 15 min	Pesticides	Removes 27–34% of residual pesticide	Wu and Kim (2007)
Onion, escarole, carrot, and spinach	12 L column	UV-C lamp with 253.7 nm wavelength and ozonated water impelled at flow rate of 115 L/h	Total aerobic mesophilic bacteria, yeasts and molds, total coliforms and *Escherichia coli*	O_3–UV: 6.6 log CFU/mL, UV: 4.0 log CFU/mL and O_3: 5.9 log CFU/mL after 60 min	Selma et al. (2008)

Table 18.2 Decontamination of Fresh Produce by Gaseous Ozone

Food Product	Scale of Operation	Operational Conditions	Microorganisms Affected	Microbial Reductions (Log Cycles)	References
Asparagus	5 L glass reactor	40 ppm ozone for 1 h	Total bacteria counts	0.75	Qiang et al. (2005)
Carrots	Medium scale	0, 300, and 1000 nL/L ozone at 10°C for 0, 1, 2 or 4 days	*B. cinerea*, *S. sclerotiorum*	1000 nL/L induced decay resistance only to *B. cinerea*	Forney et al. (2007)
Cucumber	5 L glass reactor	40 ppm ozone for 1 h	Total bacteria counts	1.04	Qiang et al. (2005)
Green bean				0.38	
Pepper (red)	5 L glass reactor	40 ppm ozone for 1 h	Total bacteria counts	1	Qiang et al. (2005)
Pepper (green)	10 L Irvine Plexiglass cylinder	2–8 mg/L for 10–40 mins	*E. coli* O157:H7	More than 5 log cycles reductions	Han et al. (2002)
Spinach	Medium scale	SanVac process—O_3 at 1.5 g/kg gas-mix (935 ppm, vol/vol), 10 psig of holding pressure, and 30 min of holding time	*E. coli* O157:H7	1.4–1.8 log CFU/g	Vurma et al. (2009)
	5 L glass reactor	40 ppm ozone for 1 h	Total bacteria counts	0.28	Qiang et al. (2005)
Lettuce	–	Neutral electrolyzed water containing 50 ppm of free chlorine	*E. coli* O157:H7, *Salmonella*, *L. innocua* and *E. carotovora*	1–2 log units	Abadias et al. (2008)
Fresh produce	–	SAEW (ca. pH 5.8 and 21 mg/L available chlorine concentration; ACC) for 90 s	*S. aureus* and *E. coli*	>5 log CFU/mL	Issa-Zacharia et al. (2010)
Celery, lettuce, and sprouts	500-mL glass beaker	AEW at pH 5.6 for 5 min	*E. coli* and *Salmonella* spp.	2.7–2.9 log CFU/g	Issa-Zacharia et al. (2011)

2–3 mins in tomatoes with no effect on product quality (Fan et al., 2012). It really appears that ozone can be an emerging application, which can overcome some drawbacks, related to the use of this technology, but requires further investigation.

In the case of fresh fruits, according to a study carried out by Bialka and Demirci (2007a) gaseous ozone has potential to be used for decontaminating blueberries on a larger scale. When treating with gaseous ozone, the maximum reductions were 3.0 and 2.2 log CFU/g of *Salmonella* and *E. coli* O157:H7, respectively. Higher reductions were obtained when blueberries were treated with aqueous ozone rather than gaseous ozone, whereby the maximum reductions achieved were 6.2 and 5.2 log CFU/g of *Salmonella* and *E. coli* O157:H7, respectively (Bialka and Demirci, 2007a).

Aqueous ozone is an effective decontaminating agent for small fruits. Since even though its efficacy is reduced in the case of microorganisms in crevices, for example, in the case of rough fruit surfaces, decontamination with aqueous ozone was more effective than washing with water. The treatment temperature did not result in a significant difference in the efficacy of the sanitizer, thus, in industries, treatment could be carried out at room temperature so as to reduce costs associated with energy expenditure (Bialka and Demirci, 2007b).

Apple is an example of a fruit that is industrially treated with aqueous ozone. A Swiss company, SAMRO Ltd, uses *Ventafresh* technology which uses a combination of ozone, ultraviolet (UV) light, and ultrasound (US) to sterilize and package food. Examples of fruits treated with this technology include apples, pears, and kiwis (Miller et al., 2013). Research was carried out by Achen and Yousef (2001) to investigate the conditions needed to effectively decontaminate whole apples inoculated with *E. coli* O157:H7 by ozone treatment. Two methods related to the delivery of ozone were investigated: dipping in ozonated water (22–24 mg O_3/L) and washing in bubbling ozone water (21, 25, and 28 mg O_3/L residual ozone). A 5-min ozone treatment is most probably impractical in industrial facilities, thus the research focused mostly on 3-min treatment. After treatment for 3 min, the maximum reductions in the surface population of *E. coli* O157:H7 were 2.6 and 3.7 log CFU/g, respectively. On the other hand, when apples were washed with water only, the surface population of the pathogens was decreased by 1.2 log CFU/g. As can be observed, ozone bubbling is more efficient than the dipping treatments due to a higher ozone concentration in the liquid layer at the gas–liquid interface compared to the surrounding solution. Therefore, even though both methods have potential applications in industries, bubbling ozone was preferred.

Temperature also affects the efficiency of this treatment since as temperature increases, the rate of reaction between ozone and the substrate increases, however, on increasing temperature, ozone solubility and stability decrease (Achen and Yousef, 2001; Khadre et al., 2001; Miller et al., 2013). The stability of aqueous ozone depends on the quality of the water; the more readily oxidizable organic and inorganic substances it contains, the less stable ozone will be (Khadre et al.,

2001; Miller et al., 2013). pH also affects the stability of aqueous ozone whereby ozone stability decreases on increasing the pH.

When whole apples were rinsed in tetrasodium pyrophosphate (a wetting agent with a low ozone demand) before the ozone wash, the population of *E. coli* O157:H7 decreased to 3–3.5 log CFU/g. Thus, ozone combined with this chemical was more effective as a disinfection technology compared to ozone on its own. This may be due to the increased exposure of the bacterial cells to ozone by the wetting agent. Thus it can be concluded that in industries, efficacy of ozone treatment can be increased by first washing apples with water so as to reduce the ozone demand by organic compounds from debris, soil, etc., followed by pretreatment with an inorganic wetting agent (in this case, tetrasodium pyrophosphate was used) and then 3-min treatment in bubbling ozone water at a temperature of 22°C (Achen and Yousef, 2001).

One of the main limitations of these methods on whole apples is that the pathogen population in the stem-calyx region did not decrease significantly (0.5 and 0.6 log CFU/g respectively), thus, it can be concluded that these methods are limited to disinfection of the surface of apples. This may be due to the decreased exposure of the bacteria to ozone due to a rougher region in the stem-calyx area (Sapers, 2001). Hence, further research is required, prior to industrial application, so as to improve the ozone delivery methods, such that ozone reaches bacterial cells in all regions of the apple (Achen and Yousef, 2001).

Overall, it can be summarized that ozone treatment is a potential technology for industrial application since it is economical, for example, the raw materials (oxygen or water) are abundant and accessible, it can be produced on-site thus reducing costs associated with transport and storage, and it does not require heat thus saving energy. The initial costs of the ozone generators may be high, however, these costs are then justified in the long term. One of the limitations is that industries must have enough space for ozone to be generated on-site due to its lack of stability. Apart from improving delivery methods, further research is also required to investigate different industrial applications of ozone and optimizing conditions so as to find a balance between obtaining disinfected products while retaining their quality (Khadre et al., 2001; Miller et al., 2013). Also, further research is needed about ozone treatment combined with other technologies.

Acidic electrolyzed water

Strongly acidic electrolyzed water (StAEW) is environmentally friendly since its generation only involves electrolysis of water and a dilute salt solution. After use, it can be converted back to ordinary water. Also, it is safe to consumers and employees since concentrated chemicals are not used during its production and it is generated on-site. It is effective, does not affect the organoleptic properties of fresh produce, is easy to handle, safer and cheaper (initial capital investment is required to purchase an electrolyzed water generator but then the operating expenses involve easily accessible raw materials (ie, water and sodium chloride), and electricity compared to other chemical sanitizers such as sodium hypochlorite). The main limitation of StAEW is that the solution rapidly loses its antimicrobial activity with time due to loss of chlorine. One way to overcome this limitation is to use slightly acidic electrolyzed water (SAEW) instead of StAEW since its pH is nearer to a neutral pH and the chlorine concentration is lower. This can result in a more stable solution compared to StAEW, therefore making the former technology more promising for industries. Electrolyzed oxidized (EO) water is produced on-site and therefore minimizes handling and exposure of dangerous chemicals. EO water has potential for industrial use (Hati et al., 2012). An issue which could be a possible limitation to large-scale use of EO water is related to the cost of the equipment and its electrical consumption (Okull and Laborde, 2004). In general, further research is required to optimize the process for individual applications.

Studies on the use of SAEW are also reported in the literature. It is considered a highly effective bactericidal decontamination method which has the potential to replace sodium hypochlorite solution (even at low available chlorine concentration (ACC) of 10–30 mg/L) (Suzuki et al., 2005) for reducing the number of pathogens on the fresh produce (Huang et al., 2008; Issa-Zacharia et al., 2010). Abadias et al. (2008) and Issa-Zacharia et al. (2010, 2011) suggested that electrolyzed water (20–55 ppm available chlorine) can effectively reduce the microbial population from 1.0 to 5.0 log CFU/g. There are a number of laboratory studies showing the potential of electrolyzed water alone or in combination with ultrasound or ozone treatment to effectively reduce the microbial levels (Guentzel et al., 2008; Hung et al., 2010; Izumi, 1999; Koseki et al., 2001, 2002, 2004a,b; Kim et al., 2003a; Olaimat and Holley, 2012; Park et al., 2001; Paola et al., 2005; Sharma and Demirci, 2003; Yang et al., 2003; Wang et al., 2004). As shown from the above studies, electrolyzed water (20–55 ppm available chlorine) with exposure times for 1–10 min is generally used for decontaminating food materials such as alfalfa seeds and sprouts, lettuce, cherry tomatoes, spinach, carrots, cilantro and bell peppers in order to reduce aerobic bacteria counts, *E. coli*, and *Salmonella* spp.

Electrostatic spraying

Another innovative technology is the electrostatic spraying in which a fine coating is applied with a substance on the surface of the food but very limited literature has so far been published on it for fresh produce decontamination (Goodburn and Wallace, 2013). Combination of malic acid and lactic acid or malic acid and grapeseed reduces the microbial population of *Salmonella* Typhimurium on spinach by 4.3 and 3.3 log CFU/g, respectively (Ganesh et al., 2010). Although log reductions by electrostatic spraying are better or comparable to ozone, chlorine, and chlorine dioxide, further research is needed to comprehend its feasibility for use in the food industry as a decontaminant.

Organic acids

Organic acids such as citric, lactic, and acetic acids are used for fresh produce decontamination and their antimicrobial action mainly depends upon environmental pH reduction, membrane permeability disruption, anion accumulation, etc. (Parish et al., 2003). Because of its strong unacceptable flavor, particularly of the acetic acid, longer exposure times (5–15 min) and impact on wastewater quality, they are not considered to be relevant for food industry applications (Ölmez and Kretzschmar, 2009).

Peroxyacetic acid

Peroxyacetic acid or peracetic acid is gaining interest as it inhibits the microorganisms at lower concentrations (about 80 ppm) than chlorine (Dell'Erbaa et al., 2007). Generally 1.0–1.7 log CFU/g reduction in microbial population occurs by using 40–80 ppm of peroxyacetic acid (peracetic acid and hydrogen peroxide) in cabbage (Hilgren and Salverda, 2000; Vandekinderen et al., 2007), celery (Hilgren and Salverda, 2000), and shredded carrots (Ruiz-Cruz et al., 2007), but not sufficient enough to obtain a significant reduction in microbial load in fresh produce (Artés et al., 2009; Rico et al., 2007; Sapers and Jones, 2006). Further studies may be required to conclude on the potential application in industrial environments.

Hydrogen peroxide

Although hydrogen peroxide is a strong oxidizing agent, as it possesses both bacteriostatic and bactericidal properties, its use is very limited in the food industry as a food decontaminant as it causes severe browning on shredded lettuce (Parish et al., 2003). It leaves no residue and its antimicrobial effectiveness is comparable to 100–200 ppm of chlorine (Parish et al., 2003) but the overall quality of food product is affected if its concentration is raised (4–5%) (Beltrán et al., 2005b; Rico et al., 2007).

Other antimicrobial substances

There are different antimicrobial substances that can be used to disinfect fresh produce, including vanillin, peroxyacetic acid, carvacrol, hydrogen peroxide, nisin, sodium lactate, citric acid, potassium sorbate, and essential oils such as thyme and oregano. A lab study was carried out on disinfection of fresh-cut apple using different antimicrobial substances, including vanillin (3, 6, and 12 g/L), carvacrol (250, 875, and 1000 mg/L), and peroxyacetic acid (40, 80, and 120 mg/L). According to this research, there is potential for the use of alternative sanitizers instead of chlorinated water, since apart from reducing pathogens during treatment, low levels were also maintained during storage. However, further research is required in order to upscale the conditions and also investigate whether the sensory qualities of the fresh produce are affected, including whether there are any chemical residues posttreatment. For example, vanillin was the most effective when the highest concentration (12 g/L) was used, however, at a high concentration, the sensory qualities of fresh-cut apples could be affected. The effectiveness of antimicrobial substances can be increased by combining with other technologies (Abadias et al., 2011).

PHYSICAL INTERVENTION TECHNOLOGIES IN FRESH VEGETABLES AND FRUITS

Thermal treatments affect the organoleptic and structural properties of food products while chemical disinfectants require a large volume of water at the industrial scale (Misra et al., 2014). There are different technologies, which are less energy-intensive, cost-efficient, and environmentally friendly, including UV, US, and pulsed electric field (Awad et al., 2012).

Ultrasound

The effective microbial reduction from the surfaces of fruits and vegetables is a challenge for the fresh produce industry (São José and Vanetti, 2012). Tables 18.1–18.3 show an overview of the major physical decontamination technologies used for reducing the microbial load in fresh produce and have been used in larger than laboratory scales. Ultrasound (US) is a safe, nontoxic and environmentally friendly technology (Bilek and Turantaş, 2013). US technology has the potential to be applied in the industry since its performance is efficient and economically viable (due to associated low costs, lack of extensive equipment, and low energy consumption). It is one of the most important decontamination technologies used and can be applied alone or with other chemical sanitizers to enhance its effectiveness (Cao et al., 2010); in addition, it is recommended for the fresh produce industry (Alegria et al., 2009; Alexandre et al., 2013; Birmpa et al., 2013; Elizaquivel et al., 2012; Huang et al., 2006; São José and Vanetti, 2012; Sagong et al., 2011, 2013; Seymour et al., 2002; Zhou et al., 2012). In the above studies, ultrasound is commonly used for decontaminating food materials such as lettuce, cherry tomatoes, spinach, carrots, and bell peppers with low frequencies (between 20 and 45 kHz) and treatment times for 1–10 min. It is also indicated that the microbial reduction of *E. coli* and *S. typhimurium* varies between 1.04–4.4 and 0.8–1.5 log CFU/g, respectively, by using ultrasound alone, while it varies between 1.3–4.0 and 2.2–3.9 log CFU/g with shorter treatment times and when ultrasound is combined with chemical sanitizers (eg, chlorine, chlorine dioxide, and organic acids).

Combined US treatment result in higher microbial reduction than the individual treatment. When ultrasound was combined with SAEW, the bactericidal activity of the latter technology was increased, which resulted in 1.29 log reductions of total aerobic bacteria, and of yeasts and molds in strawberries. On the other hand, when strawberries were treated with ultrasound treatment alone, total aerobic bacteria were reduced by 0.52 log, while yeasts and molds were reduced by 0.3 log, and when treated with SAEW alone, total aerobic bacteria were reduced by 1.07 log while yeasts and molds were reduced by 1.04 log. According to this research, the combined technology has potential as a disinfection technology of fresh produce without affecting their quality. The enhanced efficacy of the combined treatment compared to the individual treatments may be due to the ultrasound treatment increasing the penetration of the SAEW. Further research is

Table 18.3 Decontamination of Fresh Produce by Combined Technologies

Decontamination Technologies	Food Product	Scale of Operation	Operational Conditions	Microorganisms Affected	Microbial Reductions (Log Cycles)	References
Chlorination + irradiation	Shredded iceberg	–	Chlorination (200 μg/mL) + low-dose irradiation (0.55 kGy)	*E. coli* 0157:H7	5.4	Foley et al. (2002)
Aqueous ozone + chlorination	Shredded lettuce	–	Aqueous ozone (7.5 ppm or mg/L) + chlorine (150 ppm or mg/L)	Aerobic plate count	1.4 log CFU/g	Garcia et al. (2003)
Aqueous ozone + chlorination	Romaine and lollo rossa leaves	–	Sanitizer wash (Cl at 70 ppm) before cutting	*E. coli* 0157:H7	1	Nou and Lou (2010)
US alone and in combination with organic acids	Lettuce	5 L US tank	US (40 kHz) + lactic/citric/malic acid (2%) for 5 min	*E. coli* O157:H7, *S. typhimurium*, and *L. monocytogenes*	*E. coli* O157:H7: 2.7 *S. typhimurium*: 3.2 *L. monocytogenes:* 2.9	Sagong et al. (2011)
US alone, US + chlorinated water	Lettuce	5 L (small scale trial) and 40 L tanks (large-scale trial)	US alone (32–40 kHz, 10 min), US + Chlorinated water (25 ppm free chlorine)	*S. typhimurium*	US alone: 1.5, US + chlorinated water: 2.7	Seymour et al. (2002)
US alone, US + PAA	Cherry tomatoes	Medium scale	US alone (45 kHz), US + peracetic acid (40 mg/L) for 10 min	*S. enterica* Typhimurium	US alone: 0.8 and US + PAA: 3.9	São José and Vanetti (2012)
US alone or in combination with chlorine	Spinach	Pilot scale (85.5 L tank)	25, 40, and 75 kHz for 1 min	*E. coli* O157:H7	US alone: 25, 40, and 75 kHz was 0.056, 0.061, and 0.057 log CFU/(W/L). US + Chlorine: additional log reductions of 1.0 and 0.5 CFU/g	Zhou et al. (2012)
US + surfactants	Lettuce, carrots	5 L US tank	US 40 kHz + 0.1% Tween 20	*B. cereus*	Lettuce: 2.49 log CFU/g, Carrots: 2.22 log CFU/g	Sagong et al. (2013)

US alone, US + chlorinated water	Shredded carrot	45 L US tank	US alone (45 kHz, 1 min) US + chlorinated water (200 ppm free chlorine)	Total aerobic mesophilic and yeast and molds counts	TVC: 1.3, YMC: 0.9, TVC: 1.0, YMC: 0.9	Alegria et al. (2009)
US + ClO_2	Lettuce	8-L rectangular stainless steel containers	US (170 kHz, 6–10 min) + ClO_2 (5 and 10 ppm)	*Salmonella* spp., *E. coli* O157:H7	*Salmonella* spp.: 2.2–2.9 *E. coli* O157:H7: 1.3–2.2	Huang et al. (2006)
SAEW alone, US alone, SAEW + US	Cherry tomatoes	10 L water bath	US (40 kHz, 10 min)	Total aerobic bacteria, yeasts and molds	SAEW reduced approx 1.45 log CFU/g for total aerobic bacteria and 1.10 log CFU/g for yeasts and molds, US + SAEW: 1.77 and 1.29 log reductions on total aerobic bacteria, and 1.50 and 1.29 log reductions on yeasts and molds	Ding et al. (2015)
Ozonated water with 1% organic acids	Iceberg lettuce	Medium scale	1, 3, and 5 ppm ozone alone at 0.5, 1, 3, or 5 min, and combinations of 3 ppm ozone with 1% organic acids (acetic, citric, or lactic acids) for 1 min	*E. coli* O157:H7, *L. monocytogenes*	2.31 and 1.84 log reductions with 3 ppm ozone with 1% citric acid for 1 min	Yuk et al. (2006)
US alone	Red bell pepper	5.6 L	US 35 kHz for 2 min (15°C)	*L. innocua*	*L. innocua*: 1.98	Alexandre et al. (2013)
US alone	Lettuce	5.75 L US tank	US 37 kHz for 30 min	*E. coli*, *S. aureus*, *S. Enteritidis*, *L. innocua*	*E. coli*: 2.30 log CFU/g, *S. aureus*: 1.71 log CFU/g, *S. Enteritidis*: 5.72 log CFU/g, *L. innocua*: 1.88 log CFU/g	Birmpa et al. (2013)

required to optimize the conditions of US combined with SAEW in order for this technology to ensure disinfection of pathogens on fresh produce (Ding et al., 2015).

Irradiation

There are some lab studies showing the effectiveness of irradiation in reducing the microbial population by 4.1–6.7 log CFU/g at 0.5–2 KGy in different food materials (Bari et al., 2005; Foley et al., 2002, 2004; Villagomez et al., 2010), but it is still not widely used by the fresh produce industry. The commercial application of this technology requires research about its economic feasibility (initial investment may be high) and optimization of the process including radiation source, type of irradiator, etc. Also, consumers need to be more informed about the safety of food irradiation (Moy, 2005).

UV light

UV is a nonthermal technology, which is mainly effective for surface decontamination. According to the FDA and USDA, UV irradiation is safe. UV is an alternative option for commercial application in the food system. There are already commercially available systems for the decontamination of food surfaces, thus enhancing its potential application in industries. One such example is the UV tumbling process developed by C & S Equipment Company which was developed in such a way as to ensure that all sections of the products' surface are exposed to the UV source (Koutchma, 2008).

Advantages of UV treatment include its safety in use, increase of shelf-life of fresh produce, and also an improvement of the organoleptic properties of fresh produce, for example, increase in the antioxidant capacity. Limitations associated with this technology include limited food penetration, whereby the degree of penetration depends on a number of factors such as optical properties of food, including turbidity and opaqueness. UV systems are affordable since they require a low initial investment and a low operating treatment cost, therefore, this technology has potential for industries with limited capital investment. Continuous-wave UV is less expensive than pulsed UV, hence, the former technology would be more suitable for industries that have little capital for investment (Choudhary and Bandla, 2012). Further research is required to maximize UV efficacy, for example, research related to the degree of reduction in efficacy due to the shielding effects of some fruit surfaces, for example, irregular surfaces versus smooth surfaces (Koutchma, 2008).

Pulsed light

Pulsed light is an alternative nonthermal technology that can replace thermal and chemical decontamination technologies currently used in industries (Abida et al., 2014). Pulsed light rapidly inactivates the pathogenic microorganisms on the surface of food materials using short-duration intense pulses and a broad spectrum (Elmnasser et al., 2007) which induces structural changes in the pathogens and thus prevents replication of the cells (Takeshita et al., 2003). Advantages of this technology include that the lamps used are more environmentally friendly and it does not affect the organoleptic properties of the fruit. Limitations include that effectiveness depends on different factors, such as type of food (eg, food with folds may protect the pathogens from exposure to the pulsed light) and type of microorganisms (some are resistant). Also, it is limited to liquids and the surface of solid foods (Abida et al., 2014). Oms-Oliu et al. (2010) suggested that pulsed light may reduce the total aerobic mesophilic and psychrophilic microorganisms on mushrooms by about 0.6–2.2 log cycles. Pulsed UV light has potential to be used for decontaminating blueberries on a larger scale. When treated with pulsed UV light, the maximum reductions were 2.9 and 4.3 log CFU/g of *Salmonella* and *E. coli* O157:H7, respectively. According to the sensory analysis carried out, there was no significant difference between the control and treated samples. Further research needs to be carried out so as to determine whether the shelf-life is affected. Also, in the case of pulsed UV light, even though it is a potential decontamination technology that can be considered on an industrial scale, further research is required for commercialization purposes.

Nonthermal plasma treatment

Another alternative approach which does not result in extreme heat treatment is the cold gas plasmas (CAP) (or nonthermal plasma (NTP)), which show no negative effects on fresh produce and thus could replace industrial thermal treatments in the future (Kayes et al., 2007; Niemira and Sites, 2008). Generating plasma at atmospheric pressure and using it as a disinfection technology is of technical and commercial applicability to food industries since this technology can be implemented at ambient conditions, thus saving energy and reducing costs (Misra et al., 2014). This technology can be used for surface decontamination of fruit as well as other foods. For example, in research carried out by Misra et al. (2014), the total mesophilic and yeasts/mold counts for untreated strawberries were 4.99 and 4.96 log CFU/g, respectively. When treated with plasma for 5 min, the achieved reductions were of 2.4 and 3.3 log cycles, respectively. Furthermore, changes in color and firmness were statistically insignificant. A study using nitrogen CAP (Fernández et al., 2013) and atmospheric cold plasma (Ziuzina et al., 2014) are the lab-scale models which significantly reduced the microbial load in lettuce, potatoes, and cherry tomatoes.

High-energetic plasma combined with a nonthermal method makes NTP a potential application for decontamination of fresh produce since it reduces the microbial levels without adversely affecting the organoleptic properties. Furthermore, this technology is better than other potential surface treatments, such as UV light, since the ions of cold plasma can penetrate cracks and crevices. Cold plasma treatment is a promising technology since it acts fast, and it does not leave any toxic residues on the treated food. Further studies are required related to the effect of NTP on food. Cold plasma technology is not yet used commercially on a large scale, however, the equipment can be scalable; and these are already under development (Misra et al., 2011). Further research is required to show its application in the fresh produce industry including investigation of its effects on the properties of the food and on shelf-life so as to investigate the long-term effects on quality.

Photosensitization

The other novel nonthermal decontamination technology which could be a promising alternative to inactivate pathogens on the surface of fruits and vegetables is photosensitization, which is based on the interaction of photoactive compound (photosensitizer) and visible light (Aponiene et al., 2015). This treatment has been successfully applied for reducing the microbial population by 0.6–0.72 log in cauliflower (Hyp-based photosensitization) (Aponiene et al., 2015), 1.7 log in strawberries (Chl-based photosensitization) (Luksiene and Paskeviciute, 2011), 1.66–2.88 log in cherry tomatoes, nectarines, iceberg lettuce, and Chinese cabbage (Chl-based photosensitization) (Paskeviciute and Luksiene, 2009).

DECONTAMINATION TECHNOLOGIES AND ENVIRONMENT

Novel intervention and prevention strategies for fresh produce aim to improve the shelf-life of fresh produce with minimal quality deterioration. The fresh produce industry is one of the major water-intensive sectors within the food industry. Washing of fresh produce, application of disinfectants, and a sanitizing production area are the key unit operations which contribute to water consumption and wastewater stream with high concentrations of organic materials in fresh produce industry. The impetus of reducing water usage, as well as chemical emissions, changes the focus of decontamination technologies towards non-water-based or reduced water usage (eg, cold atmospheric plasma). Novel techniques appear to be environmentally friendly and studies have shown that the novel techniques can reduce water and energy consumption which has been in focus for the food industry. However, clear assessment of novel strategies compared to conventional preservation strategies is limited in the case of fresh produce. Environmental impact assessment of novel technologies based on (1) energy and water consumption and (2) reduced gas and effluent emissions has been reported for some novel technologies which have been employed at an industrial scale. In a study, Sampedro et al. (2014) carried out a cost–benefit analysis including environmental impact assessment of a commercial scale pulsed electric fields, high-pressure processing, and thermal pasteurization for orange juice as a model. This study showed that both the high-pressure processing and pulsed electric field have higher environmental impact in terms of greenhouse gas emissions (CO_2 production) compared to conventional thermal processing. Similar assessment of techniques suitable for the fresh produce sector is required to achieve the full benefits of commercial exploitation.

FUTURE PERSPECTIVES

Consumer demands for high-quality, convenient, innovative, regular, and safe fresh produce with extended shelf-life have initiated a search for innovative (dry) decontamination techniques. To meet such demands without compromising safety, the production of fresh produce is at the stage of innovative dynamics, thus stimulating a major research issue to develop and implement alternative technologies. The application of emerging techniques at various scales has the potential for producing high-quality and safe fresh produce with extended shelf-life. Some of these disinfection techniques have been exploited commercially for limited products and are not routinely employed at a commercial scale. The key processing advantages offered by these techniques, such as reduced operational costs and extended shelf-life should encourage their future adoption. These technologies will also assist food processors to meet both consumer demands for safe, nutritious and high-quality fresh produce. The applications of novel strategies and interventions are likely to grow steadily in the future because of increased consumer demands for chemical-free and minimally processed foods. There are several limitations for full commercial exploitation of these techniques including validation of disinfection techniques at commercial settings, regulatory approval and legislative requirements associated with novel disinfectants, along with the control of variables associated with the process operation have been seen as hurdles for wider adoption. An effective strategy for fresh produce is by combining emerging preservation techniques with conventional strategies. The environmental impact assessment in terms of energy efficiency, water reduction, and reduced associated emissions of the novel disinfection alternatives compared to chlorine-based techniques should be carried out for future adoption.

REFERENCES

Abadias, M., Usall, J., Oliveira, M., Alegre, I., Viñas, I., 2008. Efficacy of neutral electrolyzed water (NEW) for reducing microbial contamination on minimally processed vegetables. Int. J. Microbiol. 123, 151–158.

Abadias, M., Alegre, I., Usall, J., Torres, R., Viñas, I., 2011. Evaluation of alternative sanitizers to chlorine disinfection for reducing foodborne pathogens in fresh-cut apple. Postharvest Biol. Technol. 59 (3), 289–297.

Abida, J., Rayees, B., Masoodi, F.A., 2014. Review article pulsed light technology: a novel method for food preservation. Int. Food Res. J. 21 (3), 839–848.

Achen, M., Yousef, A.E., 2001. Efficacy of ozone against *Escherichia coli* O157: H7 on Apples. J. Food Sci. 66 (9), 1380–1384.

Akbas, M.Y., Olmez, H., 2007. Effectiveness of organic acid, ozonated water and chlorine dippings on microbial reduction and storage quality of fresh-cut iceberg lettuce. J. Sci. Food Agric. 87 (14), 2609–2616.

Alegria, C., Pinheiro, J., Goncalves, E.M., Fernandes, I., Moldao, M., Abreu, M., 2009. Quality attributes of shredded carrot (*Daucus carota* L. cv. Nantes) as affected by alternative decontamination processes to chlorine. Innovative Food Sci. Emerg. Technol. 10 (1), 61–69.

Alexandre, E.M.C., Santos-Pedro, D.M., Brandão, T.R.S., Silva, C.L.M., 2011. Influence of aqueous ozone, blanching and combined treatments on microbial load of red bell peppers, strawberries and watercress. J. Food Eng. 105, 277–282.

Alexandre, E.M.C., Brandao, T.R.S., Silva, C.L.M., 2013. Impact of non-thermal technologies and sanitizer solutions on microbial load reduction and quality factor retention of frozen red bell peppers. Innovative Food Sci. Emerg. Technol. 17, 199–205.

Alvaro, J.E., Moreno, S., Dianez, F., Santos, M., Carrasco, G., Urrestarazu, M., 2009. Effects of peracetic acid disinfectant on the postharvest of some fresh vegetables. J. Food Eng. 95 (1), 11–15.

Aponiene, K., Paskeviciute, E., Reklaitis, I., Luksiene, Z., 2015. Reduction of microbial contamination of fruits and vegetables by hypericin-based photosensitization: comparison with other emerging antimicrobial treatments. J. Food Eng. 144, 29–35.

Artés, F., Gómez, P., Aguayo, E., Escalona, V., Artés-Hernández, F., 2009. Sustainable sanitation techniques for keeping quality and safety of fresh-cut plant commodities. Postharvest Biol. Technol. 51 (3), 287–296.

Awad, T.S., Moharram, H.A., Shaltout, O.E., Asker, D., Youssef, M.M., 2012. Applications of ultrasound in analysis, processing and quality control of food: a review. Food Res. Int. 48 (2), 410–427.

Bari, M.L., Nakauma, M., Todoriki, S., Juneja, V.K., Isshiki, K., Kawamoto, S., 2005. Effectiveness of irradiation treatments in inactivating *Listeria monocytogenes* on fresh vegetables at refrigeration temperature. J. Food Prot. 68, 318–323.

Beltrán, D., Selma, M.V., Marin, A., Gil, M.I., 2005a. Ozonated water extends the shelf life of fresh-cut lettuce. J. Agric. Food Chem. 53 (14), 5654–5663.

Beltrán, D., Selma, M.V., Tudela, J.A., Gil, M.I., 2005b. Effect of different sanitizers on microbial and sensory quality of fresh-cut potato strips stored under modified atmosphere or vacuum packaging. Postharvest Biol. Technol. 37 (1), 37–46.

Bialka, K.L., Demirci, A., 2007a. Decontamination of *Escherichia coli* O157:H7 and *Salmonella enterica* on blueberries using ozone and pulsed UV-light. J. Food Sci. 72 (9), 391–396.

Bialka, K.L., Demirci, A., 2007b. Efficacy of aqueous ozone for the decontamination of *Escherichia coli* O157:H7 and *Salmonella* on raspberries and strawberries. J. Food Prot. 70 (5), 1088–1092.

Bilek, S.E., Turantaş, F., 2013. Decontamination efficiency of high power ultrasound in the fruit and vegetable industry, a review. Int. J. Food Microbiol. 166 (1), 155–162.

Birmpa, A., Sfika, V., Vantarakis, A., 2013. Ultraviolet light and ultrasound as non-thermal treatments for the inactivation of microorganisms in fresh ready-to-eat foods. Int. J. Food Microbiol. 167 (1), 96–102.

Cao, S., Hu, Z., Pang, B., Wang, H., Xie, H., Wu, F., 2010. Effect of ultrasound treatment on fruit decay and quality maintenance in strawberry after harvest. Food Control 21 (4), 529–532.

Chauhan, O.P., Raju, P.S., Ravi, N., Singh, A., Bawa, A.S., 2011. Effectiveness of ozone in combination with controlled atmosphere on quality characteristics including lignification of carrot sticks. J. Food Eng. 102 (1), 43–48.

Choudhary, R., Bandla, S., 2012. Ultraviolet pasteurization for food industry. Int. J. Food Sci. Nutr. Eng. 2 (1), 12–15.

Chung, C.C., Huang, T.C., Yu, C.-h., Shen, F.Y., Chen, H.H., 2011. Bactericidal effects of fresh-cut vegetables and fruits after subsequent washing with chlorine dioxide. In: Proceedings of International Conference on Food Engineering and Biotechnology (ICFEB 2011), vol. 9. Bangkok, Thailand, pp. 107–112.

Dell'Erbaa, A., Falsanisia, D., Libertia, L., Notarnicolaa, M., Santoroa, D., 2007. Disinfection by-products formation during wastewater disinfection with peracetic acid. Desalination 215, 177–186.

Ding, T., Ge, Z., Shi, J., Xu, Y.-T., Jones, C.L., Liu, D.-H., 2015. Impact of slightly acidic electrolyzed water (SAEW) and ultrasound on microbial loads and quality of fresh fruits. LWT—Food Sci. Technol. 60 (2), 1195–1199.

Elizaquivel, P., Sanchez, G., Selma, M.V., Aznar, R., 2012. Application of propidium monoazide-qPCR to evaluate the ultrasonic inactivation of *Escherichia coli* O157: H7 in fresh-cut vegetable wash water. Food Microbiol. 30, 316–320.

Elmnasser, N., Guillou, S., Leroi, F., Orange, A., Bakhrouf, A., Federighi, M., 2007. Pulsed-light system as novel food decontamination technology: review. Can. J. Food Microbiol. 53, 813–821.

Fan, X., Sokorai, K.J.B., Engemann, J., Gurtler, J.B., Liu, Y., 2012. Inactivation of *Listeria innocua*, *Salmonella typhimurium* and *Escherichia coli* O157:H7 on surface and stem scar areas of tomatoes using in-package ozonation. J. Food Prot. 75 (9), 1611–1618.

Fernández, A., Noriega, E., Thompson, A., 2013. Inactivation of *Salmonella enterica* serovar Typhimurium on fresh produce by cold atmospheric gas plasma technology. Food Microbiol. 33, 24–29.

Foley, D.M., Dufour, A., Rodriguez, L., Caporaso, F., Prakash, A., 2002. Reduction of *Escherichia coli* 0157:H7 in shredded iceberg lettuce by chlorination and gamma irradiation. Radiat. Phys. Chem. 63, 3–6, 2002, 391–396.

Foley, D., Euper, M., Caporaso, F., Prakash, A., 2004. Irradiation and chlorination effectively reduces *Escherichia coli* 0157:H7 inoculated on cilantro (*Coriandrum sativum*) without negatively affecting quality. J. Food Prot. 67, 2092–2098.

Food and Drug Administration (FDA), 1997. Substances generally recognized as safe, proposed rule. Fed. Regist. 62 (74), 18937–18964.

Forney, C.F., Song, J., Hildebrand, P.D., Fan, L., McRae, K.B., 2007. Interactive effects of ozone and 1-methylcyclopropene on decay resistance and quality of stored carrots. Postharvest Biol. Technol. 45, 341–348.

Ganesh, V., Hettiarachchy, N.S., Ravichandaran, M., Johnson, M.G., Griffis, C.L., Martin, E.M., et al., 2010. Electrostatic sprays of food-grade acids and plant extracts are more effective than conventional sprays in decontaminating *Salmonella typhimurium* on spinach. J. Food Sci. 75 (9), 574–579.

Garcia, A., Mount, J.R., Davidson, P.M., 2003. Ozone and chlorine treatment of minimally processed lettuce. J. Food Sci. 68 (9), 2747–2751.

Goodburn, C., Wallace, C.A., 2013. The microbiological efficacy of decontamination methodologies for fresh produce: a review. Food Control 32 (2), 418–427.

Guentzel, J.L., Lam, K.L., Callan, M.A., Emmons, S.A., Dunham, V.L., 2008. Reduction of bacteria on spinach, lettuce, and surfaces in food service areas using neutral electrolyzed oxidizing water. Food Microbiol. 25, 36–41.

Han, Y., Floros, J.D., Linton, R.H., Nielsen, S.S., Nelson, P.E., 2002. Response surface modeling for the inactivation of *Escherichia coli* O157:H7 on green peppers (*Capsicum annuum*) by ozone gas treatment. J. Food Sci. 67, 1188–1193.

Hati, S., Mandal, S., Minz, P.Z., Vij, S., Khetra, Y., Singh, B.P., et al., 2012. Electrolyzed Oxidized Water (EOW): non-thermal approach for decontamination of food borne microorganisms in food industry. Food Nutr. Sci. 03 (06), 760–768.

Hilgren, J.D., Salverda, J.A., 2000. Antimicrobial efficacy of a peroxyacetic acid/octanoic acid mixture in fresh-cut vegetable process waters. J. Food Sci. 65, 1376–1379.

Huang, T.S., Xu, C., Walker, K., West, P., Zhang, S., Weese, J., 2006. Decontamination efficacy of combined chlorine dioxide with ultrasonication on apples and lettuce. J. Food Sci. 71 (4), 134–139.

Huang, Y.-R., Hung, Y.-C., Hsu, S.-Y., Huang, Y.-W., Hwang, D.-F., 2008. Application of electrolyzed water in the food industry. Food Control 19 (4), 329–345.

Hung, Y.C., Tilly, P., Kim, C., 2010. Efficacy of electrolyzed oxidizing (EO) water and chlorinated water for inactivation of *Escherichia coli* O157:H7 on strawberries and broccoli. J. Food Qual. 33 (5), 559–577.

Issa-Zacharia, A., Kamitani, Y., Muhimbula, H.S., Ndabikunze, B.K., 2010. A review of microbiological safety of fruits and vegetables and the introduction of electrolyzed water as an alternative to sodium hypochlorite solution. Afr. J. Food Sci. 4 (13), 778–789.

Issa-Zacharia, A., Kamitani, Y., Miwa, N., Muhimbula, H., Iwasaki, K., 2011. Application of slightly acidic electrolysed water as a potential non-thermal food sanitizer for decontamination of fresh ready-to-eat vegetables and sprouts. Food Control 22 (3), 601–607.

Izumi, H., 1999. Electrolyzed water as a disinfectant for fresh-cut vegetables. J. Food Sci. 64, 536–539.

Kayes, M.M., Critzer, F.J., Kelly-Wintenberg, K., Roth, J.R., Montie, T.C., Golden, D.A., 2007. Inactivation of foodborne pathogens using a one atmosphere uniform glow discharge plasma. Foodborne Pathog. Dis. 4, 50–59.

Ketteringham, L., Gausseres, R., James, S.J., James, C., 2006. Application of aqueous ozone for treating pre-cut green peppers (*Capsicum annuum* L.). J. Food Eng. 76 (1), 104–111.

Khadre, M.A., Yousef, A.E., Kim, J.G., 2001. Microbiological aspects of ozone applications in food: a review. J. Food Sci. 66 (9), 1242–1252.

Kim, J.G., Yousef, A.E., Chism, G.W., 1999. Use of ozone to inactivate microorganisms on lettuce. J. Food Saf. 19, 17–34.

Kim, C., Hung, Y.C., Brackett, R.E., Lin, C.S., 2003a. Efficacy of electrolyzed oxidizing water in inactivating *Salmonella* on alfalfa seeds and sprouts. J. Food Prot. 66, 208–214.

Kim, J.G., Yousef, A.E., Khadre, M.A., 2003b. Ozone and its current and future application in the food industry. Adv. Food Nutr. Res. 45, 167–218.

Kim, H.J., Feng, H., Kushad, M.M., Fan, X., 2006. Effects of ultrasound, irradiation and acidic electrolyzed water on germination of alfalfa and broccoli seeds and *Escherichia coli* O157:H7. J. Food Sci. 71 (6), m168–m173.

Koseki, S., Isobe, S., 2006. Effect of ozonated water treatment on microbial control and on browning of iceberg lettuce (*Lactuca sativa* L.). J. Food Prot. 69 (1), 154–160.

Koseki, S., Yoshida, K., Isobe, S., Itoh, K., 2001. Decontamination of lettuce using acidic electrolyzed water. J. Food Prot. 64 (5), 652–658.

Koseki, S., Fujiwara, K., Itoh, K., 2002. Decontaminative effect of frozen acidic electrolyzed water on lettuce. J. Food Prot. 65, 411–414.

Koseki, S., Yoshida, K., Isobe, S., Itoh, K., 2004a. Efficacy of acidic electrolyzed water for microbial decontamination of cucumbers and strawberries. J. Food Prot. 67, 1247–1251.

Koseki, S., Yoshida, K., Kamitani, Y., Isobe, S., Itoh, K., 2004b. Effect of mild heat pre-treatment with alkaline electrolyzed water on the efficacy of acidic electrolyzed water against *Escherichia coli* O157:H7 and *Salmonella* on lettuce. Food Microbiol. 21, 559–566.

Koutchma, T., 2008. UV light for processing foods. Ozone: Sci. Eng. 30 (1), 93–98.

Long, W., Pao, S., Inserra, P., Westbrook, E., Ahn, S., 2011. Efficacy of ozone produce washers in reducing natural and artificially inoculated microorganisms on roma tomatoes and green onions. J. Food Saf. 31 (2), 268–275.

Luksiene, Z., Paskeviciute, E., 2011. Novel approach to the microbial decontamination of strawberries: chlorophyllin-based photosensitization. J. Appl. Microbiol. 110, 1274–1283.

Manganaris, G.A., Vasilakakis, M., Diamantidis, G., Mignani, I., 2006. Effect of in-season calcium applications on cell wall physicochemical properties of nectarine fruit (*Prunus persica* var. nectarina Ait. Maxim) after harvest or cold storage. J. Sci. Food Agric. 86 (15), 2597–2602.

Martin-Diana, A.B., Rico, D., Barry-Ryan, C., Frias, J.M., Mulcahy, J., Henehan, G.T.M., 2005. Comparison of calcium lactate with chlorine as a washing treatment for fresh-cut lettuce and carrots: quality and nutritional parameters. J. Sci. Food Agric. 85 (13), 2260–2268.

Martínez-Sánchez, A., Allende, A., Bennett, R.N., Ferreres, F., Gil, M.I., 2006. Microbial, nutritional and sensory quality of rocket leaves as effected by different sanitizers. Postharvest Biol. Technol. 42, 86–97.

Miller, F.A., Silva, C.L.M., Brandão, T.R.S., 2013. A review on ozone-based treatments for fruit and vegetables preservation. Food Eng. Rev. 5 (2), 77–106.

Misra, N.N., Tiwari, B.K., Raghavarao, K.S.M.S., Cullen, P.J., 2011. Nonthermal plasma inactivation of food-borne pathogens. Food Eng. Rev. 3 (3–4), 159–170.

Misra, N.N., Patil, S., Moiseev, T., Bourke, P., Mosnier, J.P., Keener, K.M., et al., 2014. In-package atmospheric pressure cold plasma treatment of strawberries. J. Food Eng. 125 (1), 131–138.

Moy, J.H., 2005. Tropical fruit irradiation—from research to commercial application. In: International Symposium "New Frontier of Irradiated food and Non-Food Products" (September), Bangkok, Thailand, pp. 22–23.

Niemira, B.A., Sites, J., 2008. Cold plasma inactivates *Salmonella* Stanley and *Escherichia coli* O157:H7 inoculated on golden delicious apples. J. Food Prot. 71, 1357–1365.

Nou, X., Lou, Y., 2010. Whole-leaf wash improves chlorine efficacy for microbial reduction and prevents pathogen cross-contamination during fresh-cut lettuce processing. J. Food Sci. 75, 283–290.

Okull, D.O., Laborde, L.F., 2004. Activity of electrolyzed oxidizing water against penicilium expansum in suspension and on wounded apples. J. Food Sci. 69 (1), 23–27.

Olaimat, A.N., Holley, R.A., 2012. Factors influencing the microbial safety of fresh produce: a review. Food Microbiol. 32, 1–19.

Ölmez, H., Akbaş, M.Y., 2009. Optimization of ozone treatment of fresh-cut green leaf lettuce. J. Food Eng. 90 (4), 487–494.

Ölmez, H., Kretzschmar, U., 2009. Potential alternative disinfection methods for organic fresh-cut industry for minimizing water consumption and environmental impact. LWT-Food Sci. Technol. 42, 686–693.

Oms-Oliu, G., Aguilo-Aguayó, I., Martín-Belloso, O., Soliva-Fortuny, R., 2010. Effects of pulsed light treatments on quality and antioxidant properties of fresh-cut mushrooms (Agaricus bisporus). Postharvest. Biol. Technol. 56, 216–222.

Paola, C.L., Rocio, C.V., Marcela, M., Milciades, D., Karina, C.A., 2005. Effectiveness of electrolyzed oxidizing water for inactivating *Listeria monocytogenes* in lettuce. Univ. Sci. 10 (1), 97–108.

Parish, M.E., Beuchat, L.R., Suslow, T.V., Harris, L.J., Garrett, E.H., Farber, J.N., et al., 2003. Methods to reduce/eliminate pathogens from fresh and fresh-cut produce. Compr. Rev. Food Sci. Food Saf. 2, 161–173.

Park, C.M., Hung, Y.C., Doyle, M.P., Ezeike, G.O.I., Kim, C., 2001. Pathogen reduction and quality of lettuce treated with electrolyzed oxidizing and acidified chlorinated water. J. Food Sci. 66, 1368–1372.

Paskeviciute, E., Luksiene, Z., 2009. Photosensitization as a novel approach to decontaminate fruits and vegetables. Časopis za procesnu tehniku i energetiku u poljoprivredi/ PTEP 13 (1), 50–53.

Qiang, Z.M., Demirkol, O., Ercal, N., Adams, C., 2005. Impact of food disinfection on beneficial biothiol contents in vegetables. J. Agric. Food Chem. 53 (25), 9830–9840.

Rico, D., Martin-Diana, A.B., Barat, J.M., Barry-Ryan, C., 2007. Extending and measuring the quality of fresh-cut fruit and vegetables: a review. Trends Food Sci. Technol. 18 (7), 373–386.

Rodgers, S.L., Cash, J.N., Siddiq, M., Ryser, E.T., 2004. A comparison of different chemical sanitizers for inactivating *Escherichia coli* O157: H7 and *Listeria monocytogenes* in solution and on apples, lettuce, strawberries, and cantaloupe. J. Food Prot. 67 (4), 721–731.

Ruiz-Cruz, S., Acedo-Felix, E., Diaz-cinco, M., Islas-Osuna, M.A., Gonzalez Aguilar, G. A., 2007. Efficacy of sanitizers in reducing *Escherichia coli O157:H7, Salmonella* spp. and *Listeria monocytogenes* populations on fresh-cut carrots. Food Control 18, 1383–1390.

Saftner, R.A., Bai, J.H., Abbott, J.A., Lee, Y.S., 2003. Sanitary dips with calcium propionate, calcium chloride, or a calcium amino acid chelate maintain quality and shelf stability of fresh-cut honeydew chunks. Postharvest Biol. Technol. 29 (3), 257–269.

Sagong, H.G., Lee, S.-Y., Chang, P.S., Heu, S., Ryu, S., Choi, Y.J., et al., 2011. Combined effect of ultrasound and organic acids to reduce *Escherichia coli* O157:H7, *Salmonella typhimurium*, and *Listeria monocytogenes* on organic fresh lettuce. Int. J. Food Microbiol. 145 (1), 287–292.

Sagong, H.G., Cheon, H.L., Kim, S.O., Lee, S.Y., Park, K.H., Chung, M.S., et al., 2013. Combined effects of ultrasound and surfactants to reduce *Bacillus cereus* spores on lettuce and carrots. Int. J. Food Microbiol. 160 (3), 367–372.

Sampedro, F., McAloon, A., Yee, W., Fan, X., Geveke, D., 2014. Cost analysis and environmental impact of pulsed electric fields and high pressure processing in comparison with thermal pasteurization. Food Bioprocess Technol. 7 (7), 1928–1937.

São José, J.F.B., Vanetti, M.C.D., 2012. Effect of ultrasound and commercial sanitizers in removing natural contaminants and *Salmonella enterica* Typhimurium on cherry tomatoes. Food Control 24 (1–2), 95–99.

Sapers, G.M., 2001. Efficacy of washing and sanitizing methods for disinfection of fresh fruit and vegetable products. Food Technol. Biotechnol. 39 (4), 305–311.

Sapers, G.M., Jones, D.M., 2006. Improved sanitizing treatments for fresh tomatoes. J. Food Sci. 71 (7), 252–256.

Selma, M.V., Beltran, D., Allende, A., Chacon-Vera, E., Gil, M.I., 2007. Elimination by ozone of *Shigella sonnei* in shredded lettuce and water. Food Microbiol. 24 (5), 492–499.

Selma, M.V., Allende, A., López-Gálvez, F., Conesa, M.A., Gil, M.I., 2008. Disinfection potential of ozone, ultraviolet-C and their combination in wash water for the fresh-cut vegetable industry. Food Microbiol. 25, 809–814.

Seymour, I.J., Burfoot, D., Smith, R.L., Cox, L.A., Lockwood, A., 2002. Ultrasound decontamination of minimally processed fruits and vegetables. Int. J. Food Sci. Technol. 37, 547–557.

Sharma, R.R., Demirci, A., 2003. Treatment of *Escherichia coli* O157:H7 inoculated alfalfa seeds and sprouts with electrolysed oxidizing water. Int. J. Food Microbiol. 86, 231–237.

Singer, P.C., 1994. Control of disinfection by-products in drinking water. J. Environ. Eng.—ASCE (American Society of Civil Engineers) 120 (4), 727–744.

Singh, N., Singh, R.K., Bhunia, A.K., Stroshine, R.L., 2002. Effect of inoculation and washing methods on the efficacy of different sanitizers against *Escherichia coli* O147: H7 on lettuce. Food Microbiol. 19, 183–193.

Smilanick, J.L., Crisosto, C., Mlikota, F., 1999. Postharvest use of ozone on fresh fruit. Perishables Handling Q. 99, 10–14.

Suzuki, K., Nakamura, T., Doi, T., Kokubo, S., Tomita, M., 2005. The disinfectant effect of slightly acidic electrolyzed water prepared with hydrochloric acid as a raw material for lettuce. J. Antibacterial Antifungal Agents 33, 589–597 (in Japanese).

Sy, K.V., McWatters, K.H., Beuchat, L.R., 2005. Efficacy of gaseous chlorine dioxide as a sanitizer for killing *Salmonella*, yeasts, and molds on blueberries, strawberries, and raspberries. J. Food Prot. 68 (6), 1165–1175.

Takeshita, K., Shibatoa, J., Sameshimaa, T., Fukunagab, S., Isobec, S., Ariharad, K., et al., 2003. Damage of yeast cells induced by pulsed light irradiation. Int. J. Food Microbiol. 85 (1–2), 151–158.

Vandekinderen, I., Van Camp, J., De Meulenaer, B., Veramme, K., Denon, Q., Ragaert, P., et al., 2007. The effect of the decontamination process on the microbial and nutritional quality of fresh-cut vegetables. Acta Hortic. 746, 173–179.

Villagomez, N.S., Herrera, E.A.N., Orozco, L.R., Wild-Padua, G., Iturriaga, M.H., 2010. Effect of storage time on the efficacy of chemical and irradiation treatments to inactivate *Salmonella* on cilantro (*Coriandrum sativum* L.). J. Food Saf. 30, 584–598.

Vurma, M., Pandit, R.B., Sastry, S.K., Yousef, A.E., 2009. Inactivation of *Escherichia coli* O157:H7 and natural microbiota on spinach leaves using gaseous ozone during vacuum cooling and simulated transportation. J. Food Prot. 72 (7), 1538–1546.

Wang, H., Feng, H., Luo, Y.G., 2004. Microbial reduction and storage quality of fresh-cut cilantro washed with acidic electrolyzed water and aqueous ozone. Food Res. Int. 37 (10), 949–956.

White, G.C., 1992. Ozone, Handbook of Chlorination and Alternative Disinfectants, third ed. Van Nostrand Reinhold, New York, NY, pp. 1046–1110.

Wu, V.C.H., Kim, B., 2007. Effect of a simple chlorine dioxide method for controlling five foodborne pathogens, yeasts and molds on blueberries. Food Microbiol. 24 (7–8), 794–800.

Yang, H., Swem, B.L., Li, Y., 2003. The effect of pH on inactivation of pathogenic bacteria on fresh-cut lettuce by dipping treatment with electrolyzed water. J. Food Sci. 68, 1013–1017.

Yuk, H.G., Yoo, M.Y., Yoon, J.W., Moon, K.D., Marshall, D.L., Oh, D.H., 2006. Effect of combined ozone and organic acid treatment for control of *Escherichia coli* O157: H7 and *Listeria monocytogenes* on lettuce. J. Food Sci. 71 (3), M83–M87.

Zhou, B., Feng, H., Pearlstein, A.J., 2012. Continuous-flow ultrasonic washing system for fresh produce surface decontamination. Innovative Food Sci. Emerg. Technol. 16, 427–435.

Ziuzina, D., Patil, S., Cullen, P.J., Keener, K.M., Bourke, P., 2014. Atmospheric cold plasma inactivation of *Escherichia coli*, *Salmonella enterica* serovar Typhimurium and *Listeria monocytogenes* inoculated on fresh produce. Food Microbiol. 42, 109–116.

CHAPTER

Microbiological risk assessments in food industry

19

Jeanne-Marie Membré

Institut National de la Recherche Agronomique, UMR1014 Sécurité des Aliments et Microbiologie, Nantes, France L'Université Nantes Angers Le Mans, Oniris, Nantes, France

INTRODUCTION

Microbiological risk assessment (MRA) is a structured process for determining the risk associated with biological hazards in a food. In the industry (defined here as a general term including primary and secondary manufacturers but also distributors), MRA approaches could be used to a broader sense, covering the food safety issue, and also the food quality issue (as quality could be altered by microbial spoilage). Covering spoilage beside safety is acknowledged by the food authority, at least in Europe: the Food Law (European Commission [EC], 2002) specifies that a food product is unacceptable for human consumption due to contamination (safety), or to putrefaction and deterioration (spoilage).

For this reason, here, the four steps of MRA are presented as they were introduced by Pujol et al. (2013): (1) hazard identification consists of the identification of the microbiological agent capable of causing an adverse health effect (for bacterial pathogens) or an inedible product (for spoilage bacteria) along the process; (2) hazard characterization is the evaluation of the nature of the adverse effect of the pathogens (type of illness) and the spoilage agents (quality default); (3) exposure assessment consists of quantifying the different pathways for contamination through the process from raw materials up to consumption; and (4) risk characterization consists of quantifying the "risk" in terms of public health (pathogen) and in terms of quality default (spoilage).

In addition, it is important to realize that industry might be more interested in estimating the level of microorganisms (concentration or prevalence) at the food product release from the manufacturer, or at the consumer's plate, rather than in estimating the number of illnesses. In that case, strictly speaking it is no longer an MRA which is performed but an MEA (microbiological exposure assessment).

P. Kotzekidou (Ed): Food Hygiene and Toxicology in Ready-to-Eat Foods.
DOI: http://dx.doi.org/10.1016/B978-0-12-801916-0.00019-4

MRA can be done in a qualitative or quantitative way. In qualitative MRA, the risk is expressed in a categorical way: low, medium, and high risk or as a binary outcome: contaminated or noncontaminated. When performing quantitative MRA (QMRA), mathematical models are used to describe the introduction of microorganisms (through raw materials or post-process recontamination) into food, the replication of microorganisms over time, the destruction of microbes during the process (eg, due to heat treatment) or during the storage (eg, due to nonthermal lethal conditions such as acidic conditions), the consumption pattern (portion size, frequency), and subsequent illness. Also, in QMRA and quantitative MEA (QMEA), both variability and uncertainty inherent to biological data are described by probability distributions. One of the main benefits of QMRA/QMEA is that it provides the flexibility of quantitatively comparing the efficiency of different risk-reduction measures which could be implemented at the operational level, by predicting their effect on the model output. Likewise, QMRA and QMEA enable to move away from defining operational process settings, food formulation, or, shelf-life based upon the worst-case scenario, as still often recommended in "traditional" food safety approaches. Indeed, for example, still recently to determine the shelf-life of chilled ready-to-eat foods, the French Food Safety Agency, commissioned by the European authority, has recommended to perform challenge-test studies with *Listeria monocytogenes* at 12°C (AFSSA, Agence Française de Sécurité Sanitaire des Aliments, 2014) considering that there are some domestic refrigerators set to this excessively high-temperature value. That is definitely a worst-case scenario which could lead directly to set unnecessarily short food shelf-life.

In addition, along with the adoption of MRA principles across the globe, since the 2000s, food safety management based on the risk has emerged (namely "risk-based food safety management"). At the operational level, this has huge consequences as this means that a hazard is defined in terms of the risk of an adverse human health consequence, rather than in terms of contamination of the food. Because of this shift towards realistic risk-based food safety management, the industry is really interested in understanding what MRA in general and MEA in particular (plus related mathematical and probabilistic concepts) are, to be then in a position of transferring risk-based food safety management concepts into practical guidelines for operational use.

This chapter will introduce risk-based food safety management before describing the mathematical and probabilistic tools developed in MEA to quantify the microbial propagation during food transformation/distribution ("farm-to-fork" continuum). The presentation of these concepts and tools is focused on industrial applications. Likewise, examples chosen to illustrate the purpose are related to industrial case studies and when possible to ready-to-eat products. The chapter concludes on current and future use of MRA in industry.

INTRODUCTION TO RISK-BASED FOOD SAFETY MANAGEMENT

The World Health Organization and the Food and Agriculture Organization, particularly through the Codex Alimentarius Commission (CAC), have encouraged the research community (academy, regulatory agencies) to perform MRA. In 1999, they published a guideline to facilitate the use of MRA (Codex Alimentarius Commission, 1999). In 2002, they set down the basis of risk-based food safety management (FAO/WHO, 2002).

The appropriate level of protection (ALOP) is a statement of the degree of public health protection that is to be achieved by the food safety systems implemented within a country (World Trade Organization, 1995). For example, an ALOP may be stated for a particular country as incidence of salmonellosis attributable to poultry to be less than 10 cases per 100,000 population. The current, and widely accepted, hypothesis is that the current number of illnesses per million inhabitants (estimated per QMRA for a given food, microorganism, and country) is deemed acceptable and could be then assimilated to an ALOP (EFSA, 2007; Gkogka et al., 2013a).

An ALOP cannot be directly translated by the industry into a practical measure. Instead, a measurable target for producers, manufacturers, and control authorities is required; this is the basis of the food safety objective (FSO) concept. An FSO corresponds to the maximum frequency and/or concentration of a hazard in a food at the time of consumption that provides or contributes to the ALOP.

Consequently and per definition, there is a quantitative link between FSO and ALOP; however there are not yet many guidelines to derive an FSO from an ALOP. Nonetheless, in the literature, there are some academic studies trying to bridge the gap between FSO and ALOP: *L. monocytogenes* in deli meat (Gkogka et al., 2013b), *Salmonella enteritidis* in egg (Stringer, 2004), *L. monocytogenes*, *Salmonella*, and *Escherichia coli* O157 in ready-to-eat food (Perni et al., 2009), etc.

Within the risk-based food safety management metrics, the performance objective (PO) is directly related to the FSO but is set at a step before the time of consumption (eg, at the point of product release from manufacture). A PO corresponds to a maximum frequency and/or concentration of a hazard in a food at a specified step in the food chain before the time of consumption that provides or contributes to an FSO or ALOP, as applicable.

To be operational, FSO and PO must be translated into criteria that can be controlled and measured in the food supply chain; this is the reason why performance criteria (PC), process criteria (PrC), and product criteria (PdC) were introduced.

The PC is the effect required of one or more control measure(s) working in concert to meet a PO; for example, a minimum log reduction required for thermal inactivation. PrC and PdC are control measures associated with the process and the formulation, respectively. For instance, time and temperature settings associated with a thermal process are PrCs, the pH of a product is a PdC (Membré, 2012).

As with FSO and ALOP, there are quantitative links between PrC, PdC, and PO/FSO. However, these links are not always straightforward, they could involve rather complex mathematical developments, to establish how a food process and/or a food formulation has an effect on the level of contamination at the moment of food consumption. The microbial propagation during food transformation/distribution ("farm-to-fork" continuum) has to be quantified accurately. That is particularly important for industrial applications as the control measures (PrC and PdC) are implemented at the operational industrial level. One may emphasize here that in the case of MRA performed for public health purposes, the detailed description of the microbial propagation during the food process steps is not always done. For instance, the FAO and the WHO have commissioned a study aiming to characterize the risk of *L. monocytogenes* in ready-to-eat foods started at retail (FAO/WHO, 2004) and then ignoring the prior food transformation steps.

Risk-based food safety management opens definitively new opportunities in terms of product innovation. Indeed, well-established thermal process settings may be revisited as a consequence of the food safety management shift towards a risk-based approach (Bean et al., 2012). That could lead to food products designed with milder heat treatment and then with an increased organoleptic quality. For instance, recently, the possibility of applying mild heat treatment in combination with curing salts to prevent any risk of *Clostridium botulinum* in shelf-stable canned meat has been discussed (Anderson et al., 2011). Likewise, a risk-based food safety framework has been set to investigate the heat treatment required to get rid of *Salmonella* in poultry products (Membré et al., 2007). The authors have considered the initial contamination in raw material, the microbial heat resistance (based upon the analysis of inactivation kinetics at different temperatures, and then calculations of time to obtain one log reduction [*D*-value], and temperature to obtain a 10-fold decrease of *D*-values [*z*-values]). This case study illustrated that the heat treatment could move from 2 min at 70°C (a widely accepted standard among industries and regulatory for food pasteurization, known as the "*Listeria* cook") down to 0.5 min at 70°C. Obviously this analysis has required specific data (raw material contamination level, thermal inactivation of *Salmonella* in poultry). Also, to be successfully implemented in factories, reducing the heat treatment required to have a reliable food safety management system in place, for example, the time and temperature have to be accurately monitored and recorded.

MATHEMATICAL AND PROBABILISTIC TOOLS DEVELOPED IN MEA

To quantify accurately the effect of the various process steps on the exposure assessment, Nauta (2001) introduced the "Modular Process Risk Model" concept where the food transformation/distribution operations are divided into smaller steps in which the contamination pathway is described. The contamination pathway includes at least (1) the introduction of microorganisms with the raw materials, (2) the inactivation (for instance by heat treatment), (3) the potential post-process recontamination of the product, and (4) the growth of microorganisms during intermediate or final storage. With ready-to-eat products, the recontamination cannot be neglected; for instance with cooked boneless ham, the slicing operation, after cooking, is a key step to control in order to avoid *L. monocytogenes* contaminating the final product (Zwietering et al., 2016).

The "Modular Process Risk Model" concept has already been used in MEA, for example by Anderson et al. (2011), Nauta et al. (2009), Ross et al. (2009), and Pujol et al. (2015a). The mathematical equations associated with each pathway are succinctly described below. For more details on modeling microbial responses, see reviews and books on predictive microbiology, for instance McKellar and Lu (2004).

MODELING INITIAL LEVEL

The first step in a QMEA study is to estimate the initial contamination level (prevalence and concentration) in the raw materials. Ideally, quantitative data from a large number of studies on the raw material could be used to assess the initial prevalence and concentration. An estimate of the batch prevalence (percentage of batches contaminated) could be derived directly from the data, and then introduced in a QMEA model through a Bernoulli distribution. When uncertainty is taken into account, the prevalence estimate is generally described by a beta distribution. An estimate of the microbial concentration in contaminated batches could be more difficult to obtain, depending on the homogeneity of the contamination. In cases of high microbial concentration, the microbial count, expressed in logarithm, is often described by a normal of a log normal distribution. In the case of low and inhomogeneous counts, negative binomial or beta Poisson distributions could be preferred (Jongenburger et al., 2012).

MODELING MICROBIAL INACTIVATION

When a heat treatment step is involved, classically, the resulting microbial inactivation is described using a first-order kinetic model (Bigelow, 1921). However, the thermal inactivation has to be interpreted as a probabilistic phenomenon (Nauta, 2001).

$$N = N_0 \times 10^{-t_{HT}/D} \tag{19.1}$$

where:

- N_0 is the bacterial concentration entering in the thermal treatment (CFU/mL or CFU/g)
- N is the microbial concentration after the heat treatment (CFU/mL or CFU/g)

- t_{HT} is the time of the heat treatment (min)
- D is the decimal reduction time (min).

The probability that one bacterium survives the heat treatment could be derived from Eq. 19.1 as follows:

$$pr_{\text{one bacterium to survive}} = 10^{-t_{HT}/D} \tag{19.2}$$

As, generally, N_0 is large and $pr^k_{\text{one bacterium to survive}}$ small, the microbial concentration after the heat treatment, N, is described by a Poisson distribution (Eq. 19.3).

$$N \sim \text{Poisson}(N_0 \times pr_{\text{one bacterium to survive}}) \tag{19.3}$$

Next, the decimal reduction time (D) is described by a secondary model which takes into account classically the heat treatment temperature but also sometimes the product characteristics (pH and water activity [a_w]). The gamma-type inactivation model (Gaillard et al., 1998; Leguérinel et al., 2000; Mafart et al., 2010) is relatively often used for industrial applications (particularly with low-acidic products where successful validations in food products have been carried out).

MICROBIAL GROWTH

One of the most used primary models in MEA to describe the growth during intermediate or final storage is the three-phase linear model (Buchanan et al., 1997):

$$\begin{aligned} \ln N_{t_{st}} &= \ln N_0 && \text{for} \quad t_{st} \le t_{lag} \\ \ln N_{t_{st}} &= \ln N_0 + \mu \times (t_{st} - t_{lag}) && \text{for} \quad t_{lag} < t_{st} < t_{max} \\ \ln N_{t_{st}} &= \ln N_{max} && \text{for} \quad t_{st} \ge t_{max} \end{aligned} \tag{19.4}$$

where:

- The natural logarithm is used (not the decimal one)
- $N_{t_{st}}$ is the population concentration at time t (CFU/mL or CFU/g)
- N_0 is the initial population concentration (CFU/mL or CFU/g)
- N_{max} is the log of the maximum population concentration supported by the environment (CFU/mL or CFU/g)
- t_{st} is the storage time (h)
- t_{lag} is the lag phase time (h)
- t_{max} is the time when the maximum population density is reached (h)
- μ is the specific growth rate (h^{-1})

Next, to describe the effect of environmental factors (such as the storage temperature, T_{st}) and formulation factor (such as pH) on the specific growth rate, the gamma model initially developed by Zwietering et al. (1992) is generally used:

$$\mu = \mu_{opt} \times \gamma(T_{st}) \times \gamma(\text{pH}) \tag{19.5}$$

where:

- μ_{opt} is the optimal specific growth rate (h^{-1}),
- $\gamma(T_{st})$ is the effect of the temperature,
- $\gamma(pH)$ is the effect of the pH,
- Both $\gamma(T_{st})$ and $\gamma(pH)$ vary between 0 and 1, being at 1 in optimal conditions.

The gamma terms are generally described by cardinal models (Rosso et al., 1995).

In Eq. 19.5, the storage temperature, T_{st}, is variable as the storage takes place at various retail and consumer places. In the case of ready-to-eat products which have to be kept in chilled conditions, the temperature could vary roughly from 2°C to 10°C or even 12°C. It is then much more appropriate to introduce the input T_{st} in the model using a probabilistic distribution. Often, temperature is described with a pert or a normal distribution.

Preferably, the pH of the product is considered as a fixed value since it corresponds to a control measure at the operational level, and more precisely to a PdC in the risk-based food safety management. The same is true for a_w.

MODELING RECONTAMINATION

Recontamination is defined as the introduction of any microorganism into the product after an inactivation step. A survey performed by the World Health Organization and Food Agricultural Organization (1995) in Europe indicated that 25% of the foodborne outbreaks could be due to recontamination. Modeling recontamination is relatively difficult.

Reij et al. (2004) and then Den Aantrekker et al. (2003) have given an overview of available models to describe recontamination. The main important point to keep in mind is that the recontamination cannot be mathematically handled in logarithm terms as the microbial growth.

Several recontamination models applied to ready-to-eat products have been presented in the literature, for example with cheese (Aziza et al., 2006) or lettuce products (Pérez Rodriguez et al., 2011). In the case of air recontamination, the concentration of microorganisms in the plant environment, C_{plant}, cannot be considered as a single, fixed value, it has to be introduced in the model as probabilistic input. To do so, data have to be collected and then analyzed. Unless a more comprehensive model linking C_{plant} with microbial concentration outside the factory, efficiency of filters, frequency of their replacement, positive pressure applied, etc. is developed. In this latter case, C_{plant} is not a model input anymore, but a latent variable, the model inputs have to be informed using data or expert knowledge. The microbial concentration outside the factory is likely to be better described by a range of values (with their probability of occurrence), it could depend on the season and the geography. On the opposite, the positive pressure applied in the factory has to be introduced into the model as a fixed value since it corresponds to a parameter (ie, a control measure) that the business operator can set.

Beside these four key microbial pathways presented here, when developing a QMEA model, mixing and partitioning steps have also to be taken into account. These are rather straightforward to quantify even if they are probabilistic per nature (Nauta, 2005).

PROBABILISTIC TECHNIQUES

Uncertainty and variability are inherent to biological process. Uncertainty arises from our lack of knowledge and may be related to the model used to characterize the risk, the parameters used to provide values for the model, or both. In some cases, we can reduce uncertainty by obtaining better information (more data, better expert elicitation), but this may not always be possible. Variability refers to real and identifiable differences between individuals within a population addressed by risk assessment. For example, variability might refer to differences between microbial species or strains, batches of manufactured products. Variability does not disappear with more data collection.

Including explicitly uncertainty and variability in QMRA/QMEA requires to perform probabilistic risk assessment (Haas et al., 2014). Probabilistic techniques enable an incorporation of inputs as probability distributions, that is, as distributions of values with their associated probability of occurrence. Uncertainty and variability are propagated into the model up to the risk estimated, using numerical method such as Monte Carlo simulation. In the last 20 years, the step-change increase in computer technology has strongly enabled the development of simulation techniques. Most of the time, Monte Carlo simulation proceeds by Latin hypercube sampling, which ensures that values from the entire range of the distribution are sampled proportionally to the probability density function, even if the number of simulations is relatively low. In principle, each simulation represents one potential true event of transmission along the food pathway.

Nowadays, most of MRAs are probabilistic QMRAs (Costa Sobrinho et al., 2014; Franz et al., 2010; Gkogka et al., 2013a; Mataragas et al., 2010; Membré et al., 2015; Romero-Barrios et al., 2013), likewise exposure assessments addressing food spoilage or food safety issues are probabilistic exposure assessments (Daelman et al., 2013; Pujol et al., 2015a; Rigaux et al., 2014; Stals et al., 2015; Tenenhaus-Aziza et al., 2014).

Nevertheless, performing probabilistic QMRA of ready-to-eat foods is difficult. Indeed, RTE foods are a broad and diverse food category, prepared and stored in different ways and under different conditions. To illustrate this diversity, let's take the study performed by the FAO/WHO (2004). The authors have evaluated the risk associated with *L. monocytogenes* in four products: pasteurized milk, ice cream, fermented meat, and cold smoked fish. The reason for their choice has been summarized as follows. Pasteurized milk is a food that is widely consumed, has very low frequencies and levels of contamination with *L. monocytogenes* but allows growth of the organism during storage. Ice cream is similar to milk but does not permit growth of *L. monocytogenes* during storage. Fermented meat products are often contaminated with *Listeria* and are produced without any lethal processing step, but their final composition prevents growth of the microbe during storage. Cold smoked fish is frequently contaminated with *L. monocytogenes*, has no lethal processing step and permits growth during an extended storage period. In these four examples, to build valuable probabilistic QMRAs, good-quality data are necessary, that could take a lot of time and effort. This has to be balanced against the real benefit of performing probabilistic QMRA. (Reducing the cost? Better product quality? Higher sustainability of the supply chain?) To illustrate furthermore the complexity/difficulty of assessing the risk or exposure level associated with ready-to-eat products, let's take the example of the probabilistic exposure assessment model developed by Daelman et al. (2013) for refrigerated and processed foods of extended durability (REPFEDs). The authors have considered the entire production and distribution process, from raw materials over pasteurization and up to the moment when the food product is consumed or discarded. The product life was divided into nine modules, each module corresponding to a basic process: (1) raw material contamination; (2) cross-contamination during handling; (3) inactivation during preparation; (4) growth during intermediate storage; (5) partitioning of batches in portions; (6) mixing portions to create the product; (7) recontamination during assembly and packaging; (8) inactivation during pasteurization; and (9) growth during shelf-life. To inform each module, a combination of newly gathered and literature data, predictive models, and expert opinions was necessary. Once built, the added value of the model was that it enabled to assess quantitatively the impact of the different processing steps and consumer behavior on the exposure to *Bacillus cereus* from REPFEDs, and then led to recommendations to food manufacturers involved in the study (Daelman, 2013).

Moreover, performing a probabilistic QMRA could be even more data-demanding if the inputs contain both uncertainty and variability. Separating uncertainty and variability has been recommended for more than a decade (Nauta, 2001). Once separated, the model is run using a second-order Monte Carlo technique, leading to building a confidence interval around a risk estimate for different realizations of variability; that enables to identify whether further investigation on uncertainty is required before making any decision (Mokhtari and Frey, 2005). There are few examples of second-order Monte Carlo techniques performed in risk or exposure assessment associated with food and spoilage issues (Cummins et al., 2008; Pouillot and Delignette-Muller, 2010; Pujol et al., 2015b). However, it is important to keep in mind that deploying second-order Monte Carlo techniques is not straightforward, it is data-demanding and can also be time-consuming.

Despite the scientific advantage of using probabilistic QMRA, it is reasonable to acknowledge that there are only few applications in industry. Several factors may explain this limited use (Membré, 2012). Among them, a crucial one is the lack of guidance to make decisions using model outputs: the risk estimates are in the format of probability distributions, while the food safety decision is a binary process (acceptable or not) (Zwietering, 2015). The second important limitation is related to the human resource demand, probabilistic techniques require definitive modeling expertise, which is generally not available in R&D departments of small and medium enterprises.

CONCLUSIONS

In industry, food safety management relies traditionally on prerequisite programs (eg, good manufacture practice, good hygiene practice) and HACCP plan in place. The use of MRA is not mandatory, even if many (large) food industry companies are using MRA to tackle risk or exposure issues due to pathogenic bacteria and probably to spoilage microorganisms, although they are less keen to communicate on this latter point, for competitor advantage reasons. MRA is acknowledged at the international level (WHO and FAO, Codex Alimentarius, EFSA, FDA, etc.) and recognized to be a transparent, documented, and scientific process. In industry MRA is applied proactively in product innovation, as an aid in process and formulation design, and/or shelf-life determination. MRA is also applied retroactively in cases of incidents, to estimate the impact on consumer safety or product quality when problems occur with products on the market.

Mathematical and probabilistic tools are now sufficiently mature to make the use of QMRA and QMEA in industry valuable. In probabilistic approaches, uncertainties generated by the lack of data/knowledge are integrated and not neglected, likewise, variability is taken into account, and consequently the output is realistic. That enables to move from worst-case scenario such as those often generated with "traditional" food safety approach (eg, when performing a challenge test at 12°C to determine the shelf-life of chilled products). The probabilistic approach is particularly valuable for refrigerated ready-to-eat food products where the temperatures at retail and consumer places have definitely to be considered as a source of variability. The authority encourages the use of QMRA for this type of product. For example, in their assessment of *L. monocytogenes* risk related to ready-to-eat foods, EFSA (2007) recommends to fix the shelf-life in relation to realistic storage temperatures and potential growth of *L. monocytogenes*, taking into account the variability of temperatures during retail and domestic storage of the foods. Likewise, the FAO and the WHO have commissioned a study aiming to characterize the risk of *L. monocytogenes* in ready-to-eat foods. The models described the growth or decline of *L. monocytogenes* between the time of purchase and consumption, using distribution of retail and home storage times and temperatures (FAO/WHO, 2004).

Overall, QMRA and QMEA enable to quantify the level of risk and exposure, respectively; they help in identifying and evaluating quantitatively (in absolute or relative terms) the efficiency of various management options to control this level of risk/exposure. That is a real asset for industry food managers as it provides tangible inputs for a cost–benefit analysis However, reanalyzing a food supply chain, gathering the data and information, fitting the model, and interpreting the outputs takes time and expertise, and it is likely that this resource effort will be made only where the expected benefit is high (eg, when they increase significantly the product quality, significant reduction of cost, a definite shift toward sustainable food manufacturing). Nevertheless, with the global adoption of risk-based food safety management, it could be predicted that MRA will be more and more used in industry.

REFERENCES

AFSSA, Agence Française de Sécurité Sanitaire des Aliments. 2014. EURL Lm Technical Guidance Document for conducting shelf-life studies on *Listeria monocytogenes* in ready-to-eat foods (Online). Available from: <http://ec.europa.eu/food/food/biosafety/salmonella/docs/technical_guidance_listeria_en.pdf> (accessed 17.07.15).

Anderson, N.M., Larkin, J.W., Cole, M.B., Skinner, G.E., Whiting, R.C., Gorris, L.G.M., et al., 2011. Food safety objective approach for controlling *Clostridium botulinum* growth and toxin production in commercially sterile foods. J. Food Prot. 74, 1956–1989.

Aziza, F., Mettler, E., Daudin, J.-J., Sanaa, M., 2006. Stochastic, compartmental, and dynamic modeling of cross-contamination during mechanical smearing of cheeses. Risk Anal. 26, 731–745.

Bean, D., Bourdichon, F., Bresnahan, D., Davies, H., Jackson, T., Membré, J.-M., et al., 2012. Risk assessment approaches to setting thermal processes in food manufacture. ILSI (International Life Sciences Institute), Belgium.

Bigelow, W.D., 1921. The logarithmic nature of thermal death time curves. J. Infect. Dis. 29, 528–536.

Buchanan, R.L., Whiting, R.C., Damert, W.C., 1997. When is simple good enough: a comparison of the Gompertz, Baranyi, and three-phase linear models for fitting bacterial growth curves. Food Microbiol. 14, 313–326.

Codex Alimentarius Commission. 1999. Principles and guidelines for the conduct of a microbiological risk assessment. In: Joint FAO/WHO Food Standards Programme (Ed.), CAC/GL-30, Rome.

Costa Sobrinho, P.D.S., Destro, M.T., Franco, B.D.G.M., Landgraf, M., 2014. A quantitative risk assessment model for *Vibrio parahaemolyticus* in raw oysters in Sao Paulo State, Brazil. Int. J. Food Microbiol. 180, 69–77.

Cummins, E., Nally, P., Butler, F., Duffy, G., O'Brien, S., 2008. Development and validation of a probabilistic second-order exposure assessment model for *Escherichia coli O157:H7* contamination of beef trimmings from Irish meat plants. Meat Sci. 79, 139–154.

Daelman, J. 2013. Quantitative microbiological exposure assessment of *Bacillus cereus* in cooked-chilled foods. Thesis submitted in fulfilment of the requirements for the degree of doctor (PhD) in Applied Biological Sciences. Faculty of Bioscience Engineering, Ghent University.

Daelman, J., Membré, J.-M., Jacxsens, L., Vermeulen, A., Devlieghere, F., Uyttendaele, M., 2013. A quantitative microbiological exposure assessment model for *Bacillus cereus* in REPFEDs. Int. J. Food Microbiol. 166, 433–449.

Den Aantrekker, E.D., Boom, R.M., Zwietering, M.H., van Schothorst, M., 2003. Quantifying recontamination through factory environments—a review. Int. J. Food Microbiol. 80, 117–130.

EFSA, 2007. Scientific opinion of the panel on biological hazards on a request from the European Commission on request for updating the former SCVPH opinion on *Listeria monocytogenes* risk related to ready-to-eat foods and scientific advice on different levels of *Listeria monocytogenes* in ready-to-eat foods and the related risk for human illness. EFSA J. 599, 1–42.

European Commission. 2002. Regulation (EC) No 178/2002 of the European Parliament and of the Council of 28 January 2002 laying down the general principles and requirements of food law, establishing the European Food Safety Authority and laying down procedures in matters of food safety. <http://data.europa.eu/eli/reg/2002/178/oj>.

FAO/WHO. 2002. Principles and guidelines for incorporating microbiological risk assessment in the development of food safety standards, guidelines and related texts, Report of a Joint FAO/WHO Consultation, Kiel, Germany.

FAO/WHO. 2004. Risk assessment of *Listeria monocytogenes* in ready-to-eat foods: interpretative summary. In: FAO (Ed.), Microbiological Risk Assessment Series; no. 4. WHO/FAO, Rome, 78 pp.

Franz, E., Tromp, S.O., Rijgersberg, H., van der Fels-Klerx, H.J., 2010. Quantitative microbial risk assessment for *Escherichia* coli O157:H7, *Salmonella*, and *Listeria monocytogenes* in leafy green vegetables consumed at salad bars. J. Food Prot. 73, 274–285.

Gaillard, S., Leguerinel, I., Mafart, P., 1998. Model for combined effects of temperature, pH and water activity on thermal inactivation of *Bacillus cereus* spores. J. Food Sci. 63, 887–889.

Gkogka, E., Reij, M.W., Gorris, L.G.M., Zwietering, M.H., 2013a. Risk assessment strategies as a tool in the application of the Appropriate Level of Protection (ALOP) and Food Safety Objective (FSO) by risk managers. Int. J. Food Microbiol. 167, 8–28.

Gkogka, E., Reij, M.W., Gorris, L.G.M., Zwietering, M.H., 2013b. The application of the Appropriate Level of Protection (ALOP) and Food Safety Objective (FSO) concepts in food safety management, using *Listeria monocytogenes* in deli meats as a case study. Food Contr. 29, 382–393.

Haas, C.N., Rose, J.B., Gerba, C.P., 2014. Quantitative microbial risk assessment, second ed. John Wiley & Sons, Hoboken, NJ.

Jongenburger, I., Bassett, J., Jackson, T., Zwietering, M.H., Jewell, K., 2012. Impact of microbial distributions on food safety I. Factors influencing microbial distributions and modelling aspects. Food. Contr. 26, 601–609.

Leguérinel, I., Couvert, O., Mafart, P., 2000. Relationship between the apparent heat resistance of *Bacillus cereus* spores and the pH and NaCl concentration of the recovery medium. Int. J. Food Microbiol. 55, 223–227.

Mafart, P., Leguérinel, I., Couvert, O., Coroller, L., 2010. Quantification of spore resistance for assessment and optimization of heating processes: a never-ending story. Food Microbiol. 27, 568–572.

Mataragas, M., Zwietering, M.H., Skandamis, P.N., Drosinos, E.H., 2010. Quantitative microbiological risk assessment as a tool to obtain useful information for risk managers—specific application to *Listeria monocytogenes* and ready-to-eat meat products. Int. J. Food Microbiol. 141, S170–S179.

McKellar, R.C., Lu, X., 2004. Modeling Microbial Responses in Food. CRC Press, London.

Membré, J.-M., 2012. Setting of thermal processes in a context of food safety objectives (FSOs) and related concepts. In: Valdramidis, V., Van Impe, J.F.M. (Eds.), Progress on Quantitative Approaches of Thermal Food Processing. Nova Science Publishers, New York, pp. 297–326.

Membré, J.-M., Bassett, J., Gorris, L.G.M., 2007. Applying the food safety objective and related standards to thermal inactivation of *Salmonella* in poultry meat. J. Food Prot. 70, 2036–2044.

Membré, J.-M., Diao, M., Thorin, C., Cordier, G., Zuber, F., André, S., 2015. Risk assessment of proteolytic *Clostridium botulinum* in canned foie gras. Int. J. Food Microbiol. 210, 62–72.

Mokhtari, A., Frey, H.C., 2005. Sensitivity analysis of a two-dimensional probabilistic risk assessment model using analysis of variance. Risk Anal. 25, 1511–1529.

Nauta, M., Hill, A., Rosenquist, H., Brynestad, S., Fetsch, A., van der Logt, P., et al., 2009. A comparison of risk assessments on *Campylobacter* in broiler meat. Int. J. Food Microbiol. 129, 107–123.

Nauta, M.J. 2001. A modular process risk model structure for quantitative microbiological risk assessment and its application in an exposure assessment of *Bacillus cereus* in REPFED (Online). Available from: <http://www.rivm.nl/bibliotheek/rapporten/149106007.html> (accessed 16.03.15).

Nauta, M.J., 2005. Microbiological risk assessment models for partitioning and mixing during food handling. Int. J. Food Microbiol. 100, 311–322.

Pérez Rodriguez, F., Campos, D., Ryser, E.T., Buchholz, A.L., Posada-Izquierdo, G.D., Marks, B.P., et al., 2011. A mathematical risk model for *Escherichia coli* O157:H7 cross-contamination of lettuce during processing. Food Microbiol. 28, 694–701.

Perni, S., Beumer, R.R., Zwietering, M.H., 2009. Multi-tools approach for food safety risk management of steam meals. J. Food Prot. 72, 2638–2645.

Pouillot, R., Delignette-Muller, M.L., 2010. Evaluating variability and uncertainty separately in microbial quantitative risk assessment using two R packages. Int. J. Food. Microbiol. 142, 330–340.

Pujol, L., Albert, I., Johnson, N.B., Membré, J.-M., 2013. Potential application of quantitative microbiological risk assessment techniques to an aseptic-UHT process in the food industry. Int. J. Food Microbiol. 162, 283–296.

Pujol, L., Albert, I., Magras, C., Johnson, N.B., Membré, J.-M., 2015a. Probabilistic exposure assessment model to estimate aseptic UHT product failure rate. Int. J. Food Microbiol. 192, 124–141.

Pujol, L., Albert, I., Magras, C., Johnson, N.B., Membré, J.-M., 2015b. Estimation and evaluation of management options to control and/or reduce the risk of not complying with commercial sterility. Int. J. Food Microbiol. 213, 124–129.

Reij, M.W., Den Aantrekker, E.D., ILSI Europe Risk Analysis in Microbiology Task Force, 2004. Recontamination as a source of pathogens in processed foods. Int. J. Food Microbiol. 91, 1–11.

Rigaux, C., André, S., Albert, I., Carlin, F., 2014. Quantitative assessment of the risk of microbial spoilage in foods. Prediction of non-stability at 55°C caused by *Geobacillus stearothermophilus* in canned green beans. Int. J. Food Microbiol. 171, 119–128.

Romero-Barrios, P., Hempen, M., Messens, W., Stella, P., Hugas, M., 2013. Quantitative microbiological risk assessment (QMRA) of food-borne zoonoses at the European level. Food Control. 29, 343–349.

Ross, T., Rasmussen, S., Fazil, A., Paoli, G., Sumner, J., 2009. Quantitative risk assessment of *Listeria monocytogenes* in ready-to-eat meats in Australia. Int. J. Food Microbiol. 131, 128–137.

Rosso, L., Lobry, J.R., Bajard, S., Flandrois, J.P., 1995. Convenient model to describe the combined effects of temperature and pH on microbial growth. Appl. Environ. Microbiol. 61, 610–616.

Stals, A., Jacxsens, L., Baert, L., Van Coillie, E., Uyttendaele, M., 2015. A quantitative exposure model simulating human norovirus transmission during preparation of deli sandwiches. Int. J. Food Microbiol. 196, 126–136.

Stringer, M., 2004. Food Safety Objectives—Role in Microbiological Food Safety Management. ILSI (International Life Sciences Institute), Belgium.

Tenenhaus-Aziza, F., Daudin, J.J., Maffre, A., Sanaa, M., 2014. Risk-based approach for microbiological food safety management in the dairy industry: the case of *Listeria monocytogenes* in soft cheese made from pasteurized milk. Risk Anal. 34, 56–74.

World Health Organization, Food Agricultural Organization, 1995. In: FAO/WHO (Ed.), Sixth Report of WHO Surveillance Programme for Control of Foodborne Infections and Intoxications in Europe. Collaborating Centre for Research and Training in Food Hygiene and Zoonoses, Berlin.

World Trade Organization. 1995. The WTO agreement on the application of sanitary and phytosanitary measures (Online). Available from: <http://www.wto.org/english/tratop_e/sps_e/spsagr_e.htm> (accessed 17.07.15).

Zwietering, M.H., 2015. Risk assessment and risk management for safe foods: assessment needs inclusion of variability and uncertainty, management needs discrete decisions. Int. J. Food Microbiol. Available from: http://dx.doi.org/10.1016/j.ijfoodmicro.2015.03.032.

Zwietering, M.H., Wijtzes, T., de Wit, J.C., van't Riet, K., 1992. A decision support system for prediction of the microbial spoilage in foods. J. Food. Prot. 55, 973–979.

Zwietering, M.H., Jacxsens, L., Membré, J.-M., Nauta, M., Peterz, M., 2016. Relevance of microbial finished product testing in food safety management. Food Control. 60, 31–43.

PART

Toxicological hazards

III

CHAPTER

Acrylamide in ready-to-eat foods

20

Bertrand Matthäus and Norbert U. Haase

Department of Safety and Quality of Cereals, Federal Research Institute of Nutrition and Food, Max Rubner-Institut, Detmold, Germany

INTRODUCTION

A completely new situation came up for consumer, food industry, and food safety authorities in April 2002 when the formation of acrylamide in food products came into the focus of interest after a communication of the Swedish National Food Administration (NFA) together with the University of Stockholm regarding findings of acrylamide in baked, fried and deep-fat fried products. In comparison to trans-fatty acids in edible oils, for the first time a heat-induced contaminant affected a large range of different food categories during food processing. Another important aspect was that total mitigation of acrylamide was linked with dramatic losses in quality. Most of the known contaminants are environmentally based such as pesticides, mycotoxins or polycyclic aromatic hydrocarbons, but in the case of acrylamide the raw material contains the precursors and during food processing the contaminant is formed. The great challenge of this finding was to avoid the formation of acrylamide during food processing while maintaining the high quality of the products.

The acrylamide story began in 2000 when Tareke et al. (2000) published their assumption that cooking of food is a major source of the background dose of acrylamide in humans. They found unexpectedly high concentrations of acrylamide adducts in a control group without exposure to acrylamide. The investigation of a number of foods showed moderate levels of acrylamide (5–50 μg/kg) in heated protein-rich food and higher amounts (150–4000 μg/kg) in carbohydrate-rich foods (Tareke et al., 2002). Already in 1991 acrylamide had been evaluated as a "genotoxic carcinogen" by the European Scientific Committee on Food (SCF), confirmed in 2002 (Anon., 2002), and in 1994 acrylamide was classified in class 2A as a "probable human carcinogen" by the International Agency for Research on Cancer (IARC) (Anon., 1994). As an unsaturated amide with an α,β-unsaturated carbonyl group acrylamide is predestined to react with nucleophilic groups such as amines, carboxylates, and thiols which are characteristic for biological molecules such as DNA (Mills et al., 2009). In June 2015 the EFSA

P. Kotzekidou (Ed): Food Hygiene and Toxicology in Ready-to-Eat Foods.
DOI: http://dx.doi.org/10.1016/B978-0-12-801916-0.00020-0

published the first full risk assessment of acrylamide in food and the experts of the EFSA Panel on Contaminants in the Food Chain (CONTAM) confirmed previous evaluations that acrylamide in food potentially increases the risk of developing cancer for consumers in all age groups (Anon., 2015). Currently a final decision of the toxicity level of acrylamide findings in foods is still absent. Therefore the ALARA principle (as low as reasonably achievable) should be used.

Until the first finding of acrylamide in food that chemical was only relevant for a small group of the population with exposure to acrylamide, but due to the large number of food categories affected the toxicological assessment of acrylamide became important for most humans.

This high relevance of acrylamide resulted in huge efforts by science and industry to find possibilities to avoid and mitigate the formation of acrylamide in food products. One prerequisite for the development of mitigation strategies was the investigation of the reaction pathway in order to find starting points for the optimization of processing towards lower amounts of acrylamide.

FORMATION PATHWAYS

For the development of mitigation strategies it is important to know how acrylamide is formed under which conditions. From this knowledge it is possible to change processing conditions selectively for reducing the content of acrylamide without impairing product quality.

Acrylamide belongs to the so-called "foodborne" contaminants as it is formed within the product during processing. The formation depends on the presence of precursors and the processing conditions such as pH-value, temperature, time, etc. Zyzak et al. (2003) showed experiments with ^{15}N-labeled asparagine and ^{13}C-labeled glucose that the three carbon atoms and the nitrogen from acrylamide derive from asparagine. Theoretically it is possible to form acrylamide directly by degradation of asparagine with elimination of carbon dioxide and water, but the formation rate is much faster in the presence of a carbonyl compound such as a reducing sugar (Yaylayan et al., 2003).

Today it is well established that the main pathway for the formation of acrylamide is within the Maillard reaction between the amino acid asparagine in the presence of certain α-dicarbonyl compounds like reducing sugars, for example, glucose or fructose (Mottram et al., 2002; Stadler et al., 2002a) (Fig. 20.1). To a much smaller extent acrylamide is also formed by thermal degradation of triacylglycerols at temperatures of $>200°C$ leading to acrolein. The following oxidation to acrylic acid leads to the formation of acrylamide in the presence of a nitrogen source such as ammonia (Izard and Libermann, 1978; Yasuhara et al., 2003).

Within the Maillard reaction the formation of acrylamide starts with the formation of a Schiff base by the reaction of the amino nitrogen from asparagine with the

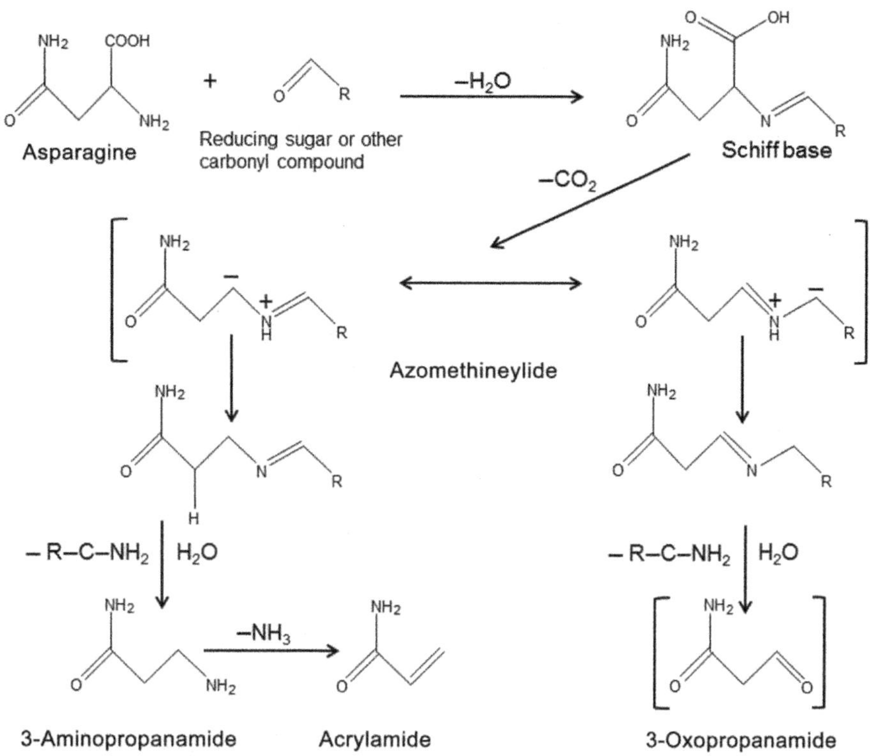

FIGURE 20.1

Reaction mechanism for acrylamide formation (Wedzicha et al., 2005; Zyzak et al., 2003).

carbonyl group of a reducing sugar (Becalski et al., 2003; Yaylayan et al., 2003). The formation of an Amadori compound is not necessary for the formation of acrylamide since on another pathway from decarboxylated Schiff base acrylamide can be formed directly by hydrolysis to 3-aminopropionamide, the Strecker amine of asparagine. Then acrylamide is formed after elimination of ammonia upon heating. Granvogl et al. (2004) showed the importance of 3-aminopropionamide as an effective precursor for the formation of acrylamide in heated food. The outstanding importance of asparagine for the formation of acrylamide has been shown by Zyzak et al. (2003) and Becalski et al. (2003). Today it is clear that glucoconjugates such as *N*-glucosides and related compounds formed from asparagine are key intermediates for the formation of acrylamide. On the other side only minor amounts of acrylamide are formed from glutamine and methionine as amino acid sources (Stadler et al., 2002a). Carbohydrate compounds for the formation of acrylamide can act on a broad range of compounds. Stadler et al. (2002b) showed that glucose, fructose, galactose, lactose, and sucrose all resulted in comparable amounts of acrylamide when heated with asparagine at 180°C. Also, sucrose as a nonreducing sugar resulted in the formation of similar amounts of acrylamide. It is assumed that thermal treatment of food results in hydrolysis of sucrose to glucose and fructose. Zyzak et al. (2003) also found in a potato snack model that other carbonyl compounds, such as 2-deoxyglucose, ribose, glyceraldehyde, glyoxal, and decanal, react with asparagine to form acrylamide.

On another pathway acrylamide is formed by the direct decomposition of the Schiff base via instable intermediates which were not yet detected. The ylide formed by decarboxylation of the Schiff base decomposes directly by splitting the C–N bond, resulting in acrylamide and a 1-amino-2-hexulose. On the other side the Schiff base can be degraded after oxidation and decarboxylation to an intermediate which decomposes to 3-aminopropionamide after enolization and hydrolysis. Finally, the separation of ammonia results in acrylamide.

The Maillard reaction also has a central importance on the quality of heated food with respect to appearance and sensory quality but the reaction is also responsible for deterioration of food during processing and storage. Another aspect is the antioxidant activity of Maillard reaction products in food products, which has been described by several authors (Lingnert and Eriksson, 1980; Natella et al., 2002; Wagner et al., 2007). Within the Maillard reaction from the Schiff base an Amadori compound can be formed by rearrangement, the degradation of which leads to a number of reaction products responsible for the development of flavor and color during food processing.

The main requirements for the formation of acrylamide are a minimum temperature of 120°C low moisture content, a pH value of about 7 or higher, and the presence of precursors, although in model systems and special products such as gingerbread acrylamide was also formed at lower temperature (Amrein et al., 2004; Biedermann and Grob, 2003a).

OCCURRENCE OF ACRYLAMIDE IN FOOD

From the appearance of acrylamide in plant-originated food a wide range of products is affected. In comparison to other heat-induced contaminants nearly all fried, baked or deep-fat-fried foods are more or less involved. Acrylamide was not found in foods prepared by boiling or microwaving (Eriksson, 2005; Törnqvist, 2005), and also in animal-based food products only very small amounts of acrylamide were found. After the first announcement of acrylamide in food a number of larger and smaller surveys were organized by governmental organizations or independent researchers all over the world to monitor the content of acrylamide in different food categories. The three main databases with information about the content of acrylamide in food are offered by the European Union, World Health Organization, and US Food and Drug Administration.

The largest dataset about the content of acrylamide in foods has been collected by the European Union together with the Member States and the European food industry. This database contains several thousand datasets which are maintained by the Institute for References Materials and Measurements (IRMM) in Geel, Belgium. The most important food matrices represented in the database are potato crisps, French fries, crispbread, breakfast cereals, pastries like biscuits, and coffee (Table 20.1). A wide range of different foods is covered by this database helping to estimate the exposure of acrylamide to humans.

Considering the data on the content of acrylamide in different foods the great variability of the concentrations not only between different types of food but also within one food group becomes obviously and also products from the same batch can contain completely different amounts of acrylamide. One reason is the susceptibility of the Maillard reaction even to small temperature differences during processing, especially at higher temperatures. That makes the development of mitigation strategies difficult.

The highest values of acrylamide were found in fried potato products, such as French fries and potato crisps, but also in cereal products such as gingerbread, crispbread, bread, or breakfast cereal high maximum levels of acrylamide can be expected. Roasted coffee and coffee substitutes are also high-risk products for acrylamide contamination. In commercial black ripe olives acrylamide levels between 176 and 1578 μg/kg of fresh weight were found in the olive pulp (Casado and Montano, 2008).

Since 2011 the European Union has established a system of indicative values for different food categories according to the German dynamic minimization concept of so-called signal values which was introduced in 2002 as part of a concerted action between the German food industry and the government. These signal

Table 20.1 Distribution of Acrylamide Levels in Different Food Categories (Anon., 2015)

Food Category	Number of Samples	Mean Value (μg/kg)	95th Percentile (μg/kg)
Potato fried products (except potato crisps and snacks)	1694	308	971
Potato crisps and snacks	34,501	389	932
Soft bread	543	42	156
Breakfast cereals	1230	161	552
Biscuits, crackers, crispbread, and similar	2065	265	1048
Coffee (dry)	1457	522	1054
Coffee substitutes (dry)	88	1499	4500
Baby foods, other than cereal-based	41	24	72
Processed cereal-based baby foods	736	73	175
Other products based on potatoes, cereals, and cacoa	569	97	370
Other products	120	330	1510

values for different food categories were calculated from acrylamide analyses of the food authorities and exceedances of these signal values started a minimization dialog between food authorities and the food manufacturer. The indicative values were based on data from EU monitoring from 2007 to 2009 (Anon., 2011) and have been adjusted to partially lower values for some food categories according to new data in 2013 (Anon., 2012) (Table 20.2). Gingerbread and coffee substitutes were introduced as food categories. No mitigation effect was found for ready-to-eat French fries and potato crisps from fresh potatoes and from potato dough.

Table 20.2 Indicative Acrylamide Values Based on the EFSA Monitoring Data From 2007 to 2012 (2013/647/EU) (Anon., 2013)

Foodstuff	Indicative Value (µg/kg)
French fries ready-to-eat	600
Potato crisps from fresh potatoes and from potato dough	1000
Potato-based crackers	
Soft bread	
1. Wheat-based bread	80
2. Soft bread other than wheat-based bread	150
Breakfast cereals (excluding porridge)	
• Bran products and whole grain cereals, gun puffed grain (gun puffed only relevant if labeled)	400
• Wheat- and rye-based products	300
• Maize-, oat-, spelt-, barley-, and rice-based products	200
• Biscuits and wafers	500
• Crackers with the exception of potato-based crackers	500
• Crispbread	450
• Gingerbread	1000
• Products similar to the other products in this category	500
Roast coffee	450
Instant (soluble) coffee	900
Coffee substitutes	
1. Coffee substitutes mainly based on cereals	2000
2. Other coffee substitutes	4000
Baby foods other than processed cereal-based foods	
1. Not containing prunes	50
2. Containing prunes	80
Biscuits and rusks for infants and young children	200
Processed cereal-based foods for infants and young children, excluding biscuits and rusks	50

MITIGATION OF ACRYLAMIDE

The main challenge for the mitigation of acrylamide is the close connection of the Maillard reaction responsible for the generation of desirable flavors and color of food to acrylamide formation. The formation of acrylamide is affected by the same factors that influence the positive quality parameters of heated food. Thus it is necessary to control and adjust the different factors influencing the Maillard reaction without changing the quality of the product: (1) content of precursors (reducing sugars and free amino acid content), (2) time–temperature conditions, (3) water content, (4) pH value of the food, and (5) presence of additives.

When looking at the processing chain it seems to be the best option to start mitigation strategies at the beginning by reducing the amount of precursors for acrylamide formation in the raw material. That may allow maintaining of the production process without any changes, which saves costs for adapting the process conditions to the new requirements. On the other side today a lot of options for minimizing acrylamide formation during food processing are available but they must be adapted to the possibilities of the process configuration and the requirements of the products. Additionally, it should be taken into account that changed composition of the raw material or process conditions not only affect the content of acrylamide but possibly also the formation of desirable compounds responsible for the quality of the ready-to-eat product. Therefore it has to be checked on an individual basis whether the different tools for the mitigation of acrylamide are useful or not for the specific process and product.

RAW MATERIAL

The content of acrylamide in heated foodstuffs is strongly affected by plant species, variety, fertilization, harvest year, and storage conditions of the raw material before processing. If it is possible to reduce the amount of precursors before processing, the potential for acrylamide formation would be reduced.

Asparagine and reducing sugars belong to the essential compounds in raw crops which are formed in several biochemical pathways. On the other side the ratio of asparagine to reducing sugars varies between crops. Consequently the relevance of the individual precursors for the development of minimizing strategies is different. The content of reducing sugars is the limiting factor for acrylamide formation in heated potatoes where asparagine is present in abundance, while the formation of acrylamide in cereals is determined by the content of asparagine with higher amounts of reducing sugars.

Variety

Amrein et al. (2003) found that glucose and fructose determine the formation of acrylamide in potato products, whereas the potential of acrylamide formation was primarily dependent upon the sugar content with a high correlation. This could be explained by a greater variation of the content of sugars in potatoes and the lower concentration in comparison to asparagine. Amrein et al. recommended that acrylamide content in potato products could be substantially reduced by selecting varieties with low concentrations of reducing sugars. Haase et al. (2004) found a strong correlation between the content of acrylamide in heated potato samples ($r = 0.75$) with the amount of reducing sugars (Fig. 20.2). A similar correlation was described by de Wilde et al. (2006) who also explained differences in the potential of acrylamide formation between different potato varieties by the content of reducing sugars ($R^2 = 0.82$, $n = 96$). Amer et al. (2014) described a strong effect of the potato variety on the potential to form acrylamide with purple-fleshed cultivars as more susceptible to storage-induced increase in acrylamide. Matthäus et al. (2004) demonstrate that different genotypes of potatoes with varying compositions of precursors for the acrylamide formation resulted in different amounts of acrylamide during processing. Fig. 20.3 shows the influence of the frying temperature on the formation of acrylamide of two different varieties of potatoes during the preparation of potato crisps. A great difference in the potential of both varieties to form acrylamide during the processing is obvious. While variety Tomensa showed only a slight increase in the acrylamide formation in dependence on the frying temperature, the increase was much more marked for potato crisps produced from the variety Saturna. Such an influence of the genotype on the potential of the acrylamide formation is also described by Amrein et al. (2003) who analyzed glucose, fructose, sucrose, free asparagines, and free

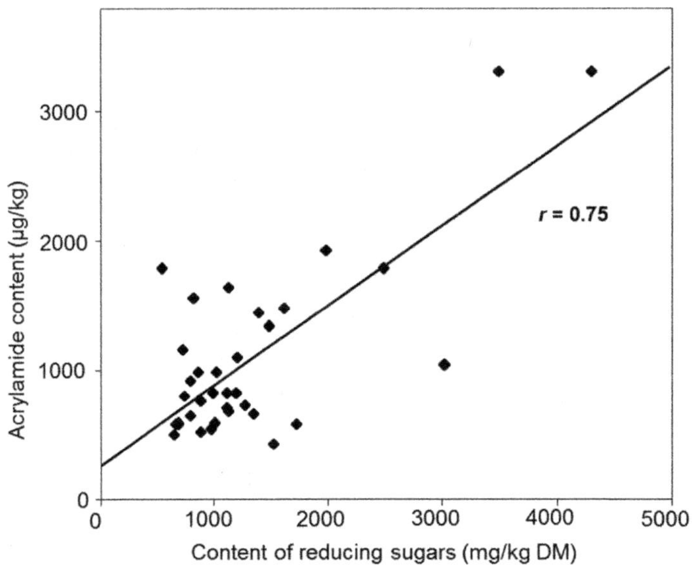

FIGURE 20.2

Correlation between acrylamide formation in potatoes and content of reducing sugars.

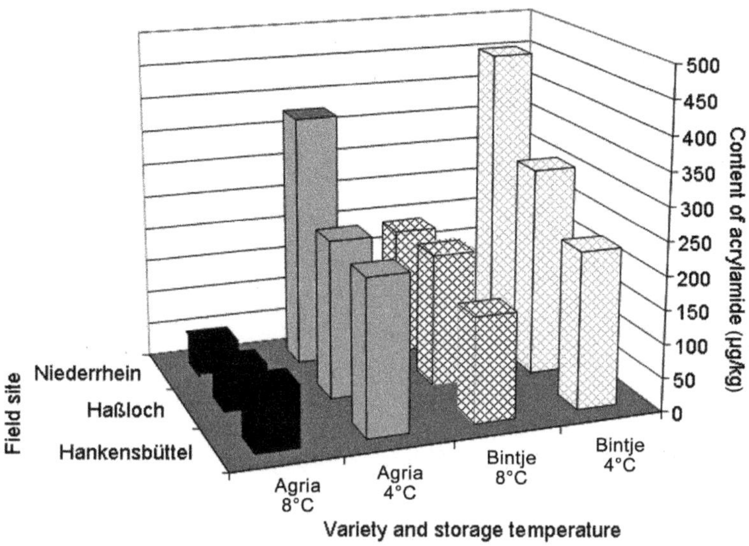

FIGURE 20.3

Influence of variety and field site on the potential for the formation of acrylamide in potatoes.

Reprinted from Matthäus, B., Haase, N.U., Vosmann, K., 2004. Factors affecting the concentration of acrylamide during deep-fat frying of potatoes. Eur. J. Lipid Sci. Technol. 106, 793–801, with permission of Wiley.

glutamine of 74 samples of potatoes from 17 cultivars. In this connection it should also be mentioned that the field site has a great influence on the potential of potato tuber to form acrylamide during processing (Haase et al., 2003).

For cereals the formation of acrylamide is determined by the content of asparagine, which is strongly influenced by the genotype, with differences of up to 200% (Martinek et al., 2009). The variation in asparagine also depends on the cereal species, with levels between 69 and 443 mg/kg for wheat and 319 and 791 mg/kg for rye, respectively (Amrein et al., 2004; Hamlet et al., 2005; Springer et al., 2003; Taeymans et al., 2004). The asparagine content in flour ranged from 0.37 to 1.89 mmol/kg resulting in acrylamide contents between 154 and 624 µg/kg (Claus et al., 2006). Also, studies in the United Kingdom, Hungary, France, and Poland demonstrated a significant variation in the content of the precursors in different varieties of wheat and rye (Curtis et al., 2009, 2010).

The great variation in the amount of reducing sugars and asparagine in potatoes and cereals allows the selection of varieties with low amounts of precursors for processing of heated food and consequently low potential for the formation of acrylamide. Additionally, varieties with low amounts of precursors present new approaches for even lower amounts by breeding.

However, unfortunately breeding of new varieties with low amounts of precursors is time-consuming and costly. After about 12 years of research the US

Department of Agriculture has approved the first genetically modified potato variety with a low potential for acrylamide formation due to low amounts of asparagine. Researchers silenced several genes associated with asparagine production (Waltz, 2015). Such an approach was described by Rommens et al. (2008) in 2008 when silencing two asparagine synthetase genes through "all-native DNA" transformation by genetic modification. They reduced the levels of free asparagine in potato tubers by up to 20 times, while the content of glutamine was a little higher. The tubers had no differences in the sensory characteristics, but the potential for acrylamide formation was as little as 5% of that of the unchanged tubers.

In a workshop in 2014, organized by the European Commission, the necessities for the development of new potato varieties with, in particular, low levels of reducing sugar and free asparagine contents as well as the need for cereal varieties with low free asparagine content were highlighted (Anon., 2014b).

Loaec et al. (2014) investigated the effects of variety on the amount of free asparagine in chicory, a raw material used for coffee substitutes and correlated these results with the formation of acrylamide during roasting. They found a variability of the content of asparagine depending on the variety and concluded that a varietal selection could be an important parameter for mitigation of acrylamide formation during processing.

For coffee two varieties, coffea *Canephora robusta* (Robusta) and coffea *Arabica* (Arabica), are known which contain relatively low concentrations of asparagine (between 0.3 and 0.9 g/kg) (Stadler and Scholz, 2004), whereas in Robusta slightly higher amounts of asparagine were reported on average (CIAA, 2004). The possibilities to reduce acrylamide formation during coffee processing by selecting or blending green beans with lower amounts of asparagine are very limited due to the narrow span within the varieties. The source of reducing sugars derives from the high amount of sucrose (30–100 g/kg) which decomposes during the roasting process into fructose and glucose.

Amrein et al. (2005b) investigated 18 samples of almonds originating from the United States and Europe for sugars and free amino acids. They found that asparagine was the major free amino acid in all samples, with amounts between 500 and 2760 mg/kg, accounting for 20–50% of the total amount of free amino acids. Almonds of European origin contained significantly less free asparagine and formed significantly less acrylamide during roasting as compared to the almonds from the United States. During roasting reducing sugars are consumed faster and to a larger extent than free asparagine, suggesting that the content of reducing sugars may be a critical factor for acrylamide formation in roasted almonds. Acrylamide was found to decrease in roasted almonds during storage at room temperature (Amrein et al., 2005a).

Climate

Another important factor for acrylamide formation are the climatic conditions, but the possibilities for influencing this variable are very limited. Different investigations have shown that the content of asparagine or reducing sugars strongly varies from harvest year to harvest year with lower potential of acrylamide formation in extremely hot summers (Haase and Weber, 2003).

Agronomy

Going further along the processing chain shows that some agronomic means may strongly influence the potential of potatoes to form acrylamide. Examples are the use of fertilizers such as nitrogen which may result in the formation of free amino acids and sugars (Muttucumaru et al., 2013) in potatoes. Martinek et al. (2009) and Muttucumaru et al. (2013) showed that the use of nitrogen increases the amount of asparagine in wheat, rye, and potatoes, while sulfur supply leads to lower amounts of glucose in potatoes (Muttucumaru et al., 2013). A similar effect was found in chicory roots, when the lowest asparagine content was found without any addition of nitrogen fertilizer with an increase in free asparagine content proportional to the amount of nitrogen added up to 60 kg fertilizer/ha and no significant difference in asparagine levels in chicory roots fertilized with higher amounts of nitrogen (Loaec et al., 2014).

Plants use free asparagine as a nitrogen store when the generation of protein is disturbed by stresses such as exposure to toxic metals, pathogen attack, and drought or salt stress. Due to its high ratio of nitrogen (N) to carbon, the relatively low relevance for enzymes and its low net charge under physiological conditions asparagine is indicated as an ideal storage compound (Lea et al., 2007). Lack of nutrients such as potassium (K), sulfur, phosphorous, and magnesium results in the accumulation of asparagine in the plant (Curtis et al., 2009; Elmore et al., 2007; Granvogl et al., 2007; Halford et al., 2012). On the other side, rising N supply increased the contents of reducing sugars in most cases, and resulted in higher contents of free amino acids, while higher K supply resulted in lower contents of reducing sugars. A frying treatment of potatoes showed the highest acrylamide content in tubers grown with high N and inadequate K supply, indicating that nutrient supply has a significant impact on the potential to form acrylamide during food processing by heat (Gerendas et al., 2007).

Storage conditions

Especially for potatoes, the storage conditions are relevant for acrylamide formation. At storage temperatures below 8°C starch breaks down, resulting in the formation of reducing sugars (Haase and Weber, 2003). Haase et al. (2004) found that storage of potatoes at 4°C led to high concentrations of reducing sugars. Matsuura-Endo et al. (2006) showed that a significant increase in reducing sugars was found at temperatures $<8°C$ when potatoes were stored at 2°C, 6°C, 8°C, 10°C, and 18°C for a period of 18 weeks. Corresponding with the increase in reducing sugar, the acrylamide level increased significantly in the heated product. A good correlation between acrylamide levels and reducing sugars for potatoes was found for a fructose/acrylamide ratio <2. At fructose/acrylamide ratio >2 the asparagine content was the limiting factor for acrylamide formation rather than the content of reducing sugars. Knutsen et al. (2009) found that the potential for acrylamide formation of potatoes cannot be predicted from the amount of reducing sugars and asparagine alone and they assumed that other factors might also change during storage influencing the potential.

Today, potatoes fit for processing are generally stored at 8–10°C to prevent excessive enrichment of reducing sugars. At those temperature regimes, tubers will rapidly lose dormancy and consequently sprout-suppressants have to be applied.

Another important aspect during storage is the storage atmosphere which also influences the formation of reducing sugars. At significantly lower oxygen levels the cold-induced enzymes which favor the degradation of starch to low-molecular-weight reducing sugars are inhibited, leading to lower amounts in the raw material (Harkett, 1971; Zhou and Solomos, 1998).

PROCESSING

Pretreatment

An interesting approach for reducing the potential of raw material for the formation of acrylamide is the removal of asparagine or reducing sugars as precursors before or during processing or the change of reaction condition of the Maillard reaction. It has to be taken into consideration that such a pretreatment can seriously affect the quality of the raw material and consequently the final product. Reducing the precursors or changing the reaction conditions by pretreatment comprises of soaking and blanching or treatment with asparaginase or antioxidants.

Soaking and blanching

An effective tool for the mitigation of acrylamide in ready-to-eat products is blanching of potatoes before processing (Fig. 20.4) with reduction rates up to 60% depending on the raw material (variety and field site) and processing conditions (blanching and frying) (Haase et al., 2003). The higher the frying temperature the higher was the effect of soaking or blanching on reduction of acrylamide in the product. The effect was based on leaching water-soluble precursors such as reducing sugars or asparagine prior to frying. Pedreschi et al. (2004) showed that blanching resulted in an average reduction of glucose and asparagine by 76% and 68%, respectively, and May et al. (2006) found an exponential dependence of the reduction from the blanching time. Similar results were found by Pedreschi et al. (2007a) who described a decreasing amount of acrylamide in potato strips with increasing soaking time with about 15% after 60 min of soaking and about 30% after 120 min, respectively, in comparison to a control.

Lukac et al. (2007) investigated the influence of blanching or soaking of almonds before roasting at 165°C for 12.5 min. They found that almonds with high initial water content formed less acrylamide than almonds with low water content. Equal temperature and time resulted in lighter color of the products at higher water content.

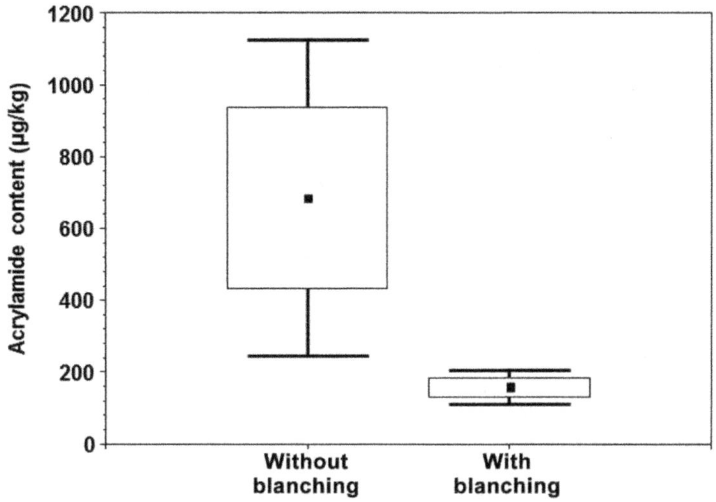

FIGURE 20.4

Influence of blanching on the potential of potato slices on the formation of acrylamide.

Reducing the pH value of the raw material during soaking has no significant effect on the content of glucose or asparagine compared to a control, but it strongly influences the potential of the raw material to form acrylamide. As acrylamide formation is part of the Maillard reaction the pH value during processing is an important factor. Maximum levels of acrylamide were found at pH = 7 or higher, while at lower pH values the potential for the formation decreased (Rydberg et al., 2003; Weisshaar and Gutsche, 2002), possibly by inhibition of acrylamide formation as a result of conversion of free non-protonated amine into non-nucleophilic protonated amine (Jung et al., 2003). From this, the treatment of potato slices with acid solutions to decrease the pH value of the material is a promising step (Kita et al., 2004; Pedreschi et al., 2007b). Pedreschi et al. (2007a) showed that soaking with citric acid resulted in a reduction of the acrylamide content for 86%, 47%, and 28% at different frying temperatures (150°C, 170°C, and 190°C, respectively) in comparison to a control. Similar results with a reduction of acrylamide formation by 90% were found by Kita et al. (2004) after soaking of potato slices in acetic acid solution (60 min/20°C). One disadvantage of such treatments was a sour taste of the final product. Ismial et al. (2013) investigated the effect of soaking potato slices and chips with different solutions on the formation of acrylamide during processing including $MgCl_2$ (0.1 M), L-cystein (0.05 M), citric acid (0.01 M), asparaginase, and phenolic acids. They found huge reduction rates of more than 90%, especially for solutions of $MgCl_2$, L-cystein, and citric acid.

Soaking of potato slices in NaCl solution after blanching before frying resulted in a significant reduction in the oil absorption and better texture parameters (Bunger et al., 2003), while simultaneously the formation of acrylamide can be reduced drastically for about 90% in comparison to a control with better assessment of the color (Pedreschi et al., 2007a).

Milling

The key determinant for the formation of acrylamide during baking is the amount of asparagine and depending on the processing commercial milling has a strong influence on the level of free asparagine in the raw material. Fredriksson et al. (2004) investigated the content of free asparagine in 11 milling fractions from wheat and rye and showed that the lowest amount was found in sifted wheat flour (0.2 g/kg) and the highest in wheat germ (4.9 g/kg). Whole grain wheat flour contained 0.5 g/kg while in whole grain rye flour 1.1 g/kg was found. Springer et al. (2003) found an impact of milling on the amount of free asparagine in different fractions and higher amounts in whole grain than in individual fractions. The highest asparagine content was not found in the starchy endosperm part of the kernel but fiber bread will form higher concentrations of acrylamide. Thus, mixing of individual fractions of grain with lower amounts of free asparagine for the preparation of dough is a promising way to reduce acrylamide formation in the final product.

Bright flours with a low mineral concentration also have a low asparagine concentration, indicating a low acrylamide formation potential. On the other hand, wholemeal flours with health benefits contain more asparagine and a relatively high acrylamide potential. To overcome this discrepancy between the positive effect of wholemeal flour on one hand and the higher potential for acrylamide formation on the other hand, Haase (2007) carried out an exemplary test to lower the asparagine level while maintaining the character of wholemeal flour. A rye sample was abrasive peeled with mass losses between 1.9% and 5%. The concentration of free asparagine dropped down from 1400 to 1250 and 900 mg/kg DM, respectively. This pretreatment ahead of milling seems to be qualified to reduce the acrylamide potential and to keep the nutritional benefits of the flour.

Additives

Asparaginase

An elegant way to remove asparagine from the raw material is treatment with asparaginase, which converts the amino acid into ammonia and aspartic acid. For commercial use the enzyme is either isolated from *Aspergillus niger* or *Aspergillus oryzae*. Ciesarova et al. (2006) reduced the amount of acrylamide in fried potato products by about 50% after treatment of fresh potato mash with 0.2 units lyophilized enzyme per gram raw material at 20°C. A reduction of acrylamide in the final product of up to 97% was reached by use of a larger amount of enzyme (2 units/g) and a higher incubation temperature (37°C). Kumar et al. (2014) reduced acrylamide formation in bakery products such as sweet bread by treatment with L-asparaginase from *Cladosporium* spp. They found no change in rheological properties and sensory characteristics of the product, but with increasing levels of asparaginase the content of acrylamide was reduced by up to 97% and 73% in the crust and crumb regions, respectively. The effect of commercially available asparaginase from *A. oryzae* was investigated in products like tortilla chips, cookies, potato tube snacks, French fries, and coffee (Hendriksen et al., 2013). In dough-based products the enzyme was either added to the dough or to pretreated potatoes with a reduction of acrylamide in the final content of up to 90%. For the preparation of French fries, potato slices were shortly dipped into the enzyme solution resulting in about 60% lower acrylamide level in the ready-to-eat product. For coffee a reduction of up to 74% was reached after treatment with asparaginase. The authors also showed that the enzyme only acts at the surface but did not penetrate into the tissue, limiting enzyme–substrate contact to the surrounding solution. For biscuits, Haase et al. (2012) demonstrated a reduction of acrylamide in a model system of about 50% while the antioxidant capacity of the products was not affected.

Although the enzymatic treatment is very effective, it has to be taken into consideration that especially for potatoes the conditions of pretreatment, such as blanching and application time, are decisive for the success of enzymatic treatment (Hendriksen et al., 2013). Another important point is that treatment with asparaginase is disproportionally expensive in comparison to other possibilities of mitigation.

Antioxidants

Different publications describe a significant effect of antioxidants and antioxidative extracts from, for example, rosemary, bamboo, or green tea on the formation of acrylamide during processing of different kinds of products. However, the results are not consistent and a little confusing, with some studies describing a reduction of the formation (Becalski et al., 2003; Hedegaard et al., 2008; Zhang and Zhang, 2008), while others found no effect or even an increase in acrylamide after addition (Cheng et al., 2010). One reason for these varied results may be the different structures of antioxidants or antioxidant compounds from extracts which can react with acrylamide precursors or intermediates of the reaction of acrylamide itself (Jin et al., 2013). They describe different pathways for the reaction of antioxidants which either promote the formation of acrylamide by triggering decomposition of sucrose, direct reaction of asparagine to acrylamide or inhibiting acrylamide formation by trapping specific Maillard reaction products, preventing lipids from oxidation, precipitation of asparagine, reaction with intermediates of the Maillard reaction, or by elimination of acrylamide after reaction with antioxidants. The different pathways strongly depend on the reaction conditions determined by the processing conditions. Due to this, for different antioxidants but also for the same antioxidant compound, different effects on the formation of acrylamide are described in the literature.

Amino acids

The addition of amino acids during processing may have a positive effect on acrylamide reduction due to a competition of the amino acid with asparagine for the carbonyl source/reducing sugars within the Maillard reaction. Different authors showed a significant reduction of acrylamide formation when soaking uncooked products in appropriate solutions of different amino acids including lysine, glycine, and cysteine with reduction rates of up to 90% depending on the food product, the processing conditions, the amino acid, and the concentration of amino acid (Kim et al., 2005; Rydberg et al., 2003). Dipping potato slices in solution of 3% lysine or glycine, respectively, for 1 min reduced the acrylamide formation in potato chips by 80% (Kim et al., 2005). For flat bread (wholemeal rye flour) and bread (white wheat flour) Brathen et al. (2005) demonstrated a significant reduction of acrylamide with 73% and 96%, respectively, after addition of 6.08 g of glycine per kg of flour.

Tareke et al. (2002) reported that acrylamide formation of potato slices can be reduced when adding chickpea protein before the frying process. Similar effects of chickpea have been published by Miskiewicz et al. (2012) who found that cookies derived from a blend of wheat and chickpea flours (1:1, w/w) formed 86.4% lower amounts of acrylamide than cookies prepared from wheat flour alone (41.9 μg/kg). The authors assumed a protective effect of the chickpea protein.

Recipe

Dough recipes often comprise different components, which significantly affect acrylamide formation during processing (Grothe et al., 2005). Additionally to other parameters like content of free asparagine and reducing sugars or low moisture content responsible for partially very high acrylamide contents, baking agents like ammonium hydrogen carbonate have been found to be one main increasing factor (Biedermann and Grob, 2003b). The assumption of Yasuhara et al. (2003) that the ammonium group may react with acrylic acid to acrylamide was not confirmed with appropriate labeling experiments (Amrein et al., 2004). Nevertheless, it was assumed that ammonium increases the potential for acrylamide formation because experiments showed that the exchange of ammonium by baking powder resulted in much lower contents in the final products (Grothe et al., 2005). A comparison between baking powder and baking agent on the basis of ammonium for the processing of brown gingerbread and shortbread showed a reduction in the acrylamide content of about 1400 μg/kg on average to below 200 μg/kg for the use of baking powder. The reduction in acrylamide was accompanied with a slightly reduced relaxation of the dough with finer pores and brighter crust and crumb browning. A simple exchange of the raising agent by baking powder eliminated most of the acrylamide, but the typical organoleptic profile of gingerbread changed significantly (Grothe et al., 2005). An optimization procedure for the raising agent with a focus on organoleptic aspects reduced the acrylamide concentration by at least 30%.

Sometimes a lightly dull aftertaste was noticed which may be covered by the application of sugar glazes, chocolate coatings, or jam fillings, which are common practice.

For shortbread, no effect of different baking agents on acrylamide formation was found, with amounts below 200 μg/kg. Only the use of ammonium-based baking agents in shortbread together with 20% of crumbs from shortbread, resulted in a slight increase in the acrylamide content in the final product up to 267 μg/kg.

Temperature

One of the major food categories responsible for acrylamide exposure to humans is the consumption of fried or deep-fat-fried potato products with respect to content and intake. During the frying process the product is exposed to temperatures between 160°C and 190°C over a period of some minutes, resulting in drying of the raw potatoes. On one hand these conditions lead to the formation of the typical taste and smell of fried potato products by the formation of desired Maillard reaction products and on the other hand the conditions during frying favor the formation of acrylamide as a result of a heat-induced reaction. Since frying is a continuous drying of wet food material water is transported from the inner part of the food to the surface where it is evaporated. As long as sufficient water reaches the surface a cooling effect takes place resulting in temperatures of about 100°C at the surface and the inner part of the product. With frying time a crust is formed as the surface dehydrates that inhibits further oil uptake and in the dehydrated crust the temperature increases resulting in the Maillard reaction and formation of acrylamide. In a temperature range between 150°C and 170°C the increase of acrylamide in the product is relatively moderate, while with increasing temperature the formation of acrylamide accelerates following a nonlinear progression (Gertz and Klostermann, 2002; Matthäus et al., 2004; Taeymans et al., 2004; Williams, 2005). On the other side it has to be taken into consideration that conventional frying at temperature below 160°C results in poor product quality with respect to color, oil content, and texture (Granda et al., 2004). Also, the moisture content of French fries should be taken into consideration with respect to consumer expectations, too dry (<38%) and too wet and soft (>45%) products will be rejected by the consumer (Fig. 20.5). Due to these results the Acrylamide Toolbox initiated by Food Drink Europe representing the European Food and Drink industry recommends a maximum temperature of 175°C for frying products (Anon., 2014a).

The major difficulty in mitigation of acrylamide is the influence of temperature not only on the formation of acrylamide but also on the development of a pleasant aroma during frying. Thus, the scope for reducing the temperature is limited. One promising possibility is reducing the temperature of frying at lower moisture content (Fiselier et al., 2006; Granda et al., 2004; Haase, 2006; Matthäus et al., 2004). Amrein et al. (2006) showed that at low moisture contents (<20%) the formation of acrylamide needs more activation energy than the

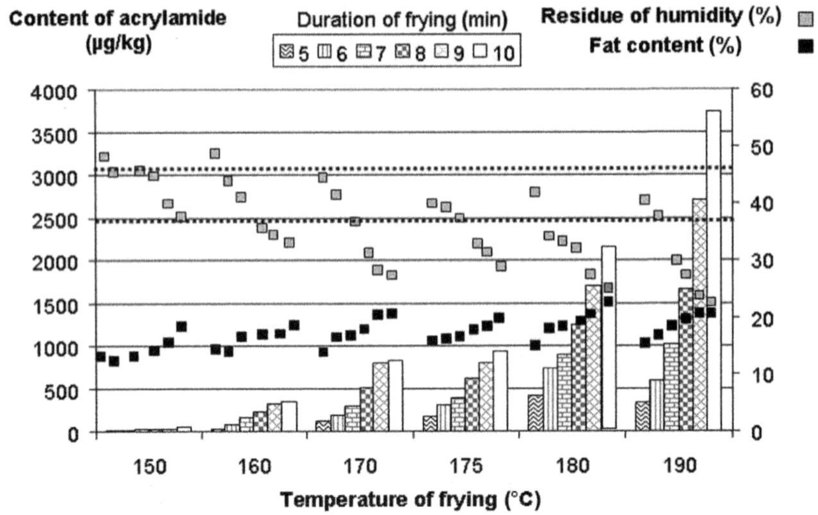

FIGURE 20.5

Influence of frying conditions on final moisture and fat content as well as acrylamide concentration of fried potato chips (area between dotted lines corresponds to the optimal moisture content of fried potato chips).

Reprinted from Matthäus, B., Haase, N.U., Vosmann, K., 2004. Factors affecting the concentration of acrylamide during deep-fat frying of potatoes. Eur. J. Lipid Sci. Technol. 106, 793–801, with permission of Wiley.

development of desirable aroma compounds leading to the conclusion that reducing the temperature in the final stage of frying results in lower amounts of acrylamide without impairing the quality of the product.

Different approaches have been developed taking this finding into consideration. One of these approaches is vacuum frying at pressure below 6.65 kPa, lowering the boiling point of oil and water resulting in utilization of lower temperature with lower final acrylamide formation (Aladedunye and Przybylski, 2009; Granda and Moreira, 2005). Vacuum frying with similar final moisture content of 1.5% ± 0.3% results in acrylamide content of 48 mg/kg at 118°C instead of 847 mg/kg for conventional frying at 165°C. Using a multistep approach, Arias-Mendez et al. (2013) showed that a process with different temperature zones during frying could be a perfect compromise between product quality regarding appearance and food safety regarding acrylamide. They suggested a short, high-temperature zone at the beginning of the deep-fat frying of potato crisps and then a progressive decrease in the temperature in the following zones. Electroporation, also known as pulsed electric field (PEF) processing, is based on the application of short pulses of high electric fields on food products. The results are pore formation in cell membrane and thus modified diffusion of intra- and extracellular media that leads to enhanced release of intracellular molecules from permeabilized tissue as well as improved uptake of low-molecular-weight substances into the sample. This technology is interesting since sugars are decreased in PEF-treated potatoes by higher release of cell liquid during drying (Janositz et al., 2011). The authors presumed that PEF is a capable assistance to thermal treatments in the processing of potato chips or French fries for the achievement of structural modifications and improved process conditions. Finally, PEF treatment results in reduced acrylamide formation during baking or deep-fat frying with lower absorption of oil.

Also in bread the main amount of acrylamide is formed in the crust as a result of higher temperature and lower moisture content during processing (Fig. 20.6). Up to 99% can be found there, while the content calculated on the whole loaves is relatively low due to low amounts in the crumb (Sadd and Hamlet, 2005). Another phenomenon is the evaporation or removal of acrylamide in dry cereal systems where the maximum level passes through at around 200°C depending on the system and the baking time and then declines (Brathen and Knutsen, 2005). Similar results were found by Becalski et al. (2003) for dry systems showing that the amount of acrylamide declined when the temperature was increased from 155°C to at least 185°C and more than 10 min heating. Brathen and Knutsen (2005) demonstrated that prolonged heating tended to reduce the amount of acrylamide in the dry systems but not in bread crusts. They assumed that water evaporation kept the effective temperature down when water was still present. In their experiments they found that the amount of acrylamide in flat breads increased only when the moisture content was below approximately 4% and lower amounts of acrylamide in flat breads than in the starch system. In bread the content of acrylamide is in most cases well correlated with the color and thus in a known process the color can be used as a parameter to determine the endpoint of the process (Stadler and Scholz, 2004).

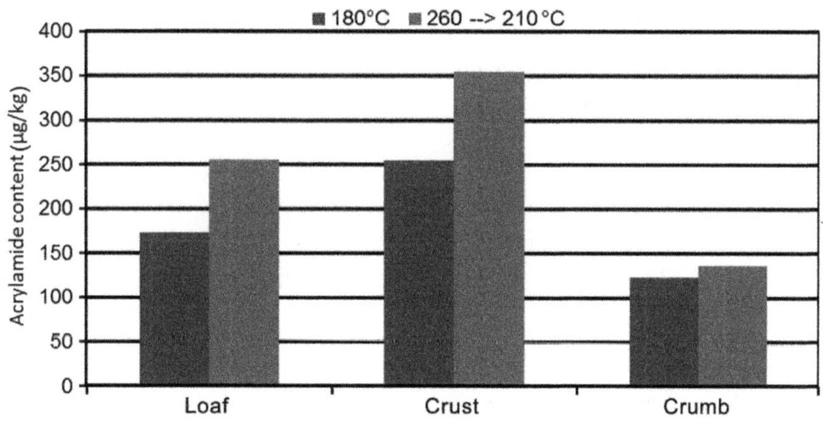

FIGURE 20.6

Influence of temperature on acrylamide formation in loaf, crust, and crumb of bread.

During the toasting of breakfast cereals the content of acrylamide increases when the moisture content falls below 5%. At this point color and flavor are formed. Some cereal products are produced by use of an extrusion cooker which gelatinizes starch but forms only a small Maillard reaction. Thus the content of acrylamide in such products is considerably below 100 μg/kg. If a toasting step is necessary for the processing the content of acrylamide may increase above this value.

Processing of coffee includes a roasting process at temperatures between 200°C and 250°C, which changes the chemical and physical properties of green coffee beans and forms the characteristic flavor of coffee by the Maillard reaction. At the beginning of roasting the amount of acrylamide increases exponentially, reaching a maximum and then decreases rapidly as the rate of degradation exceeds the rate of formation (Taeymans et al., 2004). Bagdonaite and Murkovic (2004) found that extended temperature and time lowered the acrylamide content, since temperature above 240°C and time longer than 5 min, gave a decrease in acrylamide formation.

Labeled acrylamide experiments have shown a most complete degradation during the process. Only a small percentage of the free asparagine is converted to acrylamide (Taeymans et al., 2005). Furthermore, acrylamide is not stable in the finished product, pointing to nonthermal interactions (Andrzejewski et al., 2004; Delatour et al., 2004). Consequently, light-roasted coffees may contain relatively higher amounts of acrylamide than dark-roasted ones.

As snack products, almonds are often roasted to form the typical pleasant roasted aroma that is preferred by consumers. During the roasting process acrylamide is formed in the almonds depending on the roasting temperature and time. Amrein et al. (2005b) showed that light roasting at 145°C resulted in low to moderate amounts of acrylamide (20–360 μg/kg), while temperatures at 165°C used during dark roasting led to contents up to 1500 μg/kg of acrylamide showing that the roasting temperature has a very strong influence on acrylamide formation in almonds. Lukac et al. (2007) showed that acrylamide formation in almonds started when the kernel temperature exceeded approximately 130°C and the activation energy for acrylamide formation during roasting of almonds was determined as 123 kJ/mol. They suggested that controlling of roasting temperature presents the critical factor for limiting acrylamide formation during roasting of almonds. Less acrylamide was found in almonds with higher initial moisture content at constant roasting conditions, probably due to the influence of moisture on the product temperature during roasting.

Time

In addition to the frying temperature, the time is decisive for the formation of acrylamide with a linear effect at given temperature and a higher slope at higher temperature (Matthäus et al., 2004). The frying time should be as short as possible to minimize the acrylamide formation but a higher oil uptake at lower frying temperature should be avoided. The alternative concept of temperature load (temperature × time (°Cs)) reflects the thermal input of frying. Principally, high-temperature frying will require less time to get a ready-to-eat food compared with low-temperature frying. Absolutely, temperature load values are inverse correlated to acrylamide data, also on a nonlinear level (Haase, 2006).

Surface-to-volume ratio

The formation of acrylamide in fried potato crisps and French fries mainly takes place at the surface as a result of continuous loss of water during processing, which strongly improves the conditions for acrylamide formation in this area (Matthäus et al., 2004). Gökmen et al. (2006) found 72, 2747, and 6474 μg/kg acrylamide in the surface of potato strips after frying at 150°C, 170°C, and 190°C, respectively, while in the core only 376 μg/kg were found after frying at 190°C.

An important parameter effecting the formation of acrylamide in fried potato products is the surface-to-volume ratio (SVR). Smaller pieces of potatoes comprise a greater surface, which results in higher acrylamide levels in small slices in comparison to larger slices, 1000–1500 μg/kg for 15-mm slices in comparison to 2500–13,000 μg/kg for 3-mm slices (Taubert et al., 2004). Similar results were found by Matthäus et al. (2004) for par-fried coarse-cut potato stripes (14 × 14 mm; SVR: 3.3 cm^{-1}) showing significantly ($p < 0.05$) lower amounts of acrylamide than fine cuts (8 × 8 mm; SVR: 5.4 cm^{-1}) using a constant frying time. Additionally, a constant increase in acrylamide formation with increasing temperature was only found in potato slices with low SVR (0.27 mm^{-1}) in comparison to samples with higher SVR (0.80 and 2.12 mm^{-1}). One reason for this effect is that increasing surface area for smaller slices increases the area of acrylamide formation due to water losses at the surface. Another reason could be that the heat transfer to the core may take longer in products with low SVR (Taubert et al., 2004).

Frying medium

The controversial effect of the frying medium on the formation of acrylamide has been discussed. Investigations have shown that oxidation products from degradation of fatty acids, such as α,β,γ,δ-diunsaturated carbonyl compounds, derived from the decomposition of hydroperoxides may affect the formation of acrylamide by competing effectively with carbohydrates for carbonyl-amine reactions within the Maillard reaction (Zamora and Hidalgo, 2005). Thus the authors suggested that unoxidized and oxidized lipids are able to contribute to the conversion of asparagine into acrylamide, whereby unoxidized lipids have to be oxidized in a first step (Zamora and Hidalgo, 2008). In contrast neither Matthäus et al. (2004) nor Williams (2005) found any influence of the oil quality on acrylamide formation during frying and also Mestdagh et al. (2007) stated from a model study with homogenized potato powder mixtures that acrylamide formation was independent of oxidation and hydrolysis status of the used oil. They found no evidence from the experimental results that, due to oxidative or hydrolytic oil degradation, heat transfer properties of the oil were changed to such an extent that acrylamide formation during French fries preparation would be significantly influenced. Finally, the authors concluded that the investigated oil degradation products, such as glycerol, mono-, and diacylglycerols, did not significantly influence the acrylamide formation. These results were confirmed by Kotsiou et al. (2013), showing no influence of the nature of lipids or the degree of fat oxidation on the formation of acrylamide.

Gertz and Klostermann (2002) explained higher amounts of acrylamide, especially in French fries fried in palm oil, by the relatively high amount of more polar mono- and diacylglycerols which may improve the heat transfer from the medium to the product. Zhang et al. (2015) depicted that French fries fried in oils having lower heat transfer coefficients containing lower acrylamide concentrations (913 μg/kg), whereas those fried with oils and with higher heat transfer coefficients contained higher acrylamide concentrations (1219 μg/kg). In model experiments with pure asparagine and different kinds of oil, Ehling et al. (2005) found higher amounts of acrylamide resulting from fish oils such as sardine oil (652 μg/g asparagine) and cod liver oil (435.4 μg/g), while oils such as rapeseed oil (70.6 μg/g), olive oil (73.6 μg/g), corn oil (80.7 μg/g), and soybean oil (135 μg/g) showed remarkably lower amounts. In contrast, Matthäus et al. (2004), Williams (2005), and Mestdagh et al. (2005) found no effect of the frying oil. Park et al. (2005) showed that the type of cooking oil (canola, cotton seed, peanut, safflower, shortening, soybean, and sunflower) has no important effect on the formation of acrylamide during frying of potato chip and French fries models. Only olive oil resulted in a 300% higher amount of acrylamide in comparison to corn oil (control). They explained this finding by the lower refining of olive oil in comparison to the other oils and the unusual use of olive oil for frying. Similar results were described by Becalski et al. (2003) who analyzed more acrylamide in samples fried in olive oils compared to corn oil. In contrast, Bird et al. (2009) found the lowest average acrylamide content in French fries produced in olive oil (65.12 ± 8.23 μg/kg) in comparison to other vegetable oils (260.22 ± 50.59 μg/kg) and lard (325.69 ± 85.07 μg/kg).

Oxidation products are suspicious for contributing to the formation of acrylamide. The use of carbon dioxide blanketing during frying reduces the thermo-oxidative degradation of the frying medium with the formation of total polar compounds or the anisidine value (Aladedunye and Przybylski, 2009), and also the formation of acrylamide. In a frying experiment, Aladedunye et al. (2011) showed a 58% reduction of acrylamide formation in comparison to conventional frying when using carbon dioxide blanketing at the same temperature.

Fermentation

During the production of some kinds of bread or other bakery products, such as cakes or biscuits, yeast is used for fermentation to release carbon dioxide and flavor and aroma compounds. Amino acids serve as a source of nitrogen for growth and asparagine is assimilated preferentially. Fredriksson et al. (2004)

showed that during 2 h of fermentation most of the asparagine was utilized in doughs (sifted wheat flour with whole grain wheat flour) reducing the content of free asparagine after the treatment down to less than 0.1 g/kg corresponding to 4 μg/kg acrylamide in the final product. They also found that longer fermentation times resulted in a lower potential for acrylamide formation. Similar results were found for rye bran.

CONCLUSIONS

The finding of acrylamide in a huge number of foods is a real challenge for the food industry due to the unfavorable toxicological assessment of the compound to humans. Therefore, the food industry has a duty to mitigate the contents in their products. Nowadays several approaches for the mitigation of acrylamide are available which can be applied along the production chain starting with cultivation and ending with processing. The main challenge for the food industry is reducing acrylamide in food products while maintaining the high product quality expected by the consumer. Since desirable aroma compounds are formed during the same reaction pathway of the Maillard reaction as acrylamide, changes in the reaction conditions (temperature, time, pH value, moisture, additives, etc.) directly influence the quality of the food. The approaches of choice depend on the food product and in most cases it will not be sufficient to apply only one approach. More promising is the combination of different approaches to reach an optimal reduction of acrylamide in ready-to-eat products. In the last 13 years enormous efforts have been done to mitigate acrylamide but the problem has not really been solved yet as several food categories still contain higher amounts of acrylamide. Therefore further efforts are still necessary.

REFERENCES

Aladedunye, F.A., Przybylski, R., 2009. Protecting oil during frying: a comparative study. Eur. J. Lipid Sci. Technol. 111, 893–901.

Aladedunye, F.A., Matthäus, B., Przybylski, R., 2011. Carbon dioxide blanketing impedes the formation of 4-hydroxynonenal and acrylamide during frying. A novel procedure for HNE quantification dioxide blanketing. Eur. J. Lipid Sci. Technol. 113, 916–923.

Amer, F.S., Reddivari, L., Madiwale, G.P., Stone, M., Holm, D.G., Vanamala, J., 2014. Effect of genotype and storage on glycoalkaloid and acrylamide content and sensory attributes of potato chips. Am. J. Potato Res. 91, 632–641.

Amrein, T., Bachmann, S., Noti, A., Biedermann, M., Ferraz Barbosa, M., Bidermann-Brem, S., et al., 2003. Potential of acrylamide formation, sugars, and free asparagine in potatoes: a comparison of cultivars and farming systems. J. Agric. Food Chem. 51, 5556–5560.

Amrein, T.M., Schonbachler, B., Escher, F., Amado, R., 2004. Acrylamide in gingerbread: critical factors for formation and possible ways for reduction. J. Agric. Food Chem. 52, 4282–4288.

Amrein, T.M., Andres, L., Schonbachler, B., Conde-Petit, B., Escher, F., Amado, R., 2005a. Acrylamide in almond products. Eur. Food Res. Technol. 221, 14–18.

Amrein, T.M., Lukac, H., Andres, L., Perren, R., Escher, F., Amado, R., 2005b. Acrylamide in roasted almonds and hazelnuts. J. Agric. Food Chem. 53, 7819–7825.

Amrein, T.M., Limacher, A., Conde-Petit, B., Amado, R., Escher, F., 2006. Influence of thermal processing conditions on acrylamide generation and browning in a potato model system. J. Agric. Food Chem. 54, 5910–5916.

Andrzejewski, D., Roach, J.A.G., Gay, M.L., Musser, S.M., 2004. Analysis of coffee for the presence of acrylamide by LC-MS/MS. J. Agic. Food Chem. 52, 1996–2002.

Anon., 1994. IARC International Agency for research on Cancer, Monographs on the evaluation of carcinogenic risks to human: some industrial chemicals. No. 60, pp. 389–433.

Anon., 2002. Scientific Committee on Food: Opinion of the Scientific Committee on Food on new findings regarding the presence of acrylamide in food. http://ec.europa.eu/food/fs/sc/scf/out131_en.pdf.

Anon., 2011. Commission Recommendation of 10.1.2011 on investigations into the levels of acrylamide in food (C(2010)9681 final), 2011, http://ec.europa.eu/food/food/chemicalsafety/contaminants/recommendation_10012011_acrylamide_food_en.pdf (accessed 27.08.15).

Anon., 2012. Update on acrylamide levels in food from monitoring years 2007 to 2010. EFSA J. 10, 2938.

Anon., 2013. Commission Recommendation of 8 November 2013 on investigations into the levels of acrylamide in food. Off. J. Euro.Union L 301/15.

Anon., 2014a. Food drink Europe: acrylamide toolbox 2013. <http://www.fooddrinkeurope.eu/uploads/publications_documents/FoodDrinkEurope_Acrylamide_Toolbox_2013.pdf> (accessed 25.08.15).

Anon., 2014b. Summary Report of the Standing Committee on the Food Chain and Animal Health held in Brussels on February 20, 2014. SANCO G (2014) 211687 <http://ec.europa.eu/food/committees/regulatory/scfcah/toxic/docs/sum_20140220_en.pdf>.

Anon., 2015. EFSA Panel on Contaminants in Food Chain (CONTAM), Scientific Opinion on acrylamide in food. EFSA J. 13, 4104.

Arias-Mendez, A., Warning, A., Datta, A.K., Balsa-Canto, E., 2013. Quality and safety driven optimal operation of deep-fat frying of potato chips. J. Food Eng. 119, 125–134.

Bagdonaite, K., Murkovic, M., 2004. Factors affecting the formation of acrylamide in coffee. Czech J. Food Sci. 22, 22–24.

Becalski, A., Lau, B., Lewis, D., Seaman, S.W., 2003. Acrylamide in foods: occurrence, sources, and modeling. J. Agric. Food Chem. 51, 802–808.

Biedermann, M., Grob, K., 2003a. Model studies on acrylamide formation in potato, wheat flour and corn starch; ways to reduce acrylamide contents in bakery ware. Mitt. Lebensm. Hyg. 94, 406–422.

Biedermann, M., Grob, K., 2003b. Model studies on acrylamide formation in potato, wheat flour and corn starch; ways to reduce acrylamide contents in bakery ware. Mitt. Geb. Lebensm. Unters. Hyg. 94, 406–422.

Bird, J., Guthrie, S., Flynn, N.E., 2009. Olive oil produces less acrylamide in French fries compared to lard as analyzed by HPLC and SPE. In: 237th ACS National Meeting, Salt Lake City, March 22–26, 2009.

Brathen, E., Knutsen, S.H., 2005. Effect of temperature and time on the formation of acrylamide in starch-based and cereal model systems, flat breads and bread. Food Chem. 92, 693–700.

Brathen, E.B., Kita, A., Knutsen, S.H., Wicklund, T., 2005. Addition of glycine reduces the content of acrylamide in cereal and potato products. J. Agric. Food Chem. 53, 3259–3264.

Bunger, A., Moyano, P.C., Rioseco, V., 2003. NaCl soaking treatment for improving the quality of French-fried potatoes. Food Res. Int. 36, 161–166.

Casado, F.J., Montano, A., 2008. Influence of processing conditions on acrylamide content in black ripe olives. J. Agric. Food Chem. 56, 2021–2027.

Cheng, K.W., Shi, J.J., Ou, S.Y., Wang, M.F., Jiang, Y., 2010. Effects of fruit extracts on the formation of acrylamide in model reactions and fried potato crisps. J. Agic. Food Chem. 58, 309–312.

CIAA, 2004. Acrylamide status report December 2004. A summary of the efforts and progress achieved to date by the European Food and Drink Industry (CIAA) in lowering levels of acrylamide in food, www.ciaa.be.

Ciesarova, Z., Kiss, E., Bögl, P., 2006. Impact of L-asparaginase on acrylamide content in potato products. J. Food Nutr. Res. 45, 141–146.

Claus, A., Schreiter, P., Weber, A., Graeff, S., Herrmann, W., Claupein, W., et al., 2006. Influence of agronomic factors and extraction rate on the acrylamide contents in yeast leavened breads. J. Agic. Food Chem. 54, 8968–8976.

Curtis, T.Y., Muttucumaru, N., Shewry, P.R., Parry, M.A., Powers, S.J., Elmore, J.S., et al., 2009. Evidence for genetic and environmental effects on free amino acid levels in wheat grain: implications for acrylamide formation during processing. J. Agric. Food Chem. 57, 1013–1021.

Curtis, T.Y., Powers, S.J., Balagiannis, D., Elmore, J.S., Mottram, D.S., Parry, M.A.J., et al., 2010. Free amino acids and sugars in rye grain: implications for acrylamide formation. J. Agric. Food Chem. 58, 1959–1969.

Delatour, T., Perisset, A., Goldmann, T., Riediker, S., Stadler, R.H., 2004. Improved sample preparation to determine acrylamide in difficult matrixes such as chocolate powder, cocoa, and coffee by liquid chromatography tandem mass spectroscopy. J. Agic. Food Chem. 52, 4625–4631.

De Wilde, T., De Meulenaer, B., Mestdagh, F., Govaert, Y., Ooghe, W., Fraselle, S., et al., 2006. Selection criteria for potato tubers to minimize acrylamide formation during frying. J. Agric. Food Chem. 54, 2199–2205.

Ehling, S., Hengel, M., Shibamoto, T., 2005. Formation of Acrylamide from Lipids. Springer, New York, NY.

Elmore, J.S., Mottram, D.S., Muttucumaru, N., Dodson, A.T., Parry, M.A.J., Halford, N.G., 2007. Changes in free amino acids and sugars in potatoes due to sulfate fertilization and the effect on acrylamide formation. J. Agic. Food Chem. 55, 5363–5366.

Eriksson, S., 2005. Acrylamide in Food Products: Identification, Formation and Analytical Methodology. Department of Environmental Chemistry, Stockholm University, Stockholm.

Fiselier, K., Bazzocco, D., Gama-Baumgartner, F., Grob, K., 2006. Influence of the frying temperature on acrylamide formation in French fries. Eur. Food Res. Technol. 222, 414–419.

Fredriksson, H., Tallving, J., Rosen, J., Aman, P., 2004. Fermentation reduces free asparagine in dough and acrylamide content in bread. Cereal Chem. 81, 650–653.

Gerendas, J., Heuser, F., Sattelmacher, B., 2007. Influence of nitrogen and potassium supply on contents of acrylamide precursors in potato tubers and on acrylamide accumulation in French fries. J. Plant Nutr. 30, 1499–1516.

Gertz, C., Klostermann, S., 2002. Analysis of acrylamide and mechanisms of its formation in deep-fried products. Eur. J. Lipid Sci. Technol. 104, 762–771.

Gökmen, V., Palazoglu, T.K., Senyuva, H.Z., 2006. Relation between the acrylamide formation and time-temperature history of surface and core regions of French fries. J. Food Eng. 77, 972–976.

Granda, C., Moreira, R.G., 2005. Kinetics of acrylamide formation during traditional and vacuum frying of potato chips. J. Food Process. Eng. 28, 478–493.

Granda, C., Moreira, R.G., Tichy, S.E., 2004. Reduction of acrylamide formation in potato chips by low-temperature vacuum frying. J. Food Sci. 69, E405–E411.

Granvogl, M., Jezussek, M., Koehler, P., Schieberle, P., 2004. Quantitation of 3-aminopropionamide in potatoes—a minor but potent precursor in acrylamide formation. J. Agric. Food Chem. 52, 4751–4757.

Granvogl, M., Wieser, H., Koehler, P., Von Tucher, S., Schieberle, P., 2007. Influence of sulfur fertilization on the amounts of free amino acids in wheat. Correlation with baking properties as well as with 3-aminopropionamide and acrylamide generation during baking. J. Agic. Food Chem. 55, 4271–4427.

Grothe, K.-H., Unbehend, G., Haase, N.U., Ludewig, G.-H., Matthäus, B., Vosmann, K., 2005. Einfluss von Backtriebmitteln auf die Acrylamidgehalte von Braunen Lebkuchen und Mürbkeksen. Getreidetechnologie 59, 163–167.

Haase, N.U., 2006. The formation of acrylamide in potato products. In: Skog, K., Alexander, J. (Eds.), Acrylamide and Other Hazardous Compounds in Heat-Treated Foods. Woodhead Publishing Ltd, Cambridge, pp. 41–59.

Haase, N.U., 2007. Minimization strategies: acrylamide. In: Eisenbrand, G., Engel, K.-H., Grunow, W., Hartwig, A., Knorr, D., Knudsen, I., et al., Thermal Processing of Food: Potential Health Benefits and Risks. Wiley VCH, Weinheim, pp. 155–174.

Haase, N.U., Matthäus, B., Vosmann, K., 2004. Aspects of acrylamide formation in potato crisps. J. Appl. Bot. Food Qual. 78, 144–147.

Haase, N.U., Weber, L., 2003. Variability of sugar content in potato varieties suitable for processing. Food Agric. Environ. 1, 80–81.

Haase, N.U., Matthäus, B., Vosmann, K., 2003. Minimierungsansätze zur Acrylamid-Bildung in pflanzlichen Lebensmitteln—aufgezeigt am Beispiel von Kartoffelchips. Deutsche Lebensmittel-Rundschau 99, 87–90.

Haase, N.U., Grothe, K.H., Matthäus, B., Vosmann, K., Lindhauer, M.G., 2012. Acrylamide formation and antioxidant level in biscuits related to recipe and baking. Food Addit. Contam. Part A 29, 1230–1238.

Halford, N.G., Curtis, T.Y., Muttucumaru, N., Postles, J., Elmore, J.S., Mottram, D.S., 2012. The acrylamide problem: a plant and agronomic science issue. J. Exp. Bot. 63, 2841–2851.

Hamlet, C.G., Sadd, P.A., Liang, L., 2005. Wheat variation, proving and acrylamide formation in bread. In: Poms, R.E., Cauvain, S.P. (Eds.), Cereals—The Future Challenge. ICC conference, Vienna, p 87., July 3–6, 2005.

Harkett, P.J., 1971. The effect of oxygen concentration on the sugar content of potato tubers stored at low temperature. Potato Res. 14, 305–311.

Hedegaard, R.V., Granby, K., Frandsen, H., Thygesen, J., Skibsted, L.H., 2008. Acrylamide in bread. Effect of prooxidants and antioxidants. Eur. Food Res. Technol. 227, 519–525.

Hendriksen, H.V., Budolfsen, G., Baumann, M.J., 2013. Asparaginase for acrylamide mitigation in food. Aspects Appl. Biol. 116, 41–50.

Ismial, S.A.M.A., Ali, R.F.M., Askar, M., Samy, W.M., 2013. Impact of pre-treatments on the acrylamide formation and organoleptic evolution of fried potato chips. Am. J. Biochem. Biotechnol. 9, 90–101.

Izard, C., Libermann, C., 1978. Acrolein. Mutat. Res. 47, 115–138.

Janositz, A., Noack, A.K., Knorr, D., 2011. Pulsed electric fields and their impact on the diffusion characteristics of potato slices. LWT-Food Sci. Technol. 44, 1939–1945.

Jin, C., Wu, X., Zhang, Y., 2013. Relationship between antioxidants and acrylamide formation: a review. Food Res. Int. 51, 611–620.

Jung, M.Y., Choi, D.S., Ju, J.W., 2003. A novel technique for limitation of acrylamide formation in fried and baked corn chips and in French fries. J. Food Sci. 68, 1287–1290.

Kim, E.T., Hwang, E.S., Lee, H.J., 2005. Reducing acrylamide in fried snack products by adding amino acids. J. Food Sci. 70, C354–C358.

Kita, A., Brathen, E., Knutsen, S.H., Wicklund, T., 2004. Effective ways of decreasing acrylamide content in potato crisps during processing. J. Agric. Food Chem. 52, 7011–7016.

Knutsen, S.H., Dimitrijevic, S., Molteberg, E.L., Segtnan, V.H., Kaaber, L., Wicklund, T., 2009. The influence of variety, agronomical factors and storage on the potential for acrylamide formation in potatoes grown in Norway. LWT-Food Sci. Technol. 42, 550–556.

Kotsiou, K., Tasioula-Margari, M., Fiore, A., Gokmen, V., Fogliano, V., 2013. Acrylamide formation and colour development in low-fat baked potato products as influenced by baking conditions and oil type. Eur. Food Res. Technol. 236, 843–851.

Kumar, N.S.M., Shimray, C.A., Indrani, D., Manonmani, H.K., 2014. Reduction of Acrylamide formation in sweet bread with L-asparaginase treatment. Food Bioprocess Technol. 7, 741–748.

Lea, P.J., Sodek, L., Parry, M.A.J., Shewry, R., Halford, N.G., 2007. Asparagine in plants. Ann. Appl. Biol. 150, 1–26.

Lingnert, H., Eriksson, C.E., 1980. Antioxidative Maillard reaction products. I. Products from sugars and free amino acids. J. Food Process. Preserv. 4, 161–172.

Loaec, G., Niquet-Leridon, C., Henry, N., Jacolot, P., Volpoet, G., Goudemand, E., et al., 2014. Effects of variety, agronomic factors, and drying on the amount of free asparagine and crude protein in chicory. Correlation with the acrylamide formation during roasting. Food Res. Int. 63, 299–305.

Lukac, H., Amrein, T.M., Perren, R., Conde-Petit, B., Amado, R., Escher, F., 2007. Influence of roasting conditions on the acrylamide content and the color of roasted almonds. J. Food Sci. 72, C33–C38.

Martinek, P., Klem, K., Vanova, M., Bartackova, V., Vecerkova, L., Bucher, P., et al., 2009. Effects of nitrogen nutrition, fungicide treatment and wheat genotype on free asparagine and reducing sugars content as precursors of acrylamide formation in bread. Plant Soil Environ. 55, 187–195.

Matsuura-Endo, C., Ohara-Takada, A., Chuda, Y., Ono, H., Yada, H., Yoshida, M., et al., 2006. Effects of storage temperature on the contents of sugars and free amino acids in tubers from different potato cultivars and acrylamide in chips. Biosci. Biotechnol. Biochem. 70, 1173–1180.

Matthäus, B., Haase, N.U., Vosmann, K., 2004. Factors affecting the concentration of acrylamide during deep-fat frying of potatoes. Eur. J. Lipid Sci. Technol. 106, 793–801.

May, N.J., Tong, P., Varelis, P., Paterson, J.L., 2006. Acrylamide formation in deep-fried potato products and removal of acrylamide precursors. Food Australia 58, 488–493.

Mestdagh, F.J., De Meulenaer, B., Van Poucke, C., Detavernier, C., Cromphout, C., Van Peteghem, C., 2005. Influence of oil type on the amounts of acrylamide generated in a model system and in French fries. J. Agric. Food Chem. 53, 6170–6174.

Mestdagh, F., de Meulenaer, B., van Peteghem, C., 2007. Influence of oil degradation on the amounts of acrylamide generated in the model system and in French fries. Food Chem. 100, 1153–1159.

Mills, C., Mottram, D.S., Wedzicha, B.L., 2009. Acrylamide. In: Stadler, H.R., Lineback, D.R. (Eds.), Process-Induced Food Toxicants: Occurrence, Formation, Mitigation, and Health Risks. Wiley, Hoboken, NJ, pp. 23–50.

Miskiewicz, K., Nebesny, E., Oracz, J., 2012. Formation of acrylamide during baking of shortcrust cookies derived from various flours. Czech J. Food Sci. 30, 53–66.

Mottram, D.S., Wedzicha, B.L., Dodson, A.T., 2002. Acrylamide is formed in the Maillard reaction. Nature 419, 448–449.

Muttucumaru, N., Powers, S.J., Elmore, J.S., Mottram, D.S., Halford, N.G., 2013. Effects of nitrogen and sulfur fertilization on free amino acids, sugars, and acrylamide-forming potential in potato. J. Agric. Food Chem. 62, 6734–6742.

Natella, F., Nardini, M., Giannetti, I., Dattilo, C., Scaccini, C., 2002. Coffee drinking influences plasma antioxidant capacity in humans. J. Agic. Food Chem. 50, 6211–6216.

Park, Y.W., Yang, H.W., Storkson, J.M., Albright, K.J., Liu, W., Lindsay, R.C., et al., 2005. Controlling acrylamide in French fry and potato chip models and a mathematical model of acrylamide formation—acrylamide: acidulants, phytate and calcium. In: Friedman, M., Mottram, D. (Eds.), Chemistry and Safety of Acrylamide in Food. Springer Science + Business Media, Inc., New York, NY, pp. 343–356.

Pedreschi, F., Kaack, K., Granby, K., 2004. Reduction of acrylamide formation in potato slices during frying. Lebensm.-Wiss. u.- Technol. 37, 679–685.

Pedreschi, F., Bustos, O., Mery, D., Moyano, P., Kaack, K., Granby, K., 2007a. Color kinetics and acrylamide formation in NaCl soaked potato chips. J. Food Eng. 79, 989–997.

Pedreschi, F., Kaack, K., Granby, K., Troncoso, E., 2007b. Acrylamide reduction under different pre-treatments in French fries. J. Food Eng. 79, 1287–1294.

Rommens, C.M., Yan, H., Swords, K., Richael, C., Ye, J., 2008. Low acrylamide French fries and potato chips. Plant Biotechnol. J. 6, 843–853.

Rydberg, P., Erikson, S., Tareke, E., Karlsson, P., Ehrenberg, L., Törnqvist, M., 2003. Investigation of factors that influence the acrylamide content of heated foodstuffs. J. Agric. Food Chem. 51, 7012–7018.

Sadd, P., Hamlet, C., 2005. The formation of acrylamide in UK cereal products. In: Friedman, M., Mottram, D. (Eds.), Chemistry and Safety of Acrylamide in Food. Springer, New York, NY, pp. 415–429.

Springer, M., Fischer, T., Lehrack, A., Freund, W., 2003. Development of acrylamide in baked products. Getride 57, 274–278.

Stadler, R.H., Scholz, G., 2004. Acrylamide: an update on current knowledge in analysis, levels in food, mechanisms of formation, and potential strategies of control. Nutr. Rev. 62, 449–467.

Stadler, R.H., Blank, I., Varga, N., Robert, F., Hau, J., Guy, P.A., et al., 2002a. Acrylamide from Maillard reaction products. Nature 419, 449–450.

Stadler, R.H., Verzegnassi, L., Varga, N., Grigorov, M., Studer, A., Riediker, S., et al., 2002b. Formation of vinylogous compounds in model Maillard reaction systems. Chem. Res. Toxicol. 16, 1242–1250.

Taeymans, D., Wood, J., Ashby, P., Blank, I., Studer, A., Stadler, R., et al., 2004. A review of acrylamide: an industry perspective on research, analysis, formation, and control. Crit. Rev. Food Sci. Nutr. 44, 323–347.

Taeymans, D., Andersson, A., Ashby, P., Blank, I., Gonde, P., Eijck van, P., et al., 2005. Acrylamide: update on selected research activities conducted by the European Food and Drink Industry. J. AOAC. Int. 88, 234–241.

Tareke, E., Rydberg, P., Karlsson, P., Törnqvist, S., Eriksson, M., 2000. Acrylamide—a cooking carcinogen? Chem. Res. Toxicol. 13, 517–522.

Tareke, E., Rydberg, P., Karlsson, P., Eriksson, S., Törnqvist, M., 2002. Analysis of acrylamide, a carcinogen formed in heated foodstuffs. J. Agric. Food Chem. 50, 4998–5006.

Taubert, D., Harlfinger, S., Henkes, L., Berkels, R., Schomig, E., 2004. Influence of processing parameters on acrylamide formation during frying of potatoes. J. Agric. Food Chem. 52, 2735–2739.

Törnqvist, M., 2005. Acrylamide in food: the discovery and its implications: a historical perspective. Adv. Exp. Med. Biol. 561, 1–19.

Wagner, K.H., Reichhold, S., Koschutnig, K., Chériot, S., Billaud, C., 2007. The potential antimutagenic and antioxidant effects of Maillard reaction products used as "natural antibrowning" agents. Mol. Nutr. Food Res. 51, 496–504.

Waltz, E., 2015. USDA approves next-generation GM potato. Nat. Biotechnol. 33, 12–13.

Wedzicha, B.L., Mottram, D.S., Elmore, J.S., Koutsidis, G., Dodson, A.T., 2005. Kinetic models as a route to control acrylamide formation in food. In: Friedman, M., Mottram, D.S. (Eds.), Chemistry and Safety of Acrylamide in Food. Springer, New York, NY, pp. 235–253.

Weisshaar, R., Gutsche, B., 2002. Formation of acrylamide in heated potato products—model experiments pointing to asparagine as precursor. Deutsche Lebensmittel-Rundschau 98, 397–400.

Williams, J.S.E., 2005. Influence of variety and processing conditions on acrylamide levels in fried potato crisps. Food Chem. 90, 875–881.

Yasuhara, A., Tanaka, Y., Hengel, M., Shibamoto, T., 2003. Gas chromatographic investigation of acrylamide formation in browning model systems. J. Agric. Food Chem. 51, 3999–4003.

Yaylayan, V., Wnorowski, A., Perez Locas, C., 2003. Why asparagine needs carbohydrates to generate acrylamide. J. Agric. Food Chem. 51, 1753–1757.

Zamora, R., Hidalgo, F.J., 2005. Coordinate contribution of lipid oxidation and Maillard reaction to the nonenzymatic food browning. Crit. Rev. Food Sci. Nutr. 45, 49–59.

Zamora, R., Hidalgo, F.J., 2008. Contribution of lipid oxidation products to acrylamide formation in model systems. J. Agric. Food Chem. 56, 6075–6080.

Zhang, Y., Zhang, Y., 2008. Effect of natural antioxidants on kinetic behavior of acrylamide formation and elimination in low-moisture asparagine–glucose model system. J. Food Eng. 85, 105–115.

Zhang, H., Zhang, H., Cheng, L., Wang, L., Qian, H., 2015. Influence of deep-frying using various commercial oils on acrylamide formation in French fries. Food Addit. Contam. Part A Chem. Anal. Control Expo. Risk Assess. 32, 1083–1088.

Zhou, D., Solomos, T., 1998. Effect of hypoxia on sugar accumulation, respiration, activities of amylase and starch phosphorylase, and induction of alternative oxidase and acid invertase during storage of potato tubers (*Solanum tuberosum* cv. Russet Burbank) at 1°C. Physiol. Plant. 104, 255–265.

Zyzak, D., Sandersm, R., Stojanovic, M., Tallmadge, D., Eberhart, L., Ewald, D., et al., 2003. Acrylamide formation mechanism in heated foods. J. Agric. Food Chem. 51, 4782–4787.

CHAPTER

Furan in processed foods 21

Adriana Pavesi Arisseto

Faculty of Food Engineering, State University of Campinas, Campinas, Sao Paulo, Brazil

INTRODUCTION

Food processing techniques such as fermentation, smoking and those ones involving thermal treatment (example: pasteurization, roasting, baking), have been largely employed to reduce microbiological contamination, increase shelf-life and improve the sensory characteristics of foods. From a nutritional perspective, modern food manufacturing contributes to the preservation of nutrients and/or improvement of nutrient profile by adding important vitamins and minerals, while removing toxins and antinutritional factors. Additionally, waste and product loss can be reduced, and distribution around the world can be facilitated to allow seasonal availability of many foods. Therefore, the positive impact of food processing in today's food supply is evident (Floros et al., 2010).

However, these manufacturing techniques can also cause undesirable changes in food, including the formation of toxic compounds (Stadler and Lineback, 2009). These contaminants, which are not detected in the raw material, can be found in processed foods and beverages as a result of the transformation process and, in most cases, through chemical reactions between naturally occurring compounds and/or added ingredients. A number of these substances are formed under high-temperature conditions, such as furan in roasted, baked, and jarred/canned products.

The presence of furan in the diet is a concern since this chemical is a potent hepatotoxin and hepatocarcinogen in rodents, and has been classified by the International Agency for Research on Cancer (IARC) as a possible human carcinogen (group 2B) (IARC, 1995). For this reason, furan has received special attention from the scientific community and food regulatory agencies in recent years.

FURAN

Furan (C_4H_4O) (CAS No. 110-00-9) is chemically classified as a cyclic dienic ether. It is a colorless and flammable liquid, with an ethereal odor, low molecular weight (68.08), high volatility, low boiling point (31.36°C), and a specific gravity

P. Kotzekidou (Ed): Food Hygiene and Toxicology in Ready-to-Eat Foods.
DOI: http://dx.doi.org/10.1016/B978-0-12-801916-0.00021-2

FIGURE 21.1

Furan chemical structure.

of 0.9371 at 101.1 kPa. Furan is slightly soluble in water and is soluble at greater than 10% in acetone, benzene, ether, and ethanol (NTP, 2014). All ring atoms of furan lie in a plane and form a slightly distorted pentagon (Fig. 21.1).

Furan is industrially produced by decarbonylation of furfural and primarily used as an intermediate in the synthesis and production of other organic compounds such as tetrahydrofuran, pyrrole, and thiophene. Furan is also used in the formation of lacquers, as a solvent for resins, and in the production of agricultural chemicals, stabilizers, and pharmaceuticals (NTP, 2014).

TOXICOLOGICAL ASPECTS

The toxicological aspects of furan have been reviewed in some publications (Moro et al., 2012; Mariotti et al., 2013a). A safety evaluation was carried out by the Joint FAO/WHO Expert Committee on Food Additives (JECFA) in its seventy-second meeting in 2010 (FAO/WHO, 2011).

ABSORPTION, DISTRIBUTION, METABOLISM, AND EXCRETION

Following oral administration to mice and rats, furan can easily pass through biological membranes, being rapidly and extensively absorbed from the intestines. It has been shown from studies using isotopically labeled furan (2,5-^{14}C) that the main portion of the radioactivity remaining in the tissues 24 h post-dosing (19% of the dose) was found in the liver (13%) followed by the kidney and gastrointestinal tract (together $<1\%$) (Burka et al., 1991).

In the liver, furan is rapidly metabolized by cytochrome P-450 enzymes (especially CYP2E1) via ring opening to form carbon dioxide and *cis*-2-butene-1,4-dialdehyde (BDA). BDA, which has been identified as its main metabolite, is considered to be cytotoxic and may bind irreversibly to proteins and nucleosides (Burka et al., 1991; Byrns et al., 2002). A possible endogenous formation of furan was suggested by some authors after finding the chemical in liver and blood

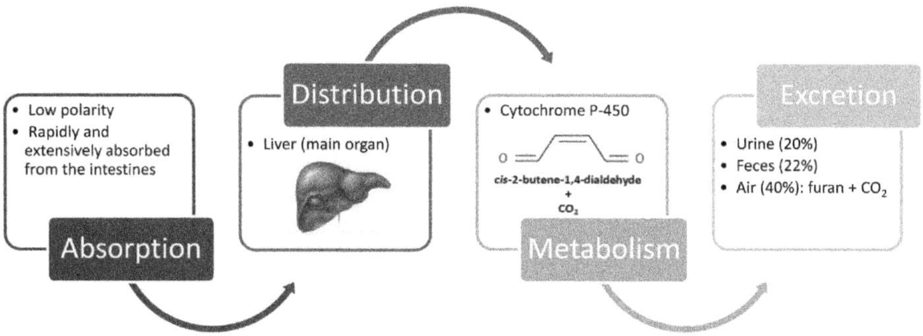

FIGURE 21. 2

Toxicokinetic aspects of furan (ADME).

tissues of rats at higher amounts than those expected from the contribution of the feed (Becalski et al., 2013).

Excretion is via feces and urine (as metabolites) and exhaled air (as unchanged furan and carbon dioxide formed as a result of ring opening). In rats, approximately 80% of the isotopically labeled furan dose was eliminated by air (40%), feces (22%), and urine (20%), 24 h following oral administration (Burka et al., 1991). The fraction eliminated by the air consisted of furan (14%) and carbon dioxide (26%). At least 10 metabolites were identified in urine. The toxicokinetic aspects of furan are illustrated in Fig. 21.2.

TOXICITY

The toxicity of orally administered furan has been studied in mice and rats considering a wide range of doses. The liver is the primary site of furan toxicity, although the kidneys and lungs are also affected at high doses (>30 mg/kg body weight (bw) per day). Changes in some hematological and hormonal parameters can occur at doses as low as 0.12 mg/kg bw per day administered 5 days/week (FAO/WHO, 2011).

Acute, subacute, and subchronic oral toxicity studies

The LD_{50} (dose with 50% lethality) of furan in solution administered by intraperitoneal injection was 6.94 mg/kg for Swiss mice (males and females) and 5.20 mg/kg for male Sprague–Dawley rats. The median lethal concentration (LC_{50}) for mice was 0.12 μg/mL (Egle and Gochberg, 1979). Acute oral exposure generally resulted in transient hepatotoxicity and no mortality was seen at doses up to 30 mg/kg per day in rats and 50 mg/kg per day in mice, which were the highest tested doses (Wilson et al., 1992). Subacute exposure (16 days) resulted in mortality at doses from 40 mg/kg per day in mice and from 80 mg/kg per day in rats (NTP, 1993).

Subchronic exposure (13 weeks) produced substantial hepatotoxicity in rats, with cholangiofibrosis and hyperplasia of the biliary tract occurring at the lowest tested dose (4 mg/kg per day) (NTP, 1993). More recent subchronic studies in Fischer-344 rat and B6C3F1 mouse were conducted extending the dose to a lower range (0.0, 0.03, 0.12, 0.5, 2.0, and 8.0 mg/kg bw per day) in order to identify a no-observed adverse effect level (NOAEL) for hepatotoxicity and to characterize non-neoplastic effects (Gill et al., 2010, 2011). A NOAEL of 0.03 and 0.12 mg/kg bw per day of furan were suggested for rats and mice, respectively. Oral treatment of furan to male rats over 90 days has also caused toxicity in the reproductive system, affecting LH (luteinizing hormone) and testosterone levels, organ weights and testicular steroidogenesis (Karacaoglu and Selmanoglu, 2010; Cooke et al., 2014). Furan does not seem to be teratogenic since it showed low toxicity in the frog embryo teratogenesis assay—Xenopus (Williams et al., 2014).

Chronic oral toxicity studies

Chronic exposure of rats and mice to furan showed a dose-dependent increase in hepatocellular adenomas and carcinomas in both sexes (NTP, 1993; Moser et al., 2009). A statistically significant increase in the incidence of these lesions was observed in animals treated with 4 and 8 mg/kg bw per day, but not in mice dosed with 0.5, 1.0, or 2.0 mg/kg bw per day (Moser et al., 2009).

For risk assessment purposes, JECFA derived a $BMDL_{10}$ (Benchmark Dose Lower Confidence Limit for a 10% Extra Risk of Tumors) of 1.34 mg/kg bw per day, which corresponds to 0.96 mg/kg bw per day when adjusted from a 5 days/week dosing schedule to an average daily dose, in female mice considering the hepatocellular adenoma and carcinoma data from the Moser et al. (2009) study as the point of departure (FAO/WHO, 2011).

The mode of action (MOA) of furan carcinogenicity may differ in rats and mice. In rats, the liver tumors occur at very high incidence and low dose, which is characteristic of a genotoxic MOA. However, the mouse tumors have been regarded by most authors as being the result of a cytotoxicity-driven proliferation in the liver (Carthew et al., 2010). More recent studies have highlighted the possible involvement of epigenetic modifications in the furan hepatotoxicity and carcinogenicity as well as evidences of a secondary genotoxic mechanism involving oxidative stress (de Conti et al., 2014, 2015). BDA, which is viewed as a potential genotoxic carcinogen, may not be directly responsible for the in vivo effects of furan on mutational spectra, since the mutation frequency did not increase in furan-treated female mice, supporting an indirect mechanism of genotoxicity (Terrell et al., 2014).

ANALYTICAL METHODS

Published methods for the analysis of furan in foods employ gas chromatography coupled to mass spectrometry (GC-MS). Due to its high volatility, headspace sampling (HS) (Fig. 21.3A) and solid-phase microextraction (SPME) (Fig. 21.3B)

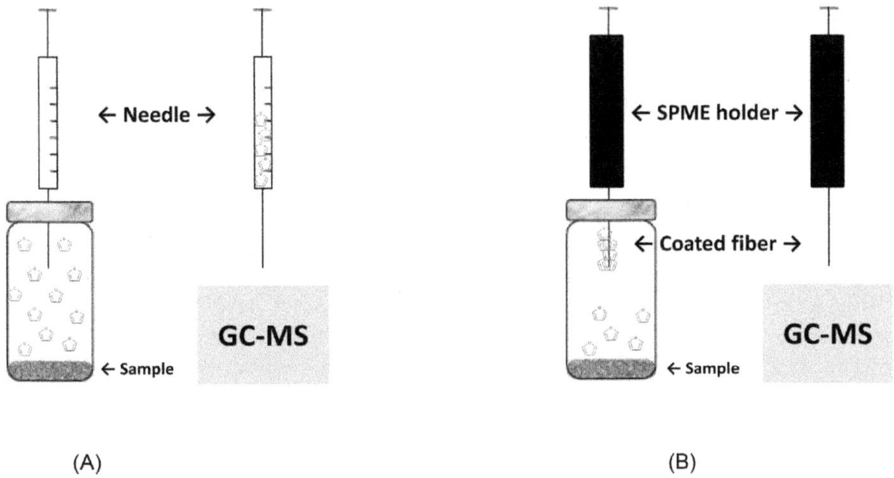

FIGURE 21.3

Extraction techniques used for furan analysis (*SPME*, solid phase microextraction; *GC-MS*, gas chromatograph—mass spectrometry). (A) Headspace; (B) SPME.

have been the most commonly used extraction techniques for different matrices. Although SPME has presented some advantages over direct HS extraction, such as lower detection limits, it has already been observed that when volatile furan precursors, including 2-butenal and furfural, are adsorbed on the CAR-PDMS fiber, additional amounts of furan can be formed on the fiber during thermal desorption, overestimating the results (Adams et al., 2012).

Other techniques have also been proposed for furan analysis including solid-phase dynamic extraction (Ridgway et al., 2010), SPME with two different coatings synthesized by sol—gel technology coupled with gas chromatography—flame ionization detector (GC—FID) (Sarafraz-Yazdi et al., 2012) and electrochemical methods (Hoveidi et al., 2013).

FORMATION AND OCCURRENCE OF FURAN IN FOODS

The occurrence of furan and furan-substituted compounds in foods was initially observed in studies aiming at identifying volatile substances generated during heating, responsible for the aroma of processed products. The first evidences of the occurrence of furan between the volatile components of coffee were raised in 1938 and confirmed only a few decades later (Merritt et al., 1963). Subsequent studies indicated the presence of the chemical in other foods such as canned beef, cooked chicken, and bread. The formation of furan was initially associated with the thermal degradation of carbohydrates (Maga, 1979).

More recently, the US Food and Drug Administration (US FDA) showed that furan may be present in a variety of thermally processed foods, including canned/jarred baby foods, sauces, soups, and vegetables (US FDA, 2004). It is well known nowadays that furan is a byproduct of thermal treatments and high-energy radiation of food.

THERMAL-INDUCED FORMATION OF FURAN

In general, higher temperatures produce higher levels of furan, especially above 120°C (Hasnip et al., 2006; Arisseto et al., 2011). Higher amounts of furan are normally formed under roasting conditions (dry heating, 200°C) compared with pressure-cooking conditions (sterilization, 121°C) (Limacher et al., 2007, 2008).

The exact mechanism of furan formation is not completely understood. The presence of the contaminant in a wide variety of foods suggests the existence of multiple formation routes rather than a single pathway. As shown in Fig. 21.4, the thermal degradation of sugars, ascorbic acid, and amino acids, the Maillard reaction between reducing sugars and amino acids, and the thermal oxidation of polyunsaturated fatty acids and carotenoids have been considered the main routes of furan formation (Locas and Yaylayan, 2004; Becalski and Seaman, 2005; Märk et al., 2006). Temperature, pH, moisture, and matrix components have been considered important factors that influenced furan formation in model systems (Fan et al., 2008; Huang et al., 2011; Owczarek-Fendor et al., 2012).

Initial investigations on furan in foods focused on heated products sealed in cans and jars, such as baby foods, infant formulas, canned vegetables, baked beans, soups, sauces, stews, and canned meats and fishes. Furan has also been found in coffee, beer, fruit and vegetable juices, soy sauce, nutritional drinks, and

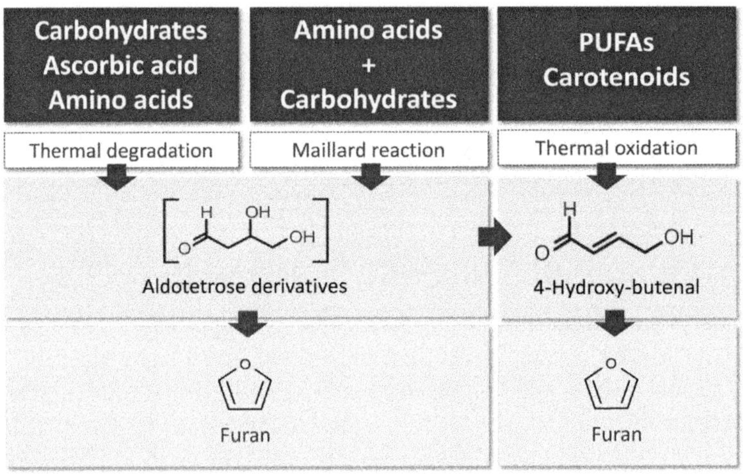

FIGURE 21.4

Main routes of furan formation (*PUFAs*, polyunsaturated fatty acids).

cereal-based foods, such as biscuits, crackers, breakfast cereals, and bread (CAC, 2011). Home-cooked foods with high levels of carbohydrates are also likely to form furan during common cooking practices (Fromberg et al., 2009).

Levels of furan in foods are available from several databases, such as those from US FDA (2009), JECFA (FAO/WHO, 2011), and European Food Safety Authority (EFSA, 2011). Monitoring studies have also been conducted by several authors in different countries (Zoller et al., 2007; Morehouse et al., 2008; Arisseto et al., 2010, 2012; Kim et al., 2009; Wegener and López-Sánchez, 2010; Scholl et al., 2012a; Waizenegger et al., 2012; Mariotti et al., 2013b; Mesias et al., 2013; Nie et al., 2013; Wu et al., 2014). Mean furan concentrations up to 4895 μg/kg have been reported, with the highest levels (up to 11,000 μg/kg) being found in roasted coffee beans (EFSA, 2011). The most important food categories in relation to furan occurrence are discussed below.

Foods processed in closed containers

Furan formed during thermal processes is easily vaporized and occupies the head-space of canned or jarred foods (Mariotti et al., 2013a). According to EFSA (2011), lower bound mean concentrations in this food category were 31 μg/kg in baby foods, 22 μg/kg in baked beans, 13 μg/kg in meat products, 23 μg/kg in soups, and 6.9 μg/kg in vegetables. Baby foods containing vegetables or vegetable and meat mixtures have consistently been reported to have higher furan levels than baby foods containing fruit only, meat only, or meat and starches.

As a consequence of its low boiling point, furan concentrations significantly decrease when the food product is reheated at moderate temperatures, especially in a saucepan or when it is stirred during reheating (Zoller et al., 2007; La Pera et al., 2009; Palmers et al., 2015). Furan volatilization during heating is limited by the food matrix, though. It has been reported that lipophilic compounds (such as oils) caused significant retention of furan (Van Lancker et al., 2009). HPHT (high-pressure/high-temperature) processing was proven to be an interesting alternative for furan reduction in vegetable-based systems, when aiming for sterilization intensities (Palmers et al., 2014). No furan formation was observed in meat, cereal, and vegetables prepared in a pressure-cooker (Arisseto et al., 2013).

Coffee

Roasted coffee (beans and grounded powder) is particularly high in furan. EFSA (2011) provided the following mean values for coffee categories: instant coffee, 394 μg/kg; roasted coffee beans, 3660 μg/kg; and roasted ground coffee, 1936 μg/kg. The high levels of furan found in coffee are due to the temperatures used in the roasting process, which are generally higher than those employed in most other thermal treatments. The amount of furan formed in beans increases with the roasting degree, but the influence of species (*Coffea arabica* and *Coffea canephora*) is not well understood (Guenther et al., 2010; Arisseto et al., 2011).

The low concentrations observed in instant coffee could be attributed to the volatilization of the contaminant during the extraction and concentration steps employed to obtain this product. The package also plays a role on furan levels of the final product, with the highest concentrations observed in vacuum-packed roasted ground coffee (Arisseto et al., 2012).

Mean and maximum furan levels in coffee brews reported by EFSA (2011) were 42 and 360 μg/kg, respectively. In contrast to roasted ground coffee, much lower concentrations are found in the beverages, which might be associated with furan volatilization due to the high temperature of water as well as its poor solubility in this solvent (Arisseto et al., 2012). It is estimated that only ~10% of the initially generated furan during roasting gets into the cup of coffee for consumption (Guenther et al., 2010). Automatic coffee machines produced coffee brews with the highest amounts of furan due to the higher ratio of coffee powder to water and because of the closed system that favors its retention. Furan content in beverages decreased from 170 up to 4 ng/mL after simulating different behaviors of coffee consumption such as standing at room temperature for 5 min, stirring for 5 min, and being cooled for 1 week in a fridge (Mesias and Morales, 2014).

Cereal products

For cereal products, EFSA (2011) reported mean and maximum furan levels of 15 and 168 μg/kg, respectively. In a survey conducted in Brazil, the highest furan concentrations were found in breakfast cereals (corn flakes), crackers (cream crackers), and biscuits (wafers), while the lowest levels were observed in loaf breads and cakes (Arisseto et al., 2012). Few studies are available on the formation of furan in this food category. In toasted bread, furan amounts varied between 3 and 12–129 ng/g for samples treated at temperatures of about 140–160°C and 170–200°C, respectively (Hasnip et al., 2006). It has been demonstrated that the color of toasted bread is directly proportional to the level of furan, with the highest concentrations observed in darker toasted bread (Fromberg et al., 2009). Other authors reported that higher levels of furan are found in the bread crust (Zoller et al., 2007).

NONTHERMAL-INDUCED FORMATION OF FURAN

Ionizing radiation has been reported to induce furan formation in fruits, fruit juices, and in simple model systems (solutions of glucose, fructose, sucrose, ascorbic acid) (Fan, 2005a,b; Fan and Sokorai, 2008). This effect is exerted through generation of free radicals from radiolysis of water, which react with food components, such as ascorbic acid and sugars, resulting in the formation of furan. UV treatment has also been reported to induce the formation of the contaminant (Fan, 2015). Interestingly, the storage of unsaturated fatty acid emulsions at 25°C for 3 days led to the formation of furan (7–11 ng/mL) even without prior thermal or nonthermal treatments (Fan, 2015). Similar observations were reported in vegetable purees after 5 months of storage at 35°C (Palmers et al., 2015).

HUMAN EXPOSURE, RISKS, AND MITIGATION STRATEGIES

Humans are constantly exposed to furan from foods, and the assessment of the risks to health from this exposure is relevant and necessary. Measured furan concentrations in selected foods have been combined with data on food consumption, by using deterministic or probabilistic approaches, to give an estimate of furan intake through the diet. Data from Europe, North America, South America, and Asia can be found in the literature (Morehouse et al., 2008; Arisseto et al., 2010; EFSA, 2011; FAO/WHO, 2011; Mariotti et al., 2013b; Scholl et al., 2012b; Mesias and Morales, 2014; Wu et al., 2014). Coffee has been considered the major contributor to dietary furan exposure for adults (40–80%) (FAO/WHO, 2011).

In its seventy-second meeting, JECFA reported that intakes of 1 and 2 μg/kg bw per day could be used to represent the dietary exposure of average and high consumers to furan, respectively. These values were considered to be sufficient to cover potential dietary exposures of infants and children. Based on the margin of exposure (MOE) approach for risk characterization, the Committee concluded that the exposure to furan at those levels may suggest a human health concern since the obtained MOEs were considered low for a carcinogenic compound that might act via a DNA-reactive genotoxic metabolite. Therefore, further efforts on developing and implementing mitigation strategies for furan in foods were highly recommended to reduce its dietary exposure (FAO/WHO, 2011).

Baby foods

Heat canned or jarred foods with stirring to allow partial volatilization and dissipation of furan.

Coffee

Choose a coffee brewing method that results in lower furan levels (instant coffee < filter drip < bean-to-cup machine).
Let coffee stand for several minutes after pouring to allow furan release before adding cream.

Bread

Toast bread to light brown, rather than dark brown, levels.

FIGURE 21.5

Mitigation options and recommendations to reduce furan exposure.

However, mitigating furan still remains a challenge and no commercial measures to achieve this objective are currently available (Anese and Suman, 2013). Reasons for this include the existence of multiple and complex pathways for furan formation, the relevance of thermal processing techniques for food safety and the development of desirable flavor and aroma, and the complex effects of the food matrix on furan retention in food (CAC, 2011). Mitigation options and recommendations are mostly targeted at the level of the consumer and include those illustrated in Fig. 21.5.

To date, there is no dose or intake below which the risk associated with furan exposure from foods is negligible. However, as indicated previously, more recent studies have supported the possible involvement of epigenetic modifications in furan hepatotoxicity and carcinogenicity, which could allow the establishment of a threshold in the dose–response relationship and the derivation of a safe intake level, such as a provisional maximum tolerable daily intake. Additionally, no regulatory maximum limits have been set for furan in foods.

CONCLUSIONS

Coffee, canned/jarred products, and a variety of other thermally processed foods may contain furan, a potent hepatotoxin in rodents, classified as a possible human carcinogen. Furan formation occurs through thermal degradation and thermal oxidation of natural food constituents, such as sugars, ascorbic acid, amino acids, and polyunsaturated fatty acids, and can be influenced by several variables including temperature, pH, moisture, and matrix components. Chronic toxicity of furan may be relevant considering the small doses continually ingested by people through the consumption of foods and beverages. Several exposure assessments have been performed and intake levels reported have suggested a potential concern to human health, especially for infant consumers of jarred baby foods. However, there are no epidemiological studies available and possible correlations between exposure to furan and cancer incidence in humans are not know so far. Exposure assessments based on a biomarker approach are highly desirable, since variations in furan levels due to its volatility until the time when food is consumed could lead to some uncertainty. Efforts have been directed at the development of mitigation strategies to reduce furan exposure, but no commercial measures to achieve this objective are currently available.

REFERENCES

Adams, A., Van Lancker, F., De Meulenaer, B., Owczarek-Fendor, A., De Kimpe, N., 2012. On-fiber furan formation from volatile precursors: a critical example of artefact formation during Solid-Phase Microextraction. J. Chromatogr. B Analyt. Technol. Biomed. Life Sci. 897, 37–41.

Anese, M., Suman, M., 2013. Mitigation strategies of furan and 5-hydroxymethylfurfural in food. Food Res. Int. 51, 257–264.

Arisseto, A.P., Vicente, E., De Figueiredo Toledo, M.C., 2010. Determination of furan levels in commercial samples of baby food from Brazil and preliminary risk assessment. Food Addit. Contam. Part A Chem. Anal. Control Expo. Risk Assess. 27, 1051–1059.

Arisseto, A.P., Vicente, E., Ueno, M.S., Amelia, S., Tfouni, V., De Figueiredo Toledo, M. C., 2011. Furan levels in coffee as influenced by species, roast degree, and brewing procedures. J. Agric. Food Chem. 59, 3118–3124.

Arisseto, A.P., Vicente, E., Furlani, R.P.Z., Ueno, M.S., Pereira, A.L.D., Toledo, M.C.F., 2012. Occurrence of furan in commercial processed foods in Brazil. Food Addit. Contam. Part A Chem. Anal. Control Expo. Risk Assess. 29, 1832–1839.

Arisseto, A.P., Vicente, E., Toledo, M.C.F., 2013. Furan levels in pressure-cooked foods. Int. J. Food Sci. 2013, Article ID 904349, 4 pp. http://dx.doi.org/10.1155/2013/904349.

Becalski, A., Seaman, S., 2005. Furan precursors in food: a model study and development of a simple headspace method for determination of furan. J. AOAC Int. 88, 102–106.

Becalski, A., Turcotte, A.-M., Cooke, G.M., Gill, S.S., 2013. Investigation of possible endogenous formation of furan in Fischer-344 rat. Toxicol. Environ. Chem. 95, 814–822.

Burka, L.T., Washburn, K.D., Irwin, R.D., 1991. Disposition of [^{14}C]-furan in the male F344 rat. J. Toxicol. Environ. Health 34, 245–257.

Byrns, M.C., Predecki, D.P., Peterson, L.A., 2002. Characterization of nucleoside adducts of cis-2-butene-1,4-dial, a reactive metabolite of furan. Chem. Res. Toxicol. 15, 373–379.

CAC (Codex Alimentarius Commission), 2011. Discussion paper on furan, In: 5th Session of Codex Committee on Contaminants in Foods (CCCF), The Hague.

Carthew, P., DiNovi, M., Setzer, R.W., 2010. Application of the margin of exposure (MoE) approach to substances in food that are genotoxic and carcinogenic example: furan (CAS No. 110-00-9). Food Chem. Toxicol. 48, S69–S74.

Cooke, G.M., Taylor, M., Bourque, C., Curran, I., Gurofsky, S., Gill, S., 2014. Effects of furan on male rat reproduction parameters in a 90-day gavage study. Reprod. Toxicol. 46, 85–90.

de Conti, A., Kobets, T., Escudero-Lourdes, C., Montgomery, B., Tryndyak, V., Beland, F. A., et al., 2014. Dose-and time-dependent epigenetic changes in the livers of Fisher 344 rats exposed to furan. Toxicol. Sci. 139, 371–380.

de Conti, A., Kobets, T., Tryndyak, V., Burnett, S.D., Han, T., Fuscoe, J.C., et al., 2015. Persistence of furan-induced epigenetic aberrations in the livers of F344 rats. Toxicol. Sci. 144, 217–226.

EFSA (European Food Safety Authority), 2011. Update on furan levels in food from monitoring years 2004–2010 and exposure assessment. EFSA J. 9 (9), 2347. Available from: http://www.efsa.europa.eu/en/search/doc/2347.pdf.

Egle, J.L., Gochberg, B.J., 1979. Respiratory retention and acute toxicity of furan. Am. Ind. Hyg. Assoc. J. 40, 310–314.

Fan, X.T., 2005a. Formation of furan from carbohydrates and ascorbic acid following exposure to ionizing radiation and thermal processing. J. Agric. Food Chem. 53, 7826–7831.

Fan, X.T., 2005b. Impact of ionizing radiation and thermal treatments on furan levels in fruit juice. J. Food Sci. 70, E409–E414.

Fan, X., 2015. Furan formation from fatty acids as a result of storage, gamma irradiation, UV-C and heat treatments. Food Chem. 175, 439–444.

Fan, X., Sokorai, K.J.B., 2008. Effect of ionizing radiation on furan formation in fresh-cut fruits and vegetables. J. Food Sci. 73, C79–C83.

Fan, X., Huang, L., Sokorai, K.J.B., 2008. Factors affecting thermally induced furan formation. J. Agric. Food Chem. 56, 9490–9494.

FAO/WHO (Food and Agriculture Organization/World Health Organization), 2011. Furan. In: Safety Evaluation of Certain Contaminants in Foods/Prepared by the Seventy-Second Meeting of the Joint FAO/WHO Expert Committee on Food Additives (JECFA). WHO Food Additives Series: 63. FAO JECFA Monographs 8, pp. 487–603.

Floros, J.D., Newsome, R., Fisher, W., Barbosa-Cánovas, G.B., Chen, H., Dunne, C.P., et al., 2010. Feeding the world today and tomorrow: the importance of food science and technology. An IFT scientific review. Compr. Rev. Food Sci. and Food Saf. 9, 572–599.

Fromberg, A., Fagt, S., Granby, K., 2009. Furan in heat processed food products including home cooked food products and ready-to-eat products. Report of the EFSA CFP/EFSA/DATEX/2007/03 project. The National Food Institute, Technical University of Denmark, Søborg. Available from: http://www.efsa.europa.eu/en/supporting/doc/1e.pdf.

Gill, S., Bondy, G., Lefebvre, D.E., Becalski, A., Kavanagh, M., Hou, Y., et al., 2010. Subchronic oral toxicity study of furan in Fischer-344 rats. Toxicol. Pathol. 38, 619–630.

Gill, S., Kavanagh, M., Barker, M., Weld, M., Vavasour, E., Hou, Y., et al., 2011. Subchronic oral toxicity study of furan in B6C3F1 Mice. Toxicol. Pathol. 39, 787–794.

Guenther, H., Hoenicke, K., Biesterveld, S., Gerhard-Rieben, E., Lantz, I., 2010. Furan in coffee: pilot studies on formation during roasting and losses during production steps and consumer handling. Food Addit. Contam. Part A Chem. Anal. Control Expo. Risk Assess. 27, 283–290.

Hasnip, S., Crews, C., Castle, L., 2006. Some factors affecting the formation of furan in heated foods. Food Addit. Contam. 23, 219–227.

Hoveidi, H., Kalal, S.H., Bidhendi, N.G., Rashedi, H., Mehr, V.D., 2013. Electrochemical behavior and the determination of furan in beverage samples using glassy carbon electrode. Int. J. Environ. Res. 7, 715–722.

Huang, X., Duan, H., Barringer, S.A., 2011. Effects of buffer and temperature on formation of furan, acetic acid and formic acid from carbohydrate model systems. LWT-Food Sci. Technol. 44, 1761–1765.

IARC (International Agency for Research on Cancer), 1995. Furan: IARC Monographs on the Evaluation of Carcinogenic Risks of Chemicals to Humans 63. IARC: Lyon, France.

Karacaoglu, E., Selmanoglu, G., 2010. Effects of heat-induced food contaminant furan on reproductive system of male rats from weaning through postpuberty. Food Chem. Toxicol. 48, 1293–1301.

Kim, T.-K., Lee, Y.-K., Kim, S., Park, Y.S., Lee, K.-G., 2009. Furan in commercially processed foods: four-year field monitoring and risk assessment study in Korea. J. Toxicol. Environ. Health Part A Current Issues 72, 1304–1310.

La Pera, L., Liberatore, A., Avellone, G., Fanara, S., Dugo, G., Agozzino, P., 2009. Analysis of furan in coffee of different provenance by head-space solid phase microextraction gas chromatography-mass spectrometry: effect of brewing procedures. Food Addit. Contam. Part A Chem. Anal. Control Expo. Risk Assess. 26, 786–792.

Limacher, A., Kerler, J., Conde-Petit, B., Blank, I., 2007. Formation of furan and methylfuran from ascorbic acid in model systems and food. Food Addit. Contam. 24, 122–135.

Limacher, A., Kerler, J., Davidek, T., Schmalzried, F., Blank, I., 2008. Formation of furan and methylfuran by Maillard-type reactions in model systems and food. J. Agric. Food Chem. 56, 3639–3647.

Locas, C.P., Yaylayan, V.A., 2004. Origin and mechanistic pathways of formation of the parent furan—a food toxicant. J. Agric. Food Chem. 52, 6830–6836.

Maga, J.A., 1979. Furans in foods. Crit. Rev. Food Sci. Nutr. 11, 355–400.

Mariotti, M.S., Granby, K., Rozowski, J., Pedreschi, F., 2013a. Furan: a critical heat induced dietary contaminant. Food Funct. 4, 1001–1015.

Mariotti, M.S., Toledo, C., Hevia, K., Pablo Gomez, J., Fromberg, A., Granby, K., et al., 2013b. Are Chileans exposed to dietary furan? Food Addit. Contam. Part A Chem. Anal. Control Expo. Risk Assess. 30, 1715–1721.

Mark, J., Pollien, P., Lindinger, C., Blank, I., Mark, T., 2006. Quantitation of furan and methylfuran formed in different precursor systems by proton transfer reaction mass spectrometry. J. Agric. Food Chem. 54, 2786–2793.

Merritt, C., Bazinet, M.L., Sullivan, J.H., Robertson, D.H., 1963. Mass spectrometric determination of the volatile components from ground coffee. J. Agric. Food Chem. 11, 152–155.

Mesias, M., Morales, F.J., 2014. Reliable estimation of dietary exposure to furan from coffee: an automatic vending machine as a case study. Food Res. Int. 61, 257–263.

Mesias, M., Guerra-Hernandez, E., Garcia-Villanova, B., 2013. Furan content in Spanish baby foods and its relation with potential precursors. Cyta-J. Food 11, 1–6.

Morehouse, K.M., Nyman, P.J., McNeal, T.P., Dinovi, M.J., Perfetti, G.A., 2008. Survey of furan in heat processed foods by headspace gas chromatography/mass spectrometry and estimated adult exposure. Food Addit. Contam. 25, 259–264.

Moro, S., Chipman, J.K., Wegener, J.-W., Hamberger, C., Dekant, W., Mally, A., 2012. Furan in heat-treated foods: formation, exposure, toxicity, and aspects of risk assessment. Mol. Nutr. Food Res. 56, 1197–1211.

Moser, G.J., Foley, J., Burnett, M., Goldsworthy, T.L., Maronpot, R., 2009. Furan induced dose response relationships for liver cytotoxicity, cell proliferation, and tumorigenicity (furan-induced liver tumorigenicity). Exp. Toxicol. Pathol. 61, 101–111.

Nie, S.-P., Huang, J.-G., Zhang, Y.-N., Hu, J.-L., Wang, S., Shen, M.-Y., et al., 2013. Analysis of furan in heat-processed foods in China by automated headspace gas chromatography-mass spectrometry (HS-GC-MS). Food Control 30, 62–68.

NTP (NationalToxicologyProgram), 1993. Toxicology and carcinogenesis studies of furan (CAS No. 110-00-9) in F344 rats and B6C3F1 mice (gavage studies). NTP Technical Report No 402. Available from: http://ntp.niehs.nih.gov/ntp/htdocs/LT_rpts/tr402.pdf.

NTP (National Toxicology Program), 2014. 13rd report on carcinogens: furan. Available from: http://ntp.niehs.nih.gov/ntp/roc/content/profiles/furan.pdf.

Owczarek-Fendor, A., De Meulenaer, B., Scholl, G., Adams, A., Van Lancker, F., Eppe, G., et al., 2012. Furan formation in starch-based model systems containing carbohydrates in combination with proteins, ascorbic acid and lipids. Food Chem. 133, 816–821.

Palmers, S., Grauwet, T., Kebede, B.T., Hendrickx, M.E., Van Loey, A., 2014. Reduction of furan formation by high-pressure high-temperature treatment of individual vegetable purees. Food Bioprocess Technol. 7, 2679–2693.

Palmers, S., Grauwet, T., Buve, C., Van de Vondel, L., Kebede, B.T., Hendrickx, M.E., et al., 2015. Furan formation during storage and reheating of sterilised vegetable purees. Food Addit. Contam. Part A Chem. Anal. Control Expo. Risk Assess. 32, 161–169.

Ridgway, K., Lalljie, S.P.D., Smith, R.M., 2010. The use of stir bar sorptive extraction-a potential alternative method for the determination of furan, evaluated using two example food matrices. Anal. Chim. Acta 657, 169–174.

Sarafraz-Yazdi, A., Abbasian, M., Amiri, A., 2012. Determination of furan in food samples using two solid phase microextraction fibers based on sot-gel technique with gas chromatography-flame ionisation detector. Food Chem. 131, 698–704.

Scholl, G., Scippo, M.L., De Pauw, E., Eppe, G., Saegerman, C., 2012a. Estimation of furan contamination across the Belgian food chain. Food Addit. Contam. Part A Chem. Anal. Control Expo. Risk Assess. 29, 172–179.

Scholl, G., Humblet, M.F., Scippo, M.L., De Pauw, E., Eppe, G., Saegerman, C., 2012b. Risk assessment of Belgian adults for furan contamination through the food chain. Food Addit. Contam. Part A Chem. Anal. Control Expo. Risk Assess. 29, 345–353.

Stadler, R.H., Lineback, D.R., 2009. Process-Induced Food Toxicants: Occurrence, Formation, Mitigation, and Health Risks. John Wiley & Sons, Inc, Hoboken, NJ.

Terrell, A.N., Huynh, M., Grill, A.E., Kovi, R.C., O'Sullivan, M.G., Guttenplan, J.B., et al., 2014. Mutagenicity of furan in female Big Blue B6C3F1 mice. Mutat. Res. Genet. Toxicol. Environ. Mutagen. 770, 46–54.

US FDA (Food and Drug Administration), 2004. Furan in food, thermal treatment; request for data and information. Available from: http://www.gpo.gov/fdsys/pkg/FR-2004-05-10/pdf/04-10588.pdf.

US FDA (Food and Drug Administration), 2009. Exploratory data on furan in food: individual food products. Available from: http://www.fda.gov/Food/FoodborneIllness Contaminants/ChemicalContaminants/ucm078439.htm.

Van Lancker, F., Adams, A., Owczarek, A., De Meulenaer, B., De Kimpe, N., 2009. Impact of various food ingredients on the retention of furan in foods. Mol. Nutr. Food Res. 53, 1505–1511.

Waizenegger, J., Winkler, G., Kuballa, T., Ruge, W., Kersting, M., Alexy, U., et al., 2012. Analysis and risk assessment of furan in coffee products targeted to adolescents. Food Addit. Contam. Part A Chem. Anal. Control Expo. Risk Assess. 29, 19–28.

Wegener, J.-W., Lopez-Sanchez, P., 2010. Furan levels in fruit and vegetables juices, nutrition drinks and bakery products. Anal. Chim. Acta 672, 55–60.

Williams, J.R., Rayburn, J.R., Cline, G.R., Sauterer, R., Friedman, M., 2014. Potential protective effect of L-cysteine against the toxicity of acrylamide and furan in exposed *Xenopus laevis* embryos: an interaction study. J. Agric. Food Chem. 62, 7927–7938.

Wilson, D.M., Goldsworthy, J.T., Popp, J.A., Butterworth, B.E., 1992. Evaluation of genotoxicity, pathological lesions, and cell proliferation in livers of rats and mice treated with furan. Environ. Mol. Mutagen. 19, 209–222.

Wu, S., Wang, E., Yuan, Y., 2014. Detection of furan levels in select Chinese foods by solid phase microextraction-gas chromatography/mass spectrometry method and dietary exposure estimation of furan in the Chinese population. Food Chem. Toxicol. 64, 34–40.

Zoller, O., Sager, F., Reinhard, H., 2007. Furan in food: headspace method and product survey. Food Addit. Contam. 24, 91–107.

CHAPTER

Biogenic amines in ready-to-eat foods

22

Ljerka Prester

Institute for Medical Research and Occupational Health, Zagreb, Croatia

INTRODUCTION

Biogenic amines (histamine, tyramine, tryptamine, putrescine, cadaverine, agmatine, phenylethylamine, spermine, and spermidine) are biologically active compounds commonly produced in many ready-to-eat foods and alcoholic beverages (Spano et al., 2010). They are mainly produced by bacterial activity (microbial decarboxylation) of the respective precursor amino acids during fermentation or spoilage (Table 22.1). According to their chemical structure, dietary amines are classified into the following categories: aromatic (tyramine, phenylethylamine) heterocyclic (histamine, tryptamine), aliphatic diamines (putrescine, cadaverine), and aliphatic polyamines (spermine, spermidine, agmatine) (Suzzi and Gardini, 2003). Biogenic amines are important in many physiological processes. On the other hand, a high dietary intake of biogenic amines can produce toxicological effects in humans (Hungerford, 2010). The highest biogenic amine content, particularly histamine, is produced in decomposed and spoiled dark-muscle fish and fermented food products (Prester, 2014). Thus, many ready-to-eat foods, especially fermented products (fish, cheese, meat) and alcoholic beverages (wine, beer) can contain a high amount of biogenic amines, chiefly histamine and tyramine (Ladero et al., 2010). On the other hand, food intolerance and food-induced migraines are associated with the consumption of moderate or low dairy amine intake (histamine, tyramine, phenylethylamine) in susceptible individuals with a less efficient detoxification system (Maintz et al., 2006). These amines can play an important role in the precipitation of headaches in children and adolescents with migraines (Millichap and Yee, 2003). Furthermore, a high dietary intake of putrescine and cadaverine can potentiate the effects of histamine and tyramine; thus from the toxicological point of view, the elevated accumulation of any biogenic amines in foods must be avoided.

Besides their adverse effects, biogenic amines are of concern in relation to food hygiene and food quality. Importantly, histamine and other biogenic

P. Kotzekidou (Ed): Food Hygiene and Toxicology in Ready-to-Eat Foods.
DOI: http://dx.doi.org/10.1016/B978-0-12-801916-0.00022-4

Table 22.1 Biogenic Amines in Foods and Their Pharmacological and Toxicological Effects

Precursor	Biogenic Amine	Physiological Role	Toxicological Effects	
			Direct	**Indirect**
Histidine	Histamine	• Vasodilatation • Hormonal mediator • Cell growth and differentiation • Attention and cognition • Allergic reactions	• Histamine poisoning • Intolerance on histamine	
Tyrosine	Tyramine	• Vasoconstriction • Hormonal mediator	• Intolerance on tyramine	
Tryptophan	Tryptamine	• Vasoconstriction • Hormonal mediator	• Less known	
Lysine Ornithine	Putrescine Cadaverine	• Precursor for spermine and spermidine • Regulator of cell growth • Tumor promotion	• Hypotension • Bradycardia • Tachycardia	• Potentiators for the toxic effects of histamine and tyramine • Carcinogenic effects
Putrescine	Spermine Spermidine	• Protein synthesis • Nucleic acid regulation	• Less known	• Carcinogenic effects
L-arginine	Agmatine	• Hormonal mediator • Inhibitor of NO synthase • Antiproliferative effect	• Less known	• Less known

amines in foods are not reduced by heat treatment; thus cooking and freezing do not decrease their level during food processing. The recovery of biogenic amines during food handling and storage is one of the greatest problems affecting the food industry. The production of biogenic amines is affected by the quality of raw fish, meat, and dairy products and it is an essential factor in its formation (Ruiz-Capillas and Jiménez-Colmenero, 2004). Except for histamine in dark-muscle fish, the regulatory limits for other biogenic amines for human consumption have not been established. Histamine/tyramine intolerance is a growing problem suggesting the importance of monitoring of amines in foods (McCabe-Sellers et al., 2006). A selective avoidance of histamine/tyramine-rich foods is the universal treatment for sensitive consumers (Millichap and Yee, 2003). Additionally, progressive steps in the food industry to certify manufactured products as histamine-free, would benefit sensitive individuals. In this chapter, a systematic overview of the presence of biogenic amines in different ready-to-eat foods and beverages is described, which is of unambiguous interest to both consumers and the industry.

BIOGENIC AMINES

In humans, histamine regulates and modulates many functions by binding to four subtypes of receptors on membranes in skin and various tissues, such as the respiratory, gastrointestinal, cardiovascular, and immunological systems (Maintz and Novak, 2007). Other vasoactive amines frequently present in ready-to-eat foods and alcoholic beverages include tyramine, tryptamine, and phenylethylamine. The vasoconstriction effect of these vasoamines can cause hypertension, migraine cases, headache, perspiration, vomiting, and midriasis (Table 22.1). Polyamines (putrescine, cadaverine, spermine, and spermidine) have been involved in a wide range of important biological reactions. They are required for cell growth, proliferation, and differentiation in all organs since they are involved in nearly every step of DNA, RNA, and protein synthesis (Wunderlichová et al., 2014). Also, polyamine can be formed de novo in the gut lumen showing that the intestinal tract plays a vital role in nutrition and preserving human health. Furthermore, polyamines are essential to cell growth of tumor cells. Multiple abnormalities in the control of polyamine metabolisms are frequently observed in cancer (Moinard et al., 2005). Based on these results, a polyamine-reduced diet has been widely studied as part of nutritional therapy in some types of cancer, such as in prostatic and colorectal carcinoma patients (Cipolla et al., 2007; Watanabe et al., 2009). Finally, the recovery of polyamines, particularly putrescine and cadaverine from fermented ready-to-eat meat and fish products can function as a precursor for the formation of carcinogenic *N*-nitrosamines in the presence of nitrites/nitrates and by heating (Al Bulushi et al., 2009; De Mey et al., 2014). Thus, the formation of nitrosamines can constitute an additional toxicological risk associated with biogenic amines in food. There is some evidence of a positive association between the intake of nitrosamine, dietary polyamines (salted/smoked fish and processed meat) and risk for certain types of cancer, particularly gastric and esophageal cancer (Jakszyn and González, 2006). Most investigations were performed in Asian countries.

The efficiency of the biogenic amine detoxification system is crucial to prevent their accumulation in humans. In healthy individuals, dietary biogenic amines are rapidly metabolized in the intestinal tract by a detoxification system based on the activities of several enzymes: monoamine oxidase (MAO; EC 1.4.3.4), diamine oxidase (DAO; EC 1.4.3.6), polyamine oxidase (PAO; EC 1.5.3.11), and histamine *N*-methyltransferase (HNMT; EC 2.1.1.8) (Maintz and Novak, 2007). Two MAO izoenzymes exist, A and B, with different human localization patterns. The metabolism of biogenic amines varies considerably among individuals due to the influence of genetic factors, disease, or drug intake. A number of drugs of different categories can inhibit MAO and DAO activity which may lead to elevated amine levels in the plasma of patients and induce histamine intolerance (Maintz and Novak, 2007). Medications such as MAO inhibitors (MAO-I) are used in psychiatric practice to serve as antidepressants (MAO-A) or as anti-Parkinsonian agents (MAO-B). These modern MAO inhibitors are reversible and selective; however, some classical and nonselective MAO-Is are also available in psychiatric practice (Fiedorowicz and Swartz, 2004). The most common DAO-inhibiting drugs include varieties of drugs such as antibiotics (isoniazide, cefuroxime, clavulanic acid), antihypertensives (verapamil, alprenolol), antiarrhythmics (propafenon), analgetics (morphine, NSAR, metamizole), narcotics (thiopental), local anesthetics (prilocaine), mucolytics (acetylcysteine), H2 receptor antagonists (cimetidine), broncholytics (aminophylline), antidepressants (amitriptyline), and cytostatics (cyclophosphamide) (Maintz and Novak, 2007; EFSA, 2011). Also, some substances (potentiators) such as alcohol and tobacco smoke slow down the metabolism of histamine and tyramine and these should be avoided during treatment with these medications.

TOXICOLOGICAL EFFECTS OF BIOGENIC AMINES

Among biogenic amines, dietary histamine and tyramine are the most toxic amines to humans. Histamine is the causative agent of histamine poisoning (HP) associated exclusively with the consumption of spoiled fish and cheese. The consumption of spoiled or decomposed dark-meat fish belonging to Scombroid fish such as tuna and mackerel commonly cause histamine fish poisoning (HFP) (scombroid poisoning) (Becker et al., 2001; Predy et al., 2003; Emborg et al., 2005). Several species of nonscombroid fish such as sardine, herring, bluefish, kahawi, fish cube, swordfish, garfish, and mahi-mahi have been implicated in incidents of HFP (Prester, 2011). Moreover, the occurrence of psychrotrophic bacteria can result in histamine production even in fish stored at refrigerated temperature and cause HFP (Kanki et al., 2004). Table 22.2 summarizes the common fish products associated with HFP. Generally, in HFP, histamine levels in the edible parts of the decomposed fish were very high (>500 mg/kg), mostly due to prolonged storage and handling of fish at an elevated temperature (>8°C) (Hungerford, 2010). The production of histamine varies widely in different parts of the fish; therefore, the consumption of the same contaminated fish may not lead to HFP in all people. Importantly, once formed, histamine and other biogenic amines are not destroyed by cooking, freezing, or canning (Lehane and Olley, 2000). Thus, decomposed or spoiled fish can cause HFP even when properly cooked, canned, or frozen. HFP occurs throughout the world and presents a common form of toxicity caused by the ingestion of decomposed fish. It can be prevented primarily by keeping fish refrigerated (Prester et al., 2009).

Table 22.2 Clinically Significant Histamine Levels in Foods (Fish and Cheese) Associated With Histamine Poisoning

Foods	Histamine (mg/kg)	Location	References
Fish product			
Billfish	>1500	Taiwan	Tsai et al. (2007b)
Canned mackerel	500–1539	Taiwan	Tsai et al. (2005)
Vacuum packed cold-smoked tuna	>7000	Denmark	Emborg et al. (2005)
Dried sardine	1700	Japan	Kanki et al. (2004)
Tuna salad	350	Canada	Predy et al. (2003)
Tuna burgers	213–3245	United States	Becker et al. (2001)
Smoked kahawi	>200	New Zealand	Fletcher et al. (1995)
Cheese			
Swiss cheese and aged Cheddar cheese	>1000	United States	Stratton et al. (1991)

Table 22.3 Health Effects of Consumption of Ready-to-Eat Foods That Contain Different Amounts of Histamine and Tyramine

Disease/ Status	Dose of BA in Foods	Symptoms		Potentiators
		Mild	Severe	
Histamine poisoning	High histamine (> 200 mg/kg)	Rash, urticaria, rhinitis, vomiting, abdominal cramps, headache, dyspnea	Tachycardia Dysrhythmia Hypotension Bronchospasm Respiratory distress	Drugs (MAO-I) Alcohol Tobacco smoke
Histamine intolerance	Low histamine (<50 mg/kg)		Hypotension Tachycardia	
Tyramine intolerance	Low tyramine (6 mg/kg)	Nausea, vomiting, migraines, rash	Hypertension Tachycardia	

BA, biogenic amine; MAO-I, monoamine oxidase inhibitor.

Besides dark-muscle fish, the ingestion of aged Cheddar and Swiss cheeses has been linked with HP (Stratton et al., 1991). The intake of foods with elevated histamine load can lead to histamine entering the systemic circulation and causing severe health adverse effects. Table 22.3 shows the typical symptoms in HP which affect the skin (rash, urticaria), the gastrointestinal tract (vomiting,

Table 22.4 Diseases/Status Associated With Increased Sensitivity to Biogenic Amines

Increased Sensitivity to Biogenic Amines	
Acquired	**Genetic Enzyme Deficit**
• MAO-I drugs • DAO-I drugs • Inflammatory GI diseases (gastritis, stomach and colon ulcer, Crohn's disease) • Neoplastic GI disease (colorectal neoplasms)	• Atopic eczema • Chronic urticaria

GI, gastrointestinal; MAO-I, monoamine oxidase inhibitor; DAO-I, diamine oxidase inhibitor.

diarrhea), neurological functions (headache, tingling), and causes hypotension (Lavon et al., 2008). The onset of symptoms in HP is within a few minutes after ingestion and usually lasts for several hours. Although HP is a relatively mild illness, severe symptoms have also been reported, mostly in elderly. In addition, some drugs, especially amine oxidase inhibitors (MAO-I and DAO-I) inhibit histamine metabolism and increase the HP risk. HP remains an important global problem for consumers and the food industry sector.

In contrast to HP, histamine and tyramine intolerance is a clinical disorder associated with the consumption of foods that contain moderate and nontoxic histamine and tyramine levels, respectively (Maintz and Novak, 2007). Typical symptoms of histamine/tyramine intolerance are similar to HP (Table 22.3). Histamine intolerance is associated with a genetic deficit of enzyme function (MAO, DAO, or HNMT) or acquired factors (drugs, diseases) (Table 22.4). In patients frequently complaining of headache, rhinitis, flushing, diarrhea, and tachycardia after food intake, allergy testing is usually performed. If true allergy on food is excluded (negative IgE on specific allergen) and no general medical disease is present, histamine/tyramine intolerance should be considered (Maintz et al., 2006). In these patients so-called "MAO-I diet" has been proposed which includes histamine- and/or tyramine-free foods (Fiedorowicz and Swartz, 2004; Rosell-Camps et al., 2013). Generally, introducing the histamine-reduced diet in patients affected by histamine/tyramine intolerance has good effects on symptoms and quality of life. If necessary, antihistamines and substitution of DAO can reduce symptoms (Maintz et al., 2006; Komericki et al., 2011). Food intolerance has been a growing problem over the last two decades. Thus, certifying fishery products as histamine-free or histamine low would be of great significance for sensitive individuals with histamine intolerance since a balanced diet has a positive effect on human health. While histamine intolerance is reported in 1% of the population, mostly female middle-aged patients, tyramine migraines can be found in 15% of the population (Maintz and Novak, 2007; EFSA, 2011).

In contrast to histamine and tyramine, dietary polyamines have low toxicological activity (Wunderlichová et al., 2014). However, a high dietary intake of putrescine and cadaverine can potentiate the effects of simultaneously ingested histamine (Rauscher-Gabernig et al., 2012).

BIOGENIC AMINES AS QUALITY INDICATORS

An increase of amine content, particularly, histamine, tyramine, putrescine, and cadaverine in foods is related to spoilage. Thus, the presence of biogenic amines in foods is a chemical indication of deterioration or spoilage of food products (Ruiz-Capillas and Jiménez-Colmenero, 2004). In most fermented food products (fish, meat, dairy, and vegetables), there is always more than one type of amine present in food. Thus, the use of several biogenic amines such as biogenic amine index (BAI) could be a better indicator of freshness/spoilage of food than single amine content. BAI is calculated as the sum of concentration of histamine, tyramine, putrescine, and cadaverine (Parente et al., 2001; EFSA, 2011). In addition, vasoactive biogenic amine index (VBAI) has been proposed as the sum of vasoactive amines (histamine, tyramine, tryptamine, and phenylethylamine) for fermented meat products (Eerola et al., 1998). Both, BAI and VBAI can indicate the quality of raw food products and hygienic conditions under which the food was processed. Thus, these parameters can be used as markers of microbial food contamination.

Histamine is the only amine with established regulatory limits in certain seafish species. Food authorities in the European Union set the histamine level in scombroid-like fish at 100 mg/kg (EC, 2006) and in the United States at 50 mg/kg (FDA, 2010). The EU regulations set the critical level of histamine at 200 mg/kg for fermented fish products. Histamine levels above 500 mg/kg in dark-muscle fish are considered toxic for human consumption (Lehane and Olley, 2000). Table 22.5 shows the regulatory limits for histamine in dark-muscle fish and recommended values for other biogenic amines in white-muscle fish, shellfish, and seafood products. BAI for dark-muscle fish of <10 and >100 mg/kg has been suggested for premium-quality fish and fish in advanced decomposition, respectively (Muscarella et al., 2005). There is no specific legislation in the EU and United States concerning the biogenic amines in other food products. Nout (1994) proposed 50–100 mg/kg for histamine and 100–800 mg/kg for tyramine as acceptable ranges in fermented foods. Other scientists stated a wide range of possible maximum tolerable concentrations for histamine, tyramine, putrescine, and cadaverine in fermented ready-to-eat foods (Shalaby, 1996; Rauscher-Gabernig et al., 2012). Based on the limited information published about patients with a histamine intolerance, even small quantities of ingested histamine may cause adverse health effects; thus, only levels below the

Table 22.5 Maximum Tolerable Concentrations of Some Biogenic Amines (mg/kg) in Fish, Shellfish, and Fermented Seafood

Seafood	Histamine	Putrescine	Cadaverine	References
Dark-muscle fish				
Tuna, mackerel	50 100	–	–	EC (2006) FDA (2010)
White muscle fish				
Carp	–	10	10	Křižek et al. (2002)
Shellfish				
Shrimp	–	3	–	Benner et al. (2003)
Fermented fish	200	–	–	EC (2006)

detectable limit of histamine can be considered safe (EFSA, 2011). Under medication with MAO inhibitors, susceptibility to histamine/tyramine can increase markedly. Tyramine levels above 50 mg/kg of food are supposed to be deleterious for those taking selective MAO-I drugs. However, a quantity of tyramine as small as 6 mg/kg for those taking classical MAO-I drugs has been associated with tyramine intolerance (Silla Santos, 1996). These levels can be easily exceeded by consumption of some fermented foods (cheese, sausages, fermented fish) which accumulate increased amounts of biogenic amines. There is no information about putrescine and cadaverine values that directly cause acute adverse health effects and/or potentiate the toxic effects of histamine.

To avoid the production of biogenic amines in ready-to-eat meat products and overall safety, several methods such as vacuum packaging and modified atmosphere packaging have been developed (Ntzimani et al., 2008). In addition, gamma-irradiation is an effective technique alone or in combination with others for reducing or eliminating microbial growth in foods. Rabie et al. (2010, 2011) reported the dramatic reduction in histamine and total biogenic amine levels in traditional processed Egyptian sausages and blue cheese by γ-irradiation. Furthermore, Baptista et al. (2014) reported that a combination of grilling, vacuum packaging, and high-dose gamma irradiation (48 kGy) significantly lowered mean levels for total biogenic amines in ready-to-eat chicken fillets through the storage (room temperature and −25°C).

BIOGENIC AMINES IN DIFFERENT TYPES OF FOOD

SEAFOOD

Among seafoods, decomposed or spoiled dark-muscle fish such as tuna, mackerel, swordfish, sardines, and bonito contain the highest amounts of histamine (Table 22.2). Besides histamine, the levels of other biogenic amines are also usually elevated, contributing to the occurrence of HFP. The production of histamine, tyramine, and polyamines in seafood is highly species-specific (Prester, 2011). In contrast to dark-muscle fish, the decomposition of white-muscle fish such as hake does not cause accumulation of histamine in its edible parts (Prester et al., 2009). Therefore, histamine is less suitable as a quality indicator for white-muscle fish. Since putrescine and cadaverine are the main amines formed in white-muscle fish, they have been suggested as markers of acceptable quality. A putrescine content <10 mg/kg (or sum of putrescine and cadaverine <20 mg/kg) was proposed as a good criterion of carp quality (Křižek et al., 2002) and BAI of 15–20 mg/kg for hake as acceptable quality (Baixas-Nogueras et al., 2005). Similarly as for white-muscle fish species, histamine formation in different shellfish species such as cephalopods (squid, octopus, cuttlefish), crustaceans (shrimp, lobster), and bivalves (mussels) is relatively low due to the low level of precursor amino acids (Prester et al., 2010). Thus, no cases of HFP have been reported as a result of the consumption of decomposed white-muscle fish and shellfish.

Seafood sauce and paste are popular ready-to-eat seafood products in the Far East and their consumption has been growing in Western countries in the last decade. These seafood products are a potentially high source of biogenic amines (Visciano et al., 2013). High quantities of histamine (>1000 mg/kg) and tyramine (>500 mg/kg) can be found in salted and fermented fish (usually sardine, anchovy, and mackerel) from European and Asian countries (Kuda et al., 2012; Koral et al., 2013). Also, polyamine content in fish and shrimp sauces can reach high values (above 100 mg/kg). In fermented shrimp paste, the total biogenic amine content varied greatly between different samples (from undetected to 976 mg/kg), meaning that few samples could be considered as possessing class 1 quality (Pilapil et al., 2015). The extremely high contents of biogenic amines in seafood sauces are the consequence of using poor-quality raw materials, contamination, and inappropriate conditions during processing and storage. The health risks following the consumption of seafood sauce have not been reported due to its relatively low level of uptake. However, due to prolonged metabolism of biogenic amines in susceptible individuals and health compromised they should pay special attention regarding the consumption of this type of ready-to-eat food. In contrast to seafood sauces, the content of biogenic amines in mullet and tuna roe was low and BAI was 90 and 15 mg/kg, respectively (Restuccia et al., 2015). Other seafood fermented products such as canned anchovy and sardine showed good overall quality and acceptable amine levels (Rabie et al., 2014a). However, canned fish products can be implicated in HFP when decomposed raw fish samples with high histamine content are used, as was reported for canned mackerel in Taiwan (Tsai et al., 2005).

CHEESE AND OTHER DAIRY PRODUCTS

After fish, cheese is the next most commonly implicated food associated with HP and biogenic amine intolerance (cheese reaction). Cheese is a leading diet factor that triggers pediatric and adolescent migraine (Millichap and Yee, 2003).

Table 22.6 Biogenic Amine (BA) Levels and Biogenic Amine Index (BAI) (mg/kg) in Different Dairy Products

Product	Country	Main BA	Histamine	BAI	References
Milk	Korea	Cadaverine 18.5 ± 0.6	nd	18.5	Min et al. (2004)
Yogurt		Cadaverine 16.7 ± 2.2	nd	16.9	
Mozzarella		Spermidine 17.5 ± 8.7	nd	2.8	
Parmesan		Spermidine 30.7 ± 1.9	1.4 ± 0.04	9.8	
Livno cheese (middle)	Bosnia and Hercegovina	Histamine	81	161	Marijan et al. (2014)
Dutch-type hard cheese	Czech Republic	Tyramine 299.8 ± 4.8	1.8 ± 0.4	364	Komprda et al. (2007)
Feta cheese	Greece	Tyramine 246	84.6	606	Valsamaki et al. (2000)
Smear-ripened cheese	Czech Republic	Cadaverine 249 (nd–967)	55 (nd–226)	604	Samková et al. (2013)
Unwashed-rind					
Washed-rind		Spermine 13.6 (nd–70.5)	nd	6.6	
Terrincho cheese	Portugal	Tyramine 283 ± 24.2	nd	770	Pintado et al. (2008)
Red Salmas Cheese	Iran	Cadaverine 701 (387–1076)	105 (12–254)	1427	Rohani et al. (2013)
Pecorino di Farindola cheese	Italy	Tyramine 1171 ± 1.4	21.8 ± 3.5	1543	Schirone et al. (2011)
Blue cheese	Portugal	Tyramine	35.7 ± 0.5	2294	Rabie et al. (2011)
Nonirradiated		2220 ± 0.6	nd	416	
Irradiated (6 kGy)		403 ± 0.5			

nd, not detected.
Results are presented as the average concentration ± standard deviation or the range.

Fermented dairy products such as smear- and mold-ripened cheeses can accumulate a large quantity of biogenic amines, particularly tyramine. Table 22.6 shows the biogenic amine content of various dairy products. In milk, yogurt, and fresh types of cheese (ricotta), histamine, tyramine, and polyamine content were very low when made from pasteurized milk (Min et al., 2004; Ladero et al., 2010). On the other hand, in most aged, ripened and raw milk cheeses, the content of tyramine, putrescine, and cadaverine commonly exceeds 100 mg/kg (Ladero et al., 2010). Moreover, in some cheeses made from ewe and goat milk more than 1000 mg/kg of tyramine has been detected (Bonetta et al., 2008; Schirone et al., 2011). Amine content can also vary within the same type of cheese and even between different sections of the same cheese, for example, between the middle and the rind of cheese (Marijan et al., 2014). Furthermore, amine levels are generally higher in cheese made from raw milk than from pasteurized milk (Linares et al., 2011). Finally, amine production differs widely and depends on several factors including storage temperature, ripening time, and ripening technology as well as microbial population (Komprda et al., 2007; Rohani et al., 2013). Washing of smear-ripened cheese surface during ripening led to a significant reduction in tyramine, histamine, and polyamine levels (Samková et al., 2013). At the moment, no specific legislation regarding biogenic amine concentrations in dairy products has been established. The upper limit for histamine and BAI of 100 and 900 mg/kg, respectively has been suggested for dairy products (Spanjer and Van Roode, 1991). According to these values, a considerable number of cheeses exceeded the recommended doses, thus potentially posing a health risk for the consumer (Table 22.6). Therefore, the regular monitoring of biogenic amines in cheese and regulative measures could improve food safety (Samková et al., 2013).

MEAT PRODUCTS

Similarly to aged cheese, fermented meat products may contain very high tyramine levels. Generally, the concentration and profile of biogenic amines varied greatly among meat products and is higher in sausages than in other meat products (Table 22.7). In most investigations, tyramine was the dominant amine in dry fermented sausages. Its concentration can reach levels higher than 400 mg/kg in French-type sausages (De Mey et al., 2014) and in *sucuk* (traditional Turkish salami) (Gençcelep et al., 2008). Lower tyramine levels (below 50 mg/kg) are usually found in sausages of small diameter and low moisture content, such as dry sausages, Boerenring, and snacks (Bover-Cid et al., 1999; De Mey et al., 2014). Other vasoactive amines such as histamine, tryptamine, and phenylethylamine were also present in sausages depending on manufacturing practices (Tasić et al., 2014). The level of histamine in most sausages was lower than that considered to generate toxic effects. However, in *sucuk* Turkish sausages, and nonirradiated Egyptian sausages, histamine was present up to 136, 263, and 768 mg/kg, respectively, which exceeded the maximum concentration of 100 mg/kg allowed in fish products. Furthermore, putrescine and cadaverine can reach values higher than 300 and 600 mg/kg, respectively in fermented sausages (Parente et al., 2001; De Mey et al., 2014; Rabie et al., 2014b). Even higher cadaverine values were reached in nonirradiated Egyptian salami (range: 28–2557 mg/kg) (Rabie et al., 2010). Interestingly, European ripened dried sausages manufactured from horse meat produced much lower putrescine levels (up to 20 mg/kg) compared to sausages manufactured from pork, beef, and turkey meat. In contrast, sausages produced from turkey meat produced the highest putrescine and histamine amount (Rabie et al., 2014b). Finally, the natural polyamines, spermine and spermidine, were found in almost all sausages, usually at low levels. Several authors suggested the BAI limit of 900 mg/kg as a quality parameter for meat

Table 22.7 Biogenic Amine (BA) Levels and Biogenic Amine Index (BAI) (mg/kg) in Different Meat Products

Product	Country	Main BA	Histamine	BAI	References
Sausages					
German type	Belgium	Tyramine 107 (16–199)	6.2 (nd–12)	215 (23–406)	De Mey et al. (2014)
Horse meats	EU	Tyramine 70	10	33–102	Rabie et al. (2014b)
Smoked turkey sausage	EU	Putrescine 285	264	207–715	
Mortadella	Spain	Tyramine nd–66.0	nd–4.8	0.6–81.7	Hernández-Jover et al. (1996)
Soppressata	Italy	Tyramine nd–557	nd–101	nd–1274	Parente et al. (2001)
Sucuk	Turkey	Tyramine 2.4–676	nd–136	4–1337	Gençcelep et al. (2008)
Salami,	Egypt	Cadverine	768	5788	Rabie et al. (2010)
Nonirradiated		2557	nd	56	
Irradiated (6 kGy)		Tyramine			
		29			
Ham					
Dry-cured ham	Italy	Tyramine	nd	40.6 ± 34.6	Virgili et al. (2007)
15-month old		40.2 (nd–110)	nd	150 ± 187	
23-month old		72.4 (1–274)			
Fillets					
Smoked turkey	Greece	Histamine	32.9 ± 1.4	62.2 ± 3.5	Ntzimani et al. (2008)
30 days/air (4°C)		32.9 ± 1.4	15.6 ± 1.2	32.6 ± 2.6	
30 days/VP (4°C)		15.6 ± 1.2			
Grilled chicken, VP	Brazil	Spermine 3.4 ± 3.1	0.07 ± 0.06	3.6 ± 2.4	Baptista et al. (2014)

nd, not detected; VP, vacuum packed.
Results are presented as the average concentration ± standard deviation or the range.

products (Eerola et al., 1998; Parente et al., 2001; EFSA, 2011). Table 22.7 shows the great variability in ranges of BAI in different types of ready-to-eat meat products. BIA is higher than the recommended value in some Italian sausages (Parente et al., 2001), sausages manufactured from turkey meat (Rabie et al., 2014b), *sucuk*, and in Egyptian salami (Rabie et al., 2010). Similarly, VBAI expressed as the sum of four vasoactive amines exceeded the recommended value of 200 mg/kg in the above sausages (Eerola et al., 1998). However, in irradiated

Egyptian sausages, BAI was almost 100 times lower compared to nonirradiated, and histamine levels decreased dramatically. Selected starter cultures used during processing of fermented sausages contribute to the reduction of biogenic amine content (ie, tyramine, tryptamine, cadaverine, and putrescine) of the product (Baka et al., 2011).

Generally, the presence of biogenic amines in other ready-to-eat meat products, such as smoked hams and some poultry products (smoked turkey fillets and grilled chicken), was low.

VEGETABLES AND SEASONINGS

Table 22.8 shows biogenic amine content in various fermented vegetables. In sauerkraut, tyramine levels can reach high values (up to several hundred mg/kg) and its concentration increased significantly with storage (Kalač et al., 2000). Furthermore, fermented soybean products, like miso and soy sauce, can contain a high amine content, particularly, histamine and tyramine (Kung et al., 2007). The histamine content in most fermented soybean products (douchi product) exceeded the 50 mg/kg USFDA guideline value (Tsai et al., 2007a). BAI (from undetected to 959 mg/kg) shows great variability in soya and miso product quality (BAI of 900 mg/kg for fermented vegetables has been suggested as acceptable) (Shalaby, 1996; Yang et al., 2014). Due to relatively small quantities of both products being consumed no reported cases of HP have been reported. Based on these results, individuals taking amine-oxidase drugs should avoid consuming soybean products and sauerkraut (Shulman and Walker, 1999). From the hygienic point of view, the content of biogenic amines in all fermented soya bean products should be studied in order to ensure consumption safety (Yang et al., 2014).

ALCOHOLIC BEVERAGES

Alcoholic beverages constitute another category of fermented products that contain substantial amounts of biogenic amines, particularly red wines and beers. The composition of amines in wine depends on climate and geological factors of wine-producing regions, as well as enological practices. Histamine, tyramine, and putrescine are the major biogenic amines in wines (Konakovsky et al., 2011). Histamine and tyramine are produced at the beginning of the aging process and their content may be changed with wine age (Ancin-Azpilicueta et al., 2008). The recommended upper limit of 2–10 mg/L for histamine in wine can be easily reached in many red wines. Histamine is considered the most important reason for wine intolerance (Konakovsky et al., 2011). However, other substances in wine, such as sulfites and acetaldehyde, may also play a part in "wine intolerance" (Maintz and Novak, 2007). In susceptible individuals, histamine intolerance was triggered due to consumption of 0.2 L of sparkling wine containing histamine at 20 mg/L. High amine levels can also be found in "high-rated" wines (Konakovsky et al., 2011). However, by controlling the critical technological

Table 22.8 Biogenic Amine Levels and Biogenic Amine Index (BAI) (mg/kg) in Some Commercial Ready-to-Eat Vegetable Products

Product	Histamine	Tyramine	Putrescine	Cadaverine	BAI	References
Ketchup	2.0–18.0	4.5–149	2.4–165	1.4–131	10–463	Kalač et al. (2002)
Tomato paste	2.0–10.1	3.0–31.1	7.9–41.1	1.4–17.6	14–100	
Soy products	nd–234	nd–237	nd–360	nd–128	nd–959	Tsai et al. (2007a)
Miso products	nd–221	nd–434	nd–12	nd–201	nd–868	Kung et al. (2007)
Pickled cucumber	27–45	–	–	–	–	Ekici and Coskun (2004)
Sauerkraut 6-month storage	nd	44–310	22–260	12–60	78–630	Kalač et al. (2000)

nd, not detected.
Results are presented as the range.

factors, it is possible to produce wine with low amine levels. Similarly to wine, beer can contain variable levels of biogenic amines. Loret et al. (2005) established BAI for beer using seven biogenic amines. While BAI < 1 mg/L presents high quality of beer and high hygienic quality of fermented process, BAI > 10 mg/L presents a poor hygienic fermentation process that could cause headache, rash, and hypertension in high-risk consumers. In spontaneously fermented beer, BAI values largely exceeded this threshold.

In summary, biogenic amines were found in fermented products such as ripened cheese, sausages, seafood, vegetables and seasonings (soy and seafood sauces), and alcoholic beverages. Consumption of fermented foods may be associated with potentially beneficial food-derived microbes and positive mental health (Selhub et al., 2014). However, these products should be avoided in individuals suffering from histamine intolerance based on reduced amine oxidase activity as a result of either treatment with drugs inhibiting the amine-oxidases or a genetically acquired inefficiency of these enzymes. Also, these food products should be avoided in individuals on polyamine-reduced diets proposed for some cancer patients. The formation of all biogenic amines can be linked to poor hygienic conditions during manufacturing, temperature abuse of raw foods during storage, and microbiota of the raw materials used for the production of ready-to-eat foods. Selected starter cultures contribute to the reduction of biogenic amine content.

CONCLUSIONS

There is increasing awareness of the risk related to dietary intake of high amine content. However, except for dark-muscle fish, the specific legislation concerning the legal limits of histamine in ready-to-eat foods has not been established. Also, there is no specific legislation concerning the maximum concentrations of other biogenic amines in foods. High levels of biogenic amines in ready-to-eat foods are of significance to public health because of their potential toxic effects. HP is a worldwide problem, generally underestimated or misdiagnosed. In the last decade, significant progress towards lower and safer amine content has been made in the seafood industry by introducing the Hazard Analysis Critical Control Point (HACCP) safety system. However, there is still high variation in the amine presence in ready-to-eat products such as cheese, sausages, fish sauces, and soya products, which requests better standardization. On the other hand, histamine/tyramine intolerance is a growing problem and susceptible groups should be adequately informed about amine-rich foods and beverages, which may contribute to more responsible consumption. In these cases, the production of high-quality foods (histamine-free products or a low-biogenic-amine diet) should be favorable, particularly for susceptible individuals. Thus, the food industry needs to adopt a more interdisciplinary approach by including scientists from the fields of medicine, nutrition, and pharmacy in order to introduce a well-balanced diet, particularly for immune-compromised and susceptible individuals.

REFERENCES

Al Bulushi, I., Poole, S., Deeth, H.C., Dykes, G.A., 2009. Biogenic amines in fish: roles in intoxication, spoilage, and nitrosamine formation-a review. Crit. Rev. Food Sci. Nutr. 49, 369–377.

Ancin-Azpilicueta, C., González-Marco, A., Jiménez-Moreno, N., 2008. Current knowledge about the presence of amines in wine. Crit. Rev. Food Sci. Nutr. 48, 257–275.

Baixas-Nogueras, S., Bover-Cid, S., Veciana-Nogués, M.T., Mariné-Font, A., Vidal-Carou, M.C., 2005. Biogenic amine index for freshness evaluation in iced Mediterranean hake (*Merluccius merluccius*). J. Food Prot. 68, 2433–2438.

Baka, A.M., Papavergou, E.J., Pragalaki, T., Bloukas, J.G., Kotzekidou, P., 2011. Effect of selected autochthonous starter cultures on processing and quality characteristics of Greek fermented sausages. LWT- Food Sci. Technol. 44, 54–61.

Baptista, R.F., Lemos, M., Teixeira, C.E., Vital, H.C., Carneiro, C.S., Mársico, E.T., et al., 2014. Microbiological quality and biogenic amines in ready-to-eat grilled chicken fillets under vacuum packing, freezing, and high-dose irradiation. Poultry Sci. 93, 1571–1577.

Becker, K., Southwick, K., Reardon, J., Berg, R., MacCormack, J.N., 2001. Histamine poisoning associated with eating tuna burgers. J. Am. Med. Assoc. 285, 1327–1330.

Benner Jr.R.A., Staruszkiewicz, W.F., Rogers, P.L., Otwell, S., 2003. Evaluation of putrescine, cadaverine, and indole as chemical indicators of decomposition in Penaeid shrimp. J. Food Sci. 68, 178–185.

Bonetta, S., Bonetta, S., Carraro, E., Coisson, J.D., Travaglia, F., Arlorio, M., 2008. Detection of biogenic amine producer bacteria in a typical Italian goat cheese. J. Food Prot. 71, 205–209.

Bover-Cid, S., Schoppen, S., Izquierdo-Pulido, M., Vidal-Carou, M.C., 1999. Relationship between biogenic amine contents and the size of dry fermented sausages. Meat Sci. 51, 305–311.

Cipolla, B.G., Havouis, R., Moulinoux, J.P., 2007. Polyamine contents in current foods: a basis for polyamine reduced diet and a study of its long term observance and tolerance in prostate carcinoma patients. Amino Acids 33, 203–212.

De Mey, E., De Klerck, K., De Maere, H., Dewulf, L., Derdelinckx, G., Peeters, M.C., et al., 2014. The occurrence of *N*-nitrosamines, residual nitrite and biogenic amines in commercial dry fermented sausages and evaluation of their occasional relation. Meat Sci. 96, 821–828.

EC, 2006. Setting maximum levels for certain contaminants in foodstuffs. Commission Regulation (EC). No 1881/2006. Off. J. Eur. Union L 364, 1–20.

Eerola, H.S., Roig Sagués, A.X., Hirvi, T.K., 1998. Biogenic amines in Finnish dry sausages. J. Food Saf. 18, 127–138.

EFSA Panel on Biological Hazards (BIOHAZ), 2011. Scientific opinion on risk based control of biogenic amine formation in fermented foods. EFSA J. 9 (2392), 1–93.

Ekici, K., Coskun, H., 2004. Histamine contents of some commercial vegetable pickles. Pak. J. Nutr. 3, 197–198.

Emborg, J., Laursen, B.G., Dalgaard, P., 2005. Significant histamine formation in tuna (*Thunnus albacores*) at 2°C- effect of vacuum—and modified atmosphere-packaging on psychrotolerant bacteria. Int. J. Food Microbiol. 101, 263–279.

FDA, 2010. Compliance program guidance manual. Chapter 3: Foodborne biological hazards. Program 7303.844. Part IV-Project 03, pp. 4–7.

Fiedorowicz, J.G., Swartz, K.L., 2004. The role of monoamine oxidase inhibitors in current psychiatric practice. J. Psychiatr. Pract. 1110, 239–248.

Fletcher, G.C., Summers, G., Winchester, R.V., Wong, R.J., 1995. Histamine and histidine in New Zealand marine fish and shellfish species, particularly kahawai (*Arripis trutta*). J. Aquat. Food Prod. Technol. 42, 53–74.

Gençcelep, H., Kaban, G., Aksu, M.I., Öz, F., Kaya, M., 2008. Determination of biogenic amines in sucuk. Food Control 19, 868–872.

Hernández-Jover, T., Izquierdo-Pulido, M., Veciana-Nogues, M.T., Vidal-Carou, M.C., 1996. Ion-pair high performance liquid chromatographic determination of biogenic amines in meat and meat products. J. Agric. Food Chem. 44, 2710–2715.

Hungerford, J.H., 2010. Scombroid poisoning: a review. Toxicon 56, 231–243.

Jakszyn, P., González, C.A., 2006. Nitrosamine and related food intake and gastric and esophageal cancer risk: a systematic review of the epidemiological evidence. World J. Gastroenterol. 12, 4296–4303.

Kalač, P., Špička, J., Křížek, M., Pelikánová, T., 2000. Changes in biogenic amine concentrations during sauerkraut storage. Food Chem. 69, 309–314.

Kalač, P., Svecova, S., Pelikánová, T., 2002. Levels of biogenic amines in typical vegetable products. Food Chem. 77, 349–351.

Kanki, M., Yoda, T., Ishibashi, M., Tsukamoto, T., 2004. Photobacterium phosphoreum caused a histamine fish poisoning incident. Int. J. Food Microbiol. 92, 79–87.

Komericki, P., Klein, G., Reider, N., Hawranek, T., Strimitzer, T., Lang, R., et al., 2011. Histamine intolerance: lack of reproducibility of single symptoms by oral provocation with histamine: a randomized, double-blind, placebo-controlled cross-over study. Wien. Klin. Wochenschr. 123, 15–20.

Komprda, T., Smělá, D., Novická, K., Kalhotka, L., Šustová, K., Pechová, P., 2007. Content and distribution of biogenic amines in Dutch-type hard cheese. Food Chem. 102, 129–137.

Konakovsky, V., Focke, M., Hoffmann-Sommergruber, K., Schmid, R., Scheiner, O., Moser, P., et al., 2011. Levels of histamine and other biogenic amines in high-quality red wines. Food Addit. Contam. Part A 28, 408–416.

Koral, S., Tufan, B., Ščavničar, A., Kočar, D., Pompe, M., Köse, S., 2013. Investigation of the content of biogenic amines and some food safety parameters of various commercially salted fish products. Food Control 32, 597–606.

Křížek, M., Pavliček, T., Vácha, F., 2002. Formation of selected biogenic amines in carp meat. J. Sci. Food Agric. 82, 1088–1093.

Kuda, T., Izawa, Y., Ishii, S., Takahashi, H., Torido, Y., Kimura, B., 2012. Suppressive effect of *Tetragenococcus halophilus*, isolated from fish-*nukazuke*, on histamine accumulation in salted and fermented fish. Food Chem. 130, 569–574.

Kung, H.F., Tsai, Y.H., Wei, C.I., 2007. Histamine and other biogenic amines and histamine-forming bacteria in miso products. Food Chem. 101, 351–356.

Ladero, V., Calles-Enriquez, M., Fernandez, M., Alvarez, M.A., 2010. Toxicological effects of dietary biogenic amines. Curr. Nutr. Food Sci. 6, 145–156.

Lavon, O., Lurie, Y., Bentur, Y., 2008. Scombroid fish poisoning in Israel, 2005–2007. Isr. Med. Assoc. J. 10, 789–792.

Lehane, L., Olley, J., 2000. Histamine fish poisoning revisited. Int. J. Food Microbiol. 58, 1–37.

Linares, D.M., Cruz Martin, M., Ladero, V., Alvarez, M.A., Fernández, M., 2011. Biogenic amines in dairy products. Crit. Rev. Food Sci. Nutr. 51, 692–703.

Loret, S., Deloyer, P., Dandrifosse, G., 2005. Levels of biogenic amines as a measure of the quantity of the beer fermentation process: data from Belgian samples. Food Chem. 89, 519–525.

Maintz, L., Novak, N., 2007. Histamine and histamine intolerance. Am. J. Clin. Nutr. 85, 1185–1196.

Maintz, L., Benfadal, S., Allam, J.P., Hagemann, T., Fimmers, R., Novak, N., 2006. Evidence for a reduced histamine degradation capacity in a subgroup of patients with atopic eczema. J. Allergy Clin. Immunol. 117, 1106–1112.

Marijan, A., Džaja, P., Bogdanović, T., Škoko, I., Cvetnić, Z., Dobranić, V., et al., 2014. Influence of ripening time on the amount of certain biogenic amines in rind and core of cow milk Livno cheese. Mljekarstvo 64, 159–169.

McCabe-Sellers, B.J., Staggs, C.G., Bogle, M.L., 2006. Tyramine in foods and monoamine oxidase inhibitor drugs: a crossroad where medicine, nutrition, pharmacy, and food industry converge. J. Food Compos. Anal. 19, 58–65.

Millichap, J.G., Yee, M.M., 2003. The diet factor in pediatric and adolescent migraine. Pediatr. Neurol. 28, 9–15.

Min, J.S., Lee, S.O., Jang, A., Lee, M., Kim, Y., 2004. Quantitative analysis of biogenic amines in raw and processed foods of animal origin on Korean domestic market. Asian-Australas. J. Anim. Sci. 17, 1764–1768.

Moinard, C., Cynober, L., de Bandt, J.P., 2005. Polyamines: metabolism and implications in human diseases. Clin. Nutr. 24, 184–197.

Muscarella, M., Iammarino, M., Centonze, D., Palermo, C., 2005. Measurement of histamine in seafood by HPLC, CE, and ELISA: comparison of three techniques. Vet. Res. Commun. 29, 343–346.

Nout, M.J.R., 1994. Fermented foods and food safety. Food Res. Int. 27, 291–296.

Ntzimani, A.G., Paleologos, E.K., Savvaidis, I.N., Kontominas, M.G., 2008. Formation of biogenic amines and relation to microbial flora and sensory changes in smoked turkey breast fillets stored under various packaging conditions at 4°C. Food Microbiol. 25, 509–517.

Parente, E., Martuscelli, M., Gardini, F., Grieco, S., Crudele, M.A., Suzzi, G., 2001. Evolution of microbial populations and biogenic amine production in dry sausages produced in Southern Italy. J. Appl. Microbiol. 90, 882–891.

Pilapil, A.R., Neyrinck, E., Deloof, D., Bekaert, K., Robbens, J., Raes, K., 2015. Chemical quality assessment of traditional salt-fermented shrimp paste from Northern Mindanao, Philippines. J. Sci. Food Agric. Available from: http://dx.doi.org/10.1002/jsfa.7167.

Pintado, A.I.E., Pinho, O., Ferreira, I.M.P.L.V.O., Pintado, M.M.E., Gomes, A.M.P., Malcata, F.X., 2008. Microbiological, biochemical and biogenic amine profiles of Terrincho cheese manufactured in several dairy farms. Int. Dairy J. 18, 631–640.

Predy, G., Honish, L., Hohn, W., Jones, S., 2003. Was it something she ate? Case report and discussion of scombroid poisoning. Can. Med. Assoc. J. 168, 587–588.

Prester, L., 2011. Biogenic amines in fish, fish products and shellfish: a review. Food Addit. Contam. 28, 1547–1560.

Prester, L., 2014. Toxicity risks associated with the recovery of bioactive compounds from marine sources. In: Hernández-Ledesma, B., Herrero, M. (Eds.), Bioactive Compounds from Marine Foods. IFT Press, Wiley Blackwell, Chichester, pp. 395–416.

Prester, L., Macan, J., Varnai, V.M., Orct, T., Vukušić, J., Kipčić, D., 2009. Endotoxin and bioganic amine levels in Atlantic mackerel (*Scomber scombrus*), sardine (*Sardina pilchaedus*) and Mediterranean hake (*Mercluccius merluccius*) stored at 22°C. Food Addit. Contam. 26, 355–362.

Prester, L., Orct, T., Macan, J., Vukušić, J., Kipčić, D., 2010. Determination of biogenic amines and endotoxin in squid, musky octopus, Norway lobster, and mussel stored at room temperature. Arch. Ind. Hyg. Toxicol. 61, 389–397.

Rabie, M.A., Siliha, H., el-Saidy, S., el-Badawy, A.A., Malcata, F.X., 2010. Effects of ϒ-radiation upon biogenic amine formation in Egyptian ripened sausages during storage. Innovative Food Sci. Emerg. Technol. 11, 661–665.

Rabie, M.A., Siliha, H.I., El-Saidy, S.M., El-Badawy, A.A., Xavier Malcata, F., 2011. Effect of γ-irradiation upon biogenic amine formation in blue cheese during storage. Int. Dairy J. 21, 373–376.

Rabie, M.A., Toliba, A.O., Sulieman, A.R., Malcata, F.X., 2014a. Changes in biogenic amine contents throughout storage of canned fish products. Pak. J. Food Sci. 24, 137–150.

Rabie, M.A., Peres, C., Malcata, F.X., 2014b. Evolution of amino acids and biogenic amines throughout storage in sausages made of horse, beef and turkey meats. Meat Sci. 96, 82–87.

Rauscher-Gabernig, E., Gabernig, R., Brueller, W., Grossgut, R., Bauer, F., Paulsen, P., 2012. Dietary exposure assessment of putrescine and cadaverine and derivation of tolerable levels in selected foods consumed in Austria. Eur. Food Res. Technol. 235, 209–220.

Restuccia, D., Spizzirri, U.G., Bonesi, M., Tundis, R., Menichini, F., Picci, N., et al., 2015. Evaluation of fatty acids and biogenic amines profiles in mullet and tuna roe during six months of storage at 4°C. J. Food Compos. Anal. 40, 52–60.

Rohani, S.M.R., Aliakbarlu, J., Ehsani, A., Hassanzadazar, H., 2013. Biogenic amine determination in some traditional cheeses in West Azerbaijan province of Iran. Vet. Res. Forum 4, 115–118.

Rosell-Camps, A., Zibetti, S., Pérez-Esteban, G., Vila-Vidal, M., Ferrés-Ramis, L., Garcia-Teresa-Garcia, E., 2013. Histamine intolerance as a cause of chronic digestive complaints in pediatric patients. Rev. Esp. Enferm. Dig. 105, 201–207.

Ruiz-Capillas, C., Jiménez-Colmenero, F., 2004. Biogenic amines in meat and meat products. Crit. Rev. Food Sci. Nutr. 44, 489–499.

Samková, E., Dadáková, E., Pelikánová, T., 2013. Changes in biogenic amine and polyamine contents in smear-ripened cheeses during storage. Eur. Food Res. Technol. 237, 309–314.

Schirone, M., Tofalo, R., Mazzone, G., Corsetti, A., Suzzi, G., 2011. Biogenic amine content and microbiological profile of Pecorino di Farindola cheese. Food Microbiol. 28, 128–136.

Selhub, E.M., Logan, A.C., Bested, A.C., 2014. Fermented foods, microbiota, and mental health: ancient practice meets nutritional psychiatry. J. Physiol. Anthropol. 32, 2–12.

Shalaby, A.R., 1996. Significance of biogenic amines to food safety and human health. Food Res. Int. 29, 675–690.

Shulman, K.I., Walker, S.E., 1999. Refining the MAOI diet: tyramine content of pizzas and soy products. J. Clin. Psychiatry 60, 191–193.

Silla Santos, M.H., 1996. Biogenic amines: their importance in foods. Int. J. Food Microbiol. 29, 213–231.

Spanjer, M.C., Van Roode, B.A.S.W., 1991. Towards a regulatory limit for biogenic amines in fish, cheese and sauerkraut. De Ware(n)-Chemicus 21, 139–167.

Spano, G., Russo, P., Lonvaud-Funel, A., Lucas, P., Alexandre, H., Grandvalet, C., et al., 2010. Biogenic amines in fermented foods. Eur. J. Clin. Nutr. 64, 95–100.

Stratton, S.S., Hutkins, R.W., Taylor, S.L., 1991. Biogenic amines in cheese and other fermented foods: a review. J. Food Prot. 54, 460–470.

Suzzi, G., Gardini, F., 2003. Biogenic amines in dry fermented sausages: a review. Int. J. Food Microbiol. 88, 41–54.

Tasić, T.A., Ikonić, P.M., Petrović, L.S., Jokanović, M.R., Tomović, V.M., Šojić, B.V., et al., 2014. Formation of vasoactive amines in dry fermented sausage Petrovská klobása during drying and ripening in traditional and industrial conditions. Int. J. Biol., Food Vet. Agric. Eng. 8, 420–423.

Tsai, Y.H., Kung, H.F., Lee, T.M., Chen, H.C., Chou, S.S., Wei, C.I., et al., 2005. Determination of histamine in canned mackerel implicated in a food born poisoning. Food Control 16, 579–585.

Tsai, Y.H., Kung, H.F., Chang, S.C., Lee, T.M., Wei, C.I., 2007a. Histamine formation by histamine-forming bacteria in douchi, a Chinese traditional fermented soybean product. Food Chem. 103, 1305–1311.

Tsai, Y.H., Hsieh, H.S., Chen, H.C., Cheng, S.H., Chai, T., Wang, D.F., 2007b. Histamine level and species identification of billfish meats implicated in two food-borne poisoning. Food Chem. 104, 1366–1371.

Valsamaki, K., Michaelidou, A., Polychroniadou, A., 2000. Biogenic amine production in Feta cheese. Food Chem. 71, 259–266.

Virgili, R., Saccani, G., Gabba, L., Tanzi, E., Soresi Bordini, C.S., 2007. Changes of free amino acids and biogenic amines during extended ageing of Italian dry-cured ham. LWT-Food Sci. Technol. 40, 871–878.

Visciano, P., Schirone, M., Tofalo, R., Suzzi, G., 2013. Biogenic amines in raw and processed seafood. Front. Microbiol. 3, 1–10.

Watanabe, S., Nagase, S., Sato, S., Ohkuma, S., 2009. Effective anticancer drug and food choices based on polyamine levels in cancer tissues. Curr. Drug Ther. 4, 214–220.

Wunderlichová, L., Buňková, L., Koutný, M., Jančová, P., Buňka, F., 2014. Formation, degradation, and detoxification of putrescine by foodborne bacteria: a review. Compr. Rev. Food Sci. Food Saf. 13, 1012–1030.

Yang, J., Ding, X., Qin, Y., Zeng, Y., 2014. Safety assessment of the biogenic amines in fermented soya beans and fermented bean curd. J. Agric. Food Chem. 62, 7947–7954.

CHAPTER

Regulated disinfection byproducts in minimally processed vegetables and beverages

23

Mercedes Gallego and María José Cardador

Department of Analytical Chemistry, Campus of Rabanales, University of Córdoba, Córdoba, Spain

INTRODUCTION

Water disinfection has been used since the early 20th century to improve the hygienic quality of drinking water by removing waterborne bacterial pathogens (Richardson, 2012). The treatment, however, generates potentially harmful disinfection byproducts (DBPs) (Hrudey and Charrois, 2012). Different disinfectants also produce different types and amounts of DBPs, trihalomethanes (THMs) and haloacetic acids (HAAs) being the two most prevalent classes (Richardson and Ternes, 2011). These chlorinated and brominated species include four THMs (trichloromethane (TCM), bromodichloromethane (BDCM), dibromochloromethane (DBCM), and tribromomethane (TBM)) and nine HAAs (monochloroacetic (MCAA), monobromoacetic (MBAA), dichloroacetic (DCAA), trichloroacetic (TCAA), bromochloroacetic (BCAA), dibromoacetic (DBAA), bromodichloroacetic (BDCAA), dibromochloroacetic (DBCAA) and tribromoacetic (TBAA) acids). The recently discovered iodinated THMs and HAAs are gaining increasing importance but they are not yet widely measured in drinking water (Ye et al., 2013). DBPs are considered to be potentially carcinogenic and mutagenic, and hence to pose a real health risk to humans (Hrudey and Charrois, 2012). Thus, some studies have found that people who drink chlorinated surface water have a higher risk of developing cancer of the bladder, rectum, and colon, and also that consumption of the water may contribute to adverse effects on reproductive health such as low-birthweight babies, premature births, miscarriages, and birth defects of the brain and spinal cord (Richardson et al., 2007). This has led the United States Environmental Protection Agency (US EPA) to establish maximum contaminant

P. Kotzekidou (Ed): Food Hygiene and Toxicology in Ready-to-Eat Foods.
DOI: http://dx.doi.org/10.1016/B978-0-12-801916-0.00023-6

levels (MCLs) in drinking water for the total concentration of four THMs and five HAAs at 80 and 60 μg/L, respectively (US EPA, 2006). Also, the European Union regulated the total concentration of four THMs to 100 μg/L after 1998 (Council Directive 98/83/EC, 1998), but has so far established no regulatory limit for HAAs. The World Health Organization (WHO) has set guideline values for the THMs TCM, BDCM, DBCM, and TBM at 200, 60, 100, and 100 μg/L, respectively, and the HAAs MCAA, DCAA, and TCAA at 20, 50, and 200 μg/L, respectively (WHO, 2006).

DBPs have recently arisen as an emergent group of contaminants in foods and beverages. This has led international bodies such as the Food and Agriculture Organization (FAO)/WHO to recommend the development of accurate, sensitive methods for their determination in beverages and foods (FAO/WHO, 2008). The ideal disinfectant for the food industry would be one with a high broad-spectrum efficacy against microorganisms and a low DBP formation potential. There may be a limited relationship to DBPs from chlorine and food contact. Among these are THMs and HAAs, which account for the largest portion of identifiable DBPs in chlorinated drinking water (up to 50%). In addition to drinking water, THMs have been found in some foods and beverages, namely: by effect of accumulation and sorption of THMs present in food packaging and wrapping materials; contact with cleaners/disinfectants used in food-processing equipment, rinses or water in food processing, storage and/or marketing; and formation during food preparation (cooking included) through reactions between residual chlorine and precursors such as carbohydrates, lipids, and proteins present in foods (Huang and Batterman, 2009).

THMs have been found at widely ranging concentrations (2–450 μg/kg) in flour, meat, dairy products, vegetables, beverages, and various other foods (US FDA, 1999; Fleming-Jones and Smith, 2003; FAO/WHO, 2008). Although HAAs can remain stable over long periods by virtue of their ionic nature and low volatility, information about their concentrations in foods and beverages is scant. This chapter provides an overview of the generation and occurrence of the regulated DBPs (THMs and HAAs) in foods potentially containing high levels of these compounds by effect of industrial treatments or water being used as an ingredient (eg, ready-to-eat foods, juices, and soft drinks).

GENERATION AND OCCURRENCE OF REGULATED DBPs IN MINIMALLY PROCESSED FRUITS AND VEGETABLES

MANUFACTURING PRINCIPLES

Fresh fruit and vegetable consumption is currently growing steadily around the world, mainly in response to the recommendations of various organizations such as WHO, FAO, USDA, and the European Food and Safety Authority based on their healthy properties (Artés and Allende, 2014). These recommendations have also increased consumers' demand for healthy, fresh, easily prepared products such as minimally processed fruits and vegetables (MPFVs), also called "fresh-cut fruit and vegetables" (Ramos et al., 2013; Artés and Allende, 2014). In line with the definition of the International Fresh-cut Produce Association (IFPA), MPFVs are "fruit or vegetables that have been trimmed and/or peeled and/or cut into 100% usable products that are bagged or prepackaged to offer consumers high nutrition, convenience, and flavor while still maintaining their freshness" (IFPA, 2005). MPFVs have similar characteristics to the whole original fruit or vegetable and retain their freshness; also, they usually need no further processing before use, which is especially attractive for consumers. The fresh-cut industry has grown enormously over the past 10 years in response to the increasing demand for MPFVs. The continuous expansion of the market for fresh-cut products reflects the prominent role of quality and safety assurance in the fresh-cut industry. One of the most important safety issues for this industry is related to bacterial pathogens and viruses. The microbiological load present in vegetables can come from all the steps in their manufacturing process including cultivation, harvest, transportation, and further processing and handling. Also, preparation of MPFVs can damage plant tissue and remove natural protective layers, thereby accelerating product decay and providing an easy entry way for microbial pathogens and chemical contaminants (Artés and Allende, 2014). For this reason, fruits and vegetables require proper handling, preparation, and storage by enforcing good hygiene practices during production, transport, and manufacturing of MPFVs.

The preparation of MPFVs in the fresh-cut industry includes precooling, sorting, washing, peeling/cutting, disinfection, rinsing, dewatering, subsequent packaging, and distribution. The general unit operations and the maximum recommended temperatures for each processing step are shown in Fig. 23.1. The most crucial steps in the process are washing, peeling/cutting, and disinfection, all of which can have a strong impact on the quality of the end-product. Thus, whole products should be precooled as soon as possible after harvesting (transport and storage steps) in order to reduce respiration and ethylene production, and delay nutritional decay as a result. The first washing step of the whole product, generally with tap water, is intended to eliminate unwanted dirt, pesticide residues, plant debris, soil, insects, and foreign matter, as well as to slow down enzymatic discoloration reactions (Soliva-Fortuny and Martín-Belloso, 2003). The peeling and cutting steps constitute a critical point in the processing line hygiene; in fact, the equipment used must be cleaned, disinfected, and sharpened at regular intervals every working day to avoid build-up of organic residues and minimize damage to the product (Codex Alimentarius Commission, 2003; US FDA, 2008). Disinfection can be achieved very simply by spraying, but generally involves immersing the product in chilled (5−10°C), sanitized water in a bath or washing tank. Modern aeration "jacuzzi" washing systems consisting of three separate washing stages in three different tanks are being increasingly frequently used for this purpose. Disinfection of the product is accomplished in the second tank, where the water is treated with a disinfectant intended to prevent cross-contamination during washing

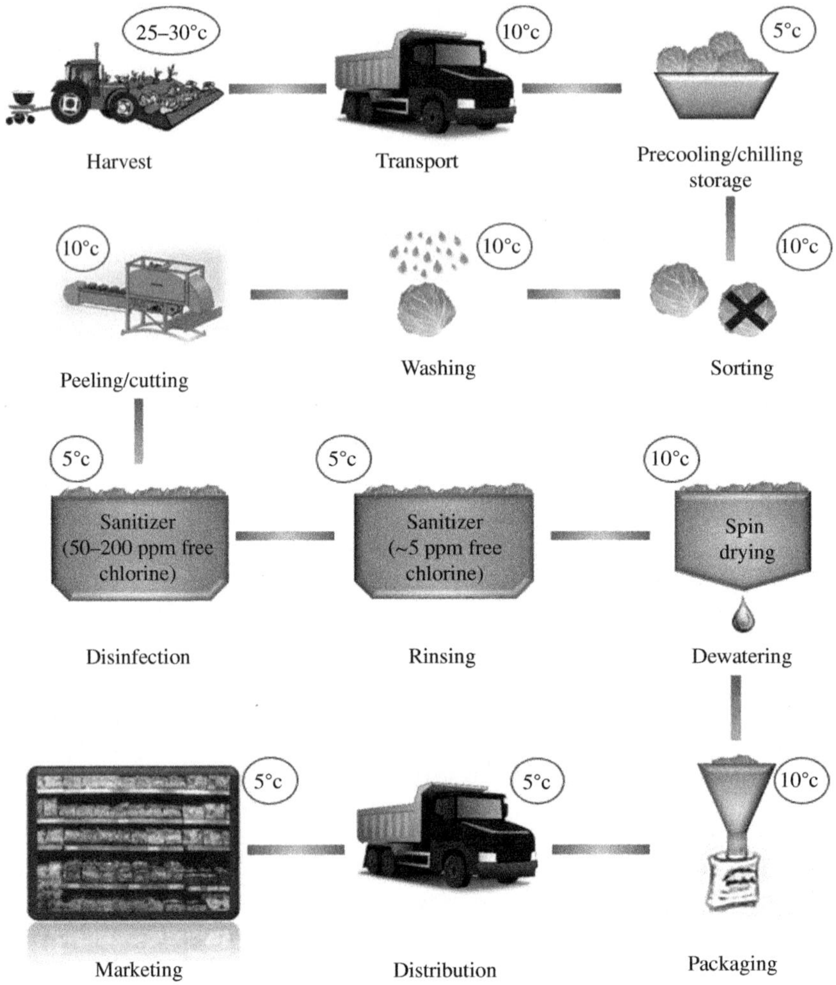

FIGURE 23.1

General unit operations of a minimally processed fruits and vegetables (MPFVs) processing plant and highest recommended temperatures for each processing step.

(FAO, 2008). A rinsing step is carried out in the third tank with a very low dose of disinfecting agent to remove residual disinfectant as far as possible (Artés and Allende, 2014). The last step before packaging is drying of the product, usually by centrifugation in stainless steel tunnels to remove excess water (Fig. 23.1).

Chlorine-based chemicals (particularly liquid chlorine and sodium hypochlorite) are the most widely used sanitizers for disinfection of fresh products and processing equipment by virtue of their simplicity, efficacy, and cost-effectiveness. The United States Food and Drug Administration (US FDA) recommends using a solution containing 50–200 mg/L (ppm) free chlorine with a contact time of 1–2 min for product surfaces and processing equipment (US FDA, 2009), followed by rinsing with a more dilute solution (~5 ppm chlorine) (Fig. 23.1). One of the main disadvantages of using chlorine-based sanitizers, especially in solutions with so high levels of free chlorine (eg, drinking water contains ~0.4 mg/L), is the potential generation of DBPs by reaction between chlorine and vegetable tissue, or between chlorine and washing water containing organic matter such as debris and exudates released by vegetables (Huang and Batterman, 2010). Because free chlorine is easily consumed by organic matter during the disinfection process, it is necessary to maintain a certain level of free chlorine in the washing water (Luo et al., 2012). For this reason, the fresh-cut industry commonly uses hyperchlorination (namely, high levels of chlorine), which can result in the formation of DBPs at high levels. Also, water from industrial processing is usually re-used to minimize consumption. This can raise organic matter and DBP levels during washing even if clean water is used for this purpose. These adverse effects have led some European countries such as Germany, the Netherlands, Denmark, Switzerland, and Belgium to ban the use of chlorine-based sanitizers to disinfect MPFVs (Artés and Allende, 2014).

There is a current trend to avoid chlorine-based sanitizers in disinfection processes in order to prevent formation of DBPs (Gil et al., 2009). Chlorine dioxide is an effective choice for this purpose as it has an oxidant capacity 2.5 times greater than that of chlorine and forms fewer halogenated DBPs (Gómez-López et al., 2009). The fresh-cut industry has explored other choices such as organic acids, hydrogen peroxide, calcium-based solutions, ozone, quaternary ammonium compounds, trisodium phosphate, bromine, iodine, peroxyacetic acid, and electrolyzed water. Some physical methods involving ultrasound, high pressure, pulsed light, ultraviolet radiation, ionizing radiation, or cold plasma have proved effective in preventing cross-contamination in washing water (Ramos et al., 2013). However, chlorine-based sanitizers remain the most effective and inexpensive disinfectants provided they are used in appropriate doses. Although the washing and disinfection steps are crucial for effective processing of MPFVs, the fresh-cut industry uses no standardized washing procedures. For example, industrial washing water sanitizers used in the United States are regulated as secondary direct food additives by the US FDA (2003a,b). In the EU, the detailed controls made under the Framework Directive have been implemented into the national law of each EU member state within the context of food additives (Council Directive 89/107/EC, 1988). Thus, legislation on sanitizers has not yet been harmonized and a global approach to controlling the use of these chemicals by the fresh-cut industry is missing.

THMs AND HAAs IN MPFVs

Studies on the generation and occurrence of DBPs in MPFVs initially focused on THMs and, mainly, on chloroform (TCM), which is the individual DBP found at the highest concentrations in treated water. Most of the scientific literature on THM formation in the washing water and its residual presence in the washed product was obtained by using simulated washing water at the laboratory scale, which does not always mimic the real situation in the fresh-cut industry. For example, Klaiber et al. (2005) used conventional industrial equipment at the pilot-plant scale to examine the effects of cold (4°C) and warm water (50°C), both chlorinated and unchlorinated, used to wash uncut peeled carrots. They found chlorine consumption and chemical oxygen demand to be low in both experiments. Washing the carrots in cold or warm chlorinated water led to an undetectable concentration of chloroform and to a 0.2 μg/L level, respectively. The chloroform concentration found in uncut peeled carrots washed with cold (4°C) and hot (50°C) water was trace levels and 2.5 μg/kg, respectively. More recently, Shen et al. (2016) found increased levels of THMs in washing water.

López-Gálvez et al. (2010) assessed the effectiveness of aqueous chlorine dioxide and sodium hypochlorite as sanitizers for fresh-cut iceberg lettuce by measuring the formation of THMs at a simulated laboratory scale. For this purpose, they analyzed both the washing water and the lettuce after washing. THMs were only found, at 217 μg/L, in the washing water treated with sodium hypochlorite, which testifies to the suitability of chlorine dioxide as a sanitizer to prevent THM formation. No THMs were detected in fresh-cut lettuce after washing with either sanitizer. Subsequently, THM formation in the washing water and in washed baby spinach was assessed by using sodium hypochlorite and electrolyzed oxidizing water with and without the addition of salt (EOW + NaCl and EOW, respectively) (Gomez-López et al., 2013). A final concentration of free chlorine within the range 2–4 mg/L was found to suffice for the authorized limits of total THMs (TTHMs) to be exceeded. Thus, the TTHM values of the washing water treated with sodium hypochlorite and EOW + NaCl (194 and 126 μg/L, respectively) exceeded the authorized limit set by European legislation and the US EPA for drinking water, whereas the TTHM levels after treatment with EOW (50 μg/L) did not. The concentration of TTHMs in washed baby spinach was low (~7 μg/kg). It is likely that THMs were weakly adsorbed to the spinach surface and hence easily removed by rinsing. These results are consistent with others previously reported (Committee on Toxicity, 2006), which revealed THM formation in commercially prepared salads to be negligible; thus, no THMs were detected in 11 commercially prepared salads and only one sample contained chloroform (16 μg/kg). Recently, the effect of free chlorine and organic load on THM production in simulated washing water was studied (Shen et al., 2016). In order to replenish free chlorine and organic load to mimic a real situation, sodium hypochlorite and lettuce juice were periodically added to the washing water. The TTHM concentrations found (20–859 μg/L) were much higher than the authorized limits established by the US EPA and European legislation on drinking water. THM production was found to depend mainly on the free chlorine concentration; however, THM levels increased slightly with increasing organic load. As a result, TTHM concentrations were efficiently reduced by using the minimum amount of free chlorine needed for disinfection of MPFVs.

As stated above, there is abundant information on THMs in MPFVs, but little is known about the formation of HAAs in washing water or end-products with a view to assessing the actual risks associated with these compounds. Similarly to THMs, the generation of HAAs in simulated washing water was recently evaluated with periodic additions of sodium hypochlorite as sanitizer and organic matter from lettuce juice (Shen et al., 2016). The total HAA (THAA) concentrations found in the washing water (88–2103 μg/L) were much higher than the authorized limits for drinking water. Repeated additions of sanitizer to the washing water to maintain appropriate levels of free chlorine is a common practice in the fresh-cut industry. On the other side, the water typically has a high load in organic compounds coming from vegetable tissue. As a result, the water usually contains increased concentrations of THAAs, residual amounts of which may remain in the end-product. The earliest report on HAA levels in commercially prepared salads, in addition to THMs, was published in 2006 (Committee on Toxicity, 2006). Various types of prepared salads from a number of manufacturers were analyzed for the presence of specific disinfectants and DBPs. The potential disinfectants and DBPs examined included chlorite, chlorate, bromate, free and total chlorine, chloramine, chloroform, total THMs, DCAA, and TCAA. An amount of 100 g of each type of salad was agitated with 100 mL of water for 1 h for analysis of the leachate. As can be seen from Table 23.1, 8 of the 12 samples studied contained DCAA and/or TCAA. The analysis levels found in most of the samples were low enough to comply with drinking water standards (2.7–30 μg/kg of THAAs). Only in one sample did the combined amounts of DCAA and TCAA (83 μg/kg) exceed the 60 μg/L permitted by US EPA drinking water regulations for THAAs. In addition, the estimated intake of each compound as calculated from salad consumption data for the different samples was at least several orders of magnitude lower than tolerable daily intakes set by WHO.

The only existing method for the determination of HAAs (specifically, MCAA, MBAA, DCAA, TCAA, BCAA, DBAA, BDCAA, DBCAA, and TBAA) in minimally processed vegetables (MPVs) is based on derivatization of the acids to volatile esters by static headspace (SHS) and determination by gas chromatography–mass spectrometry (GC–MS) (Cardador and Gallego, 2012). The method is sensitive, with limits of detection from 0.1 to 2.4 μg/kg and an average relative standard deviation of 8%. It was used to determine 100 MPVs purchased at local markets in Spain from Nov. 2011 to Feb. 2012. The samples included 40 single ingredients (3 chopped onion, 4 grated carrot, 3 green pepper, 10 chicory, 10 lettuce and 10 spinach) and 60 ready-to-eat salads containing 3–5 ingredients (lettuce, chicory, carrots, spinach, and red cabbage) each. Firstly, the stability of the nine HAAs was advocated in iceberg lettuce samples spiked with a 20 μg/kg concentration of each acid that were stored at 4°C for up to 5 days before analysis. The most common HAAs (namely, DCAA, TCAA, BCAA, DBAA, and BDCAA) remained at constant levels for up to 3 days and the others for up to 1.5 days. Table 23.1 lists the concentrations of the one-to-five HAAs found at detectable levels in the 23 positive samples (23% of all). Only DCAA

Table 23.1 Concentrations of Haloacetic Acids (HAAs) in Minimally Processed Vegetables (Range Values in μg/kg)

	n	DCAA	TCAA	BCAA	DBAA	BDCAA	Total HAAs
Salad[a]	8	nd–32	nd–51				2.7–83
MPV[b]							
Carrot	1	6.3	13	1.2	0.9	<2	21
Chicory	1	<0.4	1.6	nd	nd	nd	1.6
Green pepper	1	nd	0.7	nd	nd	nd	0.7
Iceberg lettuce	3	1.4–15	3.7–18	<0.4–3.9	nd–1.3	nd–3.2	6.8–41
Romaine lettuce	2	2.2, 3.5	1.9, 5.2	0.5, nd	nd, <0.8	nd, <2	4.6, 8.7
Salad	10	<0.4–22	0.6–24	nd–3.2	nd–1.4	nd–4.8	0.6–55
Spinach	5	nd–16	0.5–19	nd–4.1	nd– <0.8	nd	0.5–39

n, number of samples; nd, not detected.
The "less than" (<) values represent the limits of quantitation of the method.
[a]*Data from Commitee on Toxicity (2006).*
[b]*Minimally processed vegetables—Data from Cardador and Gallego (2012).*

and TCAA were detected in practically all samples, and at higher concentrations than the other HAAs; on the other hand, brominated HAAs were present in few samples and at lower levels. Most of the 23 MPVs contained HAAs at low levels; by exception, four samples (one of grated carrot (from 4), lettuce 1 (from 10), spinach 5 (from 10), and mixed salad 10 (from 60)) contained concentrations about 5–8 times higher than the other positive samples. The results were compared with those of the analysis of whole vegetables coming directly from the field that were subjected to no industrial processing—and contained no HAAs as a result. Therefore, the presence of HAAs in the MPV samples can be ascribed to the use of chlorine-based sanitizers in the washing step by the fresh-cut industry (see Fig. 23.1). The majority of the samples contained low levels of THAAs (0.5–8.7 μg/kg); by exception, four samples had THAA concentrations from 21 to 55 μg/kg but still lower than the MCL established by the US EPA and EU regulations on drinking water (60 and 80 μg/L, respectively). A person is estimated to consume ~150 g of vegetables each day; also, the most contaminated sample (salad 10, Table 23.1) contained a 55 μg/kg concentration of THAAs. Therefore, a person could ingest more than 8 μg of HAAs (roughly one-fourth of the amount usually present in tap water, 15 – 25 μg/L THAA, on the assumption of a daily consumption of 2 L) (Cardador and Gallego, 2012). Despite this evidence, scientific literature about the removal of HAAs and other DBPs from MPVs is scant. In one experiment, several portions of salads spiked with a 20 μg/kg concentration of each HAA were rinsed with tap water, tap water containing common salt, and tap water with a few drops of vinegar or sodium hypochlorite. Rinsing with salted tap water proved the most efficient choice (70 – 80%) to remove HAAs from these MPV samples (Cardador and Gallego, 2012).

GENERATION AND OCCURRENCE OF REGULATED DBPs IN BEVERAGES

MANUFACTURING PRINCIPLES

Unless juice was formerly consumed fresh, soon after pressing, fermentation was the likely, often desired consequence, until preservation techniques were developed. Codex Alimentarius defines juice as "unfermented but fermentable juice, intended for direct consumption, obtained by the mechanical process from sound, ripe fruits, preserved exclusively by physical means. The juice may be turbid or clear. The juice may have been concentrated and later reconstituted with water suitable for the purpose of maintaining the essential composition and quality factors of the juice. The addition of sugars or acids can be permitted but must be endorsed in the individual standard" (FAO/WHO, 1992). There is a great diversity of related beverages designated as fruit juices. Table 23.2 lists various juice and juice-like terms and designations (FAO, 2001).

Table 23.2 Selected Common Juice Designations

Term	Criteria	Remarks
100% Pure juice	All juice	No adjustment, not from concentrate
Fresh squeezed	Not pasteurized	Held refrigerated, food safety concerns
Chilled, ready to serve	All juice	Held refrigerated, made from concentrate or pasteurized juice
Not from concentrate	Single strength	Pasteurized after extraction
From concentrate	Made from concentrate	Reconstituted and pasteurized
Fresh frozen	Unpasteurized	Single strength, frozen after extraction
Juice blend	All juice	A mixture of pure juices
Puree	Pulp-containing	More viscous than juices, totally fruit
Nectar	Pulpy or clear	Sugar, water and acid added, 25–50% juice[a]
Nectar base	Requires reconstitution	Possesses sufficient flavor, acid, and sugar to water dilution for consumption[a]
Juice drink	Low in juice	Contains 10–20% juice[a]
Juice beverage	Low in juice	Contains 10–20% juice[a]
Juice cocktail	Low in juice	Contains 10–20% juice[a]
Fruit added	Lemonade	Contains >10% fruit juice, sugar and water[a]
Juice extract	Water extract	Fruit extracted by water, then concentrated[a]
Fruit punch	Token juice	~1% juice + natural flavors
Natural flavored	Token juice	Usually >1% juice

[a]*Differing country standards for juice solids minimum.*
Data from FAO, 2001. Principles and Practices of Small- and Medium-Scale Fruit Juice Processing. FAO Agricultural Services Bulletin 146. Food and Agriculture Organization of the United Nations, Rome.

The difference between juice, nectar, and soft drink is related to the content of fruit juice present in a packaged beverage. Worldwide, products labeled as "juice" must contain 100% fresh fruit; these are therefore pure products containing no preservatives or sweeteners and no artificial colors. This category comprises "reconstituted juices," which are basically concentrated from three to six times at the juice factory and subsequently diluted with potable drinking water at a bottling plant to return the juice to its original condition in terms of concentration of soluble solids in water at the time of bottling. One other juice category is that of "not-from-concentrate", where the juice is subjected to slight pasteurization only. Nectars constitute another type of non-carbonated soft drink made by muddling the flesh of fruits; they contain water and additives besides fruit juice. Soft drinks (also called soda, pop, soda pop, coke, fizzy drinks, seltzer, mineral, lolly water, or carbonated beverages) are beverages that typically contain carbonated water, a sweetener, and a flavor (Vaux, 2003). Overconsumption of sugar-sweetened soft drinks has been associated with obesity (Woodward-Lopez et al., 2010; Trumbo and Rivers, 2014), type 2 diabetes (Romaguera-Bosh, 2013), dental caries, and low nutrient levels (Vartanian et al., 2007; Deboer et al., 2013).

Fruit juices are widely consumed by different age groups all over the world. Their consumption has been boosted in the past two decades by claims that they reduce the risk of many chronic and degenerative diseases (Szymczycha-Madeja et al., 2014). Thus, consumption of fresh and concentrated juices increased by 1.7% and 6.8% from 2013 to 2014 (AIJN, 2014). Economical transportation and volume reduction of fruit juice is accomplished by extracting two-thirds of the water, the resulting concentrate subsequently being supplied with the same amount of previously extracted water or vitamins by reconstitution with treated water. This has fostered unlabeled dilution and adulteration, two unethical trade practices that should be rigorously avoided and condemned (Nagy and Wade, 1995).

Production of fruit juice involves several processing steps related to water, namely: fruit preparation (washing, rinsing, and trimming), reconstitution (water addition), and washing of juice tanks and packaging (Belitz et al., 2009). Vacuum concentration is the greatest contributor to the wide availability, economy, and popularity of fruit juices. Water removal by concentration greatly simplifies juice handling, storage, and shipping logistics. Some juices can be concentrated to over 70°Brix, thereby increasing the juice solid fraction up to 10-fold (FAO, 2001). The concentrate is mixed with a small amount of oil and stored in enormous bulk storage tanks or in barrels (Fellows, 2000). Refrigeration combined with pasteurization and air-tight packaging can further extend storage life with minimum quality changes; however, the resulting juice can no longer be labeled "fresh" (Crandell and Hendrix, 2001). The acid nature of most juices permits pasteurization, defined as the use of temperatures near 100°C to destroy spoilage organisms. A potentially effective process for fresh juice involves heating clean, well-sorted whole fruit at 80°C for up to 1 min. The optimum procedure depends on the particular end-product. Fig. 23.2 is a generalized flow scheme for the production of "100% pure juice" or minimally processed juice from concentrated (reconstituted and pasteurized) juice and nectar (pulpy or clear) containing sugar, water, and acid.

WATER IN RECONSTITUTED JUICES AND SOFT DRINKS

Water, which is the principal component in all reconstituted fruit juices and soft drinks (~90%), can have a crucial influence on the quality of the end-product. If water used in the manufacturing process is supplied by a reputable utility supplier and fit for drinking as such, its quality is likely to be acceptable for use in beverage manufacturing with minimal treatment solely for the purpose of removing chlorine and any other flavor taints. However, many soft drink plants have their own private water supply and the water from such sources may or may not be of suitable quality. The key factors that particularly need to be considered in relation to water for use in soft drinks are related to the water quality requirements of the

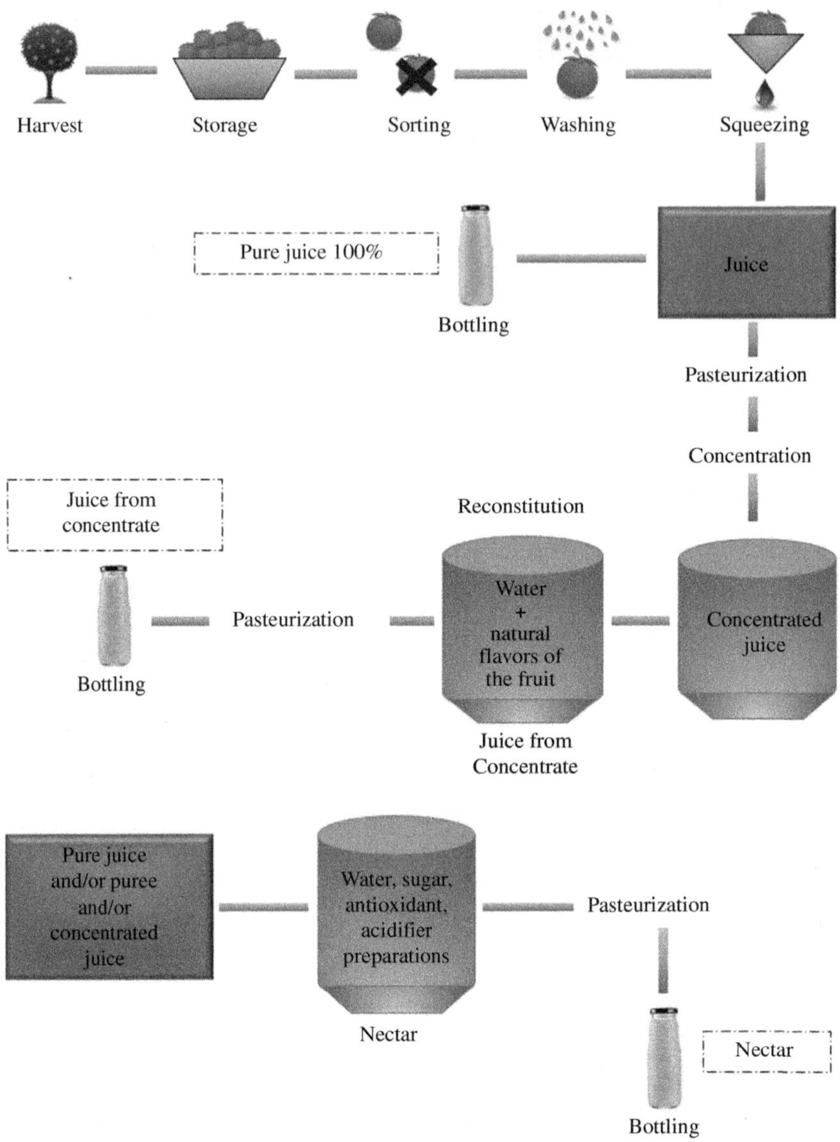

FIGURE 23.2

General unit operations in a factory producing 100% pure juice, juice from concentrate and fruit nectar.

EU Water Directive for Drinking Water (Council Directive 2001/112/CE, 2001) and US EPA (US EPA, 2006). The use of natural mineral water or spring water as an ingredient in soft drinks or reconstituted fruit juices is permitted. It should be borne in mind that, although there may be a marketing advantage in using natural mineral water or spring water, its physicochemical properties may not be ideal for manufacturing soft drinks or reconstituting fruit juice concentrates (Ashurst and Hargitt, 2009).

As stated above, unlabeled dilution and adulteration are common saving practices in the juice industry. Therefore, consideration must be given to the authenticity of juices—mostly of pure juices to be sold as such. Various methods and laboratories are available for assessing juice authenticity, so expert advice to choose the most appropriate method for this purpose may be needed (AIJN, 2014). However, because mineral water is expensive, the most common choice is tampering with treated water containing DBPs. The types of DBPs in beverages and their concentrations will depend on the particular disinfectant used to treat the tap water, the chemical constituents of the source water, and the dynamics of the water in the distribution system. The most prevalent DBPs in treated water are regulated THMs and HAAs; therefore, their control in beverages can be an indicator of the presence of treated water. Using treated water to rinse containers and packages for processing and storage at the factory can also be a source of contamination (FAO/WHO, 2008).

THMs AND HAAs IN BEVERAGES

Graham and Robertson (1988) reported the earliest reference to the content of THMs in cola soft drinks. They found chloroform at concentrations from 14 to 54 μg/L and the other THMs at 4–13 μg/L. A survey of THMs in 44 processed foods including some soft drinks and juices showed chloroform to be present at concentrations from undetectable levels to 94 μg/kg (McNeal et al., 1995). More recently, the formation of THMs during the preparation of 17 beverages (tea, coffee, chocolate, and concentrated juice) and various solid foods (potato, spinach, and corn) by using chlorinated drinking water with high concentrations of chlorine (2–12 mg/L free chlorine) was investigated. The chloroform concentration peaked in tea (3–67 μg/L), followed by coffee (3–13 μg/L), and was lowest in vegetables (<1 μg/L) (Huang and Batterman, 2009). The analysis of 74 soft drinks with polyethylene terephthalate (PET) packaging revealed concentrations of chloroform from <1 to 16.6 μg/L, and concentrations of the other THMs from undetectable to 7.2 μg/L (Silveira et al., 2011); the results, however, were not discussed. The earliest study on the influence of drinking water in contact with beverages on the recovery of HAAs was conducted on five beverages (lemonade, grape and orange juices, tea, and coffee) spiked with 10–100 μg/L concentrations of nine HAAs (Raymer et al., 2000). All HAAs except two were acceptably recovered (70–130%). No health-related information about the native presence of HAAs in the beverages was provided, however.

The only studies on the genuine presence of DBPs in beverages with different characteristics (100% pure juice, reconstituted juice, nectar juice, and soft drinks) have been conducted by our research group (Montesinos and Gallego, 2014; Cardador and Gallego, 2015). The few previous incursions into DBPs in beverages used US EPA methods based on liquid–liquid extraction and gas chromatography; because they were adapted from methodologies for water, they required manual sample handling and had a low sensitivity (Graham and Robertson, 1988; Raymer et al., 2000). Headspace analysis can be used with analytes known to be volatile; this technique has the advantage of being suitable for all matrix foods and uses few or no solvents. Whereas THMs are volatile, HAAs are not and require derivatization to volatile esters. Two new SHS methods have been proposed for this purpose: one for volatilizing THMs and the other for the simultaneous derivatization and volatilization of HAAs. This extraction technique was coupled to GC − MS for the unequivocal identification of potential chlorinated, brominated, and iodinated THMs and HAAs in beverages. The former method (Montesinos and Gallego, 2014) was used to determine the four common THMs (TCM, BDCM, DBCM, and TBM) and six emerging iodinated THMs (dichloroiodomethane (DCIM), bromochloroiodomethane (BCIM), dibromoiodomethane (DBIM), chlorodiiodomethane (CDIM), bromodiiodomethane (BDIM), and triiodomethane (TIM)). The method was sensitive enough to determine the target analytes at the nanogram-per-liter (linear range 0.03 − 100 μg/L) and highly precise (5.0 − 6.2%) in discriminating the source of the species through treated water used for beverage preparation or resulting from contamination through the industrial process (eg, washing, bottling).

Table 23.3 lists the THM levels found in the analysis of 60 juices and nectars, and 40 soft drinks (Montesinos and Gallego, 2014). Only the four common THMs were found, at TTHM concentrations from 0.2 to 38 μg/L in juices and 1.4 to 52 μg/L in soft drinks. The average TTHM concentrations in 100% pure juices (whether fresh squeezed, unpasteurized, or pasteurized) were always below 1 μg/L; also, the predominant species was chloroform (TCM), followed by BDCM. These juices contained no treated water, so the presence of the previous species can be ascribed to contamination from the industrial use of chlorinated sanitizers. In contrast, reconstituted juices contained up to four THMs in two distinct groups, with average TTHM concentrations of 0.5 ± 0.2 and 6.2 ± 3.6 μg/L, respectively. Nectars also fell in two different groups, with average TTHM concentrations of 2.6 ± 1.9 and 14.7 ± 8.7 μg/L, respectively. These differences cannot be ascribed to differences in fruit, but rather in the quality of the treated water used. This assumption is supported by a comparison of the results for the same type of product (nectar) as obtained at 12 different bottling factories; those obtained at factories M and N ($n = 5$) had extreme average TTHM concentrations (6.1 and 22 μg/L, respectively) owing to the differences in the treated water between the two. There were also two distinct groups for soft drinks (c. 90% water) containing average TTHM concentrations of 5.5 ± 2.8 and 28.8 ± 9.1 μg/L, respectively. DCIM was for the first time detected in 10 soft drinks, at levels from <0.07 (tea, isotonic beverage, orange soda) to 0.13 μg/L (cola soda). There were marked differences in TTHM concentrations between 100% pure juices, reconstituted juices, nectars, and soft drinks, with average values depending on the volume of

Table 23.3 Concentrations of Trihalomethanes (THMs) in Spanish Juices and Soft Drinks (Range Values in μg/L)

Beverage	*n*	TCM	BDCM	DBCM	TBM	Total THMs
Juices						
100%						
Apple	1	0.2	nd	nd	nd	0.2
Orange	5	0.3–0.6	nd–0.2	nd	nd	0.3–0.8
Pineapple	1	0.4	0.2	nd	nd	0.6
Reconstituted						
Apple	2	0.4, 0.9	0.1, 1.6	0.1, 1.4	<0.07, 0.3	0.6, 4.2
Cranberry	1	0.3	0.3	0.2	nd	0.8
Grape	1	0.6	0.2	<0.05	nd	0.8
Mix	11	0.2–10	nd–3.7	nd–1.3	nd–0.3	0.2–13
Orange	3	nd–3.0	nd–3.7	0.1–3.3	0.1–0.5	0.2–11
Peach	2	1.8, 3.1	0.3, 1.6	nd, 1.5	nd, 0.2	2.1, 6.4
Pineapple	1	3.6	3.3	2.9	0.4	10
Nectar						
Apple	4	0.4–6.3	0.2–2.1	0.2–0.5	0.1–0.2	0.9–9.1
Grapefruit	2	2.4, 3.9	0.6, 2.2	0.2, 1.7	nd, 0.2	3.2, 8.0
Guava	1	2.9	1.5	1.0	0.1	5.5
Mango	1	0.7	0.3	0.2	0.2	1.4
Mix	2	nd, 0.1	nd, 0.2	0.2, 0.3	0.3, 0.4	0.5, 1.0
Orange	10	0.2–18	0.1–4.8	0.1–2.9	nd–0.5	1.0–24
Passion fruit	1	4.8	2.3	1.1	0.2	8.4
Peach	3	1.5–16	1.3–2.9	0.5–1.1	nd–<0.07	3.9–20
Pear	1	2.6	2.5	1.9	0.3	7.3
Pineapple	7	0.4–32	nd–5.3	nd–2.8	nd–0.5	0.4–38
Soft drinks						
Cola soda	13	1.6–43	0.1–7.8	<0.05–1.3	nd–0.2	1.8–52
Fruit beverage	6	5.1–23	0.4–1.4	<0.05–0.2	nd–<0.07	5.5–25
Isotonic drink	4	1.8–31	0.2–4.6	0.1–0.8	nd–<0.07	2.1–35
Lemon soda	5	7.6–31	0.6–4.5	<0.05–0.8	nd–<0.07	8.3–33
Orange soda	5	1.0–31	0.2–3.4	<0.05–0.6	nd–<0.07	1.4–33
Soda	1	35	2.0	0.3	nd	37
Tea	3	1.1–27	0.5–3.5	0.2–0.7	nd–<0.07	1.8–30
Tonic water	3	3.8–20	0.4–2.1	<0.05–0.3	nd	4.2–21

n, number of samples; nd, not detected.
The "less than" (<) values represent the limits of quantitation of the method.
Data from Montesinos, I., Gallego, M., 2014. How the inclusion of treated water in beverages influences the appearance of halogenated volatile organic compounds. J. Agric. Food Chem. 62, 10240–10247.

treated water used (0.5, 3.5, 8.3, and 20 μg/L, respectively); however, there were no differences due to fruit in juices or flavor (lemon, orange, cola) in soft drinks.

An SHS–GC–MS method for determining HAAs in fruit juices and soft drinks was recently reported (Cardador and Gallego, 2015). The method is highly sensitive (limit of detection, 0.06 – 0.5 μg/L), simple, and reliable. It was used to determine the nine common HAAs (MCAA, MBAA, DCAA, TCAA, BCAA, DBAA, BDCAA, DBCAA, and TBAA) and an iodinated species monoiodoacetic acid (MIAA) in beverages. At least two HAAs (DCAA and TCAA) were present in all samples prepared with treated water; their concentrations in the beverages remained constant for 2–3 months because, unlike THMs, these species are not volatile. Table 23.4 lists the HAA levels found in the analysis of 67 juices and nectars, and 55 soft drinks (Cardador and Gallego, 2015). As can be seen, there were no significant differences in THAA concentrations between beverages of the same type packed in different containers: glass bottles, PET, cans, and cardboard boxes. Only 4 of 11 100% juices contained DCAA and/or TCAA, albeit at very low levels (<0.1 μg/L). The presence of HAAs in them cannot be ascribed to the addition of treated water (which would be a fraud), but rather to contamination through washing of packages and/or contact of squeezer/equipment with disinfectants/sanitizers, or the use of some HAA (as herbicide, eg) to boost fruit growth. Between 2 and 7 species were found in 24 reconstituted juices. No significant differences in average THAA concentrations between juices from different fruits (apple, orange, pineapple, mixed), nor between nectar flavors (apple, orange, pineapple, mixed), were observed. The average THAA concentration in Spanish nectar juices (6.7 μg/L) was higher than that in reconstituted juices (4.5 μg/L) owing to the greater amount of water used in the former. The soft drinks were found to contain the same species as the fruit juices. However, the soft drinks fell into two distinct groups, namely: (1) soda and tonic samples, with average THAA concentrations of 33.5 and 40.2 μg/L, respectively; and (2) tea, lemon, orange, apple, cola, energy drink and isotonic samples, with lower averages (namely, 2.5, 3.4, 3.7, 5.0, 3.8, 6.5, and 4.2 μg/L, respectively). The soft drinks exhibited higher HAA concentrations (average THAA 12 μg/L) than the fruit juices (5.6 μg/L) by effect of the former containing a high proportion (c. 90%) of treated water—their major ingredient.

Humans are estimated to consume about 2 L of drinking water each day including drinking water in the form of juices and other beverages containing tap water. Since the most contaminated samples (namely, cola soda in Table 23.3 and tonic water in Table 23.4) contained 52 μg/L TTHMs and 73 μg/L THAAs, a person can easily ingest more than 100 μg of THMs and/or 150 μg of THAAs each day (MCL 80–100 μg/L THMs and 60–80 μg/L HAAs) (WHO, 2011). In this situation, regular control of soft drinks for DBP content should be compulsory. A final test was performed to compare reconstituted juices of various flavors from different European countries. As can be seen from Table 23.5, the TTHM and THAA concentration ranges were similar, and not significantly different, for all countries. Also, none of the samples contained TTHM or THAA concentrations

Table 23.4 Concentrations of Haloacetic acids (HAAs) in Spanish Juices and Soft Drinks (Range Values in μg/L)

Beverage	*n*	MCAA	MBAA	DCAA	TCAA	BCAA	DBAA	BDCAA	Total HAAs
Juices									
100%									
Mango	1	nd	nd	0.1	nd	nd	nd	nd	0.1
Orange	10	nd	nd	nd–0.1	nd–<0.1	nd	nd	nd	0.1
Reconstituted									
Apple	3	nd	nd	0.6–3.9	0.2–0.6	nd–0.2	nd	nd	1.0–4.4
Cranberry	1	nd	nd	1.0	0.2	nd	nd	nd	1.2
Grape	1	nd	nd	5.6	1.0	nd	nd	nd	6.6
Mix	9	nd–1.3	nd–1.0	0.5–8.8	0.1–3.9	nd	nd	nd	0.6–15
Orange	6	nd–<0.5	nd	1.6–3.9	0.1–1.7	nd–0.1	nd	nd	1.8–4.8
Peach	2	nd	nd	0.8, 1.6	0.3, 1.1	nd	nd	nd	1.1, 2.7
Pineapple	2	<0.5, 0.6	nd, 0.5	1.7, 2.1	0.4, 0.8	0.1, 0.3	nd, 0.6	nd, 0.5	2.2, 5.4
Nectar									
Apple	4	nd–1.7	nd	0.3–6.1	0.1–3.5	nd–0.6	nd	nd–<0.5	0.4–12
Grapefruit	2	nd, 0.6	nd	3.5, 4.2	0.6, 1.8	nd, 0.5	nd	nd	4.1, 7.1
Guava	1	<0.5	nd	8.5	1.5	nd	nd	nd	10
Mix	3	nd–1.8	nd	0.2–14	nd–5.2	nd–0.6	nd–0.3	nd–<0.5	0.2–22
Orange	10	nd–1.6	nd–0.6	1.6–8.1	0.2–5.6	nd–2.7	nd–1.8	nd–1.4	2.6–18
Passion fruit	1	nd	nd	4.6	0.6	nd	nd	nd	5.2
Peach	1	nd	nd	1.8	3.3	nd	nd	nd	5.1
Pear	1	nd	nd	6.4	2.6	nd	nd	nd	9.0
Pineapple	9	nd–1.0	nd–<0.5	0.3–6.3	0.3–4.2	nd–0.6	nd	nd–<0.5	1.0–9.2

(*Continued*)

Table 23.4 Concentrations of Haloacetic acids (HAAs) in Spanish Juices and Soft Drinks (Range Values in μg/L) *Continued*

Beverage	*n*	MCAA	MBAA	DCAA	TCAA	BCAA	DBAA	BDCAA	Total HAAs
Soft drinks									
Apple	2	nd, 0.6	nd	3.0, 5.9	0.2, 0.2	nd, 0.1	nd	nd	3.2, 6.8
Cola soda	12	nd–2.1	nd	0.3–15	0.1–5.0	nd–1.5	nd	nd–1.4	0.5–20
Energy drink	4	nd–0.6	nd	1.9–3.4	0.2–0.5	0.1–7.6	nd	nd	3.4–12
Isotonic drink	5	nd–1.8	nd–0.6	1.0–5.6	0.3–18	nd–3.3	nd	nd–1.6	1.9–30
Lemon soda	8	nd–2.3	nd–<0.5	0.1–31	0.2–18	nd–7.2	nd	nd–4.1	0.6–63
Orange soda	8	nd–0.7	nd–<0.5	0.1–4.7	0.5–2.9	nd–2.5	nd	nd	0.6–7.7
Pineapple	1	nd	nd	3.3	2.1	0.8	nd	nd	6.2
Soda	4	<0.5–2.6	nd–1.5	15–18	4.8–26	0.1–2.2	nd–<0.2	nd–3.2	26–46
Tea	6	nd–0.9	nd	0.1–5.5	0.2–4.5	nd	nd	nd	0.3–7.2
Tonic water	5	nd–3.3	nd–0.5	13–33	2.8–28	nd–4.2	nd–0.7	nd–3.2	17–73

n, *number of samples; nd, not detected.*
The "less than" (<) values represent the limits of quantitation of the method.
Data from Cardador, M.J., Gallego, M., 2012. Effect of the chlorinated washing of minimally processed vegetables on the generation of haloacetic acids. J. Agric. Food Chem. 60, 7326 – 7332.

Table 23.5 Total Concentrations of Trihalomethanes (THMs) and Haloacetic Acids (HAAs) (μg/L) in European Reconstituted Juices

	European Juices			Spanish Juices		
	n	Total THMs	Total HAAs	*n*	Total THMs	Total HAAs
Apple[a]	1	0.2	3.2	2	0.6, 4.2	1.0, 4.4
Apple[b]	1	4.7	1.1			
Cranberry[c]	1	16	3.4	1	0.8	1.2
Mix[b]	2	0.4, 1.8	0.8, 1.0	9	0.2–13	0.6–15
Mix[c]	2	0.6, 18	13, 28			
Orange[a]	1	0.8	19	3	0.2–11	1.8–4.8
Orange[c]	3	0.4–18	1.8–31			
Orange[d]	2	0.2, 0.4	0.3, 1.1			
Peach[b]	1	0.6	1.2	2	2.1, 6.4	1.1, 2.7

Juices from: [a]France, [b]Portugal, [c]Belgium, [d]Poland.
n, number of samples.

above the levels allowed by US EPA or EU for drinking water. In any case, these data are only estimates owing to the small number of samples studied from all countries except Spain.

CONCLUSIONS

Existing research into the generation and occurrence of DBPs in washing water and in MPVs found that even though no THM residues remain in these products, the concentrations of THMs in washing water used in MPVs is well above the regulated limits of US EPA for drinking water. This is probably due to the high volatility of these compounds. In contrast, HAAs remain in the end-product due to their ionic nature and low volatility. Based on the foregoing, the presence of THMs in MPVs poses no special health risks, whereas HAAs are more persistent and can reach consumers, especially through hyperchlorinated products from the fresh-cut industry.

In relation to the presence of DBPs in beverages, the main conclusions are: treated water, which is used as an ingredient, is the main source of THMs and HAAs in beverages. Their concentration and speciation depend on the volume and the quality of treated water used. Soft drinks that are made with DBP-free mineral water, were negative for THMs and HAAs or contained chloroform at very low levels (<1 μg/L). Both THMs and HAAs can be used as indicators of the presence of treated water in MPFVs/beverages. But, HAAs are probably a better choice, because they are nonvolatile DBPs and can thus persist longer (several months) in the products; whereas THMs are more volatile and hence more unstable.

REFERENCES

AIJN, 2014. Liquid fruit. Market report, European Fruit Association. http://aijn.org/files/default/aijn2014-full.pdf.

Artés, F., Allende, A., 2014. Minimal processing of fresh fruit, vegetables, and juices. In: Sun, D.W. (Ed.), Emerging Technologies for Food Processing. Elsevier Ltd, Amsterdam, pp. 583–597.

Ashurst, P., Hargitt, R., 2009. Ingredients in soft drinks and fruit juices. Soft Drinks and Fruit Juices Problems Solved. Woodhead Publishing Ltd, Cambridge, pp. 20–59.

Belitz, H.D., Grosch, W., Schieberle, P., 2009. Fruit and fruit products, Food Chemistry, fourth edition Springer, Berlin, pp. 498–545.

Cardador, M.J., Gallego, M., 2012. Effect of the chlorinated washing of minimally processed vegetables on the generation of haloacetic acids. J. Agric. Food Chem. 60, 7326–7332.

Cardador, M.J., Gallego, M., 2015. Haloacetic acids content of fruit juices and soft drinks. Food Chem. 173, 685–693.

Codex Alimentarius Commission, 2003. Code of hygienic practice for fresh fruits and vegetables (new Annex III for Fresh Leafy Vegetables). CAC/RCP 53e2003. Revision 2010. www.codexalimentarius.org/input/download/standards/10200/CXP_053s_2013.pdf.

Committee on Toxicity of Chemicals in Food, Consumer Products and the Environment, 2006. COT statement on a commercial survey investigating the occurrence of disinfectants and disinfection by-products in prepared salads. http://cot.food.gov.uk/pdfs/cotstatementwashaids200614.pdf.

Council Directive 89/107/EC, 1988. Regulation on the approximation of the laws of the Member States concerning food additives authorized for use in foodstuffs intended for human consumption. Off. J. Eur. Commun. L237, 1–10.

Council Directive 98/83/EC, 1998. Regulation on the quality of water intended for human consumption. Off. J. Eur. Commun. L330, 32–54.

Council Directive 2001/112/CE, 2001. Relating to fruit juices and certain similar products intended for human consumption. Off. J. Eur. Commun. L10, 58–66.

Crandell, P.G., Hendrix, C.M., 2001. Citrus juice processing. In: Schobinger, U. (Ed.), Fruit and Vegetable Juices, third ed. Verlag Eugen Ulmer, Stuttgart.

Deboer, M.D., Scharf, R.J., Demmer, R.T., 2013. Sugar-sweetened beverages and weight gain in 2- to 5-year-old children. Pediatrics 132, 1–8.

FAO, 2001. Principles and Practices of Small- and Medium-Scale Fruit Juice Processing. FAO Agricultural Services Bulletin 146. Food and Agriculture Organization of the United Nations, Rome.

FAO, 2008. Climate change: implications for food safety. ftp://ftp.fao.org/docrep/fao/010/i0195e/i0195e00.pdf.

FAO/WHO,1992. Codex Alimentarius: Fruit Juices and Related Products. Volume 6 of Codex Alimentarius. Food and Agriculture Organization of the United Nations, Rome.

FAO/WHO, 2008. Benefits and Risks of the Use of Chlorine-Containing Disinfectants in Food Production and Food Processing. WHO Document Production Services, Ann Arbor, MI.

Fellows, P.J., 2000. Food Processing Technology: Principles and Practices, second ed. Woodhead Publishing, Ltd, Cambridge.

Fleming-Jones, M.E., Smith, R.E., 2003. Volatile organic compounds in foods: a five year study. J. Agric. Food Chem. 51, 8120–8127.

Gil, M., Selma, M.V., López-Gálvez, F., Allende, A., 2009. Fresh-cut product sanitation and wash water disinfection: problems and solutions. Int. J. Food Microbiol. 134, 37–45.

Gómez-López, V.M., Rajkovic, A., Ragaert, P., Smigic, N., Devlieghere, F., 2009. Chlorine dioxide for minimally processed produce preservation: a review. Trends Food Sci. Technol. 20, 17–26.

Gómez-López, V.M., Marín, A., Medina-Martínez, M.S., Gil, M.I., Allende, A., 2013. Generation of trihalomethanes with chlorine-based sanitizers and impact on microbial, nutritional and sensory quality of baby spinach. Postharvest Biol. Technol. 85, 210–217.

Graham, R.C., Robertson, J.K., 1988. Analysis of trihalomethanes in soft drinks. J. Chem. Educ. 65, 735–737.

Hrudey, S.E., Charrois, J.W., 2012. Disinfection By-Products and Human Health, first ed. IWA Publishing, London.

Huang, A.T., Batterman, S., 2009. Formation of trihalomethanes in foods and beverages. Food Addit. Contam. 26, 947–957.

Huang, A.T., Batterman, S., 2010. Sorption of trihalomethanes in foods. Environ. Int. 36, 754–762.

IFPA, 2005. The convenience, nutritional value and safety of fresh-cut produce. http://www.creativew.com/sites/ifpa/about.html.

Klaiber, R.G., Baur, S., Wolf, G., Hammes, W.P., Carle, R., 2005. Quality of minimally processed carrots as affected by warm water washing and chlorination. Innovative Food Sci. Emerg. Technol. 6, 351–362.

López-Gálvez, F., Allende, A., Truchado, P., Martínez-Sánchez, A., Tudela, J.A., Selma, M.V., et al., 2010. Suitability of aqueous chlorine dioxide versus sodium hypochlorite as an effective sanitizer for preserving quality of fresh-cut lettuce while avoiding by-product formation. Postharvest Biol. Technol. 55, 53–60.

Luo, Y., Nou, X., Millner, P., Zhou, B., Shen, C., Yang, Y., et al., 2012. A pilot plant scale evaluation of a new process aid for enhancing chlorine efficacy against pathogen survival and cross-contamination during produce wash. Int. J. Food Microbiol. 158, 133–139.

McNeal, T.P., Hollifield, H.C., Diachenko, G.W., 1995. Survey of trihalomethanes and other volatile chemical contaminants in processed foods by purge and trap capillary gas chromatography with mass selective detection. J. AOAC Int. 78, 391–397.

Montesinos, I., Gallego, M., 2014. How the inclusion of treated water in beverages influences the appearance of halogenated volatile organic compounds. J. Agric. Food Chem. 62, 10240–10247.

Nagy, S., Wade, R.L., 1995. Methods to Detect Adulteration of Fruit Juice Beverages. AgScience, Inc., Auburndale, FL.

Ramos, B., Miller, F.A., Brandão, T.R.S., Teixeira, P., Silva, C.L.M., 2013. Fresh fruits and vegetables—an overview on applied methodologies to improve its quality and safety. Innovative Food Science Emerg. Technol. 20, 1–15.

Raymer, J.H., Pellizzari, E., Childs, B., Briggs, K., Shoemaker, J.A., 2000. Analytical methods for water disinfection by-products in foods and beverages. J. Expo. Anal. Environ. Epidemiol. 10, 808–815.

Richardson, S.D., 2012. Environmental mass spectrometry: emerging contaminants and current issues. Anal. Chem. 84, 747–778.

Richardson, S.D., Ternes, T.A., 2011. Water analysis: emerging contaminants and current issues. Anal. Chem. 83, 4614–4648.

Richardson, S.D., Plewa, M.J., Wagner, E.D., Schoeny, R., DeMarini, D.M., 2007. Occurrence, genotoxicity, and carcinogenicity of regulated and emerging disinfection by-products in drinking water: a review and roadmap for research. Mutat. Res. 636, 178–242.

Romaguera-Bosh, D., 2013. Consumption of sweet beverages and type 2 diabetes incidence in European adults: results from EPIC-InterAct. Diabetologia 56, 1520–1530.

Shen, C., Norris, P., Williams, O., Hagan, S., Li, K., 2016. Generation of chlorine by-products in simulated wash water. Food Chem. 190, 97–102.

Silveira, M., Martendal, E., Carasek, E., 2011. Determination of THMs in soft drink by solid-phase microextraction and gas chromatography. Food Chem. 127, 290–295.

Soliva-Fortuny, R.C., Martín-Belloso, O., 2003. New advances in extending the shelf-life of fresh-cut fruits: a review. Trends Food Sci. Technol. 14, 341–353.

Szymczycha-Madeja, A., Welna, M., Jedryczko, D., Pohl, P., 2014. Developments and strategies in the spectrochemical elemental analysis of fruit juices. Trends Anal. Chem. 55, 68–80.

Trumbo, P.R., Rivers, C.R., 2014. Systematic review of the evidence for an association between sugar-sweetened beverage consumption and risk of obesity. Nutr. Rev. 72, 566–574.

US EPA, 2006. National primary drinking water regulations: stage 2 disinfectants and disinfection by-products rule. http://water.epa.gov/lawsregs/rulesregs/sdwa/stage2/regulations.cfm.

US FDA, 1999. Part 175, Indirect food additives: adhesives and components of coatings. In: Code of Federal Regulations Title 21, Food and Drugs, vol. 3. U.S. Food and Drug Administration, Washington, DC.

US FDA, 2003a. Part 173, section 173.315, Chemicals used in washing to assist in the peeling of fruits and vegetables. In: Code of Federal Regulations Title 21, Food and Drugs. Washington, DC. http://www.ecfr.gov/cgi-bin/text-idx?rgn = div5;node = 21%3A3.0.1.1.4.

US FDA, 2003b. Part 178, section 178.1010, Sanitizing solutions. In: Code of Federal Regulations Title 21, Food and Drugs. Washington, DC. http://www.ecfr.gov/cgi-bin/textidx?SID = 28346a9dced9a1f259d82d40d88e4292&mc = true&node = pt21.3.178&rgn = div5.

US FDA, 2008. Guidance for industry: guide to minimize microbial food safety hazards of fresh-cut fruits and vegetables. www.fda.gov/Food/GuidanceRegulation/GuidanceDocumentsRegulatoryInformation/ProducePlantProducts/ucm064458.htm.

US FDA, 2009. Methods to reduce/eliminate pathogens from produce and fresh-cut produce, analysis and evaluation of preventive control measures for the control and reduction/elimination of microbial hazards on fresh and fresh-cut produce. www.fda.gov/food/foodscienceresearch/safepracticesforfoodprocesses/ucm091363.htm.

Vartanian, L.R., Schwartz, M.B., Brownell, K.D., 2007. Effects of soft drink consumption on nutrition and health: a systematic review and meta-analysis. Am. J. Public Health 97, 667–675.

Vaux, B., 2003. What is your generic term for a sweetened carbonated beverage? Harv. Dialect Surv.http://www4.uwm.edu/FLL/linguistics/dialect/staticmaps/q_105.html. Retrieved July 2015.

WHO, 2006. Guidelines for Drinking Water Quality. World Health Organization, Geneva. http://www.who.int/water_sanitation_health/dwq/gdwq0506.pdf.

WHO, 2011. Guidelines for Drinking-Water Quality, fourth ed. World Health Organization, Geneva. www.who.int/water_sanitation_health/publications/2011/dwq_guidelines/en/.

Woodward-Lopez, G., Kao, J., Ritchie, L., 2010. To what extent have sweetened beverages contributed to the obesity epidemic? Public Health Nutr. 14, 499–509.

Ye, T., Xu, B., Lin, Y.L., Hu, C.Y., Lin, L., Zhang, T.Y., et al., 2013. Formation of iodinated disinfection by-products during oxidation of iodide-containing waters with chlorine dioxide. Water Res. 47, 3006–3014.

Index

Note: Page numbers followed by "*f*" and "*t*" refer to figures and tables, respectively.

B

C

D

E

HUS. *See* Hemolytic uremic syndrome (HUS)

I

T

U

V